Fertility Preservation

Emre Seli • Ashok Agarwal
Editors

Fertility Preservation

Emerging Technologies and Clinical Applications

Foreword by Charles J. Lockwood, MD

Springer

Editors
Emre Seli, MD
Department of Obstetrics,
Gynecology, and Reproductive Sciences
Yale University School of Medicine
New Haven, CT, USA
emre.seli@yale.edu

Ashok Agarwal, PhD,
HCLD (ABB), EMB (ACE)
Director
Center for Reproductive Medicine
Cleveland Clinic
Cleveland, OH, USA
agarwaa@ccf.org

ISBN 978-1-4419-1782-9 e-ISBN 978-1-4419-1783-6
DOI 10.1007/978-1-4419-1783-6
Springer New York Dordrecht Heidelberg London

Library of Congress Control Number: 2011936522

Printed on acid-free paper

Springer is part of Springer Science+Business Media (www.springer.com)

Foreword

Women diagnosed with cancer prior to or during their reproductive period often face both the uncertainty of long-term survival, and the risk of treatment-induced infertility. Treatments for common cancers occurring in younger women often necessitate extirpation of their reproductive organs, radiation therapy, and/or chemotherapy, each of which may partially or completely impair reproductive (and endocrine) function. Fortunately, during the last two decades, cancer survival in the USA has improved. This has resulted in an increased focus on maximizing the quality of life for cancer survivors. Preserving future fertility is a crucial component of the quality of life in prepubertal and reproductive age women. Similarly, cancer can also strike young men, the treatments for which pose comparable risks to future fertility. Thus, clinicians caring for such patients will find this textbook valuable in both counseling and caring for younger women and men with newly diagnosed cancers.

However, there is a third group of patients in whom fertility preservation is of increasing relevance, interest, and importance. Women in much of the industrial world have been progressively delaying initiation/completion of childbearing to later in life. More women in their late 30s and early 40s are attempting to conceive for the first time than ever before. Since the incidence of most cancers in the reproductive age group is comparatively low, intentionally postponing childbearing presents a far more common indication for fertility preservation than cancer.

The editors are well suited to the purposes of this text. Both are gifted fertility practitioners and physician scientists. Dr. Ashok Agarwal is the Director of the Center for Reproductive Medicine at Cleveland Clinic. He has authored over 150 peer reviewed publications, garnered multiple research grants, and is internationally recognized for his outstanding contributions in scientific and clinical aspects of reproductive medicine. Dr. Emre Seli is an accomplished reproductive scientist and physician at the Yale School of Medicine. Dr. Seli has conducted novel and potentially landmark research regarding basic oocyte mRNA processing, a process relevant to maternal age-associated infertility. Dr. Seli has also pioneered the development of novel technologies for assessing embryo quality in vitro. Meanwhile, Dr. Seli directs a highly successful oocyte donation program at his home institution.

Drs. Agarwal and Seli have assembled an "All Star" cast of authors, each an expert in the topics about which they write. The goal of the text is to present strategies for fertility preservation in women and men who require gonadotoxic and other fertility-impairing therapies. In addition, key epidemiologic, ethical, medico-legal, psychologic and social aspects of fertility preservation are discussed. A format that combines a concise scientific background with practical methodological information, including detailed and straightforward algorithms, is utilized. Step-by-step protocols for laboratory procedures are presented. The result is a concise, readable, and highly practical reference guide for reproductive endocrinologists, embryologists, reproductive scientists, and oncologists. This text will also greatly aid the general ob/gyn in counseling both the relatively uncommon younger patient facing fertility-compromising treatments and the increasingly common woman in her 30s or early 40s who elects to delay fertility for a myriad of professional and personal reasons. For all these reasons, I can enthusiastically recommend this textbook to you.

New Haven, CT Charles J. Lockwood, MD

Preface

Dum spiro, spero [As long as I breathe, I hope]

Marcus Tullius Cicero

Cancer is not uncommon in men and women of reproductive age. In the USA alone, over 130,000 men and women below the age of 45 were afflicted with cancer in 2008. The compromised fertility that will be faced by many of these patients can result in significant psychosocial stress and in reduced quality of life; even long after the cancer has been successfully treated.

Fortunately, the rapid pace of scientific and technological discovery has increased the options available to men and women in need of fertility preservation. More than 30 years after the birth of the first child conceived by assisted reproductive technologies, we are now able to cryopreserve embryos, spermatozoa, and oocytes as well as testicular and ovarian tissue. In addition, the effect of gonadotoxic treatments on fertility is better defined and the use of medical interventions for fertility preservation is within reach. With increasingly complex modalities available for fertility preservation comes the need for a clear, concise, readable, and practical text to serve as a resource for healthcare providers.

This textbook has been written with the aim of providing an update on options for fertility preservation for both men and women. Within this context, this text reviews current therapeutic modalities in great detail and describes the biologic basis for available interventions. A key aspect of this volume is the valuable contributions from oncologists, obstetricians, psychologists, epidemiologists, and ethicists in addition to reproductive scientists and infertility specialists. This thorough resource thus recapitulates the multidisciplinary approach that is required to adequately provide patient care with a view to fertility preservation.

Following introductory chapters on the epidemiology and ethics of fertility preservation, this book is divided into two main sections that share a similar overall format while reviewing fertility preservation in female and then male patients. Each section starts with a chapter reviewing the gender-specific biology of the reproductive system within the context of fertility preservation. Then, the impact of gonadotoxic treatment on gonadal function and on the offspring is reviewed. Additional chapters in each section summarize established and experimental methodologies currently available for fertility preservation in both sexes, providing detailed protocols for laboratory procedures. The resulting comprehensive text provides cutting edge science and evidence-based suggestions regarding potential interventions directed at fertility preservation.

This book is dedicated to our patients who, despite facing potentially lethal diagnoses and compromised quality of life, continue to hope, to live, and to pursue their dreams of parenthood.

New Haven, CT Emre Seli
Cleveland, OH Ashok Agarwal

Contents

Contributors

Ashok Agarwal, PhD, HCLD Center for Reproductive Medicine, Glickman Urological and Kidney Institute, Cleveland Clinic, Cleveland, OH, USA

Department of Obstetrics-Gynecology and Women's Health Institute, Cleveland Clinic, Cleveland, OH, USA

David F. Albertini, PhD Hall Professor of Molecular Medicine

University of Kansas Cancer Center, Kansas City, KS, USA

Desiderio Avila, Jr. MD Resident, Scott Department of Urology, Baylor College of Medicine, Houston, TX, USA

Masoud Azodi, MD Associate Professor, Department of Obstetrics, Gynecology and Reproductive Sciences, Yale University School of Medicine, New Haven, CT, USA

Mert Ozan Bahtiyar, MD Associate Professor, Section of Maternal-Fetal Medicine, Department of Obstetrics and Gynecology, Yale University School of Medicine, New Haven, CT, USA

Basak Balaban, MSc Embryologist Department of Obstetrics and Gynecology, American Hospital, Istanbul, Turkey

Gihan M. Bareh, MD, Resident, Department of Obstetrics and Gynecology, University Hospitals Case Medical Center, Case Western Reserve University, Cleveland, OH, USA

Mohamed A. Bedaiwy, MD, PhD REI Fellow, Department of Obstetrics and Gynecology, University Hospitals Case Medical Center, Case Western Reserve University, Cleveland, OH, USA

Veronica Bianchi, PhD Senior Embryologist, Tecnobios Procreazione Centre for Reproductive Health, Bologna, Italy

Andrea Borini, MD Tecnobios Procreazione Centre for Reproductive Health, Via Dante, Bologna, Italy

Clinical and Scientific Director, Tecnobios Procreazione Centre for Reproductive Health, Bologna, Italy

Hakan Cakmak, MD Resident, Department of Obstetrics, Gynecology and Reproductive Sciences, Yale University School of Medicine, New Haven, CT, USA

Ciler Celik-Ozenci, DDS, PhD Associate Professor, Department of Histology and Embryology, School of Medicine, Akdeniz University, Campus, Antalya, Turkey

Alan B. Copperman, MD Director, Division of Reproductive Endocrinology and Infertility, Mount Sinai Medical Center, New York, NY, USA

Vice-Chairman, Department of Obstetrics, Gynecology, and Reproductive Science, Mount Sinai Medical Center, New York, NY, USA

Clinical Professor, Department of Obstetrics, Gynecology, and Reproductive Science, Mount Sinai School of Medicine, New York, NY, USA

Fnu Deepinder, MD Clinical Research Fellow, Department of Endocrinology, Diabetes and Metabolism, Cedars Sinai Medical Center, Los Angeles, CA, USA

Giuseppe Del Priore, MD, MPH Director, Division of Gynecologic Oncology, Indiana University Melvin and Bren Simon Cancer Center, Indianapolis, IN, USA

Javier Domingo del Pozo, MD Gynecologist, Department of Reproductive Medicine, University of Las Palmas de Gran Canaria, Las Palmas de Gran Canaria, Spain

Ina Dobrinski, DVM, MVSc, PhD Professor, Department of Comparative Biology and Experimental Medicine, Faculty of Veterinary Medicine, University of Calgary, Calgary, AB, Canada

Abbie L. Fields, MD Medical Director, Division of Gynecologic Oncology, Washington Hospital Center, Washington, DC, USA

Bhushan K. Gangrade, PhD Director, IVF Laboratory, Center for Reproductive Medicine, Orlando, FL, USA

Paul Gittens, MD Fellow, Scott Department of Urology, Baylor College of Medicine, Houston, TX, USA

Dorothy A. Greenfeld, LCSW Clinical Professor, Department of Obstetrics and Gynecology, and Reproductive Sciences, Yale University School of Medicine, New Haven, CT, USA

Deleep Kumar Gudipudi, MBBS, DNB Resident, Department of Radiation Oncology, Basavatarakam Indo-American Cancer Institute and Research Center, Hyderabad, Andhra Pradesh, India

William W. Hurd, MD, MSc Professor, Department of Obstetrics and Gynecology, University Hospitals Case Medical Center, Case Western Reserve University, Cleveland, OH, USA

Director, Division of Reproductive Endocrinology and Infertility, Department of Obstetrics and Gynecology, University Hospitals Case Medical Center, Case Western Reserve University, Cleveland, OH, USA

Kathleen Hwang, MD Clinical Research Resident, Scott Department of Urology, Baylor College of Medicine, Houston, TX, USA

Joshua Johnson, PhD Assistant Professor, Department of Obstetrics, Gynecology and Reproductive Sciences, Yale University School of Medicine, New Haven, CT, USA

Cengiz Karakaya, PhD Rector, Department of Obstetrics and Gynecology, IVF Unit, Gazi University, Ankara, Turkey

Ertug Kovanci, MD Assistant Professor, Department of Obstetrics and Gynecology, Baylor College of Medicine, Houston, TX, USA

Maya L. Kriseman, BS Medical Student, Department of Obstetrics and Gynecology, Baylor College of Medicine, Houston, TX, USA

Christine Laky, MD Fellow, Section of Maternal-Fetal Medicine, Department of Obstetrics and Gynecology, Yale University School of Medicine, New Haven, CT, USA

Larry I. Lipshultz, MD Professor and Chair; Lester and Sue Smith Chair in Reproductive Medicine, Baylor College of Medicine, Houston, TX, USA

Chief, Division of Male Reproductive Medicine and Surgery, Scott Department of Urology, Baylor College of Medicine, Houston, TX, USA

Ana Isabel Marqués Marí, PhD Researcher, Stem Cell Bank, Prince Felipe Research Centre, Valencia, Spain

William Murk, BSc, MPH Research Assistant, Department of Obstetrics, Gynecology, and Reproductive Sciences, Yale University School of Medicine, New Haven, CT, USA

Queenie V. Neri, BSc, MSc Research Associate, The Ronald O. Perelman and Claudia Cohen Center for Reproductive Medicine, Weill Cornell Medical College, New York, NY, USA

Yuichi Niikura, PhD Research Assistant Professor, Department of Animal Sciences, Washington State University, Pullman, WA, USA

Kutluk Oktay, MD, FACOG Professor of Obstetrics & Gynecology, Medicine, and Cell Biology & Anatomy, Director, Division of Reproductive Medicine & Institute for Fertility Preservation

Ozgur Oktem, MD Specialist, Department of Obstetrics and Gynecology, American Hospital, Istanbul, Istanbul, Turkey

Gogsen Onalan, MD Associate Professor, Division of Reproductive Medicine, Department of Obstetrics and Gynecology, Baskent University, Ankara, Turkey

Carolina Ortega, MD Clinical Fellow, Centre for Reproductive Medicine, University Hospital, Dutch-speaking Brussels Free University, Brussels, Belgium

Gianpiero D. Palermo, MD, PhD Professor, The Ronald O. Perelman and Claudia Cohen Center for Reproductive Medicine, Weill Cornell Medical College, New York, NY, USA

Pasquale Patrizio, MD, MBE, HCLD Professor, Yale Fertility Center and REI Medical Practice, New Haven, CT, USA

Director, Yale Fertility Center and REI Medical Practice, New Haven, CT, USA

Antonio Pellicer Department of Obstetrics and Gynecology, Hospital Universtario Doctor Peset, Valencia, Spain

Department of Obstetrics and Gynecology, Valencia University, Valencia, Spain

Elena S. Ratner, MD Assistant Professor, Department of Obstetrics, Gynecology, and Reproductive Sciences, Yale University School of Medicine, New Haven, CT, USA

Marcia Riboldi Doctoral Student, Stem Cell Bank, Prince Felipe Research Centre, Valencia, Spain

Christine E. Richter, MD Instructor, Division of Gynecologic Oncology, Department of Obstetrics, Gynecology and Reproductive Sciences, Yale University School of Medicine, New Haven, CT, USA

Katherine J. Rodewald, MD Attending, Department of Obstetrics and Gynecology, University Hospitals Case Medical Center, Case Western Reserve University, Cleveland, OH, USA

Jose R. Rodriguez-Sosa, DVM, MAP, PhD Postdoctoral Fellow, Department of Comparative Biology and Experimental Medicine, University of Calgary, Calgary, Canada

Kenny A. Rodriguez-Wallberg, MD, PhD Head, Fertility Preservation Programme, Division of Obstetrics and Gynecology, Department of Clinical Science, Intervention and Technology, Karolinska Institute and Karolina University Hospital Huddinge, Stockholm, Sweden

Zev Rosenwaks, MD Revlon Distinguished Professor, Reproductive Medicine in Obstetrics and Gynecology, Weill Cornell Medical College, New York, NY, USA

Professor, Obstetrics and Gynecology, The Ronald O. Perelman and Claudia Cohen Center for Reproductive Medicine, Weill Cornell Medical College, New York, NY, USA

María Sánchez-Serrano, MD Gynecologist, Department of Obstetrics and Gynecology, Hospital Universtario Doctor Peset, Valencia, Spain

Stefan Schlatt, PhD Director and Professor, Center of Reproductive Medicine and Andrology, University Hospital of Münster, Münster, Germany

Peter E. Schwartz, MD John Slade Ely Professor, Vice Chair, Division of Gynecologic Oncology, Department of Obstetrics, Gynecology and Reproductive Sciences, Yale University School of Medicine, New Haven, CT, USA

Emre Seli, MD Associate Professor, Associate Director for Research, Division of Reproductive Endocrinology and Infertility; Director, Oocyte Donation and Gestational Surrogacy Program, Department of Obstetrics, Gynecology, and Reproductive Sciences, Yale University School of Medicine, New Haven, CT, USA

Dan-Arin Silasi, MD Assistant Professor, Obstetrics, Gynecology, and Reproductive Sciences, Yale University School of Medicine, New Haven, CT, USA

Carlos Simón, MD, PhD Stem Cell Bank, Pricne Felipe Research Centre, Camino de las Moreras, Valencia, Spain

Enrique Soto, MD Fellow, Minimally Invasive Gynecological Surgery Department of Obstetrics, Gynecology and Reproductive Science, Mount Sinai Hospital/Mount Sinai School of Medicine, New York, NY, USA

Pankaj Talwar, MD Lieutenant Colonel, ART Centre, Army Hospital Research and Referral, Dhaula Kuan, New Delhi, India

Herman Tournaye, MD, PhD Professor, Centre for Reproductive Medicine, University Hospital, Dutch-Speaking Brussels Free University, Brussels, Belgium

Bulent Urman, MD Associate Professor, Department of Obstetrics and Gynecology, American Hospital, Istanbul, Turkey

Hulusi Bulent Zeyneloglu, MD Professor, Department of Obstetrics and Gynecology, Baskent University, Ankara, Turkey

Part I

Overview of Fertility Preservation: Epidemiologic, Social, Ethical, and Medicolegal Aspects

1 The Epidemiology of Fertility Preservation

William Murk and Emre Seli

Abstract

Cancer survivors are living longer lives in ever greater numbers, due to significant improvements in cancer therapy and changes in population age structures. Thus, the focus from merely living to living well has highlighted the need for fertility preserving methods during cancer treatment. Pregnancy for women with a previous diagnosis of cancer is safe, with no evidence of increased risk of recurrence. However, offspring of such women may be susceptible to preterm delivery, low birth weight, and other perinatal morbidities. Both patients and providers alike may have poor knowledge about fertility preservation during cancer, which highlights the need for increased education in this area.

Keywords

Epidemiology of fertility preservation • Fertility preservation in cancer survivors • Cancer survivor fertility • Demographics of fertility preservation in cancer survivors • Systemic lupus erythematosus and fertility

In 2007, it was estimated that there were 11.2 million cancer survivors living in the USA, of whom 450,000 were of reproductive age [1]. Although cancer is commonly thought to be disease of the elderly, 9.5% of all cancers are diagnosed before the age of 45 in the USA [2]. Due to surgical removal of reproductive organs or gonadotoxic effects from cancer treatment, survivors will likely face compromised fertility. For those who do not have surgical sterility, it has been estimated that survivors of childhood cancer have an overall reduction of 46% in likelihood of ever siring a pregnancy (among men), and 19% in likelihood of ever being pregnant (among women) [3, 4]. The prospect of partial or total infertility can significantly add to anxiety and emotional strain during disease management, and may also compromise long-term quality of life [5]. To offset these risks, patients can be offered several options for fertility preservation, including conservative

E. Seli (✉)
Division of Reproductive Endocrinology and Infertility; Director, Oocyte Donation and Gestational Surrogacy Program, Department of Obstetrics, Gynecology, and Reproductive Sciences, Yale University School of Medicine, New Haven, CT, USA
e-mail: emre.seli@yale.edu

E. Seli and A. Agarwal (eds.), *Fertility Preservation: Emerging Technologies and Clinical Applications*,
DOI 10.1007/978-1-4419-1783-6_1, © Springer Science+Business Media, LLC 2012

cancer management, and cryopreservation of sperm, embryos, or ovarian and testicular tissue.

In this chapter, we describe the epidemiology of factors relevant to fertility preservation. We begin by providing a description of demographic trends that illustrates the increasing need for fertility preservation, including trends in cancer survivorship and age of first pregnancy. This is followed by a brief overview of risk factors for infertility after cancer therapy. A summary of known health risks to pregnant cancer survivors and their offspring will be presented. Finally, we consider the current state of awareness and attitudes of fertility preservation among patients and medical providers.

Demographics of Fertility Preservation

Cancer Incidence

Cancer is not an uncommon disease among reproductive men and women. In the USA, the probability of developing any cancer by the age of 40 was estimated in 2004–2006 to be 2.36% in women, and 1.51% in men [6]. The age-adjusted incidence rates (AAIRs; see Table 1.1 for a definition) of this age group in 2007 were 40.7 malignant cases per 100,000 among men, and 59.8 per 100,000 among women [7]. At the age of 49, these numbers rise to 75.7 and 121.6 cases per 100,000, respectively [7]. The top ten sites of cancer by incidence for women aged 0–40 and men aged 0–49 are shown in Fig. 1.1a, b, respectively. Reproductive tissues ranked among these top ten sites, including the cervix uteri, corpus uteri, and ovary for women; and the testis and prostate for men. Among men, the male genital system as a whole had the highest cancer incidence rate of any general tissue system (AAIR, 13.6 cases per 100,000), while genital system ranked third among women (AAIR, 7.6 cases per 100,000), within these age groups [7]. Overall, the incidence rate of cancer by any site among ages less than 65 is declining, with an age-adjusted annual percent change (APC; see Table 1.1 for definition) of −0.6 for the period 1998–2007 [2]. By site, the incidence of cancer in reproductive tissues is declining for some sites, including the cervix uteri (APC, −3.2), ovary (APC, −1.6), and prostate (APC, −0.1; not significantly different from 0) [2]. However, the incidence among other reproductive tissues is increasing, including the corpus uteri (APC, 0.4) and testis (APC, 1.0).

Childhood and adolescent cancers (i.e., those diagnosed before the age of 20) are relatively rare, with AAIRs of 17.2 malignant cases per 100,000 among males, and 16.2 per 100,000 among females [7]. Nevertheless, these cancers are a significant public health problem. As of 1 January 2007, it was estimated that there were 269,403 people living in the USA who were diagnosed with cancer before the age of 20 (not including those diagnosed more than 31 years ago, due to incomplete data before 1975) [7]. Of these, 209,957 have been alive for at least 5 years since their diagnosis. Between 2003 and 2007, there were a total of 19,257 new cases of cancer for patients below the age of 20, of which 12,424 were diagnosed at a prepubertal age (0–14) [2]. The top ten sites of cancer for girls and boys aged 0–14 are shown in Fig. 1.2a, b, respectively. Cancers of the genital system are quite rare before the age of 20, with AAIRs of 1.4 cases per 100,000 among males, and 0.5 cases per 100,000 among females [7]. Before the age of 15, the AAIRs of these cancers are 0.4 per 100,000 among boys, and effectively zero among girls [7]. Notably, while the incidence rates of most adult cancers are decreasing, the incidence rate of childhood/adolescent cancers is increasing (APC between 1975 and 2007, 0.6) [2]. AAIRs have climbed from 12.9 cases per 100,000 in 1975, to 15.7 per 100,000 in 1995, and to 16.7 per 100,000 in 2007 [2].

Cancer Survivorship During Reproductive Ages

In the USA from 1999 to 2006, the 5-year survival of women diagnosed at age 0–44 with any site of cancer was 83.7% [2]. For men, this was 75.6%. However, survival is highly dependent on site of cancer: by site at this age group, 5-year survival ranged from 21.9 (esophagus) to 99.4% (thyroid). For major cancers of the female genital

Table 1.1 Definitions of commonly used terms in epidemiology

Term	Abbreviation	Definition	Examples and interpretations
Relative risk, odds ratio, hazard ratio, or incidence rate ratio	RR, OR, HR, or IRR	These are measures of the strength of an association between an exposure and an outcome. Specifically, the ratio of the risk, odds, hazard rate, or incidence rate of an outcome in an exposed group to that in an unexposed group	e.g., RR = 2.5: an exposed individual has a 2.5 times greater *risk* of outcome, when compared with an unexposed individual RR > 1: exposure increases risk of outcome RR = 1: exposure does not affect risk RR < 1: exposure reduces risk
95% Confidence interval	95% CI	A measure of the error (uncertainty) of some statistically distributed parameter (e.g., for RR, OR, HR, IRR, prevalence, etc.). If a parameter were to be independently measured 100 times in a random sample, the parameter estimate would fall within the calculated confidence interval 95 out of those 100 times	e.g., RR = 2.5 (95% CI: 1.9–3.1). The RR is measured to be 2.5, but if repeatedly measured, it would fall within the range 1.9–3.1 95% of the time. Less formally, there is 95% certainty that the true measure of association is within the range 1.9–3.1. If the 95% CI does not include the null (1.0), then it can be concluded that the effect is significantly different than expected from chance, with a false-positive probability of 5%
Incidence rate	IR	Number of new cases per population size in a given time period	e.g., 27 cases per 100,000 people in 2007
Age-adjusted incidence rate	AAIR	Incidence rate that a population would have if it had the same age structure as some reference population (often, the US census population is used as the referent for American studies). Allows comparisons of incidence rates between populations that have different age distributions, if they are adjusted using the same reference population. Reduces the potential for confounding by age	e.g., Population 1 (mostly elderly): crude (unadjusted) IR of 78 cases per 100,000. Population 2 (mostly youth): crude (unadjusted) IR of 23 cases per 100,000. From the crude analysis, it appears that population 1 has a greater risk. However, this may be due to an effect of age (e.g., elderly are at higher risk) rather than population (e.g., population 1 is at higher risk). If this is true, then using an age-adjusted incidence rate will reveal no or reduced difference between the populations
Annual percent change	APC	An estimate of percent change in a rate over time, assuming that rates change at a constant percentage rate	APC: –8.6 for 1999–2003: there was an annual 8.6% reduction in rate between 1999 and 2003

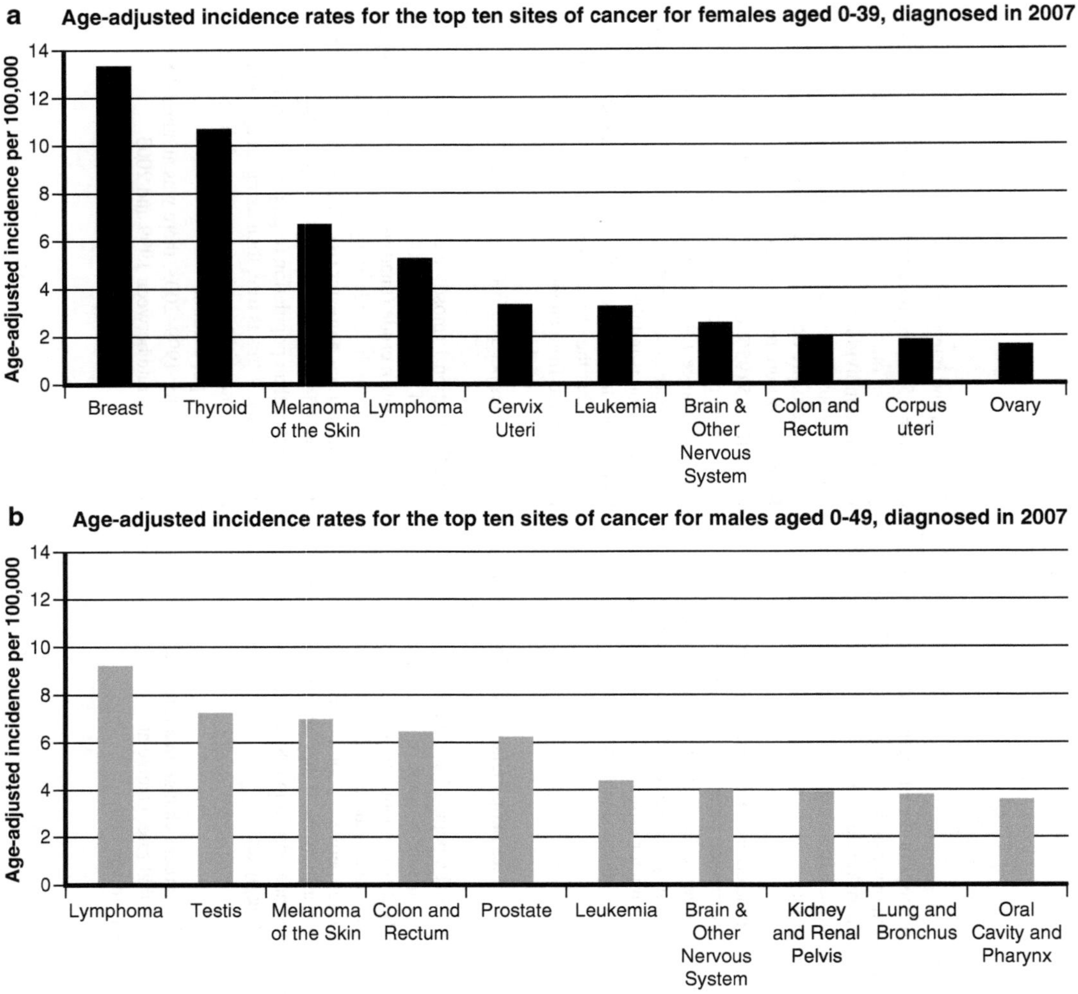

Fig. 1.1 (**a**) Age-adjusted incidence rates for the top ten sites for female subjects aged 0–39, diagnosed in 2007. (**b**) Age-adjusted incidence rates for the top ten sites of cancer for male subjects aged 0–49, diagnosed in 2007. Rates are for malignant cases and are adjusted to the 2000 US census population (data from the SEER 9 registries, including San Francisco–Oakland, Connecticut, Detroit, Hawaii, Iowa, New Mexico, Seattle, Utah, and Atlanta [7])

system, 5-year survival rates were 89.7 (corpus uteri and uterus not otherwise specified), 83.2 (cervix uteri), and 73.20% (ovary). For cancers of the male genital system, these were 96.7 (testis) and 85.3% (prostate). Overall 5-year survivorship has significantly increased over the past several decades, from 67.7% for men and women aged 0–44 in 1975–1977, to 71.6% in 1993–1995, and to 80.5% in 1999–2006. Increases in 5-year survival for women and men during these periods are shown in Fig. 1.3a, b, respectively. Similar increases in cancer survivorship have also been noted in many European countries [8, 9].

For childhood/adolescent cancers in the USA, all-site 5-year survivorship was 81.4%, for cases diagnosed in 1999–2006 [2], compared with 61.7% for cases diagnosed in 1975–1977. Increases in selected sites of cancer for children aged 0–14 are shown in Fig. 1.4. Ten-year survival has also risen, from 58.6 to 75.8% between 1975 and 1997 [2]. A study by Yeh et al. [10] estimated that the overall life expectancy of a 5-year

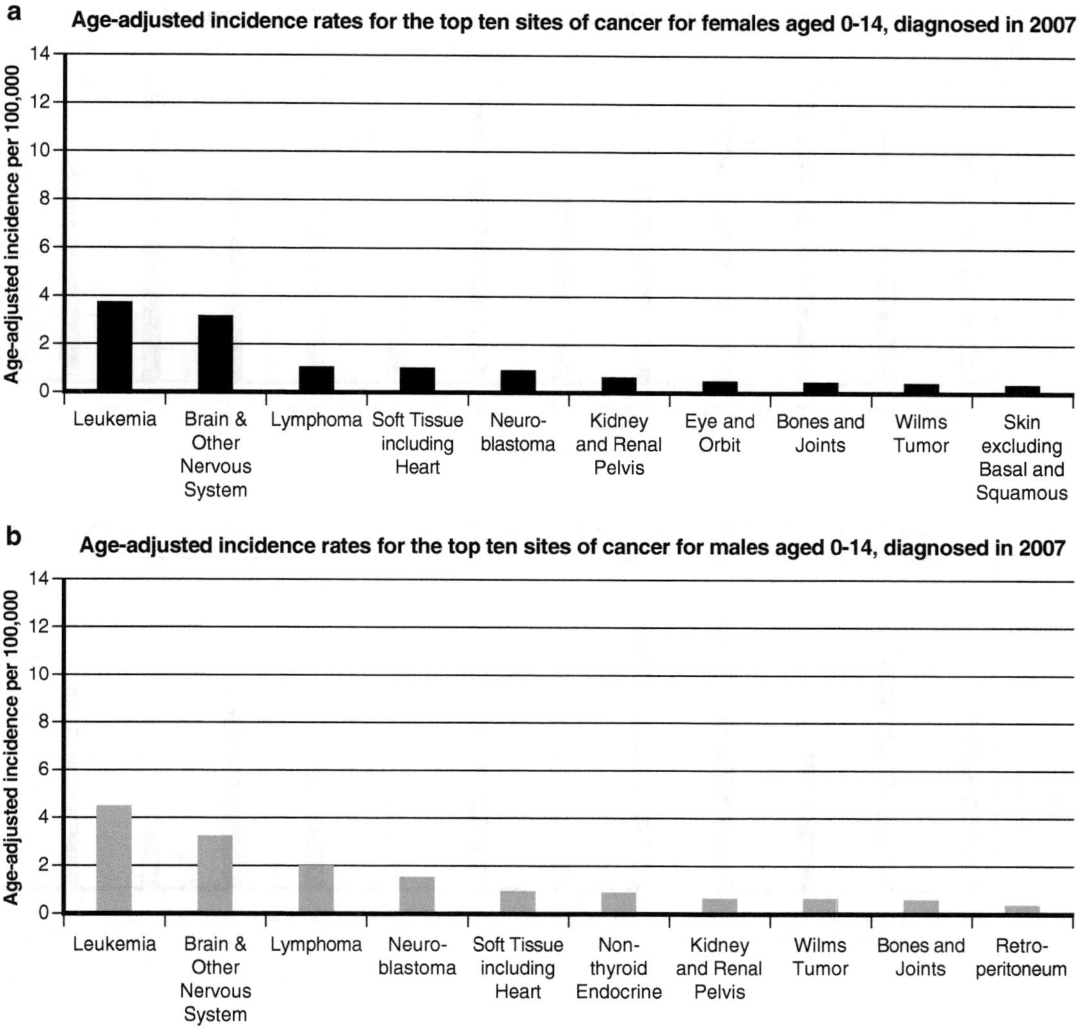

Fig. 1.2 (**a**) Age-adjusted incidence rates for the top ten sites of cancer for female subjects aged 0–14, diagnosed in 2007. (**b**) Age-adjusted incidence rates for the top ten sites of cancer for male subjects aged 0–14, diagnosed in 2007. Rates are for malignant cases and are adjusted to the 2000 US census population (data from the SEER 9 registries, including San Francisco–Oakland, Connecticut, Detroit, Hawaii, Iowa, New Mexico, Seattle, Utah, and Atlanta [7])

cancer survivor aged 15 was 65.6 years, compared with 76 years for the general US population, representing a loss of 10.4 years. However, this study was based on the data obtained over 20 years ago; life expectancy has likely increased subsequently, due to these improvements in survival rates. It has been estimated that there were 328,652 survivors of childhood cancer in the USA in 2005, of whom 27% were over the age of 39 [11].

Lupus Survivorship

In addition to cancer, other diseases may also be treated using therapies that compromise fertility. A common treatment for systemic lupus erythematosus (SLE) is the alkylating agent cyclophosphamide, which is known to cause ovarian failure and infertility in patients with this disease [12, 13]. Survival for all forms of lupus has increased

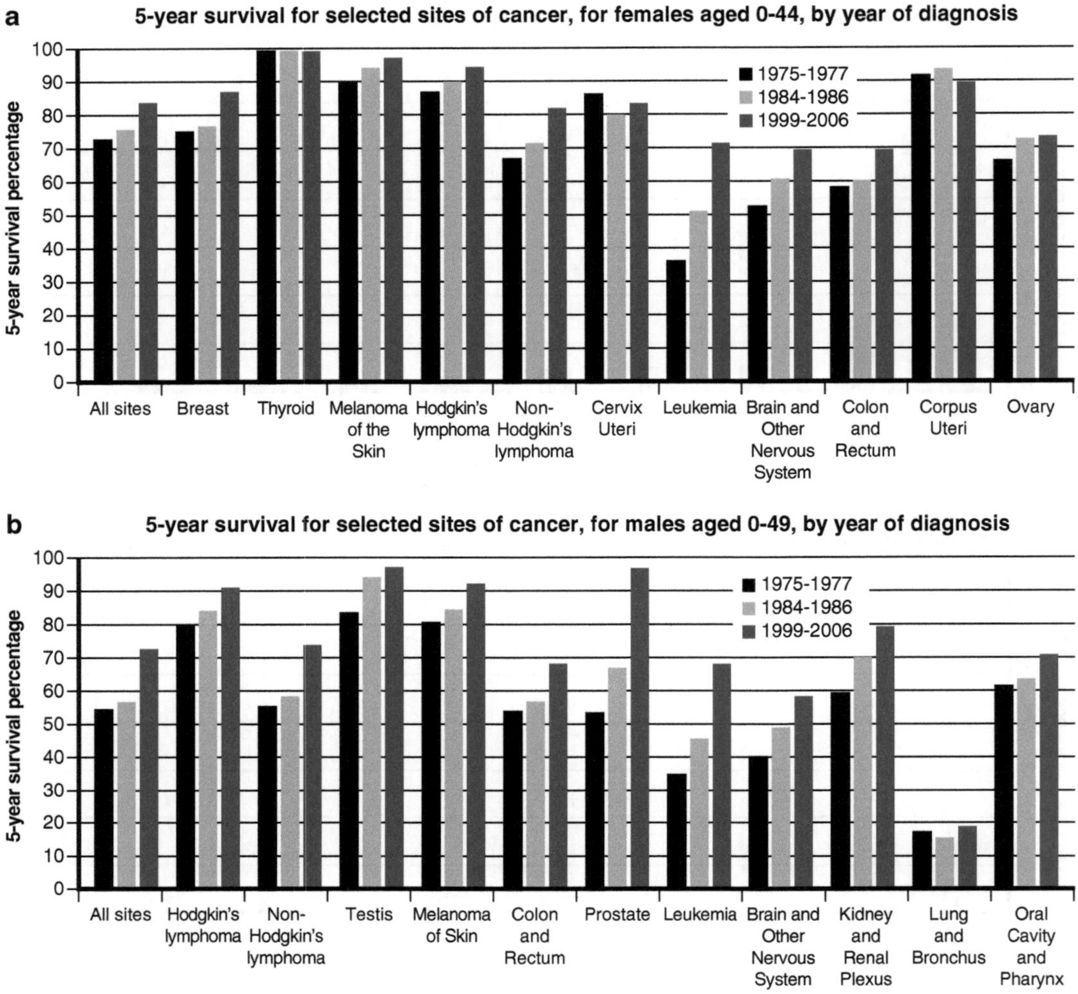

Fig. 1.3 (**a**) Five-year survival for selected sites of cancer for female subjects aged 0–44 by year of diagnosis. (**b**) Five-year survival for selected sites of cancer for male subjects aged 0–49 by year of diagnosis (data from the SEER 9 registries, including San Francisco–Oakland, Connecticut, Detroit, Hawaii, Iowa, New Mexico, Seattle, Utah, and Atlanta [7])

substantially, from a 5-year survival of 49% in 1953–1969 to 92% in 1990–1995 [14]. Today, 5-year survival rates are nearly 100%, and 10-year survival rates are almost 90% [15, 16].

Age of First Pregnancy

Among industrialized societies, various factors including prolonged education, career emphasis, longer lifespan, and availability of effective contraception have contributed toward an increasing age of first pregnancy. From 1970 to 2006, the average age of first pregnancy in the USA increased from 21.4 to 25.0 years [17]. Total birth rates and first birth rates have declined among women aged 25–29 years between 1990 and 2008, but have increased in all subsequent age groups (Fig. 1.5). One percent of first births were from women aged 35 or over in 1970, while this proportion had risen to 8.3% in 2006 [17, 18]. Although an increase in age of first pregnancy has been observed in all US states, there exists significant interstate variation, with a less than 2.5-year age increase between 1970 and 2006 in New Mexico, Mississippi, and Oklahoma, to a high of 5.0 years or more in Massachusetts, New Hampshire, and the District of Columbia [17].

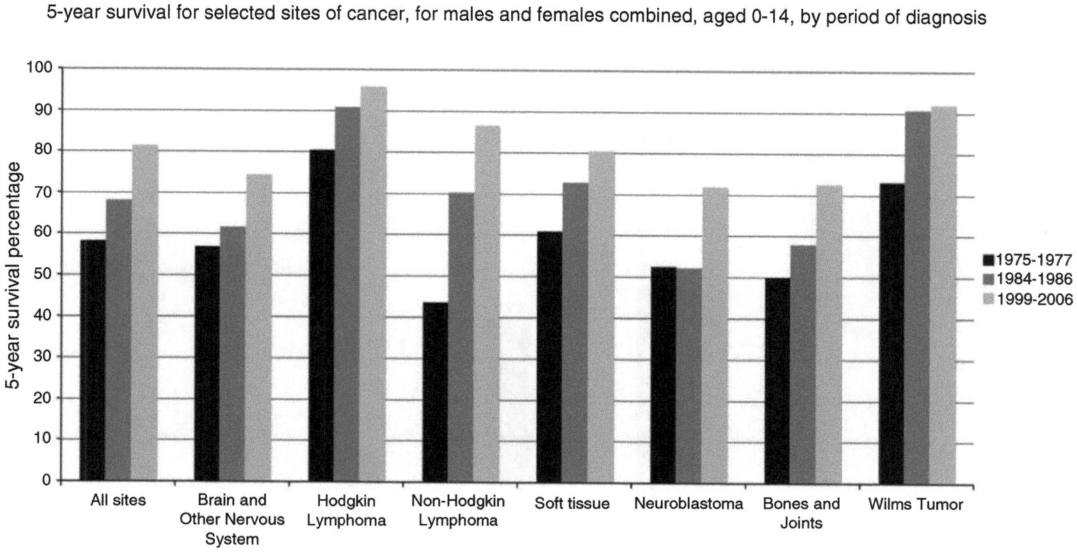

Fig. 1.4 Five-year survival for selected sites of cancer, for boys and girls combined, aged 0–14, by period of diagnosis (data from the SEER 9 registries, including San Francisco–Oakland, Connecticut, Detroit, Hawaii, Iowa, New Mexico, Seattle, Utah, and Atlanta [2])

Similar secular changes in age of first birth have been observed in other developed countries, including most European countries, Canada, and Japan (Fig. 1.6) [17, 97-101]. It has also been suggested that the delay in the age of first birth may be accompanied by a reduction in the interpregnancy interval (e.g., time between first and second birth), highlighting the increased urgency of childbearing toward the end of reproductive age [19].

Factors that Affect the Fertility of Cancer Survivors: An Overview

The risk and severity of infertility after cancer treatment are highly variable and depend on numerous patient and treatment factors. Patient factors include age of diagnosis, time since treatment, sex, pretreatment fertility, and site and stage of cancer, while treatment factors include type of drug, route of administration, site and size of radiation treatment, dose, and intensity of dose [20]. Treatment of several cancers may require sterilizing surgery. In addition, nonsterilizing surgery for cancers of the bladder, prostate, and rectum may also affect fertility by compromising potency and ejaculation [21]. Chemotherapy, and particularly alkylating agents such as cyclophosphamide and procarbazine, can significantly impact fertility, although there are clear dose-, drug-, gender- and age-dependent effects [20, 21]. Radiation at or near reproductive tissues and the pituitary gland [which produces reproductive hormones including follicle-stimulating hormone (FSH) and luteinizing hormone (LH)], or whole body radiation, may also strongly influence fertility [20, 21].

A large cohort study originating from the Childhood Cancer Survivor Study (CCSS) examined predictors of ever siring a pregnancy among 6,224 male subjects without surgical sterility aged 15–44, who were diagnosed before the age of 21 between 1970 and 1986 and survived for at least 5 years after diagnosis [3]. This study found that the likelihood of men ever siring a pregnancy was related to alkylating agent dose, hypothalamic/pituitary radiation dose, testicular radiation dose, type of cancer, and type of chemotherapeutic agent [3]. Subjects who had no exposure to alkylating agents or hypothalamic/pituitary/testes radiation had a likelihood of siring a pregnancy that was not significantly different from their control siblings [hazard ratio (HR): 0.91, 95% confidence interval (CI): 0.73–1.14]. Increasing doses of each of these exposures were clearly related a

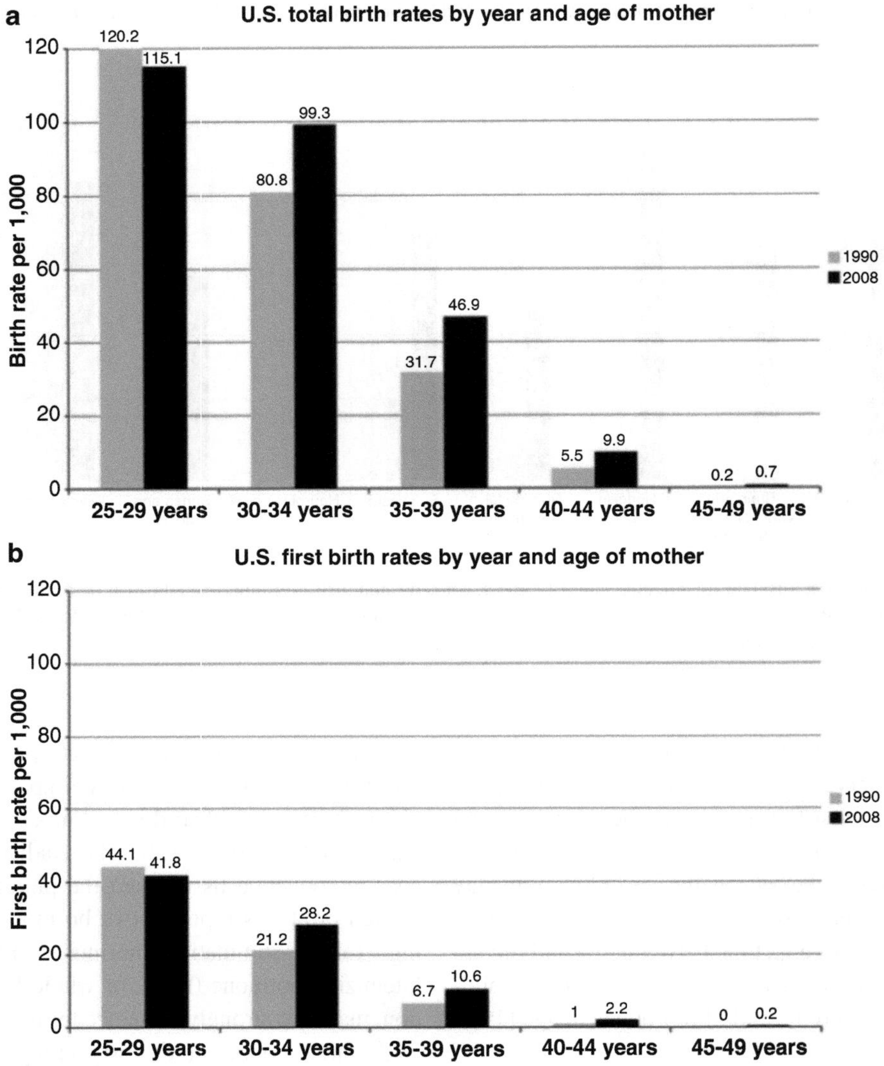

Fig. 1.5 (**a**) US total birth rates by year and age of mother. (**b**) US first birth rates by year and age of mother (data from US National Vital Statistics Reports [94–96])

decreasing likelihood of siring a pregnancy, while a significant variation of effect was observed depending on the type of chemotherapeutic agent. Younger age of diagnosis appeared to increase fertility, as those diagnosed between the age of 0 and 4 had an HR of siring of 1.80 (95% CI: 1.31–2.47), compared with those diagnosed between the age of 15 and 20.

Factors affecting fertility of female cancer survivors were also assessed in the CCSS, using 5,149 subjects without surgical sterility [4]. It was found that the likelihood of ever being pregnant 5 or more years after diagnosis was related to the increasing dose of ovarian/uterine and hypothalamic/pituitary radiation, alkylating agent dose, and type of chemotherapeutic agent. The highest dose of radiation was associated with a relative risk (RR) of infertility of 0.20 (95% CI: 0.14–0.29, for >10 Gy) for ovarian/uterine exposure, and 0.61 (95% CI: 0.44–0.85, for >30 Gy) for hypothalamic/pituitary exposure. The highest alkylating agent dose showed a RR of 0.76 (95% CI: 0.49–0.19). As with men, younger ages had an increased fertility, with those diagnosed between the age of 0 and 4 having an HR of pregnancy of 1.95 (95% CI: 1.51–2.54).

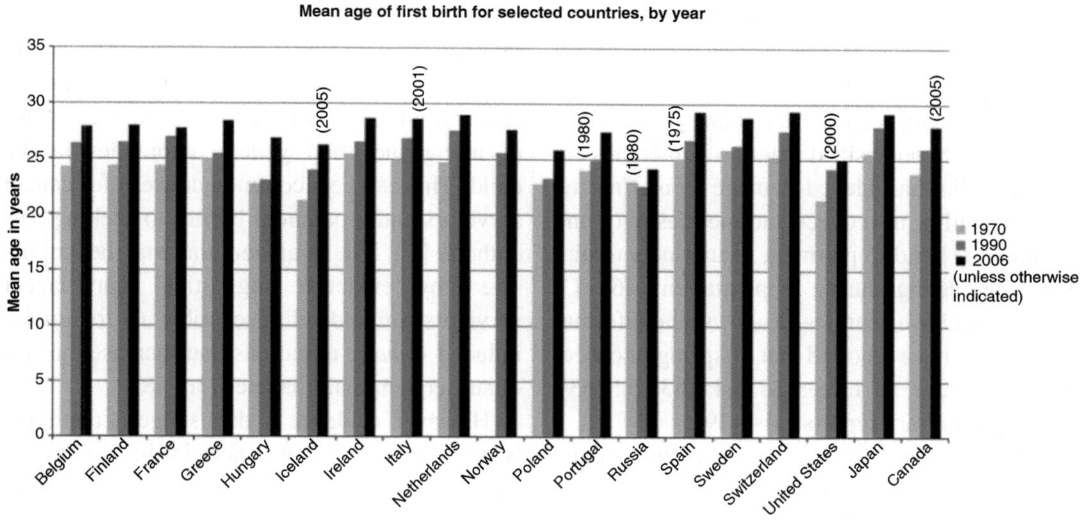

Fig. 1.6 Mean age of first birth for selected countries by year (data from [17, 97–101])

Risk of Maternal Cancer Recurrence

When considering fertility after cancer, patients and providers must discuss the possibility of recurrence of the cancer. For estrogen-associated cancers in particular, such as those of the breast, it has been hypothesized that pregnancy may promote recurrence via elevated levels of ovarian estrogens that stimulate estrogen-sensitive cell growth and survival. However, studies to date have overwhelmingly found that, for women with breast cancer who wait at least 6 months before attempting to conceive, pregnancy does not affect recurrence or survival compared with women who do not conceive [22–33]. In most cases, pregnancy has been shown to have a protective effect on breast cancer survival. This has been largely attributed to a spurious "healthy mother effect" whereby women who had a better prognosis of cancer were more likely to attempt to conceive, although it may also be due to a true protective effect of hormonal changes during pregnancy [22, 26, 34]. Breast cancer survivors may be advised to wait at least 2 years after treatment before attempting to conceive, to allow adjuvant therapy to be completed and to allow early recurrences to appear [22]. However, conflicting evidence exists regarding the effect of time after diagnosis on survival: while some studies report that time after diagnosis does not modify risk [27–29, 31], others report that the outcome is worse if conception is attempted prior to 6 months after diagnosis of breast cancer [24, 28, 30]. One study has found that, for women with a good breast cancer prognosis, there is a low risk of recurrence for all conception times after diagnosis, while for women with a poor prognosis, there is a high risk of occurrence up to, but not beyond, a wait time of 5 years before conception [34]. Information regarding pregnancy-associated recurrence of other sites of cancer is sparse. Two studies that have examined melanoma have found that pregnancy does not influence survival or recurrence [35, 36].

Health Risks for Offspring of Cancer Survivors

Perinatal Risks

Current evidence suggests that women exposed to abdominal radiation, but not chemotherapy, are at increased risk of pregnancy and neonatal complications. Numerous studies have suggested that abdominal radiation increases the risk of preterm delivery [37–43] and impaired fetal growth [37, 39, 40, 42–48]. Less frequently reported are increased risk for miscarriage

[45, 49, 50], postpartum hemorrhage [38], and fetal malposition [37]. Elevated rates of perinatal mortality have also been reported [39, 48], but conflicting evidence has been found [45, 49, 50]. It has been hypothesized that these outcomes are due to radiation-induced damage to uterine tissues, resulting in reduced uterine elasticity and volume [51] and impairment of uterine vasculature remodeling during implantation [46]. Of studies that have examined congenital malformations, most have reported that offspring of cancer survivors have no increased risk [43, 52, 53], although contrary reports exist [39].

The Childhood Cancer Survivor Study found that female survivors of childhood cancer had an overall odds ratio (OR) of preterm delivery of 1.9 (95% CI: 1.4–2.4), compared with the survivors' siblings [42]. Increasing dose of radiation to the uterus was associated with the increased risk of preterm birth, particularly among subjects who received radiation treatment before menarche; this latter affect was postulated to be due to a higher susceptibility of the immature uterus to radiation damage. Subjects with high doses of radiation to the uterus (250 cGy and above) showed an increased risk for low birth weight, but this was attributed to being born early rather than being an independent outcome. Only uterine radiation doses above 500 cGy showed a significantly increased risk of small-for-gestational-age (SGA). Radiation exposure to the ovaries or to the pituitary showed no significant effect on preterm delivery or low birth weight. Exposure to elevated doses of alkylating agents showed an increased but nonsignificant risk of preterm delivery; no effect was seen for low birth weight or SGA. Similarly, another large study, the British Childhood Cancer Survivor Study found that abdominal radiotherapy increased the risk of preterm delivery (OR 3.2, 95% CI: 2.1–4.7) and low birth weight (OR, 1.9, 95% CI: 1.1, 3.2), while no significant effects were seen for chemotherapy [40].

Germline Risks

For nonsterile cancer survivors who do not use cryopreserved sperm or embryos, radio- and chemotherapy raise the concern of treatment-related germline mutations affecting the health of offspring. Such germline mutations could theoretically manifest in the form of increased risk of malignancy and genetic anomalies in the child. This can be a common concern for cancer survivors and a potential barrier to seeking parenthood [54, 55], an anxiety that may be emphasized by evidence suggesting that cancer survivors are indeed susceptible to treatment-induced *somatic* mutations that increase the risk of subsequent malignancies [56–58].

However, evidence to date suggests that there is no increased risk of cancer in the children of cancer survivors, when cases with hereditary cancer syndromes are excluded. A population-based study in Finland of cancer patients diagnosed before the age of 35 found that children born more than 9 months after their parent's diagnosis showed no increased risk of cancer, compared with children of the cancer survivors' siblings, after removing individuals with hereditary cancer syndromes [incidence rate ratio (IRR) 1.03, 95% CI: 0.74–1.40, for all sites of cancer; median follow-up time, 14.9 years since birth] [59]. Moreover, no significant increased risks were observed when stratified by site of the cancer in the child. However, as expected, there was an increased risk of cancer if hereditary cancer cases were not removed (IRR 1.67, 95% CI: 1.29–2.12). Similarly, a population-based study in Nordic countries from 1943 to 1994 found a nonsignificant risk of all sites of cancer among offspring of cancer survivors (IRR 1.3, 95% CI: 0.8–2.0; median follow-up time, 14 years since birth), after removing likely cases of hereditary cancers [60]. The same was true when stratified by gender and age at diagnosis. The offspring of almost all subjects in this study were born at least 8 years after cancer was diagnosed in the mothers, and included follow-up to the age of 43 of the offspring. Earlier studies have likewise shown that any increased risk of cancer among offspring is due to familial aggregation rather than mutagenic effects, although these studies were limited by small case numbers and shorter follow-up time [49, 61–63].

Other manifestations of transgenerational genetic effects do not appear to be increased in offspring of cancer survivors, including no differences in cytogenetic syndromes such as Turner's and Down's syndromes, single-gene defects (Mendelian disorders), or genomic instability [64–67]. Finally, most studies have reported that there are no significant differences in the gender ratio of either male or female cancer survivors [45, 68–70]. Although one large study did find a significantly altered ratio [71], the authors postulated that this difference was not due to the increased incidence of lethal X chromosome mutations, but rather due to the lowered testosterone levels. Experiments in mice have suggested that chemotherapeutic agents including cyclophosphamide may induce aneuploidy in oocytes, early embryonic mortality, and fetal malformation, but it was postulated that oocytes maturing at the time of exposure are most susceptible to these risks, and that a sufficient interval time between exposure and pregnancy reduces these risks [72]. Chemotherapy has also been associated with aneuploidy in human spermatozoa, but these effects were also found to be transient, lasting for less than 100 days after chemotherapy [73].

The overall health of offspring born to cancer survivors does not appear to be different from expected. A population-based study in Denmark of offspring born between 1977 and 2003 found that children of cancer survivors showed no increased risk of hospitalization compared with the control population (HR 1.05, 95% CI: 0.98–1.12), after a median follow-up time to age 9.6 [74]. This was true for all discharge diagnoses, including discharges related to infections, all organ and metabolic systems, mental and behavioral problems, and injuries. There was an increased risk of malignant (HR 5.7, 95% CI: 3.0–10.8) and benign neoplasms (HR 2.0, 95% CI: 1.2–3.3), but this was attributed to hereditary cancers and increased surveillance, respectively.

Patient Awareness of Fertility Preservation Options

Patients can have significant knowledge deficits regarding options for fertility preservation and the impact of treatment on future fertility, which may be due to emotional stress, urgency of treatment, and a failure by providers to adequately discuss the issue. A systematic review of the literature using reports from 1999 to 2008 found that, among surveys addressing the issue, between 34 and 72% of patients are able to recall that they were counseled about the effect of treatment on fertility [5]. Other studies not included in this review have reported recall rates as high as 86% [75, 76].

Factors related to increased patient knowledge or recall of counseling include having treatment that was more likely to affect fertility, younger age, early stage of disease, participation in decision making, and lower anxiety [75, 77, 78]. Cancers located outside of the reproductive system and treatment without hormone therapy have been associated with the lack of patient knowledge [75]. Women may be less informed than men, possibly due to a belief by medical providers that cryopreservation methods for women are not as easy or effective as those for men [75].

Although the preceding studies are small patient surveys that are highly susceptible to selection bias (e.g., patients interested in fertility issues are more likely to respond to surveys), they do emphasize the fact that many patients have incomplete knowledge about fertility at the time of treatment, and that providers selectively provide this knowledge depending on their own beliefs and on characteristics of the patient. For childhood cancer survivors, parents may actively shield their children from learning of the consequences of treatment on future fertility, to avoid increasing the emotional burden of their children [55]. It has been noted that the lack of knowledge about fertility status may adversely affect the behavioral decisions of patients, such as whether or not to use contraception [55].

Attitudes and Awareness of Providers

Medical providers play a key role in patient awareness, decision making, and care-seeking behavior regarding fertility preservation. Patients may not raise the issue due to uncertainty or a lack of knowledge, if providers do not take the initiative in starting this conversation [79]. Surveys of physicians and nurses who are involved in the care of cancer patients report that most providers agree that fertility is an important topic worth discussing with their patients, and that general awareness of the topic is high [80–84]. A survey of oncologists, hematologists, and urologists found there was broad agreement that cryopreservation of sperm could help patients psychologically [85]. A majority of physicians report that they routinely discuss the possibility of fertility loss with their patients, particularly among patients deemed to be at high or medium risk of infertility [80–82, 86]. Oncology nurses are less likely to discuss fertility issues with their patients, even though they may believe that it is within their scope of practice [87, 88].

Although fertility may be considered to be an important topic, it is often deemed to be of low priority, both by individual physicians and medical institutions [82, 85], and providers may believe that their patients are more concerned about the topic than they are [89]. One study noted that there exists an important distinction between *discussing* and *mentioning* fertility loss, and that providers may be more likely to do the latter, which may have a significant effect on how the patient considers the issue [82].

Even if providers routinely discuss the consequences of fertility with their patients, actual referral rates for fertility preservation may be low. Several surveys have found that a large proportion of physicians do not routinely refer their patients for fertility preservation [79, 82, 84, 85]. A Dutch study conducted in 2002 reported that only 2% of female patients were referred for fertility preservation [90]. Although this low rate is likely no longer representative given subsequent improvements in cryopreservation options for women, recent studies have likewise suggested that referrals are more common for males than females [80, 81]. Moreover, postpubertal boys are more likely to be referred than early pubertal patients [80]. Some studies have suggested that female physicians are more likely to provide referrals than male physicians [79].

Disease characteristics that are commonly cited by providers to influence decisions over whether to discuss or refer for fertility preservation include urgency of treatment, disease stage, cancer site, age at diagnosis, and treatment type [82, 84, 85, 87, 89]. Additional factors cited by providers that affect likelihood of discussion include level of provider knowledge of fertility preservation options, patient HIV disease status, number of prior children, perception of high cost of fertility services, lack of convenient facilities, marital/relationship status of the patient, patient sexuality, patient interest in fertility preservation, and availability of educational materials [81, 82, 84, 87, 88]. Some providers did not refer because they believed that patients did not assign a high priority to fertility preservation [82, 86], or because they believed that success rates were too low [81]. Difficulty in accommodating parental wishes over whether or not to discuss fertility preservation may also be a factor for pediatric patients [88, 91]. Surveys among nurses have also mentioned that provider comfort level, including concerns over whether or not adolescent boys should be provided with erotic materials during semen collection, may play a role [87, 88].

Providers may lack the knowledge and training to adequately address the issue of fertility preservation. Numerous surveys have reported that oncology physicians and nurses lack knowledge regarding fertility preservation options, costs, resources, technology, efficacy, and safety, which may lead to underutilization of such services [81, 82, 84, 85, 89]. Seniority may be a factor in awareness of these issues, as more senior physicians may have less training and be less knowledgeable [82]. Physicians may have no formal training in fertility preservation, apart from observing attendings during their fellowship [82]. For training that is provided, it may only involve organ preservation methods rather than cryopreservation [82]. Lack of knowledge may also extend to awareness of the impact of cancer treatment on fertility, as one study found that only 36–62% of respondents were able to correctly ascribe the risk of gonadotoxocity for several hypothetical cases as based on American Society of Clinical Oncology guidelines [86].

While most studies cited above were surveys conducted among small numbers of physicians and nurses, they point toward a general trend that patient needs for fertility preservation are being incompletely met, and that providers are not adequately prepared or motivated to deal with these issues. With recent advances in fertility preservation methods and increasing focus on long-term quality of life for cancer survivors, this situation is likely to improve in the near future. In 2006, the American Society of Clinical Oncology published guidelines recommending that oncologists should discuss fertility preservation with reproductive age patients, and provide referrals for reproductive services if necessary [20]. Although this will help standardize practice in this area, it has been noted that physician adherence to guidelines may be compromised if physicians do not believe in the efficacy of fertility preservation [82]. Thus, the development of educational materials and training programs for physicians will be important in ensuring that patient needs are met. Social workers and nurses may also be suitable recipients of such training, as it has been suggested that they are in an ideal position to address fertility preservation needs with patients [87, 92, 93].

Finally, providers may need ethical training regarding issues surrounding the recommendation and provisioning of fertility preservation services. Expectations over reduced life span should not be a reason for denying cancer survivors access to fertility services [83]. Similarly, factors such as patient HIV status, number of prior children, marital status, sexuality, and socioeconomic status should not be determinants of whether or not a provider discusses or refers for fertility issues.

Conclusions

Cancer survivors are living longer lives in ever-greater numbers, due to significant improvements in cancer therapy and changes in population age structures. Thus, the focus from merely living to living well has highlighted the need for fertility preserving methods during cancer treatment. Pregnancy for women with a previous diagnosis of cancer is safe, with no evidence of increased risk of recurrence. However, offspring of such women may be susceptible to preterm delivery, low birth weight, and other perinatal morbidities. Both patients and providers alike may have poor knowledge about the fertility preservation during cancer, which highlights the need for increased education in this area.

References

1. Jemal A, Siegel R, Ward E, et al. Cancer statistics, 2007. CA Cancer J Clin. 2007;57:43–66.
2. Seer cancer statistics review 1975–2007.
3. Green DM, Kawashima T, Stovall M, Leisenring W, Sklar CA, Mertens AC, et al. Fertility of male survivors of childhood cancer: a report from the childhood cancer survivor study. J Clin Oncol. 2010;28:332–9.
4. Green DM, Kawashima T, Stovall M, Leisenring W, Sklar CA, Mertens AC, et al. Fertility of female survivors of childhood cancer: a report from the childhood cancer survivor study. J Clin Oncol. 2009;27:2677–85.
5. Tschudin S, Bitzer J. Psychological aspects of fertility preservation in men and women affected by cancer and other life-threatening diseases. Hum Reprod Update. 2009;15:587–97.
6. Probability of developing or dying of cancer software, version 6.4.1. Statistical research and applications branch, national cancer institute, 2009. http://srab.Cancer.Gov/devcan
7. Surveillance, epidemiology, and end results (seer) program (www.Seer.Cancer.Gov) seer*stat database: nov 2009 sub (1973–2007), national cancer institute, dccps, surveillance research program, cancer statistics branch, released april 2010, based on the november 2009 submission.
8. Berrino F, De Angelis R, Sant M, Rosso S, Bielska-Lasota M, Coebergh JW, et al. Survival for eight major cancers and all cancers combined for european adults diagnosed in 1995–99: results of the eurocare-4 study. Lancet Oncol. 2007;8:773–83.
9. Karim-Kos HE, de Vries E, Soerjomataram I, Lemmens V, Siesling S, Coebergh JW. Recent trends of cancer in Europe: a combined approach of incidence, survival and mortality for 17 cancer sites since the 1990s. Eur J Cancer. 2008;44:1345–89.
10. Yeh JM, Nekhlyudov L, Goldie SJ, Mertens AC, Diller L. A model-based estimate of cumulative excess mortality in survivors of childhood cancer. Ann Intern Med. 2010;152:409–17, W131–408.
11. Mariotto AB, Rowland JH, Yabroff KR, Scoppa S, Hachey M, Ries L, et al. Long-term survivors of childhood cancers in the united states. Cancer Epidemiol Biomarkers Prev. 2009;18:1033–40.
12. Katsifis GE, Tzioufas AG. Ovarian failure in systemic lupus erythematosus patients treated with pulsed intravenous cyclophosphamide. Lupus. 2004;13:673–8.
13. Langevitz P, Klein L, Pras M, Many A. The effect of cyclophosphamide pulses on fertility in patients with lupus nephritis. Am J Reprod Immunol. 1992;28:157–8.

14. Cameron JS. Lupus nephritis. J Am Soc Nephrol. 1999;10:413–24.
15. Nossent J, Kiss E, Rozman B, Pokorny G, Vlachoyiannopoulos P, Olesinska M, et al. Disease activity and damage accrual during the early disease course in a multinational inception cohort of patients with systemic lupus erythematosus. Lupus. 2010;19: 949–56.
16. Ravelli A, Ruperto N, Martini A. Outcome in juvenile onset systemic lupus erythematosus. Curr Opin Rheumatol. 2005;17:568–73.
17. Matthews TJ, Hamilton BE. Delayed childbearing: more women are having their first child later in life. NCHS Data Brief 2009;(21):1–8.
18. Mathews TJ, Hamilton BE. Mean age of mother, 1970–2000. Natl Vital Stat Rep. 2002;51(1):1–13.
19. Kalberer U, Baud D, Fontanet A, Hohlfeld P, de Ziegler D. Birth records from swiss married couples analyzed over the past 35 years reveal an aging of first-time mothers by 5.1 years while the interpregnancy interval has shortened. Fertil Steril. 2009;92:2072–3.
20. Lee SJ, Schover LR, Partridge AH, Patrizio P, Wallace WH, Hagerty K, et al. American society of clinical oncology recommendations on fertility preservation in cancer patients. J Clin Oncol. 2006;24: 2917–31.
21. Dohle GR. Male infertility in cancer patients: review of the literature. Int J Urol. 2010;17:327–31.
22. de Bree E, Makrigiannakis A, Askoxylakis J, Melissas J, Tsiftsis DD. Pregnancy after breast cancer: a comprehensive review. J Surg Oncol. 2010;101: 534–42.
23. Blakely LJ, Buzdar AU, Lozada JA, Shullaih SA, Hoy E, Smith TL, et al. Effects of pregnancy after treatment for breast carcinoma on survival and risk of recurrence. Cancer. 2004;100:465–9.
24. Clark RM, Chua T. Breast cancer and pregnancy: the ultimate challenge. Clin Oncol (R Coll Radiol). 1989;1:11–8.
25. Danforth Jr DN. How subsequent pregnancy affects outcome in women with a prior breast cancer. Oncology (Williston Park). 1991;5:23–30. discussion 30–21, 35.
26. Gelber S, Coates AS, Goldhirsch A, Castiglione-Gertsch M, Marini G, Lindtner J, et al. Effect of pregnancy on overall survival after the diagnosis of early-stage breast cancer. J Clin Oncol. 2001;19: 1671–5.
27. Harvey JC, Rosen PP, Ashikari R, Robbins GF, Kinne DW. The effect of pregnancy on the prognosis of carcinoma of the breast following radical mastectomy. Surg Gynecol Obstet. 1981;153:723–5.
28. Ives A, Saunders C, Bulsara M, Semmens J. Pregnancy after breast cancer: Population based study. BMJ. 2007;334:194.
29. Mignot L, Morvan F, Berdah J, Querleu D, Laurent JC, Verhaeghe M, et al. [Pregnancy after treated breast cancer. Results of a case-control study]. Presse Med. 1986;15:1961–4.
30. Mueller BA, Simon MS, Deapen D, Kamineni A, Malone KE, Daling JR. Childbearing and survival after breast carcinoma in young women. Cancer. 2003;98:1131–40.
31. Sankila R, Heinavaara S, Hakulinen T. Survival of breast cancer patients after subsequent term pregnancy: "Healthy mother effect". Am J Obstet Gynecol. 1994;170:818–23.
32. Velentgas P, Daling JR, Malone KE, Weiss NS, Williams MA, Self SG, et al. Pregnancy after breast carcinoma: outcomes and influence on mortality. Cancer. 1999;85:2424–32.
33. Weisz B, Schiff E, Lishner M. Cancer in pregnancy: maternal and fetal implications. Hum Reprod Update. 2001;7:384–93.
34. Largillier R, Savignoni A, Gligorov J, Chollet P, Guilhaume MN, Spielmann M, et al. Prognostic role of pregnancy occurring before or after treatment of early breast cancer patients aged <35 years: a get(n)a working group analysis. Cancer. 2009;115:5155–65.
35. Grin CM, Driscoll MS, Grant-Kels JM. The relationship of pregnancy, hormones, and melanoma. Semin Cutan Med Surg. 1998;17:167–71.
36. Reintgen DS, McCarty Jr KS, Vollmer R, Cox E, Seigler HF. Malignant melanoma and pregnancy. Cancer. 1985;55:1340–4.
37. Green DM, Peabody EM, Nan B, Peterson S, Kalapurakal JA, Breslow NE. Pregnancy outcome after treatment for wilms tumor: a report from the national wilms tumor study group. J Clin Oncol. 2002;20:2506–13.
38. Lie Fong S, van den Heuvel-Eibrink MM, Eijkemans MJ, Schipper I, Hukkelhoven CW, Laven JS. Pregnancy outcome in female childhood cancer survivors. Hum Reprod. 2010;25:1206–12.
39. Magelssen H, Melve KK, Skjaerven R, Fossa SD. Parenthood probability and pregnancy outcome in patients with a cancer diagnosis during adolescence and young adulthood. Hum Reprod. 2008;23: 178–86.
40. Reulen RC, Zeegers MP, Wallace WH, Frobisher C, Taylor AJ, Lancashire ER, et al. Pregnancy outcomes among adult survivors of childhood cancer in the british childhood cancer survivor study. Cancer Epidemiol Biomarkers Prev. 2009;18:2239–47.
41. Sanders JE, Hawley J, Levy W, Gooley T, Buckner CD, Deeg HJ, et al. Pregnancies following high-dose cyclophosphamide with or without high-dose busulfan or total-body irradiation and bone marrow transplantation. Blood. 1996;87:3045–52.
42. Signorello LB, Cohen SS, Bosetti C, Stovall M, Kasper CE, Weathers RE, et al. Female survivors of childhood cancer: preterm birth and low birth weight among their children. J Natl Cancer Inst. 2006;98: 1453–61.
43. Sudour H, Chastagner P, Claude L, Desandes E, Klein M, Carrie C, et al. Fertility and pregnancy outcome after abdominal irradiation that included or excluded the pelvis in childhood tumor survivors. Int J Radiat Oncol Biol Phys. 2010;76:867–73.

44. Chiarelli AM, Marrett LD, Darlington GA. Pregnancy outcomes in females after treatment for childhood cancer. Epidemiology. 2000;11:161–6.
45. Green DM, Whitton JA, Stovall M, Mertens AC, Donaldson SS, Ruymann FB, et al. Pregnancy outcome of female survivors of childhood cancer: a report from the childhood cancer survivor study. Am J Obstet Gynecol. 2002;187:1070–80.
46. Hawkins MM, Smith RA. Pregnancy outcomes in childhood cancer survivors: probable effects of abdominal irradiation. Int J Cancer. 1989;43: 399–402.
47. Salooja N, Szydlo RM, Socie G, Rio B, Chatterjee R, Ljungman P, et al. Pregnancy outcomes after peripheral blood or bone marrow transplantation: a retrospective survey. Lancet. 2001;358:271–6.
48. Li FP, Gimbrere K, Gelber RD, Sallan SE, Flamant F, Green DM, et al. Outcome of pregnancy in survivors of wilms' tumor. JAMA. 1987;257:216–9.
49. Hawkins MM, Draper GJ, Smith RA. Cancer among 1,348 offspring of survivors of childhood cancer. Int J Cancer. 1989;43:975–8.
50. Winther JF, Boice Jr JD, Svendsen AL, Frederiksen K, Stovall M, Olsen JH. Spontaneous abortion in a danish population-based cohort of childhood cancer survivors. J Clin Oncol. 2008;26:4340–6.
51. Critchley HO, Bath LE, Wallace WH. Radiation damage to the uterus – review of the effects of treatment of childhood cancer. Hum Fertil (Camb). 2002;5:61–6.
52. Sawka AM, Lakra DC, Lea J, Alshehri B, Tsang RW, Brierley JD, et al. A systematic review examining the effects of therapeutic radioactive iodine on ovarian function and future pregnancy in female thyroid cacer survivors. Clin Endocrinol (Oxf). 2008;69:479–90.
53. Winther JF, Boice Jr JD, Frederiksen K, Bautz A, Mulvihill JJ, Stovall M, et al. Radiotherapy for childhood cancer and risk for congenital malformations in offspring: a population-based cohort study. Clin Genet. 2009;75:50–6.
54. Langeveld NE, Ubbink MC, Last BF, Grootenhuis MA, Voute PA, De Haan RJ. Educational achievement, employment and living situation in long-term young adult survivors of childhood cancer in the netherlands. Psychooncology. 2003;12:213–25.
55. Zebrack BJ, Casillas J, Nohr L, Adams H, Zeltzer LK. Fertility issues for young adult survivors of childhood cancer. Psychooncology. 2004;13:689–99.
56. Goldsby R, Burke C, Nagarajan R, Zhou T, Chen Z, Marina N, et al. Second solid malignancies among children, adolescents, and young adults diagnosed with malignant bone tumors after 1976: follow-up of a children's oncology group cohort. Cancer. 2008; 113:2597–604.
57. Meadows AT, Friedman DL, Neglia JP, Mertens AC, Donaldson SS, Stovall M, et al. Second neoplasms in survivors of childhood cancer: findings from the childhood cancer survivor study cohort. J Clin Oncol. 2009;27:2356–62.
58. Sankila R, Pukkala E, Teppo L. Risk of subsequent malignant neoplasms among 470,000 cancer patients in finland, 1953–1991. Int J Cancer. 1995;60: 464–70.
59. Madanat-Harjuoja LM, Malila N, Lahteenmaki P, Pukkala E, Mulvihill JJ, Boice Jr JD, et al. Risk of cancer among children of cancer patients – a nationwide study in finland. Int J Cancer. 2010;126: 1196–205.
60. Sankila R, Olsen JH, Anderson H, Garwicz S, Glattre E, Hertz H, et al. Risk of cancer among offspring of childhood-cancer survivors. Association of the nordic cancer registries and the nordic society of paediatric haematology and oncology. N Engl J Med. 1998;338: 1339–44.
61. Bessho F, Kobayashi M. Adult survivors of children's cancer and their offspring. Pediatr Int. 2000;42: 121–5.
62. Hawkins MM, Draper GJ, Winter DL. Cancer in the offspring of survivors of childhood leukaemia and non-hodgkin lymphomas. Br J Cancer. 1995;71: 1335–9.
63. Mulvihill JJ, Myers MH, Connelly RR, Byrne J, Austin DF, Bragg K, et al. Cancer in offspring of long-term survivors of childhood and adolescent cancer. Lancet. 1987;2:813–7.
64. Bajnoczky K, Khezri S, Kajtar P, Szucs R, Kosztolanyi G, Mehes K. No chromosomal instability in offspring of survivors of childhood malignancy. Cancer Genet Cytogenet. 1999;109:79–80.
65. Byrne J, Rasmussen SA, Steinhorn SC, Connelly RR, Myers MH, Lynch CF, et al. Genetic disease in offspring of long-term survivors of childhood and adolescent cancer. Am J Hum Genet. 1998;62:45–52.
66. Tawn EJ, Whitehouse CA, Winther JF, Curwen GB, Rees GS, Stovall M, et al. Chromosome analysis in childhood cancer survivors and their offspring–no evidence for radiotherapy-induced persistent genomic instability. Mutat Res. 2005;583:198–206.
67. Winther JF, Boice Jr JD, Mulvihill JJ, Stovall M, Frederiksen K, Tawn EJ, et al. Chromosomal abnormalities among offspring of childhood-cancer survivors in denmark: a population-based study. Am J Hum Genet. 2004;74:1282–5.
68. Chow EJ, Kamineni A, Daling JR, Fraser A, Wiggins CL, Mineau GP, et al. Reproductive outcomes in male childhood cancer survivors: a linked cancer-birth registry analysis. Arch Pediatr Adolesc Med. 2009;163: 887–94.
69. Reulen RC, Zeegers MP, Lancashire ER, Winter DL, Hawkins MM. Offspring sex ratio and gonadal irradiation in the british childhood cancer survivor study. Br J Cancer. 2007;96:1439–41.
70. Winther JF, Boice Jr JD, Thomsen BL, Schull WJ, Stovall M, Olsen JH. Sex ratio among offspring of childhood cancer survivors treated with radiotherapy. Br J Cancer. 2003;88:382–7.
71. Green DM, Whitton JA, Stovall M, Mertens AC, Donaldson SS, Ruymann FB, et al. Pregnancy outcome of partners of male survivors of childhood

cancer: a report from the childhood cancer survivor study. J Clin Oncol. 2003;21:716–21.
72. Meirow D, Epstein M, Lewis H, Nugent D, Gosden RG. Administration of cyclophosphamide at different stages of follicular maturation in mice: effects on reproductive performance and fetal malformations. Hum Reprod. 2001;16:632–7.
73. Robbins WA, Meistrich ML, Moore D, Hagemeister FB, Weier HU, Cassel MJ, et al. Chemotherapy induces transient sex chromosomal and autosomal aneuploidy in human sperm. Nat Genet. 1997;16:74–8.
74. Winther JF, Boice JD Jr, Christensen J, Frederiksen K, Mulvihill JJ, Stovall M, Olsen JH. Hospitalizations among children of survivors of childhood and adolescent cancer: a population-based cohort study. Int J Cancer. 2010;127:2879–87.
75. Mancini J, Rey D, Preau M, Malavolti L, Moatti JP. Infertility induced by cancer treatment: inappropriate or no information provided to majority of french survivors of cancer. Fertil Steril. 2008;90:1616–25.
76. Partridge AH, Gelber S, Peppercorn J, Sampson E, Knudsen K, Laufer M, et al. Web-based survey of fertility issues in young women with breast cancer. J Clin Oncol. 2004;22:4174–83.
77. Duffy CM, Allen SM, Clark MA. Discussions regarding reproductive health for young women with breast cancer undergoing chemotherapy. J Clin Oncol. 2005;23:766–73.
78. Schover LR. Psychosocial aspects of infertility and decisions about reproduction in young cancer survivors: a review. Med Pediatr Oncol. 1999;33:53–9.
79. Quinn GP, Vadaparampil ST, Lee JH, Jacobsen PB, Bepler G, Lancaster J, et al. Physician referral for fertility preservation in oncology patients: a national study of practice behaviors. J Clin Oncol. 2009;27:5952–7.
80. Anderson RA, Weddell A, Spoudeas HA, Douglas C, Shalet SM, Levitt G, et al. Do doctors discuss fertility issues before they treat young patients with cancer? Hum Reprod. 2008;23:2246–51.
81. Goodwin T, Elizabeth Oosterhuis B, Kiernan M, Hudson MM, Dahl GV. Attitudes and practices of pediatric oncology providers regarding fertility issues. Pediatr Blood Cancer. 2007;48:80–5.
82. Quinn GP, Vadaparampil ST, Gwede CK, Miree C, King LM, Clayton HB, et al. Discussion of fertility preservation with newly diagnosed patients: oncologists' views. J Cancer Surviv. 2007;1:146–55.
83. Robertson JA. Cancer and fertility: ethical and legal challenges. J Natl Cancer Inst Monogr 2005;(34):104–6.
84. Schover LR, Brey K, Lichtin A, Lipshultz LI, Jeha S. Oncologists' attitudes and practices regarding banking sperm before cancer treatment. J Clin Oncol. 2002;20:1890–7.
85. Allen C, Keane D, Harrison RF. A survey of irish consultants regarding awareness of sperm freezing and assisted reproduction. Ir Med J. 2003;96:23–5.
86. Forman EJ, Anders CK, Behera MA. A nationwide survey of oncologists regarding treatment-related infertility and fertility preservation in female cancer patients. Fertil Steril. 2010;94:1652–6.
87. King L, Quinn GP, Vadaparampil ST, Gwede CK, Miree CA, Wilson C, et al. Oncology nurses' perceptions of barriers to discussion of fertility preservation with patients with cancer. Clin J Oncol Nurs. 2008;12:467–76.
88. Vadaparampil ST, Clayton H, Quinn GP, King LM, Nieder M, Wilson C. Pediatric oncology nurses' attitudes related to discussing fertility preservation with pediatric cancer patients and their families. J Pediatr Oncol Nurs. 2007;24:255–63.
89. Zapzalka DM, Redmon JB, Pryor JL. A survey of oncologists regarding sperm cryopreservation and assisted reproductive techniques for male cancer patients. Cancer. 1999;86:1812–7.
90. Jenninga E, Hilders CG, Louwe LA, Peters AA. Female fertility preservation: practical and ethical considerations of an underused procedure. Cancer J. 2008;14:333–9.
91. de Vries MC, Bresters D, Engberts DP, Wit JM, van Leeuwen E. Attitudes of physicians and parents towards discussing infertility risks and semen cryopreservation with male adolescents diagnosed with cancer. Pediatr Blood Cancer. 2009;53:386–91.
92. Clayton H, Quinn GP, Lee JH, King LM, Miree CA, Nieder M, et al. Trends in clinical practice and nurses' attitudes about fertility preservation for pediatric patients with cancer. Oncol Nurs Forum. 2008;35:249–55.
93. King L, Quinn GP, Vadaparampil ST, Miree CA, Wilson C, Clayton H, et al. Oncology social workers' perceptions of barriers to discussing fertility preservation with cancer patients. Soc Work Health Care. 2008;47:479–501.
94. Hamilton BE, Sutton PD, Ventura SJ. Revised birth and fertility rates for the 1990s and new rates for Hispanic populations, 2000 and 2001: United States. Natl Vital Stat Rep. 2003;51:1–94.
95. Martin JA, Hamilton BE, Sutton PD, Ventura SJ, Menacker F, Kimeyer S, Matthews TJ. Births: final data for 2006. Natl Vital Stat Rep. 2006;57:1–104.
96. Hamilton BE, Martin JA, Ventura SJ. Births: preliminary data for 2008. Natl Vital Stat Rep. 2010;58:1–17.
97. Vienna Institute of Demography. European demographic data sheet 2008. Retrieved from: http://www.oeaw.ac.at/vid/datasheet/download/sources_notes_datasheet2008.pdf. [Retrieved 7/20/11].
98. Statistics Canada. Report on the demographic situation in Canada, 2005 and 2006. Retrieved from: http://dsp-psd.pwgsc.gc.ca/collection_2008/statcan/91-209-X/91-209-XIE2004000.pdf. [Retrieved 7/20/11].
99. Council of Europe Publishing. Recent demographic developments in Europe, 2002. Retrieved from: http://www.coe.int/t/e/social_cohesion/population/d%C3%A9mo211960EN.PDF. [Retrieved 7/20/11].
100. United Nations economic commission for Europe. Trends in Europe and North America. The statistical yearbook of the economic commission for Europe 2003.
101. Mathews TJ, Hamilton BE. Mean age of mother, 1970–2000. Natl Vital Stat Rep. 2002;51:1–13

Ethical Discussions in Approaching Fertility Preservation

2

Pasquale Patrizio

Abstract

Cancer survivors may wish to become parents, if they have lost their reproductive function, by using previously stored gametes or gonadal tissue. Fertility preservation serves such a wide range of medico-social circumstances, some quite unique, that patient care requires an individualized and multidisciplinary approach. In particular, fertility specialists offering fertility preservation options to cancer patients should be properly trained and knowledgeable to discuss patient's treatment plan, prognosis, as well as unusual health risks for future offspring and the potential harmful effects of pregnancy. Overall, there should be no ethical objections to offer these services since they are offered with the scope of preserving future fertility. However, in practice, there are objections: many options are still experimental; posthumous use of stored tissue or gametes carries some legal issues; concerns exist about the welfare of offspring resulting from an expected shortened life span of the parent; concerns exist about the welfare of children born using gametes frozen after chemotherapy already started; and reseeding of cancer is possible after transplanting cryopreserved tissue.

Keywords

Autonomy • Beneficence • Nonmaleficence • Justice • Veracidity • Informed consent • Ethical principles in fertility preservation

Ethical Principles: General Considerations

Before discussing the ethical dilemmas associated with fertility preservation, it is necessary to describe the ethical constructs most commonly used to formulate guidelines. It is also important

P. Patrizio (✉)
Yale Fertility Center and REI Medical Practice,
150 Sargent Drive (2nd Floor), New Haven,
CT 06511, USA
e-mail: Pasquale.Patrizio@yale.edu

E. Seli and A. Agarwal (eds.), *Fertility Preservation: Emerging Technologies and Clinical Applications*,
DOI 10.1007/978-1-4419-1783-6_2, © Springer Science+Business Media, LLC 2012

to realize that any policy or guideline needs to have a certain degree of built-in flexibility, because natural rights and human dignity, in the context of creating families, cannot always be clearly defined. It is understandably difficult to balance an individual right to reproductive autonomy and privacy with the societal obligations to protect the potentiality of life.

The most common ethical norms used to conduct ethical analyses are based on the five principle-based ethics. These principles are autonomy, beneficence, nonmaleficence, justice, and veracidity.

Autonomy

Autonomy or *respect* for persons, acknowledges an individual's right to hold views, makes choices and takes actions on the basis of personal values and beliefs. This principle is at the base of informed consent and respect for privacy. In the reproductive field, this principle is the basis for reproductive rights and reproductive choices. Some examples: Should fertility preservation be offered to a woman who knows that her prognosis for long-term survival is uncertain? Should a widower be permitted to use embryos frozen while his wife was alive, and to use a member of his wife's family as a gestational carrier?

As to each question posed, the formulation of guidelines must take into account the right of an individual to decide, with the understanding that procreative liberty has some limits when these rights conflict with the child's best interest. In formulating an informed consent, it is therefore important to anticipate these scenarios and request disposal directives for cryopreserved reproductive tissue or gametes or embryos.

Beneficence

Beneficence represents the obligation to promote the patient's well-being. Beneficence represents the balancing of risks and benefits of an intervention. It is the principle that dictates that subjects be protected from harm and that efforts be taken to secure their well-being. It may conflict with the principle of autonomy. For example, the woman with breast cancer that requests controlled ovarian hyperstimulation despite being positive for estrogen and progesterone receptors and carrying breast cancer gene mutations.

Nonmalficence

Nonmaleficence is the obligation to "do no harm" also known by its original Latin expression of "primum non nocere," which needs to be balanced with the principle of autonomy. Both beneficence and nonmaleficence may overlap, for example, a patient who wants to postpone the beginning of chemotherapy treatments and, contrary to the oncologist opinion, insists in undergoing fertility preservation.

Justice

Justice concerns fairness and equity, i.e., the need to be fair in sharing the burden (costs) and the distribution (benefits) of resources to all members of the community. In particular, the concept of *distributive justice* is often applied to situations requiring a decision about the equitable allocation of resources. The current practice of IVF in general and fertility preservation in particular however, is not fair. Since many options to preserve fertility are experimental, insurance companies are not covering these services. Particularly for low income people, these services are only offered on institutional grants or on charitable basis; consequently, most patients are unable to have access to these services.

Veracidity

The principle of *veracidity* stipulates that a provider always tells the truth to his/her patient and avoids the exploitation of vulnerable populations.

Fertility Preservation for Cancer: Ethical Considerations

Ethical Consideration for Fertility Preservation in Adults

Fertility preservation developed first with the intent of preserving the potential for genetic parenthood in adults or children at risk of sterility due to chemo- or radiotherapy. Today, many young patients with cancer are surviving (the 5-year survival rates for Caucasian and Hispanic American women have increased for Hodgkin's lymphoma from 86 to 98% in the quarter century to the year 2000, and for breast cancer from 78 to 91%) [1]. At the same time, diagnoses of some malignant diseases have become more prevalent (e.g., breast and testicular cancer) [2]. The net effect has been an increase in numbers of patients in their reproductive years at risk of sterilization or early menopause [1]. As a result of this progress, quality of life issues after cancer is becoming increasingly more important and protection of fertility is a preeminent quality of life paradigm.

Although there are many strategies to preserve fertility, embryo freezing and sperm freezing are the only established options while all the others are still considered experimental; experimental procedures include oocyte freezing, ovarian tissue or whole ovary freezing, and in vitro maturation of oocytes or in vitro folliculogenesis.

Likewise, for men, when the option of semen cryopreservation is not available as for prepubertal boys, the harvesting and isolation of spermatogonial cells from testicular biopsies or the freezing of testicular tissue for later transplantation or even xenografting, are being tested but remain highly experimental.

When using experimental techniques, the informed consent is essential and both women and men have the right to know all options concerning fertility preservation and their implications including the risks and costs involved. In addition, experimental procedures are considered under the umbrella of research protocols and thus should also be reviewed and approved by institutional review boards.

Providing thorough informed consent in recruiting persons to participate in research is the foundation for the ethical conduct of research. It is based on three components: adequate, comprehensible information; a competent decision-maker; and a voluntary decision process.

Patients have the right to know what will happen if they or any children that are created are injured or disabled, including issues related to health insurance and compensation. Providers obtaining consents for experimental techniques must focus not simply on disclosure but also on effective communication and comprehension. The use of quizzes and documenting responses to questions after information is presented are effective tools to assist in demonstrating that patients understand the experimental or innovative nature of some modes of fertility preservation. Consent to the use of one of the many therapeutic strategies may require involving a surrogate decision maker in the case of young children or mentally impaired persons.

From an ethical standpoint, the key reason for pursuing fertility preservation is to restore personal autonomy to those who might, in the future, become unable to conceive [3, 4]. The presentation of risk information is complicated by the fact that both the adult and their offspring may be involved. A core principle of medical ethics is to do no harm. Ideally, the decision about who is candidate for fertility preservation should be rendered by a team including a medical oncologist, a reproductive endocrinologist, a pathologist and a psychologist, all guided by written protocols that can be shared with patients [4]. Patients should not be provided with false hopes. Alternative plans including no intervention with the prospect of adoption or childlessness should also be a part of the discussion. Equity or ownership interests of caregivers in novel technologies utilized in research must be disclosed to potential subjects. It is reasonable in the absence of grant funds to seek reimbursement from patients to cover the expenses of the research, but there should be no charge for clinical fees until the experimental options have been proven safe and effective.

Concerns about the children of mothers affected by cancer fall into three categories: first, the possible shortened life expectancy of mothers should cancer recur meaning that children could be orphaned at an early age. Another is the health of the mothers' posttreatment and the fear that they will not have the energy and stamina to care for their children. The third is the health of the children born after the thawing and retransplantation of ovarian cortical tissue or from the in vitro maturation of follicles or oocytes. Therefore, in the absence of any long-term follow-up studies or registries, it is imperative that those involved with these techniques continue monitoring each of these issues. It also means that novel forms of fertility preservation involving, for example, ovarian and testicular harvesting for freezing should be performed only in a few specialized centers working with proper IRB permission, the capacity for follow-up with subjects, adequate social service support, and subjects signing ethics committee approved consents.

Ethical Consideration for Fertility Preservation in Children

Impaired future fertility is a possible consequence of exposure to cancer therapies even for children. This risk may be difficult for children to conceptualize, but potentially traumatic to them when they become adults.

Since the modalities that are available to children for preserving their fertility are limited by their sexual immaturity, they are all considered experimental. For prepubertal boys who cannot produce mature sperm, harvesting and cryopreservation of testicular stem cells with the hope of future autologous transplantation, or in vitro maturation, represent potential methods of fertility preservation. For girls, isolation and cryopreservation of ovarian cortical strips/primordial follicles followed by in vitro maturation of gametes, when fertility is desired, is a possible option. Extensive research is still required to refine these modalities to safely offer them to patients as therapies [4].

Again, assisted reproductive technologies must be scrutinized on the basis of efficacy and safety and they must be subjected to rigorous ethical deliberation by independent review board committees before they can be offered. The modalities involved in fertility preservation of young children are no exception to these rules.

In addition to ensuring that the basis for offering the intervention is scientifically sound, the execution of the intervention must be deemed ethically sound. This determination requires that the intervention in question be evaluated within an ethical framework that considers it in terms of beneficence, respect for persons (autonomy), and justice [5].

It can be argued that fertility preservation aimed at children is ethical because it prevents morbidity (reproductive and psychosocial) and it safeguards their reproductive autonomy [6]. Therefore, the main ethical question concerns the process and the techniques necessary to protect fertility. The special situation of children as research subjects and at the same time patients makes the provider open to potential abuse of the technologies in an impetus to have a breakthrough [6, 7]. To avoid this risk, it is prudent to have multiple caregivers involved in the consent process.

Programs must make every effort to minimize financial barriers to access for children and to work with patient advocacy groups to seek coverage for children and families who cannot afford to participate in fertility treatment or research. Research involved in childhood fertility preservation should be conducted on patients who could experience personal benefit from the research, eliminating the prospect of exploitation for the gain of others.

Children represent a unique and vulnerable population with respect to medical research. They have diminished autonomy, diminished capacity to understand the risks and benefits of research objectives and lack the ability to provide consent for research studies. As a result, they require special protection against potential violation of their rights that may occur during research investigations [4, 5]. Until very recently, institutional attitudes impeded significant participation by children in medical research for the fear of exploitation.

Children should not be exploited to participate in pediatric research, nor should they be deprived of the benefits research has to offer because of their vulnerable status. With respect to childhood fertility preservation, proper attainment of informed consent from a legally authorized representative (i.e., parent or guardian) and of childhood assent must be ensured [4, 5]. Assent – the active affirmation by the research subject – can be obtained from incompetent minors and it should be obtained from children whenever possible. While the benefits of gamete cryopreservation are promising, they are largely unquantifiable because human data on the survival of gametes after the freeze-thaw-transplant process are limited. Until more data become available providers cannot tell patients what percentage of gametes will survive and what the probability of conception is, and must not provide patients with false hope. Alternatives to gamete cryopreservation should be discussed and patients should be given the option of no intervention [6]. Barriers to the consent process for fertility preservation interventions may develop. While parents may be competent to consent for their children, the scenario is very complex clinically and emotionally [7].

It has been suggested that to overcome some of the practical obstacles involved in the consent process, it should be performed in stages [8]. If a two-stage process is adopted, the issues of gonadal harvesting/storage and gamete manipulation can be handled as two separate topics at distinct time points. The decision to harvest gametes would be made at the time of cancer diagnosis and consent for the procedure would be left to parents/guardians. The decision of whether to use the gametes after they have been isolated can then be made at a future point by the child, when adulthood is reached. At such a point in time, the young patient would be better able to express personal preferences about the handling of the tissue based on an enhanced capacity to understand the ramifications of the possible medical interventions available at that time.

Summary of Clinical Outcomes Within the Context of Informed Consent

Health of Children Born from Oocyte Cryopreservation

Many patients with cancer are choosing the option of oocyte cryopreservation as a fertility preservation strategy. However, despite the birth of hundreds of children, the ASRM still considers this technology as experimental [9, 10]. Recent summary reports have documented that babies born from the use of cryopreserved oocytes by vitrification methods (outcome of 200 infants born) are not at higher risk for congenital malformations and not at an increased risk of adverse perinatal outcome [9].

Another study summarized 58 reports (1986–2008) that included 609 live born babies (308 from slow freezing, 289 from vitrification, and 12 from both methods) [10]. In addition, 327 other live births were verified. Of the total 936 infants, 1.3% [12] had a birth anomaly: three ventricular septal defects, one choanal and one biliary atresia, one Rubinstein-Taybi syndrome, one Arnold-Chiari syndrome, one cleft palate, three clubfoot, and one skin hemangioma. On the whole, these observations demonstrate that, so far, children conceived from oocyte freezing are healthy and not at an increased risk of adverse outcome.

Health of Children Born from Cryopreserved Embryos

The first prospective study aimed at assessing the postnatal growth and development of children born from cryopreserved embryos was carried out in 1995 [11]. The findings of that study were that children conceived from cryopreservation had a lower mean birth weight and mean gestational age, but the incidence of minor and major congenital malformations was similar to that of a control group and, furthermore, these children performed on a similar functional level as the control group.

A recent systematic review evaluated the medical outcome of children born after cryopreservation, slow freezing and vitrification of early cleavage stage embryos, blastocysts and oocytes during the years 1984–2008 [12]. Most studies found comparable malformation rates between frozen and fresh ART cycles and overall data concerning infant outcome and psychological well-being after cryopreservation of embryos were reassuring. As for oocyte cryopreservation data, the number of properly controlled follow-up studies of neonatal outcome after embryo cryopreservation is still somewhat limited. Long-term follow-up studies for all cryopreservation techniques are also essential.

Health of Children Born from Frozen/Retransplanted Human Ovarian Tissue

Ovarian cryopreservation and transplantation techniques, either as heterotopic or orthotopic allografts, are becoming steadily more successful. So far, 14 children have been born worldwide as a result of transplanting frozen/thawed ovarian tissue [13–17]. The very first was born in Belgium in 2004 [13] and the subsequent births have been reported in Israel in 2005 [14], Denmark [15, 16], Spain (birth of twins) [17], and in the USA [18]. One of the patients from Denmark gave birth to two children [19]. Recently, [20] birth from a noncancer patient with thawed ovarian transplants grafted in the pelvic sidewall was reported in France.

Many births have also been reported by using fresh ovarian transplants between monozygotic twins [18]. Ten monozygotic twin pairs requested ovarian transplantation and nine have undergone the procedure (some after failing oocyte donation from their sisters) with cryopreservation of spare tissue. All recipients reinitiated ovulatory menstrual cycles and showed normal day 3 serum FSH levels by 77–142 days. Seven conceived naturally (three twice). Currently, seven healthy babies have been delivered out of ten pregnancies using fresh ovarian tissue transplants. The oldest transplant ceased functioning by 3 years, but this patient conceived again after a second transplant using spare frozen-thawed tissue. Very recently, a birth from the transplant of a whole fresh ovary between two sisters HLA-compatible has also been reported [21].

In summary, when providing an informed consent, it is perfectly legitimate to offer these encouraging but still preliminary results. International fertility preservation society and national special interest groups in both USA and Europe are also closely monitoring the field with follow-up registry.

Conclusions

Cancer survivors may wish to become parents, if they have lost their reproductive function, by using previously stored gametes or gonadal tissue. Fertility preservation serves such a wide range of medico-social circumstances, some quite unique, that patient care requires an individualized and multidisciplinary approach. In particular, fertility specialists offering fertility preservation options to cancer patients should be properly trained and knowledgeable to discuss patient's treatment plan, prognosis, as well as unusual health risks for future offspring and the potential harmful effects of pregnancy.

Overall, there should be no ethical objections to offer these services since they are offered with the scope of preserving future fertility.

However, in practice, there are objections:

1. The options available, except sperm storage and embryo cryopreservation, are all experimental. There is a lack of extended follow-ups about their safety.
2. Posthumous use of stored tissue or gametes. When gametes or tissue is stored for later use, written directives for posthumous use may be given effect, and subsequently born children may be recognized as legal offspring of the deceased. Postmortem reproduction with stored gametes or tissue should be honored when the deceased has given specific consent; programs storing gametes, embryos, or gonadic tissue from cancer patients should inform the options for making advance directives for future use. Whether posthumously conceived or implanted offspring will inherit property from the deceased or will qualify for government benefits will depend on the law of the jurisdiction in which death occurs [22].
3. Concerns about the welfare of offspring resulting from an expected shortened life span of the parent. This concern, however, should not be a sufficient reason to deny cancer survivors assistance in reproducing. Although the effect of the early loss of a parent on a child is regrettable, many children experience stress and sorrow from other circumstances of their birth.

The risk that a cancer survivor will die sooner than other parents does not impose an appreciably different burden than the other causes of suffering and unhappiness that persons face in their lives. Protecting such children by preventing their birth altogether is not a reasonable ground for denying cancer survivors' chance to reproduce [22].

4. Concerns about the welfare of children born using gametes frozen after chemotherapy already started.
5. Reseeding of cancer after transplanting cryopreserved tissue.

Future successful production of germ cells de novo could have applications in fertility preservation. Sterile gonads would no longer limit reproduction as it will be possible to produce artificial gametes by dedifferentiation of somatic cells.

References

1. Gosden RG. Fertility preservation-definition, history and prospect. Semin Reprod Med. 2009;27(6):433–7.
2. SEER Cancer Statistics Review, 1975–2000, National Cancer Institute, Bethesda, MD, (http://seer.cancer.gov)
3. Bromer JG, Patrizio P. Gynecologic management of fertility. In: Wingard JR, Gastineau D, Leather H, Snyder EL, Szczepiorkowski ZM, editors. Hematopoietic stem cell transplantation: a clinician's handbook. Bethesda, MD: AABB Press; 2009.
4. Patrizio P, Butt S, Caplan A. Ovarian tissue preservation and future fertility: emerging technologies and ethical considerations. J Natl Cancer Inst Monogr. 2005;34:107–10.
5. Hirtz DG, Fitzsimmons LG. Regulatory and ethical issues in the conduct of clinical research involving children. Curr Opin Pediatr. 2002;14:669–75.
6. Grundy R, Larcher V, Gosden RG, Hewitt M, Leiper A, Spoudeas HA, et al. Fertility preservation for children treated for cancer (2): ethics of consent for gamete storage and experimentation. Arch Dis Child. 2001;84:360–2.
7. Grundy R, Larcher V, Gosden RG, Hewitt M, Leiper A, Spoudeas HA, et al. Fertility preservation for children treated for cancer (1): scientific advances and research dilemmas. Arch Dis Child. 2001;84:355–9.
8. Bahadur G. Ethics of testicular stem cells medicine. Hum Reprod. 2004;19:2702–10.
9. Chian RC, Huang JY, Tan SL, et al. Obstetric and perinatal outcome in 200 infants conceived from vitrified oocytes. Reprod Biomed Online. 2008;16:608–10.
10. Noyes N, Porcu E, Borini A. Over 900 oocyte cryopreservation babies born with no apparent increase in congenital anomalies. Reprod Biomed Online. 2009; 18(6):769–76.
11. Sutcliffe AG, Dsouza SW, Cadman J, et al. Minor congenital anomalies, major congenital malformations and development in children conceived from cryopreserved embryos. Hum Reprod. 1995;10: 3332–7.
12. Wennerholm UB, Soderstrom-Anttila V, Bergh C, Aittomaki K, Hazekamp J, Nygren KG, et al. Children born after cryopreservation of embryos or oocytes: a systematic review of outcome data. Hum Reprod. 2009;24(9):2158–72.
13. Donnez J, Dolmans MM, Demylle D, et al. Livebirth after orthotopic transplantation of cryopreserved ovarian tissue. Lancet. 2004;364(9443):1405–10.
14. Meirow D, Levron J, Eldar-Geva T, et al. Pregnancy after transplantation of cryopreserved ovarian tissue in a patient with ovarian failure after chemotherapy. N Engl J Med. 2005;353(3):318–21.
15. Demeestere I, Simon P, Emiliani S, Delbaere A, Englert Y. Fertility preservation: successful transplantation of cryopreserved ovarian tissue in a young patient previously treated for Hodgkin's disease. Oncologist. 2007;12(12):1437–42.
16. Andersen CY, Rosendahl M, Byskov AG, et al. Two successful pregnancies following autotransplantation of frozen/thawed ovarian tissue. Hum Reprod. 2008;23:2266–72.
17. Sanchez-Serrano M, Crespo J, Mirabet V, et al. Twins born after transplantation of ovarian cortical tissue and oocyte vitrification. Fertil Steril. 2010;93:268. e11–3.
18. Silber SJ, DeRosa M, Pineda J, Lenahan K, Grenia D, Gorman K, et al. A series of monozygotic twins discordant for ovarian failure: ovary transplantation (cortical versus microvascular) and cryopreservation. Hum Reprod. 2008;23:1531–7.
19. Ernst E, Bergholdt S, Jørgensen JS, Andersen CY. The first woman to give birth to two children following transplantation of frozen/thawed ovarian tissue. Hum Reprod. 2010;25:1280–1.
20. Roux C, Amiot C, Agnani G, Aubard Y, Rohrlich PS, Piver P. Live birth after ovarian tissue autograft in a patient with sickle cell disease treated by allogeneic bone marrow transplantation. Fertil Steril. 2010;93: 2413.e15–9.
21. Silber SJ, Grudzinskas G, Gosden RG. Successful pregnancy after microsurgical transplantation of an intact ovary. N Engl J Med. 2008;359:2617–8.
22. Robertson JA. Procreative liberty, harm to offspring, and assisted reproduction. Am J Law Med. 2004; 30:7–40.

Part II

Fertility Preservation in Women

Ovarian Follicle Development and Fertility Preservation

3

Yuichi Niikura and Joshua Johnson

Abstract

There are strong incentives to better understand the basic physiology of the mammalian ovary. Functioning ovaries have an enormous impact on the health and well-being of women beyond their role in conception. Further, survivorship after treatment for cancer and other life-threatening illnesses is increasing. Maintaining ovarian function during and after treatment is therefore critical, given the health benefits of functional ovaries and the potential desire to conceive. Basic research efforts have resulted in increased access to human ovary tissue, and a number of key technologies moving into widespread clinical use in less than a decade. In this chapter, we review the basic biology of the ovary and ovarian follicle, including recent data on stem cell support of the organ, and clarify the difference between oocyte and follicle maturation in vivo with the clinical goal of "oocyte maturation." Breakthroughs will continue to result from the interplay among infertility treatment, efforts to preserve fertility, and advances in our understanding of basic ovarian biology.

Keywords

Ovarian follicle development • Fertility preservation in women • Ovarian function after cancer treatment • Oocyte and follicle maturation • Infertility treatment in women

Optimally functioning ovaries, with their cohort of immature egg cells (oocytes) within follicles, produce the premenopausal hormonal and signaling milieu [1]. Development of an immature primordial follicle to a mature (preovulatory) state can be followed by ovulation of a mature egg, its fertilization, and subsequent offspring development. Over time, the total number of primordial follicles decreases for all women. The loss of follicles below a threshold amount results in ovarian failure, and thus menopause. Menopause that begins prior to age 40 is referred to as premature ovarian failure (POF). Unexplained POF occurs in up to 1% of the world's female population (and in 1 in

J. Johnson (✉)
Assistant Professor, Department of Obstetrics, Gynecology and Reproductive Sciences, Yale University School of Medicine, New Haven, CT, USA
e-mail: josh.johnson@yale.edu

E. Seli and A. Agarwal (eds.), *Fertility Preservation: Emerging Technologies and Clinical Applications*,
DOI 10.1007/978-1-4419-1783-6_3,

1,000 women between 15 and 29 years of age) [2], and results in significant health and well-being compromises for the remainder of a woman's life.

Ovarian failure poses health challenges far greater than the loss of fertility. Bone density, cardiovascular health, muscle mass, body fat composition, and many self-reported well-being measures [3] worsen upon ovarian failure. The earlier a woman experiences ovarian failure, the longer she faces such health and well-being compromises. Moreover, the safety of hormone replacement therapy (HRT) has recently been reassessed in a large multicenter trial. At least in its current form, HRT results in limited improvements in the above parameters, and also worsens a woman's risk of stroke [4] and cardiovascular disease [5]. Thus ovarian hormones, at least as currently administered, cannot replace functioning ovaries. Overall health care costs of managing ovarian failure are so high that they are not readily calculated. Annual sales of a single commercial hormone replacement preparation (Premarin) exceed two billion dollars in the USA. In light of such health risks, mixed benefits to women who undergo treatment, and costs, there is great benefit to be found in protecting ovarian function and fertility.

Recent strides in cancer treatment have resulted in steadily increasing numbers of women who survive their disease. When comparing all sites of disease, the average survival rate has increased by 11% when comparing data from 1975 to 1977 and 1996 to 2003. Survivorship from reproductive tract cancers (including that of the breast) has increased by approximately 8% over the same time frame [6]. We have reached the point where about 90% of girls and women will survive their cancer and treatments. Thus we can now increasingly focus on the "side effects" of treatment once a patient becomes well. Compromised ovary function and infertility that can result from cancer (and other) treatments are major concerns as patients seek to regain maximal quality of life.

It is well established that certain chemo- or radiotherapeutic regimens correlate with increased risk of gonadal failure. Female cancer survivors face compromised reproductive function as the very treatments that save lives are often toxic to the ovaries and the population of oocytes within. Chemotherapies that target rapidly growing cancer cells can be exquisitely toxic to oocytes [7] and surrounding granulosa cells [8]. Pelvic radiotherapy can also result in destruction of the oocyte population [9]. Here again, if the number of oocytes drops below a threshold of a few thousand, ovarian failure or "premature menopause" results.

There are two primary goals of Reproductive Endocrinology and Infertility research as applied to women: (1) the consistent generation of mature eggs capable of giving rise to healthy offspring when desired and (2) the discovery of methods to protect oocytes from damage and loss, so that they may support ovary function or be used for (1). The latter goal is now a key subdiscipline, referred to as "Fertility Preservation." We maintain that addressing the clinical goals of this field requires increased understanding of the basic mechanisms that control oocyte and follicle development. Here, we discuss the developmental history and adult function of oocytes and ovarian follicles. Special emphasis is placed on "oocyte maturation" as it is referred to in the clinical setting vs. the continuum of oocyte meiotic maturation within the growing follicle.

The Mammalian Oocyte and Ovarian Follicle

Ovarian and Oocyte Development

The early stages of oocyte development begin before birth. These stages include the critical stages of meiosis entry and follicle formation (Fig. 3.1, I and II). It is thought that, in humans, all primordial oocytes develop at this time and stay in a dormant stage for years, until sexual maturity is reached and some follicles begin to grow and mature. This means that decades may pass before an immature oocyte develops into a fertilization-competent egg. There are some data that suggest that the laboratory mouse and other vertebrates harbor ovarian stem cells that can produce new oocytes during adult life, a prospect we briefly consider next.

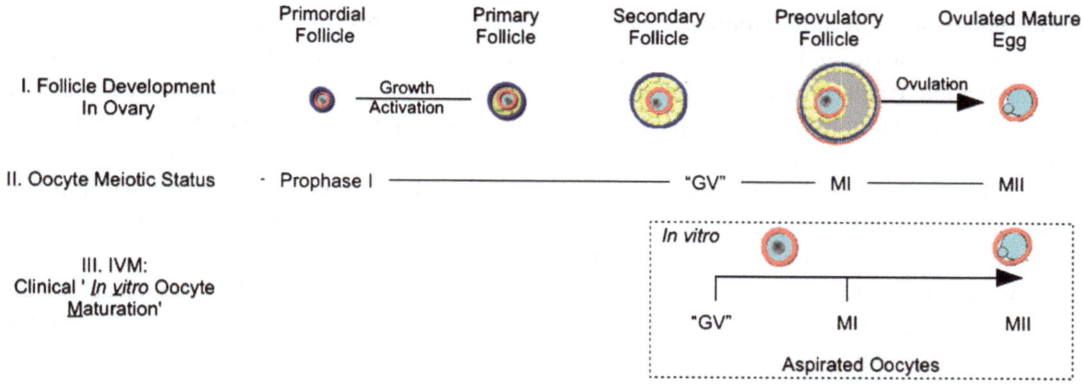

Fig. 3.1 Simplified staging of follicle development (I) and oocyte meiotic maturation (II) overlaid with clinical in vitro maturation (III): After follicle formation during fetal life, the pool of primordial follicles is essentially dormant. At puberty, cohorts of arrested *primordial follicles* undergo growth activation and transition to the *primary follicle* state. Subsequent follicle growth consists of proliferation of cumulus cells (*yellow*), growth of the oocyte (*blue*), and acquisition of the fluid-filled antral cavity (*gray*), culminating in the mature *preovulatory follicle*. Oocyte growth is accompanied by progression through the stages of meiosis (II). Arrest in prophase I ends with the development of the germinal vesicle ("GV") nucleus followed by meiotic progression through meiosis I (MI) and II (MII). Thus, coordinated follicle and oocyte development produces the ovulated mature egg. (III). Clinically speaking, in vitro maturation refers to the collection of oocytes in different meiotic stages during aspiration. If very few or no mature MII eggs are available, any remaining immature oocytes may be cultured in attempts to allow maturation to fertilization competence at MII

Until quite recently, all mammals were thought to have a fixed pool of oocytes endowed at birth that represent all those available for ovarian function and potential conception throughout life. Instead, multiple groups have detected the production of new oocytes in adult mice, generated by proposed female germ stem cells (fGSCs) [10–12] (see [13–15] for reviews). Further, it has been shown that fGSCs can be isolated from the adult ovary, can be cultured long-term in vitro, and that are capable of supporting offspring production after transplantation [11]. Since those mouse studies, adult production of new oocytes via stem cells has been directly visualized in a vertebrate, the *Medaka* fish [16]. Whether these data are relevant to humans remains controversial and is limited to the realm of experimental models for the foreseeable future.

Whatever their origins, mammalian ovarian follicles consist of an oocyte and intimately associated somatic cells that surround them, including granulosa cells. Maturation of this unit (see [17, 18] for reviews) occurs when the primordial follicle enters a period of growth that includes increased oocyte volume, granulosa cell proliferation, and the maturation of each cell type (Fig. 3.1). Development continues until the follicle either reaches a large, multilaminar preovulatory state or dies via a regulated apoptotic process termed follicle atresia. Fully mature ovulatory follicles contain perhaps tens of thousands of granulosa cells and surrounding theca cells that have proliferated and differentiated to support the mature oocyte. The signals that control follicle growth continue to be exhaustively characterized (see Sect. "Growth of the Mammalian Ovarian Follicle").

Oocyte Development and Meiosis

Oocytes arise from the primordial germ cells (PGCs) during embryo development. After PGC migration from the yolk sac and colonization of the urogenital ridge at embryonic stage (E, equivalent to days *post coitum*) 9.5–11.5 [19], they undergo a period of proliferation and initiate the specialized meiotic cell cycle. This begins with a single round of DNA synthesis known as premeiotic DNA replication, thereby entering prophase of the first meiotic division (Prophase I) at E13.5. Subsequent chromosome condensation and recombination result in the dictyate stage, at which point the chromosomes become dispersed and oocytes are surrounded by a single layer of flattened epithelial-like somatic cells (pregranulosa cells). Such Prophase I-arrested oocytes are referred to as primordial oocytes, which reside in primordial follicles [20].

Although a cohort of follicles begins to grow during prepubertal life, Prophase I arrest remains in effect until puberty. At puberty, the entire developmental process of oocyte maturation (compare to clinical "Oocyte Maturation," below and Fig. 3.1) can thus take place. In response to the luteinizing hormone (LH) surge, fully grown oocytes present in preovulatory antral follicles resume meiosis. Oocytes arrested at prophase I have an intact nuclear envelope (referred to as the germinal vesicle, or "GV"), and germinal vesicle break down (GVBD) is the first visible sign of the resumption of meiosis. Following GVBD, a Metaphase I (MI) spindle forms and when all chromosome bivalents have established stable microtubule–kinetochore interactions, Anaphase I occurs. Following completion of MI, oocytes enter directly into Meiosis II (MII) without an intervening S-phase, at which point they arrest for the second time at the MII. Fertilization triggers resumption and completion of MII, at which time the haploid oocyte accepts the haploid paternal genome, producing the zygote. We now pause to put these developmental events into context with the clinical concept of in vitro oocyte maturation.

Developmental Oocyte Maturation In Vivo vs. Clinical "Oocyte Maturation/In Vitro Maturation"

Confusion can arise in discussions of oocyte maturation as the label has taken on a specific clinical connotation (Fig. 3.1, III), separate from the continuum of oocyte development in vivo. "Oocyte Maturation" and "In Vitro Maturation (IVM)" in the Assisted Reproduction clinic often refer specifically to attempts to foster the development of retrieved immature oocytes to a mature metaphase II state in vitro. This is an area of intense research interest, as increasing the number of fertilizable eggs per retrieval while potentially lessening the need for gonadotrophin stimulation is almost always desirable. Summarizing, one center's report of oocytes retrieved and their eventual disposition is instructive when considering the potential of IVM to add to the number of available oocytes.

In a relatively large study, Vanhoutte et al. [21] evaluated the potential of MI oocytes to develop to MII, and if so, their rates of production of embryos, clinical pregnancies, and implantations. Of 1,208 total oocytes collected from a cohort of patients (mean age 35.9), 300 were at the MI stage, and 132 were at the GV stage, excluding those that were degenerating. Thus approximately one-third of all oocytes retrieved were immature, representing a significant potential resource. The authors stated that "…in each cycle of [the] study, at least on MI oocyte was present." MI oocytes were allowed to mature in vitro for a minimum of 2 h and a maximum of 26-h post retrieval (IVM group), after which ICSI was performed on those that matured to MII (as evidenced by the presence of the first polar body). Forty-three percent of MI oocytes matured to MII in the time frame indicated.

Subsequent development after ICSI was then compared with the development of eggs retrieved at MII (in vivo matured). Significantly fewer IVM eggs fertilized and developed as far as day 3 post-ICSI compared with in vivo matured oocytes. Further, of live embryos, significantly more IVM embryos were scored "poor quality" in the IVM group, consistent with a number of prior studies. In the end, one clinical pregnancy (and one healthy baby) resulted from 13 transfers of exclusively IVM oocytes; keeping the low number of transfers in mind, this performance was worse that achieved when exclusively MII oocytes were used (17 clinical pregnancies of 63 transfers).

Progress is being made in the clinic and experimental models of IVM. For example, attempts to mature both retrieved GV stage oocytes and MI oocytes to the MII stage are becoming more common. However, caution is warranted due to reports of increased aneuploidy of IVM oocytes vs. in vivo matured MII oocytes

[22, 23]. Sophisticated attempts to regulate the biochemical properties of immature oocytes to improve their potential are being attempted. In a macaque model [24], increasing intracellular glutathione (GSH) using a cell-permeable GSH donor resulted in improved fertilization rates and embryo development compared with oocytes matured in media alone. In the human, supplementation of culture media with the epidermal growth factors amphiregulin and epiregulin resulted in a near doubling of the number of GV oocytes that matured to MII in 24 h, and improved embryo development after fertilization [25]. Whether these types of biochemical strategies can also reduce the aneuploidy seen in IVM produced oocytes remains to be determined. If so, greater numbers of oocytes can be retrieved per cycle, and the reliance on gonadotrophins should be reduced. We now return to our consideration for the development of oocytes in vivo.

Transcription Factor Networks Guide Development of Oocytes from PGCs

In developing fetal ovaries, PGCs enter meiosis in a retinoid-dependent manner. A characteristic feature of this event is the onset of oocyte DNA replication, which has been shown to be driven by *St*imulated by *R*etinoic *A*cid *G*ene 8 (STRA8) (reviewed by Bowles and Koopman [26]). STRA8 is the only known molecule that controls meiotic entry of PGCs in mammals. Disruption of the *Stra8* gene in mice results in a meiotic entry block in germ cells in both females and males, a significant gametogenesis defect resulting in infertility [27, 28].

Conversion of PGCs into oocytes occurs through the orchestrated regulation of a transcription factor network that includes *S*permatogenesis and *O*ogenesis basic *H*elix-*L*oop-*H*elix (bHLH) transcription factor 1 and 2 (SOHLH1 and SOHLH2) and the downstream targets, *L*im *H*omeobox protein 8 (LHX8) and *N*ewborn *O*vary home*obox* (NOBOX). Depletion of these genes in mice results in POF, again due to an oogenesis defect [29–31]. Interestingly, in neonatal ovaries of these mutant mice, *Stra8* is highly expressed compared to levels in wild-type mice, suggesting that PGCs fail to transition into oocytes and thus undifferentiated germ cells are present even after birth [29–31].

Recent studies have discovered mutations in the NOBOX and LHX8 genes in women with POF [32–34]. A Sohlh1 target molecule, *F*actor *i*n the *G*ermline α (Figla), is known to control folliculogenesis in neonatal mouse ovaries. *Figla*-deficient mice are infertile due to failure in primordial follicle formation and correspondingly massive oocyte depletion [35]. Mutations in the *FIGLA* gene have also been reported in POF patients [36]. Thus genes that control the developmental specification of oocyte fate in the mouse are likely to control that process in humans, and thereby control the size of the primordial follicle pool in women.

Primordial Follicle Growth Arrest and Activation

The majority of primordial follicles exist in a quiescent state with the oocyte arrested at prophase I of meiosis. In sexually mature animals, follicles leave the arrested pool and undergo the primordial-to-primary follicle transition. One of the initial events in this process is a change in the shape of granulosa cells from squamous to cuboidal. Key among these gene products is the TGF-beta family protein Mullerian Inhibitory Substance (MIS; also referred to as Anti-Mullerian Hormone/AMH).

MIS was initially identified as the hormone produced by Sertoli cells of the fetal testis that promotes regression of the Mullerian ducts during differentiation of the male reproductive tract [37]. More recent studies have demonstrated that MIS is also expressed by granulosa cells [38, 39]. Interestingly, mice that bear a targeted deletion of the *MIS* gene show an increased rate of primordial follicle growth activation, resulting in a premature depletion of their ovarian reserve [40]. Since MIS is exclusively expressed in granulosa cells of growing follicles but not of primordial follicles, it was hypothesized that growing follicles exert an inhibitory feedback influence on resting primordial follicles. This was later shown to be true in an in vitro system, where MIS was found to block the initiation of growth of human primordial follicles [41]. After it was found that MIS levels correlate with the size of the primordial pool of follicles in mice [42], efforts to predict the size of the human primordial reserve occurred. MIS level measurement is now becoming more common as a measure of ovarian function and of the size of the ovarian reserve [43].

Significant progress has been made recently in understanding the mechanisms that enforce growth arrest within primordial follicles. A striking initial observation that knockout of the FoxO3a gene in mice results in the simultaneous growth activation of all primordial follicles [44] opened the door to how the process is controlled by upstream signals. Investigation of the potential role(s) of the mTOR/Akt signaling pathway (see Hay [45], and Wullschleger et al. [46] for reviews) in follicle development revealed that factors in the pathway are critical regulators of the growth activation of primordial follicles [47–52]. Increased mTOR activity is associated with protein translation, an active cell cycle, and tissue growth [46]. When mTOR activity was increased specifically in the oocyte by using a tissue-specific knockout of its negative regulators *Tsc1* or *Tsc2*, the entire pool of primordial follicles growth-activated around puberty. This resulted in nearly complete follicle loss by early adulthood, and thus POF [49]. The expression pattern of a phosphorylated form (Phospho Serine 2448-mTOR) of mTOR associated with the active enzyme is shown in Fig. 3.2.

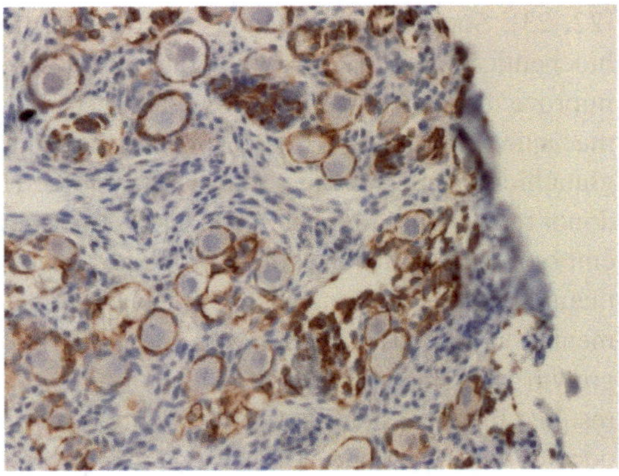

Fig. 3.2 *Expression of Phospho-Ser 2448-mTOR in the fetal human ovary.* mTOR kinase is controlled by nutritional and stress signals and controls protein translation, organelle production, the cell cycle, and a number of other critical cell processes. This phosphorylated form of mTOR, associated with the active kinase, is highly expressed in the granulosa cells of primordial follicles in this late second trimester sample

Experiments targeting Akt further confirmed the importance of mTOR/Akt signaling in the control of primordial follicle growth activation [52]. In this case, investigators manipulated signaling by the Phosphatase with TENsin homology deleted in chromosome 10 (PTEN) phosphatase, known to act as a negative regulator of Akt. Inhibition of PTEN, itself a negative regulator of Akt, resulted in increased primordial follicle growth activation in mouse and human ovarian grafts. The PTEN/Akt data are consistent with the initial result with FoxO3a knockout mice [44], as PTEN inhibition was found to result in Fox03a export from the nucleus, the presumed site of its transcriptional enforcement of primordial follicle arrest. It is likely that an intersection between paracrine MIS signaling and oocyte and granulosa cell Akt/mTOR activity will be identified. It seems that a relatively complete model of the control of the growth activation of the primordial pool is forthcoming.

Growth of the Mammalian Ovarian Follicle

Follicle growth and development (termed “folliculogenesis”) consists of the recruitment, selection and growth of ovarian follicles from the primordial stage to ovulation and subsequent corpus luteum (CL) formation. Both oocyte-generated signals and extrinsic somatic signals from the granulosa and theca cells regulate this complex process. After their squamous to cuboidal transition, granulosa cells engage in a period of proliferation. Follicle growth also involves granulosa cell cytodifferentiation into distinct populations [53, 54] and the recruitment of theca cells (below) to the follicles. There are three populations of granulosa cells that become delineated with advancing follicle growth. First, the *mural* granulosa cells are the outermost layer, in contact with the exterior basement membrane of the follicle. Mural granulosa cells convert androgens produced by the surrounding theca cells (below) into estradiol. Second, *antral* granulosa cells surround the so-called antrum, an acentric fluid-filled cavity that develops within growing follicles as they acquire multiple layers of granulosa cells. Last, the *cumulus* granulosa cells, those in close contact with the oocyte, contribute to the proper environment for successful oocyte maturation and fertilization. Cumulus granulosa cells supply growth factors and energy sources to the oocyte. Since cumulus granulosa cells also interact with mural cells directly, gonadotropin stimuli can be transmitted from peripheral mural granulosa cells, through cumulus cells, to the oocyte at the center of the follicle. Cumulus granulosa cells are ovulated along with the oocyte and contribute greatly to the process of fertilization.

A large number of studies have shown that oocytes are responsible for driving early follicle growth [55–57]. Oocyte-derived growth factors of the TGF-β superfamily (in particular, GDF-9 and BMP-15) are required for the early rounds and continued granulosa cell proliferation [58]. The TGF-beta superfamily is one of the largest groups of extracellular growth factors that are developmentally conserved through insect to vertebrate. Members of this superfamily control multiple aspects of tissue development, homeostasis, and repair by directing cell proliferation, differentiation and death. While they share a common structure, homo- or heterodimers as bioactive forms, their physiological roles are diverse due to tissue-dependency and developmental stage-dependency. Several members of the TGF-beta superfamily are considered as master regulators of ovarian follicle development: Bone Morphogenetic Protein (in particular, BMP-15), MIS, the growth and differentiation factors (in particular, GDF-9), activins, inhibins, glial cell line-derived neurotrophic factor (GDNF), and likely others that have yet to be identified.

GDF-9 has been found to be selectively expressed by oocytes from early stage follicles [59, 60], and is now considered a NOBOX down-stream target molecule due to NOBOX binding sites found in its promoter [61]. Its receptor subunits are expressed by granulosa cells of the corresponding early follicle stages, making these cells potential targets for paracrine signaling [62]. *Gdf-9*-deficient mice are infertile and exhibit arrested follicle development at the primary stage, thus indicating that oocyte-derived GDF-9 is essential for further follicle progression [62]. Indeed, GDF-9 treatment in vitro enhanced the progression of early to late stage primary follicles in ovaries in the mouse [63] and human [64]. Interestingly, GDF-9 action appears to occur in a synergistic fashion with the effects of BMP-15 [65, 66].

BMP-4 and -7 have also been shown to regulate follicle development. This was originally revealed by detecting ovarian expression of these factors (and their cognate receptors in the ovary), and by culturing rat granulosa cells in the recombinant proteins [67]. Later work showed that BMP-4 and -7 promote the primordial-to-primary follicle transition to enhance follicle survival. Intrabursal injection of BMP-7 decreases the numbers of primordial follicles but increases the numbers of primary, preantral, and antral follicles [68]. Similarly, BMP-4 treatment raises the proportion of developing follicles in cultured ovary in vitro. On the other hand, exposing a neutralizing antibody to BMP-4 resulted in smaller ovaries accompanied by a progressive loss of oocytes and primordial follicles due to increased cellular apoptosis [69]. In sum, many factors (notably TGF-beta signaling pathway molecules) control the early development and survival of ovarian follicles.

Follicle development discussed to this point is thought to occur without being influenced by circulating hormones. In contrast, much of subsequent follicle growth and development to the preovulatory state is dependent on the circulating gonadotropin hormones follicle-stimulating hormone (FSH) and luteinizing hormone (LH). These hormones are secreted from the anterior pituitary gland under the control of pulses of gonadotropin-releasing hormone (GnRH) from the hypothalamus. Low-frequency GnRH pulses stimulate a slight increase in FSH levels, which enhances the proliferation of granulosa cells. Later, high-frequency pulses of GnRH lead to a sharp rise in levels of LH (referred to as LH surge), which triggers resumption of meiosis in oocytes [70]. We now consider intra-oocyte signaling events that control meiosis resumption in response to the LH surge.

Meiosis Resumption Within the Oocyte

In growing follicles, endogenously produced cyclic AMP (cAMP) in resting oocytes actively represses re-entry of prophase I (reviewed by Solc et al. [71]). Production of cAMP in resting oocytes depends on the activity of G-protein-coupled receptor 3 (GPR3), a major oocyte-G-protein-coupled orphan receptor. Stimulation of GPR3 activates adenylate cyclase 3 (ADCY3) that catalyzes transformation of ATP to cAMP. Cyclic AMP in turn activates protein kinase A (PKA), which indirectly inhibits the activity of meiosis-promoting factor, MPF (cyclin-dependent kinase 1 (CDK1)/cyclin B complex) through Wee2 kinase, a PKA substrate. The activation of Wee2 kinase causes inhibitory phosphorylation of CDK1 and thus prevents resumption of meiotic arrest. In fact, deletion of *Gpr3* gene in mice causes spontaneous resumption of prophase I arrest in oocytes [72]. Also, leaky prophase I arrest has been reported by RNAi gene knockdown of Wee2 [73].

Phosphodiesterase 3a (PDE3a) is an essential PDE isoform for promoting cAMP degradation. Oocytes derived from Pde3a-deficient mice cannot resume meiosis even after the LH surge in vivo [74]. Since LH signaling does not terminate the GPR3-ADCY3 pathway, the increase in PDE3a activity is likely sufficient to promote the maturation-associated decrease in oocyte cAMP, which leads to MPF activation for meiotic spindle formation and degradation of GV (reviewed by Brunet and Maro [75]). A pivotal role of LH in cAMP homeostasis in oocytes and the intermediate signaling pathway is discussed in the following section.

Importantly, LH secreted from the pituitary targets mural granulosa cells. LH activation stimulates cAMP signaling in mural cells that in turn induces the expression of the EGF-like growth factors epiregulin and amphiregulin, which then act upon cumulus granulosa cells. At the same time, granulosa cell levels of cyclic GMP (cGMP), an inhibitor of meiosis resumption, decrease in response to LH treatment. cGMP in cumulus granulosa cells passes through gap junctions into the oocyte, where it blocks hydrolysis of cAMP by masking phosphodiesterase 3a [76]. This inhibition keeps the intra-oocyte concentration of cAMP high and thereby blocks meiotic progression.

It is well known that the epidermal growth factor (EGF) signaling pathway regulates follicle development (including meiosis resumption) via coordinated signaling between the oocyte and granulosa cells (reviewed by Sun et al. [77]). Cumulus granulosa cells were found to be the main targets for EGF-induced meiotic resumption as oocytes enclosed with cumulus cells initiate GVBD in response to EGF, and inhibition of EGF receptor prevents MAP kinase activation in cumulus-enclosed oocytes but not denuded oocytes.

The EGF-like growth factors: Amphiregulin, Epiregulin, and Betacellulin were initially identified as autocrine mitogens secreted from MCF-7, NIH-3T3, and pancreatic beta cell tumor cell lines, respectively [78–80]. Ten years later, biological functions of these EGF-like growth factors in oocyte maturation after the LH surge had been revealed [81]. EGFs are secreted from mural granulosa cells in response to the LH surge via cAMP activation and diffuse to act on cumulus granulosa cells (reviewed by Conti et al. [82]).

When granulosa cells were cultured in the presence of neutralizing antibodies against Amphiregulin, Epiregulin, and Betacellulin, EGF receptor phosphorylation and subsequent MAP kinase activation were inhibited [83]. Similarly, selective inhibitors against EGF receptor kinase blocked LH-induced oocyte maturation in follicles in vitro [81]. Since MAP kinase activation in cumulus granulosa cells results in reduced intra-oocyte levels of cGMP/cAMP (due to altered gap junction transport, above), activation of EGF signaling through EGF receptor/MAP kinase pathway is favored as a key mechanism in meiosis resumption at the LH surge.

MAP kinase (MAPK) signaling in granulosa cells is an essential part of the events downstream of the activation of the EGF receptor. Specific deletion of MAPK genes in mouse granulosa cells results in inhibited meiotic resumption and a failure to ovulate even after exogenous hormone treatment [84]. Accordingly, activated MAPK in cumulus granulosa cells phosphorylates connexin 43, a major component of the gap junctions, that couple cumulus granulosa cells with their oocyte. This causes a closure of the gap junctions between these cells, resulting in decreased cGMP in oocytes [85] and increased meiotic maturation.

Oocyte Control of Follicle Development

Granulosa cells play a crucial role in maintaining meiotic arrest and providing the oocyte with nutrients and metabolic support (in particular, amino acids uptake, glycolysis, and cholesterol biosynthesis [reviewed by Sugiura et al. [86] and Su et al. [87]]). As will be seen, however, the oocyte plays an active role in controlling the fate of its surrounding somatic cells and thus the entire follicle.

Initial studies analyzing amino acid uptake in oocytes had been reported using amino acid labeled with radioisotopes [88, 89]. These studies showed that denuded oocytes exhibit much less radioactivity compared with oocytes enclosed with cumulus granulosa cells. In fact, Slc38a3, a neutral amino acid transporter harboring high substrate preference for L-alanine, is highly expressed in cumulus granulosa cells but neither in oocytes nor in mural granulosa cells [57]. Similarly, glucose metabolism has also been examined in granulosa cell-associated oocytes and denuded oocytes. Denuded oocytes failed to utilize glucose as an energy source [90]. Cumulus granulosa cells metabolize glucose into pyruvate and transfer the glycolysis substrate to oocytes through gap junctions and/or paracrine secretions. The molecular basis of this metabolic cooperation has recently been provided by a study in which transcripts encoding key enzymes in the glycolytic pathway were found to be robustly expressed in cumulus granulosa cells, but barely detectable in oocytes or mural granulosa cells (Sugiura et al. [86]; reviewed by Sugiura and Eppig [91]).

Another example of metabolic cooperation between oocytes and cumulus granulosa cells involves cholesterol biosynthesis. It has been reported that the gene expression of enzymes in the cholesterol biosynthesis pathway are highly expressed in cumulus granulosa cells but neither in oocytes nor in mural granulosa cells [92]. This suggests that mouse oocytes lack the full enzymatic system required for synthesizing cholesterol. Moreover, when mouse oocytes were incubated with radio-labeled acetate as the precursor for cholesterol biosynthesis in culture, the levels of radio-labeled cholesterol were barely detected in denuded oocytes, as opposed to oocytes that were enclosed by cumulus granulosa cells during culture [92].

Oocytes within follicles use paracrine signals to actively control gene expression involved in the biosynthesis and/or uptake of nutrients in cumulus granulosa cells [57, 92, 93]. As described earlier, the expression of genes encoding key enzymes for glycolysis, cholesterol biosynthesis, and amino acid transporters is exclusive to granulosa cells in follicles. However, such gene expression was dramatically reduced when oocytes were microsurgically removed from the follicles [57, 86, 92, 93]. Consistent with this, functional analysis of glycolytic and cholesterol biosynthesis activities and amino acid uptake demonstrated less glucose metabolism, cholesterol biosynthesis and L-alanine uptake in granulosa cells compared with those before oocyte removal [57, 86, 92, 93].

Interestingly, co-culture of cumulus granulosa cells with maturing oocytes from large antral follicles increased expression levels of genes encoding enzymes for glycolysis and cholesterol biosynthesis as well as amino acid transporters [57, 86, 92, 93]. This suggests that soluble factors released by the oocyte can at least in part substitute for those that cross gap junctions in intact follicles. Thus oocytes promote, at least, metabolic activities in granulosa cells via paracrine signaling mechanisms. Taken all together, oocytes actively control follicle metabolism by promoting the expression of genes in cumulus granulosa cells that are "missing" from oocytes so that follicle growth is controlled as a single unit.

Theca Cells and the Production of the Preovulatory Follicle

Theca cells are another important cell type in the follicle. Interactions among theca cells, the different granulosa cell types, and oocytes are the primary local factors that control folliculogenesis (see Magoffin [94] for a review). The theca cell layer surrounds the follicular basement membrane, where it has a key role in the production of steroid hormones. In response to the LH surge, theca cells start the production of androgens from cholesterol. These androgens are then converted into estrogen or progesterone by the aromatase enzyme in mural granulosa cells. Beyond their local roles in follicle development, intra-follicular communication, and corpus luteum function, estrogen and progesterone can both positively and negatively control GnRH secretion from hypothalamus during the menstrual cycle (estrous cycle in rodents), providing feedback to the greater hypothalamic–pituitary axis.

Fertility Preservation: A Brief Overview

Damage to Oocytes, Follicles, and the Ovary Occurs During Therapeutic Treatment

As mentioned, genetic and developmental factors that compromise ovarian function in mammals have been identified. Indeed, oocytes are lost throughout life due to baseline follicle atresia and the rate of loss may be affected by these factors. In addition, data from clinical and experimental studies have also led to a better understanding of factors that damage the ovary. Radiation therapy and chemotherapy are two factors that can greatly accelerate oocyte loss. While the inarguable first priority is the use of therapies most likely to save a patient's life, increasing survivorship after many cancers has meant increased attention in protecting the fertility of these women, and the rise of fertility preservation as a medical discipline.

Methods of Fertility Preservation

Two basic strategies are employed to protect ovaries from the insults discussed earlier to preserve fertility (reviewed in Bianchi and Woodruff [95]). First, methods are used to protect intact ovaries from damage. Second, ovarian tissue containing oocytes or oocytes themselves can be isolated, cryopreserved, and replaced later for attempts at resumption of ovarian function and possible conception. If immature oocytes are isolated, in vitro maturation to fertilization-competence is an option (Sect. "Developmental Oocyte Maturation In Vivo vs. Clinical Oocyte Maturation/In Vitro Maturation"), possibly after cryopreservation and thawing.

Protecting the Ovary In Situ

Ovarian transposition consists of a (most often) laparoscopic surgical relocation of ovaries from their normal position(s) either medially or laterally so that they are removed from the radiation field during treatment. Perhaps unsurprisingly, this technique is very effective in protecting the ovarian reserve from damage due to radiation. Very recently, Al-Badawi et al. [96] published a case report discussing their use of robotic assistance to perform ovarian transposition, and argue that increased surgical precision may improve postrecovery outcomes. While ovarian transposition is a proven technique, it is clearly not an option for women who require total body irradiation, chemotherapeutic treatment, or who are afflicted with ovarian cancer.

Removal, Storage, and Reintroduction of Ovarian Tissue or Oocytes

Transplantation of pieces of ovarian cortex containing oocytes has been performed in a wide range of mammals and the growth of follicles to ovulation is generally successful. In humans,

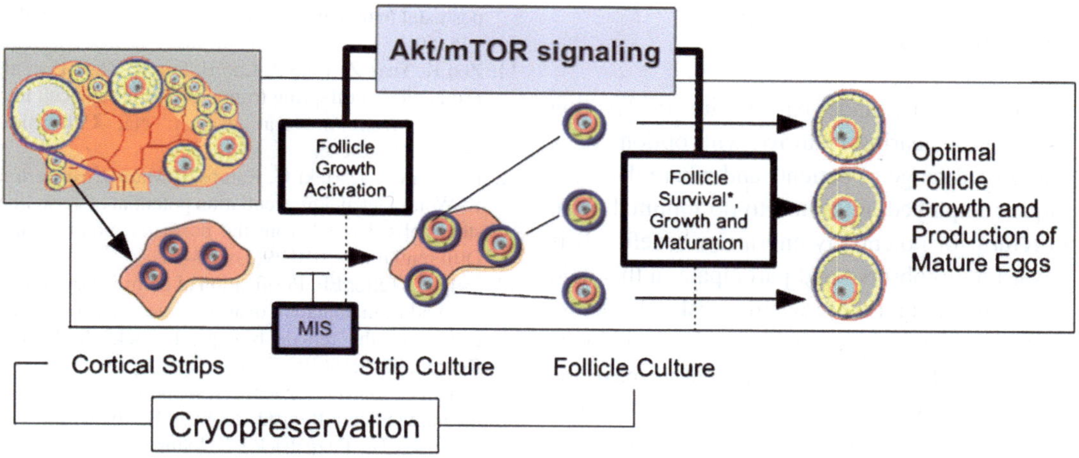

Fig. 3.3 *Controlling follicle development*: Akt/mTOR signaling has been shown to be critical in the control of the arrest of primordial follicles and the proliferation and differentiation of granulosa cells during follicle growth. This pathway can be taken advantage of infertility preservation efforts using ovarian cortical strips. Here, cortical strips are isolated and optionally cryopreserved. Pharmacologic manipulation of Akt/mTOR signaling or the action of MIS may allow the growth activation of increased numbers of primordial follicles per strip. After initial follicle growth (*center panel*) individual follicles can be isolated for a second stage of culture. Targeting Akt/mTOR may allow for fine control of the rate of follicle growth and maturation, increasing the likelihood of producing fertilizable eggs (*right panel*)

where transplantation of a patient's own tissue is necessarily the most important factor, the key has been cryopreservation of ovarian cortex until the patient recovers. Figure 3.3 summarizes the key approaches in in vitro ovarian cortex and follicle cryopreservation, and how to manipulate the key pathways above may lead to optimal follicle growth and the production of normal eggs.

Ovarian cortical tissue ("cortical strips") or individual follicles can be stored for later use via cryopreservation until the patient recovers from her illness. These pieces of ovarian cortex can contain thousands of the most immature follicles, each harboring an immature egg. After thawing, this tissue can either be transplanted back into the patient's ovaries or the tissue can be cultured in vitro in hopes of generating healthy, mature eggs. Proof of principle of the latter strategy has been shown by Telfer et al. [97], who have shown that a two-step procedure can result in the generation of large antral follicles. Here, cortical strips are first cultured intact, and then growing follicles are mechanically isolated and cultured singly. In combination with cryopreservation, this technique holds additional promise for a day, when even ovarian cancer survivors at risk for reintroduction of the disease if orthotopic transplantation was performed, might have the entire continuum of oocyte development from the primordial oocyte to the mature MII egg take place in vitro, with only an embryo transferred back for attempted pregnancy.

Equally as exciting are recent successes seen in culturing immature human follicles using bioengineering approaches. In particular, the use of alginate matrices [98] has resulted in promising follicle development in vitro, up to large antral stages. Such matrices are thought to provide an improved mechanical environment better approximating growth in vivo. When cryopreserved follicles were thawed and cultured in such a system, the growth of those follicles and their cytoskeletal properties were approximated as those seen in noncryopreserved follicles. [99]. Bioengineering approaches are truly at their dawn in fertility preservation, and it is certain that such approaches will help in the quest for the consistent, predictable development of immature follicles into fertilization-competent eggs in the reasonably near future.

Conclusions

The case can be made that fertility preservation as a field is nearing maturity, with options for a very broad range of patients and where the techniques employed are found to do minimal harm in return for potentially enormous benefits. It is rewarding to observe and participate in this field due to the high ratio of reward to risk. This being said, if the goal of preserving ovarian function is to maintain fertility, it is paramount to make the production of (chromosomally and metabolically) normal oocytes the priority regardless of which techniques are employed. Producing more oocytes increase the likelihood of compromised offspring is not acceptable. Interplay between basic approaches and clinical demand will allow us to be up to this task.

References

1. Conway GS. Premature ovarian failure. Curr Opin Obstet Gynecol. 2007;9:202–6.
2. Skillern A, Rajkovic A. Recent developments in identifying genetic determinants of premature ovarian failure. Sex Dev. 2008;2:228–43.
3. Hadji P. Menopausal symptoms and adjuvant therapy-associated adverse events. Endocr Relat Cancer. 2008;15:73–90.
4. Billeci AM, Paciaroni M, Caso V, Agnelli G. Hormone replacement therapy and stroke. Curr Vasc Pharmacol. 2008;6:112–23.
5. Mattar CN, Harharah L, Su LL, Agarwal AA, Wong PC, Choolani M. Menopause, hormone therapy and cardiovascular and cerebrovascular disease. Ann Acad Med Singap. 2008;37:54–62.
6. Jemal A, Siegel R, Ward E, Hao Y, Xu J, Murray T, et al. Cancer statistics, 2008. CA Cancer J Clin. 2008;58(2):71–96.
7. Reh A, Oktem O, Oktay K. Impact of breast cancer chemotherapy on ovarian reserve: a prospective observational analysis by menstrual history and ovarian reserve markers. Fertil Steril. 2008;90(5):1635–9.
8. Jeruss JS, Woodruff TK. Preservation of fertility in patients with cancer. N Engl J Med. 2009;360:902–11.
9. Wo JY, Viswanathan AN. Impact of radiotherapy on fertility, pregnancy, and neonatal outcomes in female cancer patients. Int J Radiat Oncol Biol Phys. 2009;73:1304–12.
10. Johnson J, Canning J, Kaneko T, Pru JK, Tilly JL. Germline stem cells and follicular renewal in the postnatal mammalian ovary. Nature. 2004;428(6979):145–50.
11. Zou K, Yuan Z, Yang Z, Luo H, Sun K, Zhou L, et al. Production of offspring from a germline stem cell line derived from neonatal ovaries. Nat Cell Biol. 2009;11(5):631–6.
12. Pacchiarotti J, Maki C, Ramos T, Marh J, Howerton K, Wong J, et al. Differentiation potential of germ line stem cells derived from the postnatal mouse ovary. Differentiation. 2010;79(3):159–70.
13. Tilly JL, Telfer EE. Purification of germline stem cells from adult mammalian ovaries: a step closer towards control of the female biological clock? Mol Hum Reprod. 2009;15(7):393–8.
14. Abban G, Johnson J. Stem cell support of oogenesis in the human. Hum Reprod. 2009;24(12):2974–8.
15. Thomson TC, Fitzpatrick KE, Johnson J. Intrinsic and extrinsic mechanisms of oocyte loss. Mol Hum Reprod. 2010;16(12):916–27.
16. Nakamura S, Kobayashi K, Nishimura T, Higashijima S, Tanaka M. Identification of germline stem cells in the ovary of the teleost medaka. Science. 2010;328(5985):1561–3.
17. Erickson GF. in Endotext.com. (2008).
18. Gougeon A. Human ovarian follicular development: from activation of resting follicles to preovulatory maturation. Ann Endocrinol (Paris). 2010;71:132–43.
19. Richardson BE, Lehmann R. Mechanisms guiding primordial germ cell migration: strategies from different organisms. Nat Rev Mol Cell Biol. 2010;11:37–49.
20. Mehlmann LM. Stops and starts in mammalian oocytes: recent advances in understanding the regulation of meiotic arrest and oocyte maturation. Reproduction. 2005;130:791–9.
21. Vanhoutte L, De Sutter P, Van der Elst J, Dhont M. Clinical benefit of metaphase I oocytes. Reprod Biol Endocrinol. 2005;3:71.
22. Requena A, Bronet F, Guillén A, Agudo D, Bou C, García-Velasco JA. The impact of in-vitro maturation of oocytes on aneuploidy rate. Reprod Biomed Online. 2009;18(6):777–83.
23. Zhang XY, Ata B, Son WY, Buckett WM, Tan SL, Ao A. Chromosome abnormality rates in human embryos obtained from in-vitro maturation and IVF treatment cycles. Reprod Biomed Online. 2010;21(4):552–9.
24. Curnow EC, Ryan JP, Saunders DM, Hayes ES. Primate model of metaphase I oocyte in vitro maturation and the effects of a novel glutathione donor on maturation, fertilization, and blastocyst development. Fertil Steril. 2011;95:1235–40.
25. Ben-Ami I, Komsky A, Bern O, Kasterstein E, Komarovsky D, Ron-El R. In vitro maturation of human germinal vesicle-stage oocytes: role of epidermal growth factor-like growth factors in the culture medium. Hum Reprod. 2010;26:76–81.
26. Bowles J, Koopman P. Retinoic acid, meiosis and germ cell fate in mammals. Development. 2007;134:3401–11.

27. Baltus AE, Menke DB, Hu YC, Goodheart ML, Carpenter AE, de Rooij DG, et al. In germ cells of mouse embryonic ovaries, the decision to enter meiosis precedes premeiotic DNA replication. Nat Genet. 2006;38:1430–4.
28. Anderson EL, Baltus AE, Roepers-Gajadien HL, Hassold TJ, de Rooij DG, van Pelt AM, et al. Stra8 and its inducer, retinoic acid, regulate meiotic initiation in both spermatogenesis and oogenesis in mice. Proc Natl Acad Sci USA. 2008;105: 14976–80.
29. Pangas SA, Choi Y, Ballow DJ, Zhao Y, Westphal H, Matzuk MM, et al. Oogenesis requires germ cell-specific transcriptional regulators Sohlh1 and Lhx8. Proc Natl Acad Sci USA. 2006;103:8090–5.
30. Choi Y, Ballow DJ, Xin Y, Rajkovic A. Lim homeobox gene, lhx8, is essential for mouse oocyte differentiation and survival. Biol Reprod. 2008;79:442–9.
31. Rajkovic A, Pangas SA, Ballow D, Suzumori N, Matzuk MM. NOBOX deficiency disrupts early folliculogenesis and oocyte-specific gene expression. Science. 2004;305:1157–9.
32. Zhao XX, Suzumori N, Yamaguchi M, Suzumori K. Mutational analysis of the homeobox region of the human NOBOX gene in Japanese women who exhibit premature ovarian failure. Fertil Steril. 2005;83:1843–4.
33. Qin Y, Choi Y, Zhao H, Simpson JL, Chen ZJ, Rajkovic A. NOBOX homeobox mutation causes premature ovarian failure. Am J Hum Genet. 2007;81: 576–81.
34. Qin Y, Shi Y, Zhao Y, Carson SA, Simpson JL, Chen ZJ. Mutation analysis of NOBOX homeodomain in Chinese women with premature ovarian failure. Fertil Steril. 2009;91(4 Suppl):1507–9.
35. Soyal SM, Amleh A, Dean J. FIGalpha, a germ cell-specific transcription factor required for ovarian follicle formation. Development. 2000;127:4645–54.
36. Zhao H, Chen ZJ, Qin Y, Shi Y, Wang S, Choi Y, et al. Transcription factor FIGLA is mutated in patients with premature ovarian failure. Am J Hum Genet. 2008;82(6):1342–8.
37. Donahoe PK, Ito Y, Price JM, Hendren 3rd WH. Müllerian inhibiting substance activity in bovine fetal, newborn and prepubertal testes. Biol Reprod. 1977; 16:238–43.
38. Ueno S, Takahashi M, Manganaro TF, Ragin RC, Donahoe PK. Cellular localization of müllerian inhibiting substance in the developing rat ovary. Endocrinology. 1989;124:1000–6.
39. Weenen C, Laven JS, Von Bergh AR, Cranfield M, Groome NP, Visser JA, et al. Anti-Müllerian hormone expression pattern in the human ovary: potential implications for initial and cyclic follicle recruitment. Mol Hum Reprod. 2004;10:77–83.
40. Durlinger AL, Kramer P, Karels B, de Jong FH, Uilenbroek JT, Grootegoed JA, et al. Control of primordial follicle recruitment by anti-Müllerian hormone in the mouse ovary. Endocrinology. 1999;140: 5789–96.
41. Carlsson IB, Scott JE, Visser JA, Ritvos O, Themmen AP, Hovatta O. Anti-Müllerian hormone inhibits initiation of growth of human primordial ovarian follicles in vitro. Hum Reprod. 2006;21:2223–7.
42. Kevenaar ME, Meerasahib MF, Kramer P, van de Lang-Born BM, de Jong FH, Groome NP, et al. Serum anti-mullerian hormone levels reflect the size of the primordial follicle pool in mice. Endocrinology. 2006;147:3228–34.
43. Seifer DB, Maclaughlin DT. Mullerian Inhibiting Substance is an ovarian growth factor of emerging clinical significance. Fertil Steril. 2007;88:539–46.
44. Castrillon DH, Miao L, Kallipara R, Horner J, Depinho RA. Suppression of ovarian follicle activation in mice by the transcription factor Foxo3a. Science. 2003;301:215–8.
45. Hay N. The Akt-mTOR tango and its relevance to cancer. Cancer Cell. 2005;8:179–83.
46. Wullschleger S, Loewith R, Hall MN. Tor signaling in growth and metabolism. Cell. 2006;124:471–84.
47. Liu L, Rajareddy S, Reddy P, Du C, Jagarlamudi K, Shen Y, et al. Infertility caused by retardation of follicular development in mice with oocyte-specific expression of Foxo3a. Development. 2007;134:199–209.
48. Adhikari D, Flohr G, Gorre N, Shen Y, Yang H, Lundin E, et al. Disruption of Tsc2 in oocytes leads to overactivation of the entire pool of primordial follicles. Mol Hum Reprod. 2009;15:765–70.
49. Adhikari D, Zheng W, Shen Y, Gorre N, Hämäläinen T, Cooney AJ, et al. Tsc/mTORC1 signaling in oocytes governs the quiescence and activation of primordial follicles. Hum Mol Genet. 2010;19:397–410.
50. Adhikari D, Liu K. mTOR signaling in the control of activation of primordial follicles. Cell Cycle. 2010;9:1673–4.
51. Reddy P, Zheng W, Liu K. Mechanisms maintaining the dormancy and survival of mammalian primordial follicles. Trends Endocrinol Metab. 2010;21:96–103.
52. Li J, Kawamura K, Cheng Y, Liu S, Klein C, Liu S, et al. Activation of dormant ovarian follicles to generate mature eggs. Proc Natl Acad Sci USA. 2010;107:10280–4.
53. Latham KE, Bautista FD, Hirao Y, O'Brien MJ, Eppig JJ. Comparison of protein synthesis patterns in mouse cumulus cells and mural granulosa cells: effects of follicle-stimulating hormone and insulin on granulosa cell differentiation in vitro. Biol Reprod. 1999;61(4): 82–92.
54. Johnson J, Espinoza T, McGaughey RW, Rawls A, Wilson-Rawls J. Notch pathway genes are expressed in mammalian ovarian follicles. Mech Dev. 2001; 109:355–61.
55. Eppig JJ, Wigglesworth K, Pendola FL. The mammalian oocyte orchestrates the rate of ovarian follicular development. Proc Natl Acad Sci USA. 2002;99(5): 2890–4.
56. Matzuk MM, Burns KH, Viveiros MM, Eppig JJ. Intercellular communication in the mammalian ovary: oocytes carry the conversation. Science. 2002; 296(5576):2178–80.

57. Eppig JJ, Pendola FL, Wigglesworth K, Pendola JK. Mouse oocytes regulate metabolic cooperativity between granulosa cells and oocytes: amino acid transport. Biol Reprod. 2005;73(2):351–7.
58. Pangas SA. Growth factors in ovarian development. Semin Reprod Med. 2007;25:225.
59. McGrath SA, Esquela AF, Lee SJ. Oocyte-specific expression of growth/differentiation factor-9. Mol Endocrinol. 1995;9:131–6.
60. Carabatsos MJ, Elvin J, Matzuk MM, Albertini DF. Characterization of oocyte and follicle development in growth differentiation factor-9-deficient mice. Dev Biol. 1998;204:373–84.
61. Choi Y, Rajkovic A. Characterization of NOBOX DNA binding specificity and its regulation of Gdf9 and Pou5f1 promoters. J Biol Chem. 2006;281: 35747–56.
62. Mazerbourg S, Klein C, Roh J, Kaivo-Oja N, Mottershead DG, Korchynskyi O, et al. Growth differentiation factor-9 signaling is mediated by the type I receptor, activin receptor-like kinase 5. Mol Endocrinol. 2004;18:653–65.
63. Hayashi M, McGee EA, Min G, Klein C, Rose UM, van Duin M, et al. Recombinant growth differentiation factor-9 (GDF-9) enhances growth and differentiation of cultured early ovarian follicles. Endocrinology. 1999;140:1236–44.
64. Hreinsson JG, Scott JE, Rasmussen C, Swahn ML, Hsueh AJ, Hovatta O. Growth differentiation factor-9 promotes the growth, development, and survival of human ovarian follicles in organ culture. J Clin Endocrinol Metab. 2002;87:316–21.
65. Su YQ, Wu X, O'Brien MJ, Pendola FL, Denegre JN, Matzuk MM, et al. Synergistic roles of BMP15 and GDF9 in the development and function of the oocyte-cumulus cell complex in mice: genetic evidence for an oocyte-granulosa cell regulatory loop. Dev Biol. 2004;276:64–73.
66. McNatty KP, Juengel JL, Reader KL, Lun S, Myllymaa S, Lawrence SB, et al. Bone morphogenetic protein 15 and growth differentiation factor 9 co-operate to regulate granulosa cell function. Reproduction. 2005;129:473–80.
67. Shimasaki S, Zachow RJ, Li D, Kim H, Iemura S, Ueno N, et al. A functional bone morphogenetic protein system in the ovary. Proc Natl Acad Sci USA. 1999;96(13):7282–7.
68. Lee WS, Otsuka F, Moore RK, Shimasaki S. Effect of bone morphogenetic protein-7 on folliculogenesis and ovulation in the rat. Biol Reprod. 2001;65(4):994–9.
69. Tanwar PS, O'Shea T, McFarlane JR. In vivo evidence of role of bone morphogenetic protein-4 in the mouse ovary. Anim Reprod Sci. 2008;106(3–4):232–40.
70. Shupnik MA. Gonadotropin gene modulation by steroids and gonadotropin-releasing hormone. Biol Reprod. 1996;54(2):279–86.
71. Solc P, Schultz RM, Motlik J. Prophase I arrest and progression to metaphase I in mouse oocytes: comparison of resumption of meiosis and recovery from G2-arrest in somatic cells. Mol Hum Reprod. 2010; 16(9):654–64.
72. Mehlmann LM. Oocyte-specific expression of Gpr3 is required for the maintenance of meiotic arrest in mouse oocytes. Dev Biol. 2005;288(2):397–404.
73. Han SJ, Chen R, Paronetto MP, Conti M. Wee1B is an oocyte-specific kinase involved in the control of meiotic arrest in the mouse. Curr Biol. 2005;15(18): 1670–6.
74. Masciarelli S, Horner K, Liu C, Park SH, Hinckley M, Hockman S, et al. Cyclic nucleotide phosphodiesterase 3A-deficient mice as a model of female infertility. J Clin Invest. 2004;114(2):196–205.
75. Brunet S, Maro B. Cytoskeleton and cell cycle control during meiotic maturation of the mouse oocyte: integrating time and space. Reproduction. 2005;130(6): 801–11.
76. Norris RP, Ratzan WJ, Freudzon M, Mehlmann LM, Krall J, Movsesian MA, et al. Cyclic GMP from the surrounding somatic cells regulates cyclic AMP and meiosis in the mouse oocyte. Development. 2009; 136(11):1869–78.
77. Sun QY, Miao YL, Schatten H. Towards a new understanding on the regulation of mammalian oocyte meiosis resumption. Cell Cycle. 2009;8(17):2741–7.
78. Shoyab M, McDonald VL, Bradley JG, Todaro GJ. Amphiregulin: a bifunctional growth-modulating glycoprotein produced by the phorbol 12-myristate 13-acetate-treated human breast adenocarcinoma cell line MCF-7. Proc Natl Acad Sci USA. 1988;85(17): 6528–32.
79. Toyoda H, Komurasaki T, Uchida D, Takayama Y, Isobe T, Okuyama T, et al. Epiregulin. A novel epidermal growth factor with mitogenic activity for rat primary hepatocytes. J Biol Chem. 1995;270(13): 7495–500.
80. Shing Y, Christofori G, Hanahan D, Ono Y, Sasada R, Igarashi K, et al. Betacellulin: a mitogen from pancreatic beta cell tumors. Science. 1993;259(5101): 1604–7.
81. Park JY, Su YQ, Ariga M, Law E, Jin SL, Conti M. EGF-like growth factors as mediators of LH action in the ovulatory follicle. Science. 2004;303(5658): 682–4.
82. Conti M, Hsieh M, Park JY, Su YQ. Role of the epidermal growth factor network in ovarian follicles. Mol Endocrinol. 2006;20(4):715–23.
83. Panigone S, Hsieh M, Fu M, Persani L, Conti M. Luteinizing hormone signaling in preovulatory follicles involves early activation of the epidermal growth factor receptor pathway. Mol Endocrinol. 2008;22(4): 924–36.
84. Fan HY, Liu Z, Paquet M, Wang J, Lydon JP, DeMayo FJ, et al. Cell type-specific targeted mutations of Kras and Pten document proliferation arrest in granulosa cells versus oncogenic insult to ovarian surface epithelial cells. Cancer Res. 2009;69(16):6463–72.
85. Sela-Abramovich S, Chorev E, Galiani D, Dekel N. Mitogen-activated protein kinase mediates luteinizing

hormone-induced breakdown of communication and oocyte maturation in rat ovarian follicles. Endocrinology. 2005;146(3):1236–44.
86. Sugiura K, Pendola FL, Eppig JJ. Oocyte control of metabolic cooperativity between oocytes and companion granulosa cells: energy metabolism. Dev Biol. 2005;279(1):20–30.
87. Su YQ, Sugiura K, Eppig JJ. Mouse oocyte control of granulosa cell development and function: paracrine regulation of cumulus cell metabolism. Semin Reprod Med. 2009;27(1):32–42.
88. Cross PC, Brinster RL. Leucine uptake and incorporation at three stages of mouse oocyte maturation. Exp Cell Res. 1974;86(1):43–6.
89. Colonna R, Mangia F. Mechanisms of amino acid uptake in cumulus-enclosed mouse oocytes. Biol Reprod. 1983;28(4):797–803.
90. Donahue RP, Stern S. Follicular cell support of oocyte maturation: production of pyruvate in vitro. J Reprod Fertil. 1968;17(2):395–8.
91. Sugiura K, Eppig JJ. Society for Reproductive Biology Founders' Lecture, Control of metabolic cooperativity between oocytes and their companion granulosa cells by mouse oocytes. Reprod Fertil Dev. 2005; 17(7):667–74.
92. Su YQ, Sugiura K, Wigglesworth K, O'Brien MJ, Affourtit JP, Pangas SA, et al. Oocyte regulation of metabolic cooperativity between mouse cumulus cells and oocytes: BMP15 and GDF9 control cholesterol biosynthesis in cumulus cells. Development. 2008; 135(1):111–21.
93. Sugiura K, Su YQ, Diaz FJ, Pangas SA, Sharma S, Wigglesworth K, et al. Oocyte-derived BMP15 and FGFs cooperate to promote glycolysis in cumulus cells. Development. 2007;134(14):2593–603.
94. Magoffin DA. Ovarian theca cell. Int J Biochem Cell Biol. 2005;37(7):1344–9.
95. Woodruff TK. Preserving fertility during cancer treatment. Nat Med. 2009;15(10):1124–5.
96. Al-Badawi I, Al-Aker M, Tulandi T. Robotic-assisted ovarian transposition before radiation. Surg Technol Int. 2010;19:141–3.
97. Telfer EE, McLaughlin M, Ding C, Thong KJ. A two-step serum-free culture system supports development of human oocytes from primordial follicles in the presence of activin. Hum Reprod. 2008;23(5): 1151–8.
98. West ER, Xu M, Woodruff TK, Shea LD. Physical properties of alginate hydrogels and their effects on in vitro follicle development. Biomaterials. 2007; 28(30):4439–48.
99. Barrett SL, Shea LD, Woodruff TK. Noninvasive index of cryorecovery and growth potential for human follicles in vitro. Biol Reprod. 2010;82(6):1180–9.

Impact of Chemotherapy and Radiotherapy on the Ovary

4

Maya L. Kriseman and Ertug Kovanci

Abstract

Chemotherapy and radiation can both significantly decrease fertility potential. For chemotherapy, factors such as dose, age of patient, type of cancer, and type of drug being used are the most significant prognostic factors in determining future reproductive potential. As the dose and age of a patient increases, the chance of pregnancy decreases. Alkylating agents are the biggest culprits of decreased fertility. Some cancers that are more associated with decreased fertility potential are Hodgkin's lymphoma and breast cancer. For radiotherapy, the most significant factors for future reproductive potential are dose and location. Radiotherapy to the ovaries is the most significant risk factor for acute ovarian failure. There is a direct inverse relationship between radiation dose and fertility potential. Considering the high cure rate of childhood cancers and the deleterious long-term effects of chemotherapy and radiation, it is important to be familiar with both the treatment options for cancer and future fertility.

Keywords

Ovarian function • Chemotherapy • Dosing • Location • Faddy–Gosden Model • Fertility

Hematologic malignancies and solid tumors that occur in children and reproductive-aged women are treated more successfully today than ever before [1]. The 5-year survival rates for all childhood cancers combined have increased from 58.1% in 1975–1977 to 79.6% in 1996–2003 [2]. As a result, the lasting adverse effects of radiation and chemotherapy are receiving increased attention. We find ourselves dealing with newly emergent issues as the ovary has become an unintended target. Although the majority of girls and adolescents with cancer are cured with nonsterilizing regimens, there remains a concern with regard to a shortened reproductive period [3]. In premenopausal women diagnosed with breast cancer and requiring chemotherapy, risk of menopause and infertility has been shown to be a major concern [1]. In fact, the risk of nonsurgical premature

E. Kovanci (✉)
Department of Obstetrics and Gynecology, Baylor College of Medicine, 1 Baylor Plaza, Houston, TX 77030, USA

E. Seli and A. Agarwal (eds.), *Fertility Preservation: Emerging Technologies and Clinical Applications*, DOI 10.1007/978-1-4419-1783-6_4,

menopause in childhood cancer survivors is 13-fold higher than that of their siblings, with a cumulative incidence of 8% by 40 years [4]. Likewise, animal studies showed a reduction in the resting ovarian follicle pool in a dose-related manner [2]. As a consequence, significant and in some cases, permanent reproductive dysfunction can create a distressing long-term sequel to therapy in young women who make an otherwise good recovery [5].

Events in the Ovary

The nonrenewable pool of ovarian primordial follicles declines, by atresia, from two million at birth to 500,000 at menarche [1, 3]. By the approximate age 38, the total number declines to about 25,000 and spontaneous and assisted conceptions become increasingly difficult [3]. At this point, follicle loss accelerates until ovarian senescence at, on average, 51 years of age [1]. Therefore, any insult as a result of chemotherapy and/or radiation that hastens the decline in follicle numbers may lead to a decreased reproductive potential [3].

Folliculogenesis

In folliculogenesis, there is continual recruitment of small numbers of primordial follicles to differentiate and grow prior to birth and throughout life. However, regular recruitment of primordial follicles (as in steady and more predictable) into a pool of growing follicles only begins once puberty is reached. The number of follicles that enter the growth phase in each cycle appears to be a fixed proportion of the primordial follicles remaining in the ovary [5]. An experiment on animal ovaries determined that the fewer the number of follicles at birth, the more rapidly they move into growth phase and are depleted [1]. This is a critical characteristic of the regulation of the ovarian follicular pool. The pool of functional primordial follicles is the ovarian reserve [5].

Assessing Ovarian Function

There is no reliable measure of ovarian reserve available for the individual woman. Assessment of ovarian function relies on the use of surrogate markers such as follicle-stimulating hormone, inhibin B, and anti-mullerian hormone as well as ultrasound assessment of ovarian volume and antral follicle count (AFC) [6]. Table 4.1 shows recommended cut-points for several tests in regards to fertility assessment.

Menstrual History

Presence or absence of menstruation is an inaccurate assessment of ovarian function. At a mean age of 37 years, there is an accelerated loss of follicle numbers with corresponding decline in fertility, even in the presence of normal menses [6]. The age of 43 is considered to be the point when fertility stops and sterility starts. The actual menopause occurs approximately 10 years after the substantial loss of conception potential [7]. Therefore, women can have regular menstrual cycles for several years, yet have close to zero chance of attaining a pregnancy. Likewise in young women with cancer who receive radiation and/or chemotherapy, while their menses may return posttreatment, their fertility is unknown as their mean age of menopause is drastically decreased. In a study by Byrne et al. [8], 1,067 women in whom cancer was diagnosed before

Table 4.1 Tests and cut-points for assessing ovarian function[a]

Menstrual cycle day-3 serum FSH level: >12 mIU/mL
Serum inhibin B level: <50 pg/mL
Serum AMH level: <1 ng/mL
Combined cycle day 3 or day 10 serum FSH levels in a clomiphene citrate challenge test: >10 mIU/mL
Ovarian volume per ovary: <3 cm^3
Total number of antral follicles (<10 mm): <4/ovary or <8 total

[a] Values are consistent with reduced ovarian reserve
Data from Stovall and McGee [1]

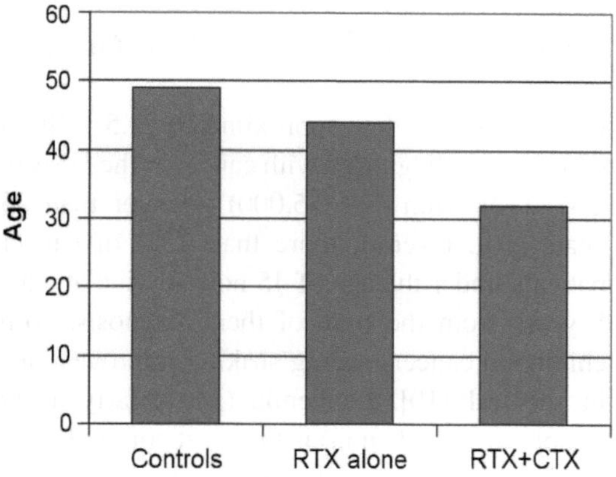

Fig. 4.1 Median age at menopause (from Byrne et al. [8], with permission. Copyright Elsevier 1992)

age 20, who were at least 5-year survivors, and who were still menstruating at age 21, had a four times greater risk of menopause compared with the control group. The cancer survivors also had significantly increased relative risks of menopause during the early 20s after treatment with either radiotherapy alone or alkylating agents alone. For women aged 21–25 treated with both radiation below the diaphragm and alkylating agent chemotherapy, the risk of menopause increased 27-fold for women, and by age 31, 42% of these women had reached menopause compared with 5% for controls (Fig. 4.1). In another study by Byrne [9], cancer survivors diagnosed after puberty and treated with radiation below the diaphragm were nearly ten times more likely to reach menopause during their 20s than controls, regardless of their primary diagnosis (relative risk=9.6 for Hodgkin disease and 8.56 for all other cancers). Therefore, even in the presence of menses, the fertility window can be drastically decreased.

Follicle-Stimulating Hormone

In normal menstruating women, ovarian function depends on pituitary gonadotropin production. Follicle-stimulating hormone (FSH) stimulates the granulosa cells of the growing follicle to proliferate and produce estradiol. Ovulation occurs in response to the mid-cycle surge of luteinizing hormone (LH), following which the corpus luteum produces progesterone, which prepares the endometrium to receive the fertilized ovum [5, 6]. In regards to assessing ovarian function, a rise in early follicular phase FSH with maintained estrogen production is a well-recognized feature of perimenopause. Increases in both FSH and estradiol on menstrual cycle day 3 are predictors of poor ovarian response to exogenous gonadotropin treatment [1]. This elevation is frequently seen in women of advanced reproductive age, reflecting poor follicle development. Before age 40 years, persistent FSH concentrations above 30 IU/L in the setting of amenorrhea, regardless of inciting events, suggests a diagnosis of premature ovarian failure [10]. An elevated estradiol level may suppress cycle day 3 FSH level into the normal range, and as such, an elevated FSH level reflects a more severe loss of ovarian function [1].

Inhibin B

Inhibin B is produced by small growing follicles after cyclic recruitment and thus is at its highest in the early follicular phase of the menstrual cycle. Inhibin B is involved in the feedback regulation of FSH secretion from the pituitary [6], and as such early follicular inhibin B levels decrease during reproductive aging leading to increasing FSH concentrations [11]. Reduced inhibin B levels have been reported in women 35 years and older with normal menstrual cycles and cycle day 3 FSH levels, compared with women less than 35 years of age. This suggests that serum inhibin B levels might be an earlier predictor of ovarian reserve loss in comparison with serum FSH level [1]. However, inhibin B levels vary across the menstrual cycle and must be correlated to the cycle day making interpretation of findings more complicated.

Anti-mullerian Hormone

Anti-mullerian hormone (AMH), produced by the granulosa cells of growing follicles, is involved in the regulation of primordial follicle recruitment [6]. In contrast to FSH, estradiol, inhibin B, and AFC, AMH levels do not appear to vary with cycle day. Moreover, AMH has a superior cycle-to-cycle reproducibility compared with inhibin B and FSH [11]. AMH is a relatively good predictor of the number of small FSH-sensitive follicles and thus of the number of oocytes recovered following controlled ovarian stimulation. As AMH is the product of the smallest growing follicles, it may more accurately reflect the ovarian reserve than AFC and basal and stimulated inhibin B [12]. AMH shows a marked decline in levels with advancing age [6] and has been judged to be the best endocrine marker for assessing the age-related decline of reproductive capacity [1].

Biophysical Measures

The pool of primordial follicles in the ovary is related to the number of growing antral follicles. Antral follicles are responsive to gonadotropin stimulation and the measure of ovarian reserve can be defined as the total number of follicles that can be stimulated to grow under maximal stimulation [13]. Biophysical measures including ovarian volume and AFC have been shown to correlate with reproductive potential [6], as there is a gradual decline in the number of primordial follicles with increasing age [13]. AFC may be measured by ultrasound when they reach 2 mm in size. The number of small antral follicles measuring 2–10 mm on transvaginal ultrasound has been shown to have the best correlation with chronological age in healthy female volunteers with normal menstrual cycles and proven fertility [6]. In a study analyzing depletion of ovarian reserve following cancer treatment in childhood, ovarian volume, but not AFC, was reduced in cancer survivors compared with controls [12]. However, both ovarian volume measurement and AFC have been successfully used to predict the response to FSH and subsequent pregnancy in IVF [1]. AFC has a significant association with the number of eggs collected and likewise is associated with the likelihood of clinical pregnancy [13].

Cancer During Reproductive Years

It is estimated that approximately 1.5 million people were diagnosed with cancer in the USA in 2009 [14], with 4% (55,000) younger than 35 years [10]. Overall, more than 75% of cancer patients under the age of 45 now survive at least 5 years from the time of their diagnosis, with childhood cancers seeing striking improvements in survival [10]. Leukemia (particularly acute lymphocytic leukemia) is the most common cancer in children (aged 0–14 years), followed by brain cancer and other nervous system tumors, soft tissue sarcomas, renal (Wilms) tumors, and non-Hodgkin lymphoma [14]. Women of reproductive age are most often afflicted with breast or gynecological malignancies, with up to 15% of breast and 43% of cervical cancer diagnoses in patients younger than 45 years [10].

Chemotherapy

One of the best established and effective treatments for malignancy is chemotherapy. However, the varied nature of the gonadal insult after chemotherapy makes it often difficult to predict whether a patient about to undergo treatment will subsequently have impaired fertility. Chemotherapy can lead to both infertility and loss of sex steroid hormone production in females. It is known that chemotherapy causes depletion of the primordial follicle pool in a drug- and dose-dependent manner [15]. That along with patient age can predict the extent of damage. There is an interplay among the three factors as the exact dose of chemotherapy associated with gonadal failure is not always predictable, but depends on the agent given and the age at time of administration [10]. Histological studies of human ovaries have demonstrated that the end result of chemotherapy is ovarian atrophy and global loss of primordial follicles [16, 17]. Although the cytotoxic-induced damage is reversible in other tissues of rapidly dividing cells such as bone marrow, gastrointestinal tract, and thymus; it appears to be progressive and irreversible in the ovary, where the number of germ cells is limited, fixed since the fetal life, and cannot be regenerated [18].

Mechanism

The exact mechanism of the effect of chemotherapy on the ovary is unknown [4]. Several reports have shown that apoptosis is a primary factor in ovarian follicle depletion post chemotherapy [16, 19]. However, other mechanisms by which chemotherapy affects the ovary may also be implicated. In a study by Meirow et al. [16], injury to blood vessels and focal fibrosis of the ovarian cortex were present in ovaries of patients previously exposed to chemotherapy. These modes of injury were present in nonatrophic ovaries of patients that were not sterilized by chemotherapy. The injury to blood vessels led to ischemic regions in the ovary and was therefore hypothesized to result in ovarian follicle depletion.

Age

The risk of developing premature ovarian failure or primary ovarian insufficiency is directly dependent on the number of primordial follicles, and thus on the age of the patient. Prepubertal ovaries appear to be more resistant to cytotoxic agents than postpubertal ovaries, possibly because they have more follicles and/or because of the absence of active folliculogenesis in prepubertal ovaries [4].

Drug Dosage

With standard doses of chemotherapy, ovarian function is often retained or recovered. However, premature menopause may occur [4]. It is difficult to analyze the impact of different treatment practices on the incidence of amenorrhea in view of the multitude of confounding factors [which drug(s) is/are being used, cumulative dose, dose-intensity]. However, it has been shown that the number of surviving primordial follicles following exposure to chemotherapy is in reverse correlation with the dose of chemotherapy [16]. In an unpublished study by Meirow [17], exposure to all doses of cyclophosphamide tested caused follicular destruction. However, the primordial follicle number decreased progressively in direct linear proportion to increasing dose of cyclophosphamide (significance, $p = 0.0001$).

Hormones

Chemotherapy has been shown to result in decreased ovarian reserve and higher baseline follicular FSH levels, indicating accelerated oocyte atresia and decreased oocyte quality [10]. Reduced AMH concentrations, but not reduced inhibin B concentrations, in young women treated for cancer during childhood in whom ovarian function is maintained has also been shown [15]. In Bath et al. [12], changes in AMH concentrations were shown to indicate gonadal toxicity during chemotherapy for breast cancer more clearly than those in E_2 or inhibin B, supporting a role for AMH as a marker of ovarian damage during such therapies.

Chemotherapeutic Agents

The effect of chemotherapy on the ovaries depends greatly on the agent used to treat a patients' disease. Table 4.2 shows different classes of chemotherapy, their mechanism of action, and their attributable risk for premature ovarian failure.

Alkylating Agents

The chemotherapeutic agents that are most toxic to the gonads are the "alkylators." Alkylating agents include the nitrogen mustards, nitrosoureas, busulfan, and thiotepa. They covalently bind an alkyl group to the DNA template and induce apoptosis [4]. Alkylating agents have also been shown to cause follicular and oocyte depletion, as well as being mutagenic to preovulatory oocytes [17]. Univariate associations have been shown among the chemotherapeutic agents (cyclophosphamide, busulfan, CCNU, chlorambucil, nitrogen

Table 4.2 Classes of chemotherapy, their action, and premature ovarian failure risk

Class of agent	Examples	Mechanism of action	Risk
Alkylating agents	Cyclophosphamide Mechlorethamine Chlorambucil Busulfan Melphalan	Crosslinks DNA strands, and inhibits RNA formation	High
Platinum derivatives	Cisplatin Carboplatin	Crosslinks DNA strands	Intermediate
Antimetabolites	Methotrexate 5-Fluorouracil Cytarabine	Inhibits pyrimidine or purine synthesis or incorporation into DNA	Low
Vinca alkaloids	Vincristine Vinblastin	Dissociation of microtubules leading to disruption of spindle	Low
Antiobiotics	Daunorubicin Bleomycin Adriamycin (doxorubicin)	Various, e.g., DNA intercalation, inhibition of transcription	Low (except adriamycin: intermediate)
Designer drugs	Imatinib	Tyrosine kinase inhibitor	Unknown

mustard, procarbazine) and primary ovarian insufficiency [4]. Multivariate associations were seen with the alkylating agents, cyclophosphamide and procarbazine, as significant risk factors for acute ovarian failure (AOF). Although exposure to procarbazine was an independent risk factor for AOF, regardless of age at treatment, cyclophosphamide significantly increased that risk only in subjects treated at an older age [20]. Cyclophosphamide at a dose of 5 g/m^2 is likely to cause amenorrhea in women over 40, while many adolescents will continue to menstruate after >20 g/m^2 [21]. Ifosfamide is a newer analogue of cyclophosphamide and also appears to be toxic to the ovary [4]. Although young women may not become amenorrheic after cytotoxic therapy, the risk of early menopause, defined as cessation of menses younger than age 40, is significant [21, 22]. In a review by Green et al. [22], the incidence of nonsurgical premature menopause in 2,819 survivors of childhood cancer compared with 1,065 female siblings was significantly higher (8% vs. 0.8%, $p<0.01$) (Fig. 4.2). For survivors who were treated with alkylating agents plus abdominal pelvic radiation, the cumulative incidence of nonsurgical premature menopause approached 30% [22] (Fig. 4.3).

Platinum Derivatives

Platinum derivatives have been shown to have an intermediate risk of premature ovarian failure.

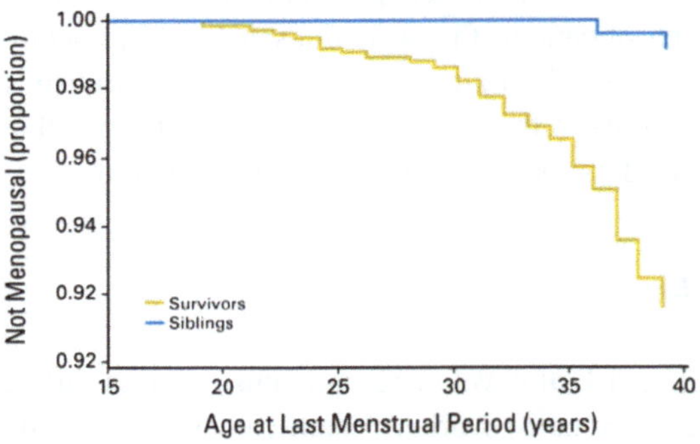

Fig. 4.2 Cumulative incidence curves of nonsurgical premature menopause in survivors compared with siblings

While the chance of retaining fertility is much higher than with the use of alkylating agents, the organoplatinum compounds (platinol cisplatin agents) are also female mutagens. They have been shown to cause different types of chromosomal damage (deletions, ring formation, and rearrangements) that induce genetic defects in oocytes, which may result in early embryonic mortality [17].

Antimetabolites/Vinca Alkaloids

Little gonadotoxicity is noted with these classes of chemotherapeutic agents [21] and are therefore classified as low-risk therapeutic agents Shamberger et al. [23]. Treated seven women (aged 13–31) with high-dose methotrexate for osteosarcoma and found no gonadal dysfunction. Likewise, Koyama et al. [24] observed no ovarian

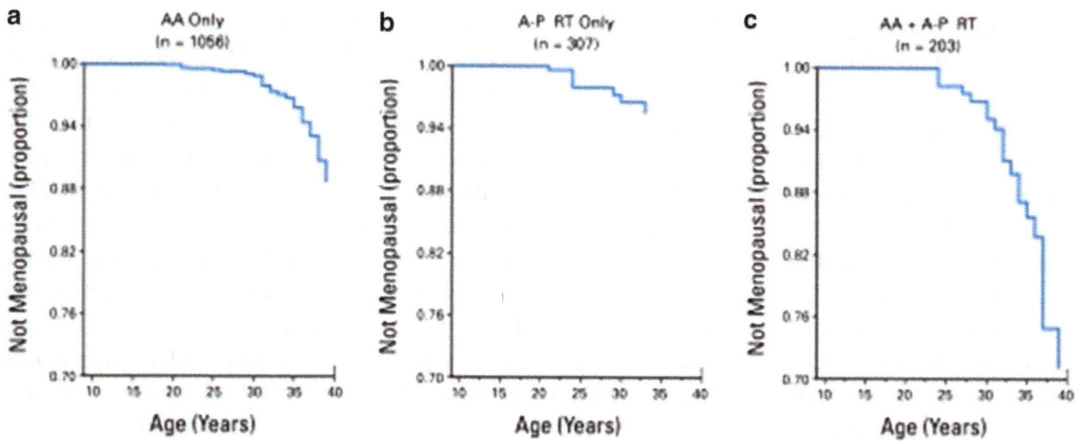

Fig. 4.3 Cumulative incidence curves of nonsurgical premature menopause in survivors according to treatment exposures. (**a**) Survivors treated with alkylating agents (AA) but not with abdominal-pelvic radiation therapy (A-P RT). (**b**) Survivors treated with A-P RT but not AA. (**c**) Survivors treated with AA and A-P RT

Table 4.3 Fertility assessment after current treatment for common cancers in childhood and adolescence

Low risk (<20%)	Medium risk	High risk (>80%)
ALL	AML (difficult to quantify)	Whole-body irradiation
Wilms' tumor	Hepatoblastoma	Localized radiotherapy: pelvic or testicular
Soft-tissue sarcoma: stage I	Osteosarcoma	Chemotherapy conditioning for bone-marrow transplantation
Germ-cell tumors (with gonadal preservation and no radiotherapy)	Ewing's sarcoma: nonmetastatic	Hodgkin's disease: treatment with alkylating-drugs
Retinoblastoma	Soft-tissue sarcoma: stage II or III	Soft-tissue sarcoma: stage IV (metastatic)
Brain tumor: surgery only, cranial irradiation <24 Gy	Neuroblastoma Non-Hodgkin lymphoma Hodgkin's disease: alternating treatment Brain tumor: craniospinal radiotherapy, cranial irradiation >24 Gy	Ewing's sarcoma: metastatic

From Green et al. [22]. Reprinted with permission.

toxicity among nine breast cancer patients treated with adjuvant fluorouracil [25].

Antibiotics

Bleomycin and adriamycin (a topoisomerase II inhibitor) are also proposed to be female specific mutagens. They both have been shown to induce dominant lethal mutations in maturing/preovulatory oocytes of female mice [17].

Type of Cancer

The risk for ovarian failure can be classified according to the type of malignant disease and its associated treatment (Table 4.3). For example, conditioning cytotoxic treatment before bone-marrow transplantation with high-dose chemotherapy and whole-body irradiation has a substantial risk of gonadotoxic effects, as well as treatment of metastatic sarcoma. However, current treatment for acute lymphoblastic leukemia, the most common malignant disease in childhood and adolescence, has a low risk of severe gonadotoxic effects [26]. Hodgkin's disease is the most common malignancy in the population aged 15–24 years. Prolonged survival of almost 90% of patients is now expected for young patients treated with cytotoxic chemotherapy for Hodgkin's disease [25]. Since treatment for Hodgkin's often involves alkylating agents, Hodgkin's also carries a high risk of gonadotoxic effects [27].

Breast Cancer

Breast cancer, the most common malignancy in women, affects approximately 185,000 women per year in the USA. Almost 25% of the cases occur premenopausally [25]. Breast cancer that develops at a younger age is more likely to be poorly differentiated and is typically associated with less favorable outcome [28]. Although adjuvant chemotherapy prolongs survival, it has been shown to have a high risk for premature ovarian failure [25, 28]. While newer treatment regimens employed are less gonadotoxic, most chemotherapeutic regimens employed in breast cancer treatment still interfere with some aspect of the cell cycle, thus affecting all proliferating cells in the body including healthy cells [29]. Furthermore, drugs are chosen to attack neoplastic cells that share a significant degree of functional, histological, and behavioral homology to germinal cells. Therefore, many of these drugs also destroy germinal tissue such as steroid-producing cells of the ovary (granulosa and theca cells), as well as oocytes, with resultant POF and early menopause [29].

Unfortunately, research determining the risk of developing premature ovarian failure does not take into account the recurrence rate of these diseases, and therefore, the additive gonadotoxic effects of repeated and likely more aggressive treatment. In a study by Arya et al. [30] analyzing 254 children treated for ALL with one current standard protocol, 17.9% relapsed, where bone marrow was the most common site of relapse (n=29, 72.5%). Once recurrence occurs, stronger doses of chemotherapy and alternative therapies are subsequently used. In instances of bone marrow recurrences, bone marrow transplantation (BMT) becomes a good treatment option. However, a patient must be conditioned before transplant. A common conditioning regimen consists of 40 mg/kg cyclophosphamide, thoracoabdominal irradiation of 4.5 Gy, and five doses of antithymocyte globulin [31]. Several studies describing the impact of high-dose chemotherapy plus total body irradiation, as preparative therapy for BMT, on gonadal function noted that more than 90% of young women developed ovarian failure [32, 33]. On the contrary, in a study by Sarafoglou et al. [32] on BMT during childhood, only 44% of girls went on to develop ovarian failure. Several studies have demonstrated that younger age (<25 years) at BMT is an important predictor of ovarian function recovery [33]. However, the short follow-up and overall increased ovarian reserve in children may account for the discrepancy. Factors that are associated with a higher fertility potential are having a single transplant procedure rather than a double transplant, as well as using less gonadotoxic conditioning chemotherapy. As such, cyclophosphamide and irradiation are known to significantly decrease fertility potential, and the previously low risk of infertility with treatment of ALL can become a high risk. Likewise, diseases in the other low and medium risk groups may increase to high risk in instances of recurrences as additional radio- or chemotherapy with higher doses and/or more gonadotoxic agents may be required. In these patients, previous treatment with any kind of chemo- or radiotherapy may decrease the probability of successful fertility preservation compared with the probability prior to receiving any cancer treatment.

Radiotherapy

Radiotherapy produces lethal cell damage, mostly due to its production of double-strand DNA breaks that are not amenable to repair [5].

Dose

The radiation dose necessary to induce ovarian failure is age dependent. Single-dose radiotherapy also seems to be more toxic than fractionated doses [10]. The effective sterilizing dose (ESD: dose of fractionated radiotherapy [Gy] at which premature ovarian failure occurs immediately after treatment in 97.5% of patients) decreases with increasing age at treatment. ESD at birth is 20.3 Gy (~2,030 rads; 1 Gy=100 rads), at 10 years 18.4 Gy, at 20 years 16.5 Gy, and at 30 years 14.3 Gy [34]. In women under age 40, doses in excess of 5 Gy directly to the pelvis (5–10 Gy) are required to produce amenorrhea in more than 95% of women. In women over age

40, a dose of 3.75 Gy produces amenorrhea in almost 100% of women [34]. When desiring complete ovarian estrogen production cessation, as in the management of breast cancer, almost 100% of women with a dose of 10–15 Gy in 4–5 fractions will become amenorrheic [35]. In a study by Chemaitilly et al. [20], doses of radiotherapy to the ovary of at least 2,000 cGy were associated with the highest risk of AOF with more than 70% of patients exposed to such doses developing AOF.

Location

The location of radiation can have a significant impact on fertility potential. Radiation involving the pelvis and the cranial region are the two most responsible for adversely affecting future childbearing. In a study showing the effects of abdominal irradiation secondary to childhood malignancies on ovarian function, primary or secondary amenorrhea and elevated pituitary gonadotropin levels occurred in 68% of the girls who had both ovaries included in the radiation field (estimated ovarian dose of 12–50 Gy), 14% of the girls who had at least one ovary at the edge of the radiation field (estimated ovarian dose of 0.9–10 Gy), and none of the girls who had at least one ovary outside the treatment field (estimated ovarian dose of 0.05–1.5 Gy) or did not receive abdominal radiation [35]. In another study on gonadal function status posttreatment for acute lymphocytic leukemia (ALL), 93% of girls receiving craniospinal plus abdominal irradiation showed elevated gonadotropins and arrested or delayed puberty, while only 49% of girls receiving craniospinal irradiation alone showed the same results [5].

These findings show that the effects of radiation on the ovary can be minimized with low doses of radiation and with either shielding the ovaries from radiation or moving the ovaries away from the radiation field, such as with oophoropexy [17, 35, 36].

Uterus

Radiation has several effects on the uterus. First, uterine vasculature is altered, potentially impairing cytotrophoblast invasion and resulting in decreased fetal-placental blood flow, and fetal growth restriction. Second, uterine elasticity and volume can be decreased from radiation-induced myometrial changes, which can lead to preterm labor and delivery [10]. Even with standard estrogen replacement, the uteri of young girls are often reduced to 40% of normal adult size. The uterine volume also correlates with the age at which radiation is received [37]. Third, radiotherapy can injure the endometrium, preventing normal decidualization and causing disorders of placental attachment, such as placenta accreta [10]. In one study of women receiving abdominal irradiation, the endometrium became unresponsive to physiological serum levels of estradiol and progesterone attained by exogenous administration [36]. In another study by Bath et al. [38], however, following physiological sex steroid replacement for 3 months, all measures of uterine function improved such that there was no significant difference in uterine blood flow and endometrial thickness from the comparison group.

This radiation damage to the uterine musculature and vasculature may adversely affect prospects for pregnancy in these women. However, there are no studies that have shown a significant decline in fertility as a result of uterine damage.

Ovaries

Radiotherapy to the ovaries is the most significant risk factor for AOF [20]. After irradiation, damaged oocytes either undergo repair or eliminate from the ovary by phagocytosis. The time at which degenerative changes occur is independent of the dose administered, but dose influences the number of oocytes affected [36]. In a study by Sarafo-glou et al. [32], at least 50% of girls treated prepubertally retained adequate ovarian function to enter puberty and menstruate regularly. However, while ovarian function may be preserved in women treated prepubertally with total body irradiation, the risk of early ovarian failure remains [38]. On the contrary, postpubertal treatment with total body irradiation carries a significant risk of permanent ovarian failure [32, 38]. As would be expected, the women who had radiation-induced ovarian failure had undetectable levels of inhibin B, a measure of follicular reserve. However, the woman whose ovarian function was preserved, inhibin B was also undetectable in the early follicular phase of an apparently ovulatory cycle [38]. This reflects that the resumption of menses does not guarantee normal ovarian reserve.

Central Nervous System

Cranial radiation can lead to disruption of the Hypothalamic–Pituitary–Gonadal (HPG) axis, resulting in dysregulation of the hormonal pathways responsible for menstruation and fertility. This radiation can affect the timing of the onset of puberty [36] as well as cause endocrinopathies such as hypogonadism and hyperprolactinaemia, which result in amenorrhea and infertility [10].

Faddy–Gosden Model

The Faddy–Gosden model produces a biologically realistic model of ovarian follicle disappearance with increasing age [39]. This model is the only one that takes into account all available histologic data and combines these with known ranges for the age of menopause [40].

Radiation, depending on the dose and location as discussed earlier, significantly decreases the oocyte population resulting in a much earlier menopause. Unfortunately, the precise age of resultant menopause is not known.

However, this model can be applied to patients who receive irradiation, showing what happens to an oocyte population after irradiation (Fig. 4.4). Given the solution to the Faddy–Gosden equation

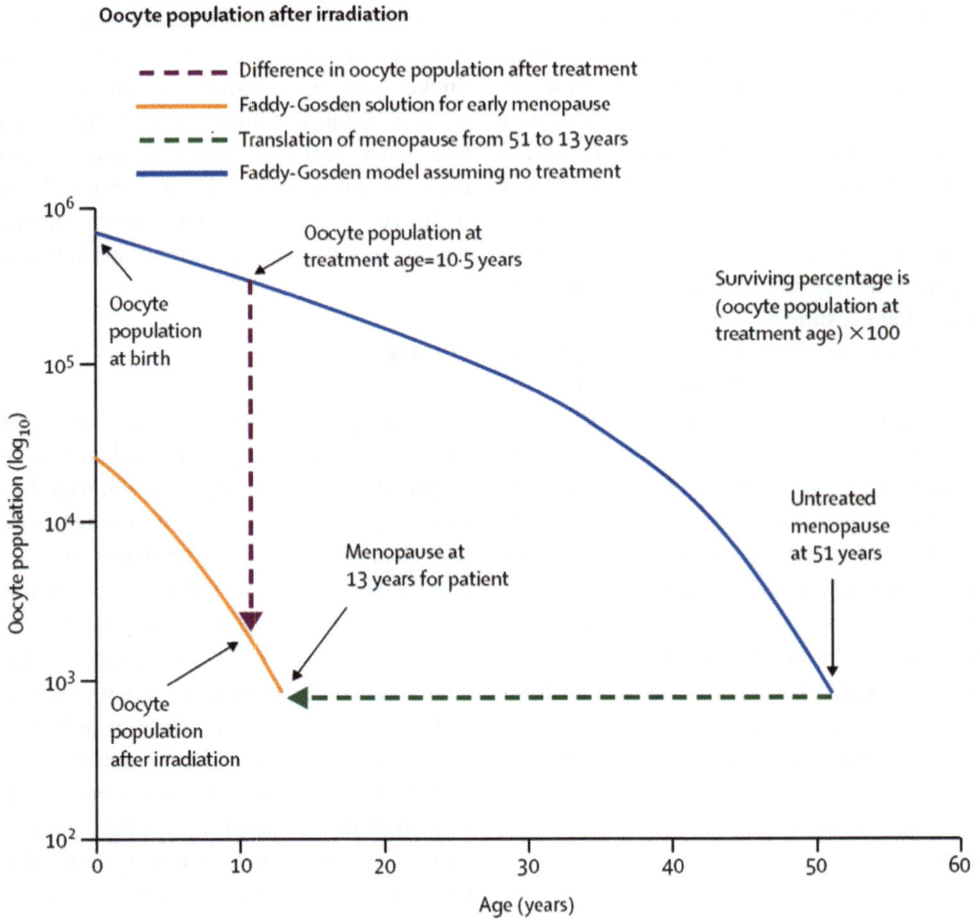

Fig. 4.4 Oocyte population after irradiation

and the surviving percentage function, we can predict the age of ovarian failure for a patient given a known dose (z Gy) at a known chronologic age (x_{chron} years) [40].

While the Faddy–Gosden model calculates the shift of menopause to age 13, both the location and quantity of radiation used can cause a later onset of menopause around age 21. Therefore, it cannot be assumed that a female who begins her menses at a normal age has not suffered from ovarian loss.

When to Discuss Fertility

Only 50% of patients are counseled regarding the option of fertility preservation and the potential risks associated with pregnancy and parenthood after cancer [10]. Unfortunately, in patients who have cancer treatment without a prior fertility preservation procedure, reproduction seems to be substantially reduced [10]. Therefore, if the option of fertility is important to the patient, then it must be discussed prior to cancer treatment.

Conclusions

Today, cancer is treated more successfully than ever before. As a result, new issues have arisen addressing the long-term outcomes of these treatments. While chemo- and radiotherapy are very effective in treating several common cancers such as Hodgkin's lymphoma and breast cancer, the consequences of these treatments on the ovary are significant. In a study by Chemaitilly et al. [20], compared with survivors who did not develop AOF, survivors with AOF were older at diagnosis, were more likely to have been diagnosed with Hodgkin's lymphoma, and were more likely to have been exposed to an alkylating agent and abdominal/pelvic radiotherapy. Therefore, when treating a patient with childbearing potential, it is important to be aware of the various treatment options both for the cancer and fertility.

References

1. Stovall D, McGee E. How chemotherapy harms ovarian function and how to assess your patients' risk and reproductive status. Srm-ejournal.com. Accessed September 2, 2010.
2. Childhood cancers: National Cancer Institute. http://www.cancer.gov/cancertopics/factsheet/Sites-Types/childhood. Accessed September 2, 2010.
3. Larsen EC, Müller J, Schmiegelow K, Rechnitzer C, Andersen AN. Reduced ovarian function in long-term survivors of radiation- and chemotherapy-treated childhood cancer. J Clin Endocrinol Metab. 2003;88(11):5307–14.
4. Cohen L. Cancer treatment and the ovary: the effects of chemotherapy and radiation. Ann N Y Acad Sci. 2008;1135:123–5.
5. Salooja N, Reddy N, Apperley J. Vulnerability of the reproductive system to radiotherapy and chemotherapy. In: Tulandi T, Gosden R, editors. Preservation of fertility. London: Taylor & Francis; 2004. p. 39–64.
6. Johnston RJ, Wallace WH. Normal ovarian function and assessment of ovarian reserve in the survivor of childhood cancer. Pediatr Blood Cancer. 2009;53(2):296–302.
7. Tarlatzis BC, Zepiridis L. Perimenopausal conception. Ann N Y Acad Sci. 2003;997:93–104.
8. Byrne J, Fears TR, Gail MH, Pee D, Connelly RR, Austin DF, et al. Early menopause in long-term survivors of cancer during adolescence. Am J Obstet Gynecol. 1992;166(3):788–93.
9. Byrne J. Infertility and premature menopause in childhood cancer survivors. Med Pediatr Oncol. 1999;33(1):24–8.
10. Knopman JM, Papadopoulos EB, Grifo JA, Fino ME, Noyes N. Surviving childhood and reproductive-age malignancy: effects on fertility and future parenthood. Lancet Oncol. 2010;11(5):490–8.
11. Knauff EA, Marinus JC, Eijkemans MJ, et al. Anti-müllerian hormone, inhibin B, and antral follicle count in young women with ovarian failure. J Clin Endocrinol Metab. 2009;94(3):786–92.
12. Bath LE, Wallace WHB, Shaw MP, Fitzpatrick C, Anderson RA. Depletion of ovarian reserve in young women after treatment for cancer in childhood: detection by anti-Mullerian hormone, inhibin B and ovarian ultrasound. Hum Reprod. 2003;18(11):2368–74.
13. Muttukrishna S, McGarrigle H, Wakim R, Khadum I, Ranieri DM, Serhal P. Antral follicle count, anti-mullerian hormone and inhibin B: predictors of ovarian response in assisted reproductive technology. BJOG. 2005;112(10):1384–90.
14. Jemal A, Siegel R, Ward E, Hao Y, Xu J, Thun MJ. Cancer statistics. CA Cancer J Clin. 2009;59(4):225–49.
15. Anderson RA, Themmen APN, Qahtani AA, Groome NP, Cameron DA. The effects of chemotherapy and

long-term gonadotrophin suppression on the ovarian reserve in premenopausal women with breast cancer. Hum Reprod. 2006;21(10):2583–92.
16. Meirow D, Dor J, Kaufman B, Shrim A, Rabinovici J, Schiff E, et al. Cortical fibrosis and blood-vessels damage in human ovaries exposed to chemotherapy. Potential mechanisms of ovarian injury. Hum Reprod. 2007;22(6):1626–33.
17. Meirow D. Ovarian injury and modern options to preserve fertility in female cancer patients treated with high dose radio-chemotherapy for hemato-oncological neoplasias and other cancers. Leuk Lymphoma. 1999;33(1–2):65–76.
18. Blumenfeld Z. Gynaecologic concerns for young women exposed to gonadotoxic chemotherapy. Curr Opin Obstet Gynecol. 2003;15(5):359–70.
19. Ben-Aharon I, Bar-Joseph H, Tzarfaty G, Kuchinsky L, Rizel S, Stemmer SM, et al. Doxorubicin-induced ovarian toxicity. Reprod Biol Endocrinol. 2010; 8:20.
20. Chemaitilly W, Mertens AC, Mitby P, Whitton J, Stovall M, Yasui Y, et al. Acute ovarian failure in the childhood cancer survivor study. J Clin Endocrinol Metab. 2006;91(5):1723–8.
21. Schwartz C. Long-term survivors of childhood cancer: the late effects of therapy. Oncologist. 1999;4(1): 45–54.
22. Green DM, Sklar CA, Boice Jr JD, Mulvihill JJ, Whitton JA, Stovall M, et al. Ovarian failure and reproductive outcomes after childhood cancer treatment: results from the Childhood Cancer Survivor Study. J Clin Oncol. 2009;27(14):2374–81.
23. Shamberger RC, Rosenberg SA, Seipp CA, et al. Effects of high-dose methotrexate and vincristine on ovarian and testicular functions in patients undergoing postoperative adjuvant treatment of ostosarcoma. Cancer Treat Rep. 1981;65:739–746.
24. Koyama H, Wada T, Nishizawa Y, et al. Cyclophosphamide-induced ovarian failure and its therapeutic significance in patients with breast cancer. Cancer. 1977;39:1403–1409.
25. Blumenfeld Z. Preservation of fertility and ovarian function and minimalization of chemotherapy associated gonadotoxicity and premature ovarian failure: the role of inhibin-A and -B as markers. Mol Cell Endocrinol. 2002;187(1–2):93–105.
26. Wallace WHB, Anderson RA, Irvine DS. Fertility preservation for young patients with cancer: who is at risk and what can be offered. Lancet Oncol. 2005;6(4): 209–18.
27. Blumenfeld Z, Avivi I, Ritter M, Rowe JM. Preservation of fertility and ovarian function and minimizing chemotherapy-induced gonadotoxicity in young women. J Soc Gynecol Investig. 1999;6(5):229–39.
28. Kovacs P. Preserving ovarian function after breast cancer treatment. http://www.medscape.com/viewarticle/701901. Accessed September 2010.
29. Anchan RM, Ginsburg ES. Fertility concerns and preservation in younger women with breast cancer. Crit Rev Oncol Hematol. 2010;74(3):175–92.
30. Arya LS, Kotikanyadanam SP, Bhargava M, Saxena R, Sazawal S, Bakhshi S, et al. Pattern of relapse in childhood ALL: challenges and lessons from a uniform treatment protocol. J Pediatr Hematol Oncol. 2010;32(5):370–5.
31. Veitia RA, Gluckman E, Fellous M, Soulier J. Recovery of female fertility after chemotherapy, irradiation, and bone marrow allograft: further evidence against massive oocyte regeneration by bone marrow-derived germline stem cells. Stem Cells. 2007;25(5): 1334–5.
32. Sarafoglou K, Boulad F, Gillio A, Sklar C. Gonadal function after bone marrow transplantation for acute leukemia during childhood. J Pediatr. 1997;130(2): 210–6.
33. Liu J, Malhotra R, Voltarelli J, Stracieri AB, Oliveira L, Simoes BP, et al. Ovarian recovery after stem cell transplantation. Bone Marrow Transplant. 2008;41(3): 275–8.
34. Wallace WH, Thomson AB, Saran F, et al. Predicting age of ovarian failure after radiation to a field that includes the ovaries. Int J Radiat Oncol Biol Phys. 2005;62:738–44.
35. Thomas GM. Female genital tract. In: Cox J, Ang K, editors. Radiation oncology: rationale, technique, results. 8th ed. St. Louis, MO: Mosby; 2003. p. 758.
36. Ogilvy-Stuart AL, Shalet SM. Effect of radiation on the human reproductive system. Environ Health Perspect. 1993;101 Suppl 2:109–16.
37. Critchley HO, Bath LE, Wallace WH. Radiation damage to the uterus – review of the effects of treatment of childhood cancer. Hum Fertil (Camb). 2002;5(2):61–6.
38. Bath LE, Critchley HOD, Chambers SE, Anderson RA, Kelnar CJ, Wallace WH. Ovarian and uterine characteristics after total body irradiation in childhood and adolescence: response to sex steroid replacement. Br J Obstet Gynaecol. 1999;106(12):1265–72.
39. Faddy MJ, Gosden RG. A model conforming the decline in follicle numbers to the age of menopause in women. Hum Reprod. 1996;11(7):1484–6.
40. Wallace WHB, Thomson AB, Saran F, Kelsey TW. Predicting age of ovarian failure after radiation to a field that includes the ovaries. Int J Radiat Oncol Biol Phys. 2005;62(3):738–44.

Impact of Chemotherapy and Radiotherapy on the Uterus

5

Abbie L. Fields, Deleep Kumar Gudipudi, and Giuseppe Del Priore

Abstract

Survivorship issues surrounding conception and gestational outcome have gained recognition as the management and outcome of childhood malignancies has become increasingly successful. Issues surrounding future fertility are likely multifactorial including both direct and indirect effects of therapy on the pituitary system as well as the end organs. Radiation therapy impairs normal function of both the ovaries and the uterus. The magnitude of risk is related to the age and menarchal status at the time of treatment. Although all treatment for malignant disease is a balance between successful therapeutic outcome and minimized toxicity, it appears that the threshold for radiation is considerably lower than initially perceived. As cancer treatment and cures improve, assisted reproductive technology strives to keep pace. Oocyte cryopreservation, in vitro maturation, ovarian tissue cryopreservation, and uterus and ovary transplantation are all evolving to treat secondary infertility in cancer patients. Because of the rapidly changing field, it has become of paramount importance to educate both the parents and the patient with pediatric malignancy as to the impact their treatment may have on fertility and pregnancy outcome.

Keywords

Chemotherapy and fertility • Uterus and chemotherapy • Radiotherapy effects on the uterus • Cancer survivorship and fertility • Cancer and fertility

G.D. Priore (✉)
Division of Gynecologic Oncology,
Indiana University Melvin and Bren Simon
Cancer Center, Indianapolis, IN, USA

E. Seli and A. Agarwal (eds.), *Fertility Preservation: Emerging Technologies and Clinical Applications*,
DOI 10.1007/978-1-4419-1783-6_5,

It is anticipated that in 2010, 739,940 women will be diagnosed with cancer, of whom 1 in 48 will be less than 40 years of age [1]. One of every 640 individuals between the ages of 20 and 39 is a survivor of childhood malignancy [2]. Prior to 1970, the overall 5-year survival for these malignancies was less than 50%. Presently, overall 5- and 10-year survivorship for invasive cancers in this age group is reported as 80 and 75%, respectively [3]. Given continued advances in early detection, treatment and improved outcomes, issues of extended survivorship have become even more significant.

Primary malignancies differ among various age groups. In the pediatric population (<20 years old) nearly 75% of diagnosis were leukemias (31%), central nervous system malignancies (21.3%), lymphomas (8.1%), neuroblastoma (7.1%), or Wilms tumor (5.2%) [1]. In the reproductive age group, up to 15% of breast cancer and 43% of cervical cancers occur in women under the age of 45 [4]. Therapeutic modalities require surgery, radiation, chemotherapy, or a combination of the above, all of which are known to significantly impact future fertility outcome.

The impact of cancer treatment on ovarian function is well known and is discussed at length elsewhere. The purpose of this chapter is to detail the impact of chemotherapy and radiation therapy on uterine function, physiology, pregnancy, and neonatal outcomes in female survivors of childhood malignancies.

The Normal Uterus

The uterus is a thick-walled muscular organ with embryologic origin from the paramesonephric duct (uterus, uterine tubes) and the surrounding undifferentiated mesenchyme (connective tissue and muscle) [5]. Sagittal section finds an endometrial cavity lined by columnar epithelial glands and surrounding stroma. The thickest middle layer is 1–2 cm of interdigitating smooth muscle, and the organ is surrounded by the visceral peritoneum, serosa.

Blood supply to the uterus is from the uterine artery, which branches off the anterior division of the internal iliac artery. Sympathetic innervation is from the inferior hypogastric plexus and parasympathetic from the pelvic splanchnics (S2–4). Recognizing considerable variation in size, a nulliparous adult uterus generally measures $8.0 \times 4.5 \times 3.0$ cm in size and weighs approximately 90 g.

During puberty, the uterus undergoes significant change. Serial ultrasound evaluations of healthy young girls and women age 6–25 have demonstrated that uterine growth occurs from approximately the age of 8 and continues through age 20 [6, 7]. Onset of uterine growth precedes breast development by about 2 years. It has also been shown that during this time, the shape of the uterus changes from a tubular to a pear-shaped organ, and the endometrial thickness increases. During pubertal maturation, there is also a significant increase in uterine artery blood flow velocity with an associated decline in vascular resistance measured as pulsatility index [8].

Vascular supply to the uterus is through the gonadal vessels and uterine vessels. There is a minor component though ascending vaginal arteries. The uterus is resilient to complete interruption of any of these vascular pedicles with almost no change in perfusion or oxygenation [9]. This may be important for its recovery after injury from chemotherapy or radiation.

The adult, post adolescent uterus is an organ relatively resistant to radiation injury. Exposure of 40–60 Gy is tolerable before radiation fibrosis and endometrial atrophy becomes clinically significant [10].

Effect of Chemotherapy on the Uterus

To complete the process of conception, term gestation and successful delivery requires not only an intact hypothalamic–pituitary–ovarian axis but also an end-organ (uterus) that retains capability for implantation, adequate distensibility, and sufficient blood flow to support growth and development of the fetus to term. Although the effects of chemotherapy on the ovary result in

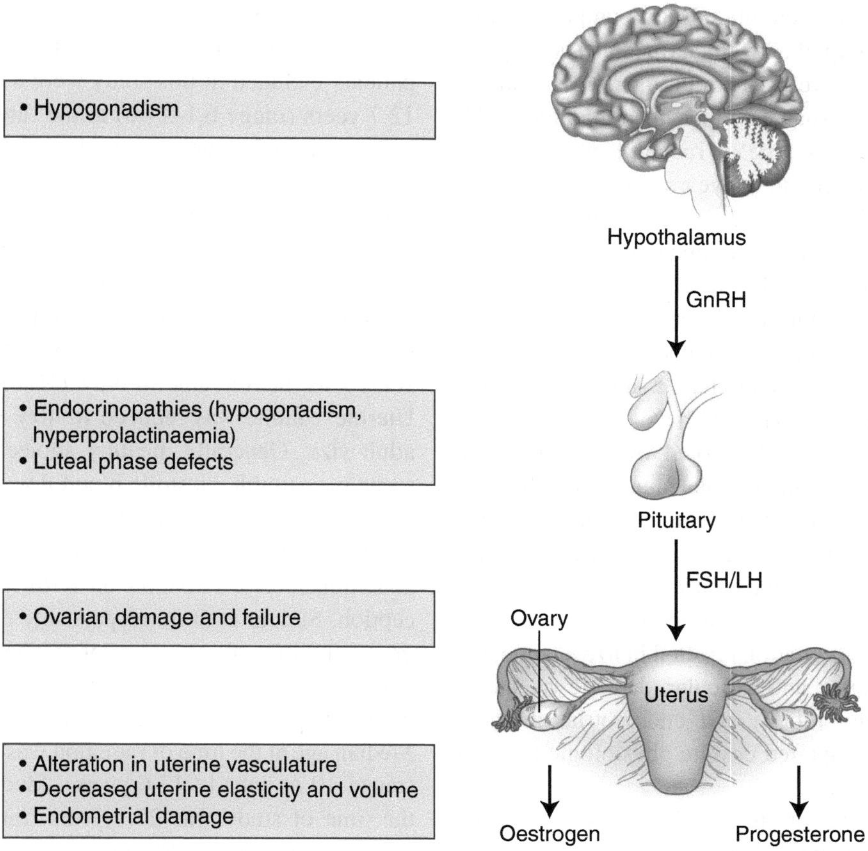

Fig. 5.1 The effects of radiation on fertility and reproductive outcome

impaired oocyte reserve, function and quality are well described, there seems to be little evidence that chemotherapy contributes directly to uterine damage [11].

Effect of Radiation on the Uterus

The effects that radiation bears on the uterus are both direct and indirect. Direct effects include impaired uterine distensibility due to myometrial fibrosis: uterine vascular damage and impaired decidualization and placentation due to endometrial injury [4, 12–14]. Indirect effects are attributed to disorders of the hypothalamic–pituitary–ovarian axis resulting in endocrinopathies associated with luteal phase defects and other factors contributing to infertility [15]. Radiation effects on fertility and reproductive outcome are seen in Fig. 5.1.

The degree of uterine damage depends on the radiation field, total radiation dose, fractionation schedule, and patient age at the time of treatment [12, 16, 17]. Studies suggest that the prepubertal uterus is more vulnerable than the adult uterus to the effects of pelvic irradiation, with doses of 14–30 Gy likely to cause uterine dysfunction [16, 17].

This is because the prepubertal uterus has not yet undergone the growth and shape changes recognized to occur in association with pubescence. These changes appear to be unique to the process of irradiation on uterine musculature and vasculature as opposed to the hormone deprivation effect associated with premature ovarian failure, also recognized to occur with pelvic radiation.

A comparative observational study was reported on a small group of 10 women (median age 24 years; range 15–31 years) with premature ovarian failure due to whole abdominal radiation

(20–30 Gy) in childhood and compared with a control group of 22 women (median age 31 years; range 23–37 years) with premature ovarian failure due to sources other than radiation [16]. Median age at time of radiation was 2.5 years (range 5 weeks to 11 years). Both groups were treated with cyclic exogenous hormone replacement therapy and experienced withdrawal bleeding. Women were followed by serial ultrasounds documenting uterine length, uterine blood flow (pulsatility index), and endometrial thickness. The radiated women had a uterine length of 4.1 ± 0.8 cm compared to that of the control group of 7.3 ± 0.6 cm. Also noted were absent/diminished uterine artery doppler signals in 8/10 treated women compared to 0/22 in the control group, as well as linear increase in endometrial thickness vs. no change in the control vs. radiated groups, respectively. This study concluded that early childhood irradiation resulted in irreversible uterine muscular and vascular damage as uterine length, blood flow, and endometrial thickness remained impaired despite physiologic hormone replacement.

A more recent study of 100 survivors of childhood cancer treated by radio- and/or chemotherapy addressed the same issues [12]. Median age at diagnosis was 5.4 years, range 0.1–15.3. This study also confirmed reduced uterine volume to be significantly associated with prepubertal age at the time of treatment with radiotherapy. The ability to stimulate the endometrium with exogenous hormone replacement therapy sufficient to induce withdrawal bleeding does not connote normal fertility. It has been shown that when postpubertal women treated with <30 Gy of radiation are given physiologic doses of estradiol, uterine function may be improved. However, patients receiving radiotherapy prior to puberty have seemingly irreversible damage to the uterus (uterine size, shape, blood flow, endometrial thickness) despite estradiol supplementation [12, 17, 18].

In addition to restricting uterine volume, radiation likely also results in uterine vessel damage. Holm et al. evaluated the effect of total body irradiation (TBI) and allogenic bone marrow transplant for childhood leukemia or lymphoma on uterine volume and uterine blood flow by ultrasound and Doppler [13]. The 12 female patients included in this study were a median of 12.7 years (range 6.1–17.6) at the time of treatment and 21.5 years (range 11.6–25.6) at the time of follow-up evaluation. At follow-up, 11/12 had achieved menarche, eight of which required hormone replacement. Despite the lower radiation dose associated with TBI (14 Gy) as compared to the dose of whole abdominal radiation (20–30 Gy), uterine diastolic blood flow in this study was measureable in only 11% of patients. Uterine volume was reduced to 40% of normal adult size. Generally, healthy subjects demonstrate measurable diastolic blood flow in 35% of prepubertal and 100% of adult females [14].

To better elucidate the clinical impact of abdominal/pelvic radiation on fertility and conception, Sudour et al. retrospectively evaluated a group of 84 female survivors of childhood malignancy [18]. Fifty of these patients were treated while prepubertal and 34 were postpubertal. Median age at the time of radiation was 11.3 years (range 10 months to 17.6 years). Median age at the time of study was 27.1 years (range 18–45 years). Actual dosimetry curves were not available. The radiation dose delivered to the reproductive organs was estimated on the basis of theoretic points and dosimetry was then calculated on the basis of the actual dose, fractionation schedule, and treatment field as detailed in medical records. Appropriate consideration was given to those with ovarian transposition and pelvic shielding. The radiation field included or excluded the pelvis in 27 and 57 patients, respectively. All patients received multiagent chemotherapy, 60% of which included alkylators. Patients were then allocated as being fertile (normal puberty regular spontaneous menses, at least one recorded pregnancy), potentially fertile (normal puberty, regular spontaneous menses, no recorded pregnancy at the time of study), or difficult fertility (abnormal puberty, absence of spontaneous menses, no recorded pregnancy). This study found that direct pelvic irradiation, total radiation dose, and dose delivered to reproductive organs all significantly impacted fertility (Table 5.1). They also concluded

Table 5.1 Prognostic factors for fertility

Group	Age at time of RT	Abdominal RT alone ($n=57$)	RT including Pelvis ($n=27$)	Mean total dose	Mean dose in ovaries (R/L)	Mean dose in uterus	Chemotherapy with alkylating drugs	No. of pregnancies	No. of children
Fertile ($n=28$)	10.2 years	23	5	33.4 Gy	2.1/3.8 Gy	4.9 Gy	13	67	49
Potentially fertile ($n=34$)	9.6 years	29	5	26.7 Gy	3.5/2.2 Gy	3.5 Gy	19	0	0
Difficult fertility ($n=22$)	10.1 years	5	17	32.5 Gy	14.9/14 Gy	18.1 Gy	18	0	0
P value (analysis of varience)	NS		<0.0005	0.05	<0.0005	<0.0005	0.04		

RT radiotherapy; *NS* not significant
From Sudour et al. [18], with permission from Elsevier

that RT doses between 4 and 15 Gy put patients at risk for subfertility (also dependent on age and associated treatment modalities) and that only radiation doses less than 4 Gy appear to be associated with normal conception/fertility.

Effect of Radiation on Placentation

As already discussed, pelvic irradiation has an irreversible effect on uterine development, musculature, elasticity, and vasculature. Reports have been made at the time of cesarean section in these women of a particularly thin, fibrotic wall concerning for risk of rupture [18]. Radiation damage to the endometrium could potentiate issues of abnormal decidulization and placentation resulting in placenta accreta, percreta, and uterine rupture [18–20]. One report details a 23-year-old patient who at 13 weeks gestation suffered spontaneous incomplete abortion complicated by hemoperitoneum and subsequent supracervical hysterectomy for presumed uterine rupture [20]. This patient received a total of 70 Gy pelvic RT for a gluteal tendon sheath sarcoma at the age of 7. The hysterectomy specimen found a uterine wall thickness of 2–5 mm, with absence of decidua and an implantation site with villi extending to the serosal surface.

A similar occurrence in a 23-year-old patient treated with 8.75 Gy whole body irradiation in a single fraction at age 5 for leukemia, suffered a uterine rupture requiring supracervical hysterectomy at 17 weeks gestation [19]. Pathologic examination of this specimen described a small, atrophic uterus with wall thickness of 1–6 mm along with full thickness cytotrophoblast invasion infiltrating the serosa.

Rates of survivorship in childhood malignancies continue to increase. Innovative methods of fertility preservation and assisted reproductive technology are allowing these women realistic opportunities for conception. It is becoming clear that appropriate counseling should be provided to these women regarding concerns and possible risk of abnormal placentation as well as the other potential complications that accompany their pregnancies.

Effects of Radiation on Pregnancy Outcome

Impaired uterine distensibility and irreversible vascular damage cause concerns for potential issues of miscarriage, preterm delivery, and reduced blood flow to the fetus and placenta resulting in growth impairment and low birth weight. Given substantial improvement in survival after cancer over the past 30 years, the value of documenting long-term morbidity and mortality issues, e.g., uterine function, associated with treatment of childhood malignancies has become clear.

The Childhood Cancer Survivor Study (CCSS) was designed to investigate survivorship parameters of a cohort of 5-year childhood/adolescent cancer survivors diagnosed between 1970 and 1986 [21]. Evaluable data from this cohort included more than 14,000 survivors from 25 centers in the USA and Canada. Tumors included were leukemia, CNS tumors, Hodgkin disease, non-Hodgkin lymphoma, Wilms tumor, neuroblastoma, soft tissue sarcoma, and malignant bone tumors. The control group used as the comparison cohort was a random sample of nearest-age siblings of the survivors who were not surgically sterile.

For the subgroup analysis designed to assess overall fertility and miscarriage rates, data were abstracted from 1,915 women. Among this group, there were 4,029 conceptions resulting in 63% live births. Also reported were 17% elective terminations, 15% spontaneous miscarriages, 3% unknown, and 1% stillbirth. The male:female ratio of offspring born was 1.09:1.00, consistent with US population trends [22].This finding argues against the likelihood of transmission of lethal X-linked mutations among female survivors of childhood cancers [21].

Radiation doses in the CCSS cohort were estimated on the basis of radiation treatment records for the individual patients. Oophoropexy and pelvic shielding were accounted for in dose estimates. Methodology for the dosimetry estimates have been fully described elsewhere [23]. The CCSS study found that the female survivors age 15–30 years were significantly less likely than their sibling controls (at the same age of pregnancy) to have a live birth [20]. The relative risk (RR) of miscarriage among survivors age 21–30 years was also increased, although not statistically significant. Patients with tumors of the Central Nervous System (CNS) reported more miscarriages than their sibling counterparts [RR = 1.65 (95% CI 1.16–2.34), $P = 0.006$].

Subgroup analysis did not indicate that this outcome was related to cranial irradiation. In fact, to assess the possible effect of cranial vs. spinal RT on miscarriage rates, various radiation regimens were compared among survivors of acute lymphoblastic leukemia (ALL) and CNS tumors (Table 5.2). Statistical significance was noted only among those receiving cranial and spinal radiation compared with those receiving no irradiation with RR = 3.63(95% CI 1.70–7.78) ($P < 0.001$).

This finding may suggest spinal radiation (which likely involves the uterus) to be the adverse effect on pregnancy outcome. The risk of stillbirth (>20 weeks gestation) and neonatal death rates (within 28 days following birth) have also been evaluated through cohort analysis based on CCSS data [23]. There were 1,657 women in this study who had survived childhood cancer. One thousand-forty-two (63%) of these women have been treated with radiotherapy.

Radiation doses received by the ovaries and uterus were exactly concordant in 92% of the radiation-exposed data points thus complicating the ability to differentiate the independent effects of these organs. Age at menarche was known in 80% of participants and estimated in an additional 10%. The adjusted RR of stillbirth and neonatal death is shown in Table 5.3. For the group with the highest risk of fetal loss (>10.0 Gy), the meanRT dose to the uterus was 17.52 Gy (SD 12.03) and to the ovaries was 18.08 (SD 9.75). Table 5.4 describes the impact of radiation premenarche and post menarche. As noted, the risk of stillbirth/neonatal death continues to rise as the dose of RT is increased. Although not presented in tabular form, the death rates continued to increase in those treated premenarche with a rate of 4/49 (8%) offspring with RT doses of 2.50–9.99 Gy [RR = 5.8 (95% CI 1.2–28.2)]; and to 5/23 (22%) with RT doses >10.00 Gy [RR = 19.0 (95% CI 5.6–65.2)].

Given the variety of cancer diagnoses in this group of patients, a sub group of 510 women treated with high-dose pituitary irradiation could be evaluated. The lack of effect noted [RR = 1.1 (95% CI 0.5–2.4) for >20 Gy vs. no irradiation] supports the conclusion made by Green [21] earlier. It is the direct radiation effects on the target organs (uterus and ovaries), which are more significant in pregnancy outcome than are the indirect effects from cranial or pituitary radiation on the hypothalamic – pituitary–ovarian axis, which can generally be managed by appropriate hormonal manipulation.

The low birth weight of offspring born to female survivors of childhood cancer has been thoroughly documented in both Wilms tumor survivors as well as those receiving abdominal/pelvic RT for many different tumors [18, 21, 24, 25]. Clarification of whether the low birth weight was secondary to preterm birth or simply small for gestational age (SGA) was lacking. Signorello and colleagues evaluated a cohort of 2,201 singleton births to 1,264 survivors and 1,175 children born to 601 females in the sibling cohort from the CCSS database [26]. Logistic regression estimated odds ratios (OR) to assess association between modes of therapy and preterm birth (<37 weeks), low birth weight (<2.5 kg) and SGA (lowest tenth percentile). This study confirmed that children of survivors were more likely to be born preterm than their siblings' children (21.1% vs. 12.6%; OR = 1.9, 95% CI = 1.4–2.4; $P < 0.001$). They did not note an increase in SGA between children of survivors and their siblings' offspring. It was demonstrated that >500 cGy RT to the uterus significantly increased the risk of preterm birth ($P = 0.003$), low birth weight ($P = 0.001$) and SGA ($P = 0.003$) as compared to those not having undergone RT (Tables 5.5–5.7). The effect of RT on the uterus was greater and the threshold lower for girls who were premenarchal at the time of treatment. Preterm birth and low birth weight were noted to be particularly high among children of survivors from Wilms tumor.

Earlier evaluation reported by the National Wilms Tumor Study Group evaluated the impact of flank irradiation and other associated treatment for Wilms tumor on pregnancy outcome [24].

Table 5.2 Frequency of miscarriage by cranial or spinal irradiation

Received central nervous system radiation	All CCSS members			Acute lymphoblastic leukemia			All CCSS members central nervous system		
	No.	Miscarriage	RR (95% CI)	No.	Marriage	RR (95% CI)	No.	Miscarriage	RR (95% CI)
No radiation therapy	1,147	160	1.00	171	17	1.00	104	21	1.00
Cranial + spinal radiation therapy	110	28	2.22 (1.36–3.64)	72	20	3.63 (1.70–7.78)	13	2	0.86 (0.18–4.18)
Cranial radiation therapy only	499	91	1.40 (1.02–1.94)	333	48	1.60 (0.85–3.00)	77	18	1.33 (0.61–2.93)
Spinal radiation therapy only	5	1	2.10 (0.18–24.5)	0	0	–	2	1	4.52 (0.27–76.4)
No cranial/spinal radiation therapy	1,483	215	1.06 (0.82–1.36)	2	0	–	19	4	1.21 (0.50–2.97)
Unknown	785	128	1.23 (0.92–1.64)	178	30	1.80 (0.88–3.68)	61	13	0.79 (0.30–2.03)

From Green et al. [21], with permission from Elsevier

Table 5.3 Association between organ-specific radiotherapy doses and risk of stillbirth or neonatal death in offspring of survivors of childhood cancer

	All pregnancies lasting at least 20 weeks[a]	Stillbirth or neonatal death	Relative risk (95% CI) of stillbirth or neonatal death		Relative risk (95% CI) of stillbirth
			Crude	Adjusted[b]	Adjusted[b]
Women					
Not treated with radiation	1,075	21 (2%)	Reference	Reference	Reference
Radiation dose to uterus and ovaries (Gy)					
0.01–0.99	1,404	24 (2%)	0.8 (0.4–1.4)	0.7[c] (0.4–1.4)	0.7[d] (0.3–1.5)
1.00–2.49	155	5 (3%)	2.1 (0.8–5.7)	1.9[c] (0.7–5.4)	2.4[d] (0.8–7.3)
2.50–9.99	126	5 (4%)	1.6 (0.4–6.0)	1.6[c] (0.4–6.5)	1.9[d] (0.5–7.6)
≥10.00	28	5 (18%)	9.2 (3.3–25.4)	9.1[c] (3.4–24.6)	7.3[d] (2.3–23.0)
Men					
Not treated with radiation	734	12 (2%)	Reference	Reference	Reference
Radiation dose to testes (Gy)					
0.01–0.09	692	8 (1%)	0.7 (0.3–1.8)	0.8 (0.3–2.0)	1.1 (0.4–3.0)
0.10–049	337	5 (1%)	0.9 (0.3–2.7)	0.8 (0.3–2.3)	0.7 (0.2–2.8)
≥0.50	241	3 (1%)	0.8 (0.2–2.8)	0.6 (0.2–1.9)	0.9 (0.2–3.2)

Data are number or number (%), unless otherwise indicated

[a]Stillbirths and livebirths

[b]Adjusted for calendar year of birth and maternal age (for analyses of uterus or ovaries), and paternal age (for analyses of testes)

[c]Adjusted relative risks, with exposure defined as uterine and maximum ovarian radiation dose, were 0.7 (95% CI 0.4–1.4) for 0.01–0.99 Gy, 1.5 (0.5–4.6) for 1.00–2.49 Gy, 1.3 (0.3–6.4) for 2.50–9.99 Gy, and 7.8 (3.1–19.4) for 10.00 Gy or more with outcomes noted in 24 (2%), four (3%), four (3%), and six (14%) offspring, respectively

[d]Adjusted relative risks, with exposure defined as uterine and maximum ovarian radiation dose, were 0.7 (0.3–1.5) for 0.01–0.99 Gy, 1.9 (0.6–6.2) for 1.00–2.49 Gy; 1.6 (0.3–7.5) for 2.50–9.99 Gy, and 6.9 (2.5–19.5) for 10.00 Gy or more with outcomes noted in 16 (1%), four (3%), three (2%), and four (10%) offspring, respectively

From the Lancet, Signorello et al. [33], with permission of Elsevier

Table 5.4 Association between radiotherapy doses to uterus and ovaries and risk of stillbirth or neonatal death in offspring of survivors of childhood cancer

	Treatment before menarche		Treatment after menarche	
	Risk of stillbirth or neonatal death	Relative risk[a,*] (95% CI)	Risk of stillbirth or neonatal death	Relative risk[a,**] (95% CI)
No radiation	5/494 (1%)	Reference	13/447 (3%)	Reference
0.01–0·99 Gy	11/636 (2%)	1.3 (0.5–3.9)	7/599 (1%)	0.3 (0.1–1.0)
1.00–2.49 Gy	3/69 (4%)	4.7 (1.2–19.0)	2/70 (3%)	1.2 (0.2–6.4)
≥2.50 Gy	11/82 (13%)	12.3 (4.2–36.0)	1/85 (1%)	0.2 (0.0–1.4)

Data are n/N (%), unless otherwise indicated. Data are for the offspring of only 1,481 (89%) of 1,657 female survivors for whom timing of treatment in relation to menarche could be established. For the 160 women in whom age at menarche was missing and needed to be estimated, we assumed they were treated before menarche if they were treated at age 9 or younger, and after menarche if they were treated at age 18 or older

[a]Adjusted for calendar year of birth and maternal age

**P* value for trend was 0.006

***P* value for trend was 0.32

From the Lancet, Signorello et al. [33] with permission from Elsevier

Table 5.5 Risk of preterm birth among live-born children of female childhood cancer survivors, by radiation treatment and organ dose

	Preterm birth, *N* (%)	Full-term birth, *N* (%)	OR (95%CI)	*P*
Not treated with any radiation	121 (19.6)	496 (80.4)	1.0 (referent)	
Radiation dose to uterus, cGy (treatment at all ages)				
0–10	81 (17.3)	386 (82.7)	0.9 (0.6–1.4)	0.72
10–50	53 (19.9)	214 (80.1)	1.2 (0.7–2.0)	0.53
50–250	74 (26.1)	209 (73.9)	1.8 (1.1–3.0)	0.03
250–500	21 (39.6)	32 (60.4)	2.3 (1.0–5.1)	0.04
>500	23 (50.0)	23 (50.0)	3.5 (1.5–8.0)	0.003
Radiation dose to uterus, cGy (treatment premenarche)				
0–10	29 (16.9)	143 (83.1)	0.9 (0.5–19)	0.85
10–50	23 (27.4)	61 (72.6)	2.2 (1.0–4.8)	0.05
50–250	21 (26.3)	59 (73.8)	2.1 (1.0–4.6)	0.05
>250	15 (48.4)	16 (51.6)	4.9 (1.7–13.9)	0.003
Radiation dose to uterus, cGy (treatment postmenarche)				
0–10	39 (21.2)	145 (78.8)	1.2 (0.6–2.4)	0.69
10–50	20 (15.4)	110 (84.6)	0.8 (0.3–1.7)	0.52
50–250	39 (27.5)	103 (72.5)	1.8 (0.8–4.3)	0.18
>250	24 (40.7)	35 (59.3)	1.9 (0.7–4.9)	0.21
Radiation dose to ovary, cGy				
0–10	75 (17.7)	349 (82.3)	0.9 (0.6–1.5)	0.81
10–20	25 (21.2)	93 (78.8)	1.2 (0.7–2.4)	0.51
20–50	27 (17.1)	131 (82.9)	0.9 (0.4–1.7)	0.66
50–100	36 (26.3)	101 (73.7)	15 (0.8–3.0)	0.22
>100	9 (25.0)	27 (75.0)	1.2 (0.4–3.8)	0.76
Radiation dose to pituitary, cGy				
0–50	74 (30.5)	169 (69.5)	1.6 (1.0–2.7)	0.05
50–250	96 (21.1)	360 (78.9)	1.0 (0.6–1.9)	0.90
250–2,000	27 (23.9)	86 (76.1)	1.4 (0.7–2.7)	0.34
>2,000	54 (18.7)	235 (81.3)	10 (0.6–1.6)	0.96

From Signorello et al. [26], with permission of Oxford University Press

Table 5.6 Risk of low birth weight (LBW) among the live-born children of female childhood cancer survivors, by radiation treatment and organ dose

	LBW, *N* (%)	Non-LBW, *N* (%)	OR (95% CI)	*P*
Not treated with any radiation	47 (7.6)	575 (92.4)	1.0 (referent)	
Radiation dose to uterus, cGy				
0–10	31 (6.5)	443 (93.5)	1.5 (0.7–34)	0.35
10–50	19 (7.1)	250 (92.9)	1.2 (0.5–3.2)	0.66
50–250	25 (8.7)	262 (91.3)	1.2 (0.5–3.2)	0.66
250–500	14 (25.5)	41 (74.5)	4.3 (14–12.8)	0.01
>500	17 (36.2)	30 (63.8)	6.8 (2.1–22.2)	0.001
Radiation dose to ovary, cGy				
0–10	30 (7.0)	401 (93.0)	1.6 (0.7–36)	0.30
10–20	6 (5.2)	109 (94.8)	0.5 (0.1–2.1)	0.37
20–50	12 (7.4)	151 (92.6)	2.3 (0.7–7.0)	0.15
50–100	12 (8.8)	124 (91.2)	0.9 (0.2–3.1)	0.81
>100	5 (14.7)	29 (85.3)	1.7 (0.3–9.6)	0.52
Radiation dose to pituitary, cGy				
0–50	38 (15.6)	205 (84.4)	1.7 (0.7–3.9)	0.23
50–250	28 (6.1)	434 (93.9)	2.1 (0.8–5.9)	0.15
250–2,000	10 (8.7)	105 (91.3)	1.4 (0.4–4.7)	0.62
>2,000	30 (10.1)	266 (89.9)	1.5 (0.6–3.8)	0.36

From Signorello et al. [26], with permission of Oxford University Press

Table 5.7 Risk of a small-for-gestational-age (SGA) birth among the live-born children of female childhood cancer survivors, by radiation treatment and organ dose

	SGA, *N* (%)	Non-SGA, *N* (%)	OR (95% Cr)	*P*
Not treated with any radiation	46 (7.8)	547 (92.2)	1.0 (referent)	
Radiation dose to uterus, cGy				
0–10	39 (8.7)	409 (91.3)	1.1 (0.6–2.1)	0.66
10–50	21 (8.2)	236 (91.8)	1.3 (0.6–2.8)	0.46
50–250	20 (7.3)	256 (92.8)	1.0 (0.4–2.2)	0.92
250–500	3 (5.7)	50 (94.3)	1.3 (0.3–4.8)	0.70
>500	8 (18.2)	36 (81.8)	4.0 (1.6–9.8)	0.003
Radiation dose to ovary, cGy				
0–10	36 (8.9)	369 (91.1)	1.2 (0.6–2.2)	0.61
10–20	8 (7.0)	106 (93.0)	0.8 (0.3–2.5)	0.75
20–50	14 (9.2)	138 (90.8)	1.4 (0.6–3.3)	0.46
50–100	7 (5.2)	127 (94.8)	0.7 (0.2–2.2)	0.57
>100	3 (8.8)	31 (91.2)	1.2 (0.2–67)	0.81
Radiation dose to pituitary, cGy				
0–50	20 (8.6)	213 (91.4)	1.7 (0.8–3.4)	0.14
50–250	35 (7.9)	408 (92.1)	1.7 (0.7–4.7)	0.27
250–2,000	5 (4.7)	10 I (95.3)	0.3 (0.1–1.4)	0.12
>2,000	30 (106)	252 (89.4)	1.1 (0.6–2.1)	0.69

From Signorello et al. [26], with permission of Oxford University Press

This group evaluated 427 pregnancies (>20 weeks) resulting in 409 singleton live births and 12 live-born twin gestations. Statistical analysis was based on available medical records from 309 pregnancies. Findings of fetal malpresentation, umbilical cord complications, preterm delivery (<36 weeks), and low birth weight (<2,500 g) were all increased among irradiated female survivors of Wilms tumor. Many of these findings had been previously observed and presumed to be associated with the flank radiation as well as the recognized congenital malformations of the genitourinary tract frequently associated with Wilms tumor [24, 25, 27, 28].

Potential Mechanisms to Improve Reproductive Function

As discussed earlier, despite success in eliciting physiologic cycling of the endometrium with hormone replacement, residual myometrial and vascular damage remains a concern. An expanding body of literature now exists regarding the utility of selective estrogen receptor modulators (SERMs) in both management of malignancy and assisted reproduction. Clomiphen is a SERM introduced into clinical care before the concept of SERM was widely used. SERMs are typically a variation of 17 beta-estradiol with subclassification based on their side chain.

Tamoxifen is classified as a triphenylethylene with recognized potential for endometrial stimulation and has been evaluated as a means of ovulation induction since the early 1970s. The mechanism of action relies on the agonistic effect of tamoxifen when endogenous estrogen levels are low. Tamoxifen administration in anovulatory women with depleted estrogen receptors in the hypothalamus can have normalization of GnRH secretion leading to optimized FSH secretion with resultant follicular development and subsequent ovulation. Tamoxifen has been successful in ovulation induction both in clomiphene citrate failures and as combination therapy with clomiphene citrate [29, 30]. There is also data to support an increased endometrial thickness to accompany tamoxifen ovulation induction [31]. Tamoxifen is associated with the subsequent risk of endometrial cancer (<1 case/1,000 women treated/year) and sarcoma very rarely.

Raloxifene is a benzothiophene with little, if any, endometrial stimulation effect. Although raloxifene could be similarly considered as an agent for ovulation induction, due to its known antagonist properties at the uterine level, it has not been further evaluated. Raloxifene's apparent safer side effect profile may be bias due to its relatively short and limited long term data.

GnRH agonist/antagonist, progestins, antiprogestins, and antiestrogens are all used in the treatment of various malignancies of their side effects. The short-term effects of these agents are well documented and predictable. However, the long-term effect of the uterus is not known. For now, the use and exposure to these agents should be elicited and documented in the patients' charts for possible future consideration.

Conclusions

Survivorship issues surrounding conception and gestational outcome have gained recognition as the management and outcome of childhood malignancies has become increasingly successful. Issues surrounding future fertility are likely multifactorial including both direct and indirect effects of therapy on the pituitary system and the end organs.

Radiation therapy impairs normal function of both the ovaries and the uterus. The magnitude of risk is related to the age and menarchal status at the time of treatment. Total radiation dose, radiation field, fractionation schedule, ovarian position, and extent of shielding are all factors that significantly contribute to the extent of tissue damage secondary to treatment. Although physiologic regimens of hormone replacement are able to elicit cycling of the endometrium, this does not necessarily insure a favorable pregnancy outcome.

Although all treatment for malignant disease is a balance between successful therapeutic outcome and minimized toxicity, it appears that the threshold for radiation is considerably lower than initially perceived. Present literature supports only radiation doses of less than 4 Gy to the uterus to be associated with normal fertility and pregnancy outcome.

Beyond this dose, issues of diminished uterine volume/distensability, myometrial fibrosis, irreversible uterine vascular damage, and endometrial atrophy contribute to associated complications of miscarriage, abnormal placentation, stillbirth/neonatal death, preterm delivery, and impaired fetal growth.

As cancer treatment and cures improve, assisted reproductive technology strives to keep pace. Oocyte cryopreservation, in vitro maturation, ovarian tissue cryopreservation, uterus, and ovary transplantation are all evolving as options for managing secondary infertility in cancer patients. Because of the rapidly changing field, it has become paramount importance to educate both the parents and the patient with pediatric malignancy as to the impact their treatment may have on fertility and pregnancy outcome. They should understand the importance of continuing their care at specialized centers, aware of the risks associated with their pregnancies, and equipped to manage any complication to ensure a successful outcome [32].

References

1. American Cancer Society: Cancer Facts and Figures, 2010.
2. Robison LL, Green DM, Hudson M, et al. Long-term outcomes of adult survivors of childhood cancer. Cancer. 2005;104(11 Suppl):2557–64.
3. Rowland JH, Baker F. Resilience of cancer survivors across the lifespan. Cancer. 2005;104(11 Suppl): 2543–8.
4. Knopman JM, Papadopoulos EB, Grifo JA, et al. Surviving childhood and reproductive-age malignancy: effects on fertility and future parenthood. Lancet Oncol. 2010;11:490–8.
5. Copeland LJ, editor. Textbook of gynecology. 2nd ed. Philadelphia: W.B. Saunders; 2000. p. 8–9.
6. Holm K, Mosfeldt L, Brocks V, et al. Pubertal maturation of the internal genitalia: an ultrasound evaluation of 166 healthy girls. Ultrasound Obstet Gynecol. 1995;6:175–81.
7. Bridges NA, Cooke A, Healy MJR, et al. Growth of the uterus. Arch Dis Child. 1996;75(4):330–1.
8. Mosfeldt Laursen E, Holm K, Brocks V, et al. Doppler assessment of flow veocity in the uterine artery during pubertal maturation. Ultrasound Obstet Gynecol. 1996;8:341–5.
9. Sieunarine K, Boyle DCM, Corless DJ, Noakes DE, Ungar L, Marr CE, et al. Pelvic vascular prospects for uterine transplantation. Int Surg. 2006;91(4):217–22.
10. Arrive L, Chang YGF, Hricak H, et al. Radiation-induced uterine changes: MR imaging. Radiology. 1989;170:55–8.
11. Critchley HO, Wallace WH. Impact of cancer treatment on uterine function. J Natl Cancer Inst Monogr. 2005;34:64–8.
12. Larsen EC et al. Radiotherapy at a young age reduces uterine volume of childhood cancer survivors. Acta Obstet Gynecol Scand. 2004;83(1):96–102.
13. Holm K et al. Ultrasound B-mode changes in the uterus and ovaries and Doppler Changes in the uterus after total body irradiation and allogenic bone marrow transplantation in childhood. Bone Marrow Transplant. 1999;23(3):259–63.
14. Laursen EM et al. Doppler assessment of flow velocity in the uterine artery during pubertal maturation. Ultrasound Obstet Gynecol. 1996;8(5):341–5.
15. Wo JY, Viswanathan AK. Impact of radiotherapy on fertility, pregnancy and neonatal outcomes in female cancer patients. Int J Radiat Oncol Biol Phys. 2009;5:1304–12.
16. Critchley HO et al. Abdominal irradiation in childhood; the potential for pregnancy. Br J Obstet Gynaecol. 1992;99(5):392–4.
17. Bath LE et al. Ovarian and uterine characteristics after total body irradiation in childhood and adolescence: response to sex steroid replacement. Br J Obstet Gynaecol. 1999;106(12):1265–72.
18. Sudour H et al. Fertility and pregnancy outcome after abdominal irradiation that included or excluded the pelvis in childhood tumor survivors. Int J Radiat Oncol Biol Phys. 2010;76(3):867–73.
19. Norwitz ER, et al. Placenta percreta and uterine rupture associated with prior whole body radiation therapy. Obstet Gynecol. 2001;98(5 Pt 2):929–31. [PubMed: 10084257].
20. Pridjian G, Rich NE, Montag AG. Pregnancy hemoperitoneum and placenta percreta in a patient with previous pelvic irradiation and ovarian failure. Am J Obstet Gynecol. 1990;162(5):1205–6.
21. Green DM, Whitton JA, Stovall M, et al. Pregnancy outcome of female survivors of childhood cancer: a report from the childhood cancer survivor study. Am J Obstet Gynecol. 2002;187(4):1070–80.
22. Ventura SJ, Martin JA, Curtin SC, et al. Births: final data for 1999. Natl Vital Stat Rep. 2001;49:1–100.
23. Stovall M et al. Dose reconstruction for therapeutic and diagnostic radiation exposures: use in epidemiologic studies. Radiat Res. 2006;166:141–57.
24. Green DM et al. Pregnancy outcome after treatment for Wilms tumor: a report from the National Wilms Tumor Study Group. J Clin Oncol. 2002;20(10): 2506–13.
25. Chiarelli AM, Marrett LD, Darlington GA. Pregnancy outcomes in females after treatment for childhood cancer. Epidemiology. 2000;11(2):161–6.
26. Signorello LB et al. Female survivors of childhood cancer: preterm birth and low birth weight among their children. J Natl Cancer Inst. 2006;98(20):1453–61.

27. Li FP et al. Outcome of pregnancy in survivors of Wilms tumor. JAMA. 1987;257:216–9.
28. Byrne J et al. Reproductive problems and birth defects in survivors of Wilms tumor and their relatives. Med Pediatr Oncol. 1988;16:233–40.
29. Borenstein R et al. Tamoxifen treatment in women with failure of clomiphne citrate therapy. Aust N Z J Obstet Gynecol. 1989;29:173–5.
30. Suginami H et al. A clomiphene citrate and tamoxifen citrate combination therapy: a novel therapy for ovulation induction. Fertil Steril. 1993;59:976–9.
31. Reynolds K et al. Comparison of the effect of tamoxifen on endometrial thickness in women with thin endometrium (<7mm) undergoing ovulation induction with clompiphene citrate. Fertil Steril. 2010;93:2091–3.
32. Del Priore G, Stega J, Sieunarine K, Ungar L, Smith JR. Human uterus retrieval from a multi-organ donor. Obstet Gynecol. 2007;109:101–4.
33. Signorello LB et al. Stillbirth and neonatal death in relation to radiation exposure before conception: a retrospective cohort study. Lancet. 2010;376:624–30.

Ovarian Transposition

6

Elena S. Ratner, Dan-Arin Silasi, and Masoud Azodi

Abstract

Preservation of ovarian function is a critical concern in premenopausal cancer patients whose treatments include radiation therapy to pelvic structures and incorporate the ovaries in the radiated field. Ovarian transposition is a surgical maneuver used to protect ovarian function prior to delivery of gonadocidal doses of radiation therapy.

Keywords

Ovarian transposition • Hodgkin's disease • Cervical cancer • Radiation in ovarian cancer • Fertility in women • Menopause

Ovarian transposition was first described in 1958 as a means of maintaining ovarian function in patients irradiated for cervical carcinoma [1]. Over half the patients with cervical cancer are premenopausal, so maintaining ovarian function and potential future fertility is important to this cohort [2].

Malignant disease indications for ovarian transposition are broad, ranging from cervical cancer, upper vaginal cancer, and Hodgkin's disease to anal cancer [3].

Historically, premenopausal patients undergoing pelvic irradiation for Hodgkin's disease frequently underwent ovarian transposition at the time of a staging laparatomy. With various new disease managements and the proliferation of laparoscopic surgical techniques, ovarian transposition can now take place in other clinical settings, including gastrointestinal cancers that require pre-operative radiation [4]. Ovarian transposition is no longer used for Hodgkin's disease because radiation therapy is no longer applied systemically.

Literature reviews suggest multiple techniques for ovarian transposition. One of the methods described is the medial transposition of the ovaries behind the uterus [5]. Lateral transposition of the ovaries to the paracolic gutters is an ulterior technique [6, 7]. A laparoscopic approach to the above methods is favored because it allows the patient to return to normal function and start radiation treatment without delay [8, 9]. In addition to preserving premenopausal hormonal status, ovarian transposition with subsequent pelvic irradiation has been reported to preserve a woman's (with normal genital tract morphology) ability to become pregnant and give birth [4]. No increased risk of fetal anomalies has been found in such women [4].

M. Azodi (✉)
Department of Obstetrics, Gynecology, and Reproductive Sciences, Yale University School of Medicine, New Haven, CT, USA

E. Seli and A. Agarwal (eds.), *Fertility Preservation: Emerging Technologies and Clinical Applications*, DOI 10.1007/978-1-4419-1783-6_6,

Radiation Threshold Doses

The results of radiotherapy are dependent upon both the radiation dose delivered to the ovaries and the patient's age [10]. "Minimal tolerance dose" and "maximal tolerance dose" are definitions that refer to sterility complication rates of 5% and 50%, respectively, within 5 years of completion of radiation treatment. For the ovaries, these values are approximately 300 and 1,200 cGy, respectively [3]. Radiation doses received by the ovaries ranging from 3.2 to 20 Gy have been found to induce menopause [7]. Cessation of ovarian function has been observed after an ovary receives direct radiation of approximately 2,000 cGy, which is significantly less than the dose needed for treatment of cervical cancer [11]. Bieler et al. studied gonadotropin, estradiol, and progesterone levels during radiation therapy. Patients with nontransposed ovaries began losing ovarian function between 560 and 1,130 cGy. Estradiol levels decreased first, followed successively by rising FSH and LH levels [12].

Patient's age also plays a role in the extent of functional loss following radiotherapy to the ovaries. Whereas a dose of 6 Gy was sufficient to induce menopause in patients older than 40 years of age, the ovaries of patients younger than 20 years of age withstood doses of 20 Gy [13]. When intracavitary brachytherapy alone is used, menses resume in 76% of patients [14].

Procedure Methods

The literature discusses numerous surgical techniques for ovarian transposition. When ovarian transposition was implemented during treatment of Hodgkin's disease, the ovaries transported to the midline behind the uterus in order to avoid radiation directed at the lymph node chains [15]. When ovarian transposition is utilized in the management of cervical cancer, however, a more radical approach is needed because the entire pelvis has to be radiated. Among the techniques that have been described are lateral subcutaneous placement, positioning intra- or retroperitoneally on the psoas muscle, intraperitoneal placement lateral to the psoas muscle, and intraperitoneal placement high in the pericolic gutters above the pelvic brim [16–19].

Laparotomy

When laparatomy is performed, ovaries are examined macroscopically to assure the absence of metastatic disease. The ovaries are then mobilized after the ureters are identified. Next, the uteroovarian ligament is transected and the fallopian tubes are separated from the ovaries. The peritoneum is then incised along the infundibulopelvic ligament to further mobilize the ovaries. Subsequently, the ovarian vessels are dissected to the level of the aortic bifurcation. After this, the ovaries are transposed bilaterally to the paracolic gutters at or above the level of the bifurcation. Finally, metallic clips are applied in a specific fashion for future imaging localization.

Laparoscopy

When laparoscopy is utilized, the ovarian transposition is performed in a similar fashion to the laparotomy approach described previously. Three or four port incisions are made and the trocars are introduced into the peritoneal cavity at the umbilicus, both lower quadrants, and/or the suprapubic region. Under direct visualization of the ureters, the utero-ovarian ligaments are separated. The peritoneum under and lateral to the ovarian vessels is incised to an area outside of the true pelvis. The ovaries and tubes are then fixed high in the paracolic gutters, below the spleen and the liver, and are sutured in 3 points to prevent torsion. Good blood supply to the ovaries is confirmed. Clips are placed [4].

Placing the ovaries intraperitoneally has been suggested in order to avoid the development of ovarian cysts. Intraperitoneal pericolic gutter ovarian transposition has been modified to include retroperitoneal placement of the ovarian vascular pedicles while maintaining an intraperitoneal ovarian location. This modification has been shown to decrease the risk of ovarian vascular tension and ovarian devascularization as well as bowel herniation.

Success of Ovarian Transposition

Ovarian transposition has been reported to be a successful approach to preserving ovarian function in patients requiring radiation therapy for gynecologic, gastrointestinal, and Hodgkin's disease malignancies. Historically, exploratory laparatomy performed for staging or debulking procedures for these malignancies provided an opportunity to perform ovarian transposition. As laparoscopic procedures became more widespread, surgical procedures dedicated to ovarian transposition alone became more common. A laparoscopic approach has been shown to be safe and not to delay the proposed treatment for the malignancy.

Preservation of ovarian function following ovarian transposition and pelvic irradiation for cervical cancer varies at a rate between 17 and 71% [20–22]. Ovarian function can be measured in three ways: by serum FSH levels, assessment of estrogenization of vaginal epithelium, or the presence of postmenopausal signs and symptoms [23]. A serum FSH greater than 40 miU/ml is consistent with menopause. Ovarian function preservation after ovarian transposition and irradiation appears likely to be age related. If the treatment is performed before age 40, 85% of women will retain ovarian function. If, however, the hysterectomy is performed between ages of 40 and 55, only 68% will maintain normal ovarian function [24].

Haie-Meder et al. reported that ovarian transposition success is influenced by two factors: the patient's age at the time of treatment and the radiation dose administered to the ovaries. The authors used the cut-off of 25 years of age as measure of successful ovarian preservation. The patients in this cohort were treated for either Hodgkin's disease or dysgerminoma and the median age was younger than those encountered in cervical cancer [7].

Morice et al., found an 83% ovarian function preservation rate among the 107 patients who underwent ovarian transposition during surgical treatment for cervical cancer at the Institute Gustave Roussy from 1985 to 1998. These patients were subdivided into three categories: (1) patients treated exclusively with surgery (100% of these patients preserved ovarian function), (2) patients treated with surgery and brachytherapy (90% preserved ovarian function), and (3) patients treated with surgery, external radiation therapy, and brachytherapy (60% preserved ovarian function). Sixteen patients went into menopause during the initial treatment, nine patients experienced menopause at the end of the treatment, and seven patients presented with menopausal symptoms after a median delay of 48 months. Additionally, 11 patients reached menopause at a "physiologic" age (>45 years) and more than 5 years after cessation of the treatment [25].

Covens et al. assessed quantitatively the preservation of ovarian function in three patients who underwent ovarian transposition. They found that serum FSH was within normal limits in two of three patients, 24–32 months after radiation. Additionally, all three patients menstruated regularly [26]. Tulandi et al. reported initial postmenopausal symptoms and elevated FSH in a 34-year-old woman who underwent radiation for rectal carcinoma after ovarian transposition. However, normal menstruation resumed and FSH levels returned to premenopausal range 8 months after treatment. This patient subsequently had a normal spontaneous pregnancy [27, 28].

Chambers et al. reported a 29% ovarian failure rate in patients who underwent lateral ovarian transposition with subsequent radiation for treatment of cervical cancer [20]. When lateral transposition is not successful, ovarian failure is presumed to be caused by vascular compromise, due to either erroneous surgical technique or radiation exposure to the vascular pedicle [29].

Complications

Chambers et al., reported that 24% of patients who underwent ovarian transposition developed symptomatic ovarian cysts. Sixteen percent of patients in this cohort required an oophorectomy.

A history of endometriosis or pelvic inflammatory disease increased these risks [22]. Patients receiving external radiation therapy had a lower incidence of ovarian cysts, but this was offset by a higher rate of menopause. Subcutaneous ovarian transposition yielded a 7% rate of ovarian failure and a high incidence of ovarian cysts. Seventeen percent of patients required repeated cyst aspirations and 7% underwent subsequent oophorectomy [12].

Morice et al. described delayed complications in 27 out of 107 patients. One patient with bulky stage IB cervical squamous carcinoma re-presented with metastasis to the ovary 3 years after initial treatment with surgery and brachytherapy and subsequently died from her disease [4]. Benign ovarian cysts were observed in 22 (23%) patients. Three patients underwent surgical exploration for persistent cysts. Three patients had chronic abdominal pain at the site of transposed ovaries, without evidence of cysts. One patient developed small bowel obstruction from adhesions at the site of transposed ovary.

Metastasis from cervical squamous or adenocarcinoma to the transposed ovary is rare in early stage disease, but has been reported to occur in 5% of patients with bulky, stage IIB tumors [30–32].Tabata et al. reported a significantly higher rate of ovarian metastasis in patients treated for stage Ib to III cervical adenocarcinoma, compared to patients with squamous cervical carcinoma (28% vs. 17–20%) [33]. Sutton et al., however, did not confirm these findings. The incidence of primary ovarian cancer is not increased in patients with cervical carcinoma and preservation of the ovarian function does not adversely influence the course of squamous cervical cancer.

Fertility Options

After successful completion of treatment in patients desiring fertility, the ovaries can be stimulated to produce eggs that can be retrieved, fertilized, and possibly re-implanted into the uterus. This may allow the patient to have a biological child either herself or through a surrogate. When doses of external beam radiation in the range of 8,500 cGy combined with intracavitary brachytherapy have been used, endometrial damage has been reported, thereby precluding successful implantation. Other fertility-sparing procedures, including cryopreservation, grafting of ovarian tissue, and oocyte freezing, have been reported to be effective in preserving fertility in patients treated with irradiation [34–36].

Conclusions

Ovarian transposition is a relatively easy and effective surgical procedure used to preserve ovarian function for patients at risk of iatrogenic ovarian failure from radiation therapy. Patient selection needs to be appropriate, such as choosing patients who are younger than 40 years old. For cervical carcinoma, important parameters include early stage of the disease, a tumor size <3 cm, the tumor being confined to the cervix, and the absence of macroscopic extracervical disease. The laparoscopic approach is favored.

References

1. McCall M, Keatye C, Thompson JD. Conservation of ovarian tissue in the treatment of carcinoma of the cervix with radical surgery. Am J Obstet Gynec. 1958;75:590.
2. Coppleson M, Managhan J, Morro C, Tattersall M. Gynecologic Oncology. 2nd ed. Edinburgh: Churchill Livingstone; 1992. p. 11–29.
3. Ray GR, Trueblood HW, Enright LP, Kaplan HS, Nelsen TS. Oophoropexy: a means of preserving ovarian function following pelvic megavoltage radiotherapy for Hodgkin's disease. Radiology. 1970;96(1): 175–80.
4. Farber L, Ames J, Rush GD. Laparoscopic ovarian transposition to preserve ovarian function before pelvic radiation and chemotherapy in a young patient with rectal cancer. MedGenMed. 2005;7(1):66.
5. Scott S, Schlaff W. Laparoscopic medial oophoropexy prior to radiation therapy in an adolescent with Hodgkin's disease. J Pediatr Adolesc Gynecol. 2005; 18:355–7.
6. Howard F. Laparoscopic lateral ovarian transposition before radiation treatment of Hodgkin disease. J Am Assoc Gynecol Laparosc. 1997;4(5):601–4.
7. Morice P, Castaigne D, Haie-Meder C, et al. Laparoscopic ovarian transposition for pelvic

malignancies: indications and functional outcomes. Fertil Steril. 1998;70:956–60.
8. Tinga DJ, Dolsma WV, Tamminga RY, et al. Preservation of ovarian function in 2 young women with Hodgkin disease by laparoscopic transposition of the ovaries prior to abdominal irradiation. Ned Tijdschr Geneeskd. 1999;143:308–12.
9. Classe JM, Mahe M, Moreau P, et al. Ovarian transposition by laparoscopy before radiotherapy in the treatment of Hodgkin's disease. Cancer. 1998;83: 1420–4.
10. Haie-Meder C, Mlika-Cabanne N, Michel G, Briot E, Gerbaulet A, Lhomme C, et al. Radiotherapy after ovarian transposition: ovarian function and fertility preservation. Int J Radiat Oncol Biol Phys. 1993; 25(3):419–24.
11. Serber W, Amendola B. Principles and practice of radiation oncology. Philadelphia: Lippincott; 1992.
12. Bieler E, Schnabel T, Knobel J. Persisting cyclic ovarian activity in cervical cancer after surgical transposition of the ovaries and pelvic irradiation. Br J Radiol. 1976;49:875–9.
13. Lushbaugh CC, Casarett GW. The effects of gonadal irradiation in clinical radiation therapy. A review. Cancer. 1976;37:1111–20.
14. Krebs C, Blixenkrone-Moller N, Mosekilde V. Preservation of ovarian function in early cervical cancer after surgical lifting of the ovaries and radiation therapy. Acta Radiol Ther Phys Biol. 1963;1: 176–82.
15. Ray G, Trueblood H, Enright L. Oophoropexy: a means of preserving ovarian function following pelvic mega-voltage radiotherapy for Hodgkin's disease. Radiation. 1970;96:175–80.
16. Kovacev M. Exteriorization of ovaries under the skin of yound women operated upon for cancer of the cervix. Am J Obstet Gynecol. 1968;101(6):756–9.
17. Nahhas W, Nisce L, D'Angio G, Lewis J. Lateral ovarian transposition: ovarian relocation in patients with Hodgkin's disease. J Obstet Gynecol. 1971;38: 785–8.
18. Hodel K, Rich W, Austin P, DiSaia P. The role of ovarian transposition in conservation of ovarian function in radical hysterectomy followed by pelvic radiation. Gynecol Oncol. 1982;13:195–202.
19. Gaetini A, De Simone M, Urgesi A, et al. Lateral high abdominal ovariopexy: an original surgical technique for protection of the ovaries during curative radiotherapy for Hodgkin's disease. J Surg Oncol. 1988; 39:22–8.
20. Chambers S, Chambers J, Kier R, Perscel R. Sequelae of lateral ovarian transposition in irradiated cervical cancer patients. Int J Rad Oncol Biol Phys. 1991; 20:1305–8.
21. Feeney DD, Moore DH, Look KY, et al. The fate of the ovaries after radical hysterectomy and ovarian transposition. Gynecol Oncol. 1995;56:3–7.
22. Chambers SK, Chambers JT, Holm C, et al. Sequelae of lateral ovarian transposition in unirradiated cervical cancer patients. Gynecol Oncol. 1990; 39:155–9.
23. Anderson B, LaPolla J, Turner D, Chapman G, Buller R. Ovarian transposition in cervical cancer. Gyn Onc. 1993;49:206–21.
24. Ranney B, Abu-Ghazaleh S. The future function and fortune of ovarian tissue which is retained in vivo during hysterectomy. Am J Obstet Gynecol. 1977;128:626–32.
25. Morice P, Juncker L, Rey A, El-Hassan J, Haie-Meder C, Castaigne D. Ovarian transposition for patients with cercial carcioma treated by radiosurgical combination. Fertil Steril. 2000;74(4):743–8.
26. Covens AL, van der Putten HW, Fyles AW, et al. Laparoscopic ovarian transposition. Eur J Gynaecol Oncol. 1996;17:177–82.
27. Tulandi T, Al-Took S. Laparoscopic ovarian suspension before irradiation. Fertil Steril. 1998;70:381–3.
28. Tulandi T. Profile: Dr. Togas Tulandi. McGill University Health Centre Health Perspectives. August/September 2002, p 5.
29. Feeney D, Moore D, Look K, Stehman F, Sutton G. The fate of the ovaries after radical hysterectomy and ovarian transposition. Gynec Oncol. 1995;56:3–7.
30. Kjorstad k, Bond B. Stage IB adenocarcinoma of te cervix: Metastatic potetial and patterns of dissimination. Am J Obstet Gynecol. 1984;15:297–9.
31. Toki N, Tsukamoto N, Kaku T. Microscopic ovaries metastasis of the uterine cervical cancer. Gynecol Oncol. 1991;41:46–51.
32. Sutton GP, Budny BN, Delgado G, Sevin BU, Creasman WT, Major FJ, et al. Ovarian metastases in stage IB carcinoma of the cervix: a GOG study. Am J Obstet Gynecol. 1992;166:50–3.
33. Tabata M, Ichinoe K, Sakuragi N, Shiina Y, Yamaguchi T, Mabuchi Y. Incidence of ovarian metastasis in patients with cancer of the uterine cervix. Gynecol Oncol. 1987;28:255–6.
34. Newton H, Fisher J, Arnold JR, Pegg DE, Faddy MJ, Gosden RG. Permeation of human ovarian tissue with cryoprotective agents in preparation for cryopreservation. Hum Reprod. 1998;13(2):376–80.
35. Schmidt KL, Ernst E, Byskov AG, Nyboe Andersen A, Yding Andersen C. Survival of primordial follicles following prolonged transportation of ovarian tissue prior to cryopreservation. Hum Reprod. 2003;18(12): 2654–9.
36. Donnez J, Martinez-Madrid B, Jadoul P, Dolmans M. Ovarian tissue cryopreservation and transplantaton: a review. Hum Reprod Update 12: 519–35.

Embryo Cryopreservation and Alternative Controlled Ovarian Hyperstimulation Strategies for Fertility Preservation

7

Bulent Urman, Ozgur Oktem, and Basak Balaban

Abstract

Fertility preservation with embryo cryopreservation is a safe and effective option in women with cancer and other disorders requiring gonadotoxic treatment. Most patients presenting have surgically treated breast cancer, who are about to undergo chemotherapy. Other less commonly encountered indications include hematological malignancies and autoimmune diseases such as severe SLE with organ involvement and ITP. Women who are undergoing surgery for endometriosis as well as women with genetic disorders such as Turner syndrome and fragile-X premutation who face similar risks further contribute to the population of women who need fertility preservation procedures. In women with an established partner, fertility preservation entails ovarian stimulation, egg collection, fertilization, and embryo cryopreservation. Several concerns have been voiced over ovarian stimulation and resulting hyperestrogenism in women with surgically treated breast cancer. Alternative ovarian stimulation protocols using aromatase inhibitors combined with gonadotropins have been developed with the aim to obviate some of the potential adverse effects. Collection of immature oocytes followed by in vitro maturation and subsequent fertilization appears to be a viable and effective method that may prevent most of the drawbacks of classical ovarian stimulation.

Keywords

Fertility preservation with embryo cryopreservation • Embryo cryopreservation • Alternative ovarian hyperstimulation strategies for fertility preservation • Cancer and fertility preservation • Hyperestrogenism in women with surgically treated breast cancer

B. Urman (✉)
Department of Obstetrics and Gynecology,
American Hospital, Guzelbahce Sok No 20 Nisantasi,
34365 Istanbul, Turkey
e-mail: burman@superonline.com

E. Seli and A. Agarwal (eds.), *Fertility Preservation: Emerging Technologies and Clinical Applications*,
DOI 10.1007/978-1-4419-1783-6_7,

Cryopreservation of human embryos is a time honored technology that has been proven to be safe and efficient in couples undergoing assisted reproduction. The application of assisted reproductive techniques often results in surplus embryos that can be cryopreserved for later use. Since the first pregnancy that was reported in 1983, laboratory techniques, instrumentation and culture media, and endometrial preparation protocols evolved constantly that has permitted the achievement of acceptable pregnancy rates with the transplantation of frozen-thawed embryos [1, 2]. Embryo cryopreservation enabled the storage of excess embryos derived from in vitro fertilization/intracytoplasmic sperm injection (IVF/ICSI) thus obviating ethical concerns associated with embryo destruction, paved the way to single embryo transfer and prevention of multiple pregnancies, and increased the safety of assisted reproductive technology (ART) treatment in patients facing the iatrogenic risk of ovarian hyperstimulation syndrome [3–8]. Furthermore, cryopreservation provides the couple with multiple attempts at embryo transfer after a single ovarian stimulation cycle with IVF, thus improving cumulative pregnancy rates while decreasing exposure to gonadotropins and reducing treatment costs. Today successful cryopreservation of embryos should be an integral part of every IVF program. The dilemma between safety and high success in IVF can be only resolved with the establishment of an efficient cryopreservation program. Cryopreservation has enabled the implementation of single embryo transfers obviating ethical concerns regarding the disposition of excess embryos and furthermore boosting cumulative pregnancy rates [9–11].

Outcome of Cryopreserved-Thawed Embryo Transfer in Infertile Couples

Transfer of cryopreserved embryos is associated with implantation rates that are still inferior to those achieved by the transfer of their fresh counterparts. Most recent data from the Society for Assisted Reproductive Technology and the European IVF Monitoring Program report a pregnancy rate of 34% following frozen embryo transfer (FET) in women younger than 35 years and an overall pregnancy rate of 19%, respectively [12, 13]. Several variables may influence the outcome including quality of the gametes, quality of the embryos, laboratory conditions, technology applied, endometrial preparation protocols, outcome of the fresh transfer cycle, and whether all or remaining embryos after fresh transfer were cryopreserved [14–21].

Techniques of Cryopreservation

There are basically three techniques that are employed today for embryo cryopreservation; namely slow freezing, ultrarapid freezing, and vitrification. Ultrarapid freezing appears to be inferior to the other two techniques. Clinical outcome of slow freezing and vitrification appear to be similar and dependent upon the experience of the embryologist performing the procedure and the characteristics of the couple and the derived embryo [2, 22]. Data indicate higher post-thaw survival rates in vitrified embryos; however, this does not reflect upon the pregnancy rates [23]. A very recent meta-analysis, however, showed higher pregnancy rates after the transfer of vitrified embryos [24]. Besides pregnancy rates any comparison of slow freezing with vitrification should also take into account the workload each procedure imposes on the laboratory and more importantly safety. A recent review assessing the medical outcome of ART children born after cryopreservation reported reassuring results. The rate of preterm birth, birth defects, and chromosomal abnormalities was not significantly different between children born after transfer of fresh or cryopreserved embryos. Similarly, these children demonstrated similar growth and mental development [25, 26].

Cryopreservation for Fertility Preservation

Following the introduction of embryo cryopreservation for excess embryos generated through ART, it became apparent that the technique may also be used for fertility preservation in married couples or in couples with an established partner. Preservation of fertility may be indicated in women with curable cancer where conception has to be postponed until the resolution of the primary disease, in couples who wish to postpone childbearing and may face the risk of age-related decline in fertility when conception is desired, and in men with cancer or other serious illnesses where conception is best delayed until a more healthy environment for fatherhood is achieved. Administration of gonadotropin-releasing hormone analogues or inhibitors of apoptosis, cryopreservation of unfertilized oocytes following in vitro fertilization (IVF) or in vitro maturation (IVM), cryopreservation of embryos following IVF or IVM and ovarian tissue cryopreservation are the current methods available for fertility preservation. When the threat to gonads is limited to direct radiotherapy, surgically positioning the ovaries out of the irradiation field is an effective method and can be combined with other methods if deemed necessary. At present, embryo cryopreservation following IVF is the only method endorsed by the American Society of Clinical Oncology (ASCO) and the American Society of Reproductive Medicine (ASRM), while the other methods are still considered experimental. This chapter will mainly focus on embryo cryopreservation as a fertility preserving procedure in women with cancer and other diseases that require gonadotoxic therapy or surgical removal of the ovaries. Fertility preservation through embryo cryopreservation for social indications (i.e., storage of embryos to overcome age related decline in fertility) is discussed in Chap. 17.

The Effect of Gonadotoxic Treatment on Fertility

Women are born with a finite number of oocytes that will eventually perish at the time of menopause. However, fertility potential is diminished approximately 10 years preceding this life event. The natural decline in fertility is markedly accelerated following gonadotoxic therapy for cancer and other non-oncologic conditions. Cancer continues to be a major health problem despite advances in its diagnosis and treatment. It is estimated that in 2009 approximately 713,220 women in the United States will be diagnosed with cancer [27]. The most common cancers in females under age 40 are breast cancer, cancers of the lung and bronchus, colon and rectum, leukemia and lymphomas, and cervical cancer. The probability of being diagnosed with an invasive cancer for women under the age 40 is 2%. This rate increases to 9% by the age 60. Among females, the leading cause of cancer death before age 20 years is leukemia. Breast cancer ranks first at age 20–59 years, and lung cancer ranks first at age 60 years and older. When cancers of all sites and all races are considered, the survival rate increased from 50% between 1975 and 1977 to 66 in 1996 and 2004 period [28]. Survival rates are more encouraging in children [29]. The 5-year relative survival rate among children for all cancer sites combined improved from 58% for patients diagnosed in 1975–1977 to 80% for those diagnosed in 1996–2004 [30]. Furthermore, significant progress has been made in reducing cancer death rates during the same period resulting mostly from reductions in tobacco use, increased screening allowing early detection of several cancers and modest to large improvements in treatment for specific cancers [31]. Overall cancer incidence rates decreased in the most recent time period in both men (1.8% per year from 2001 to 2005) and women (0.6% per year from 1998 to 2005), largely because of decreases in the three major cancer sites in men (lung, prostate, and colon and rectum) and in two major cancer sites in women (breast and colorectal). Among women, overall cancer death rates between 1991 and 2005 decreased by 11.4%, with decreases in breast (37%) and colorectal (24%) cancer rates accounting for 60% of the total decrease. The reduction in the overall cancer death rates has resulted in the avoidance of about 650,000 deaths from cancer over the 15-year period [28].

Fertility Preservation in Cancer Patients Who Are About to Undergo Chemo/Irradiation

Effective treatments of cancer have resulted in increased life expectancy in these patients and childhood cancers have been particularly favorably affected [32]. Although initially often ignored, loss of fertility potential will inevitably become a long-term health problem in survivors of childhood cancers [33, 34]. Modern combination chemotherapy and radiotherapy regimens have a substantial negative impact on reproduction. Premature ovarian failure and other poor reproductive outcomes subsequent to cancer therapies are being recognized. Patients who are exposed to gonadotoxic agents for the treatment of non-oncologic diseases such as systemic lupus erythematosus, those who are undergoing surgery for endometriosis as well as women with genetic disorders such as Turner syndrome and fragile-X premutation face similar risks, further contributing to the population of women who need fertility-preservation procedures [35–37]. Therefore, preservation of gonadal function and fertility has become one of the major quality of life issues for cancer survivors at reproductive ages. Accordingly, clinical guidelines, encouraging fertility preservation among all young cancer survivors with interest in fertility have been issued by American Society of Clinical Oncology [38].

Young women in their reproductive life period should be counseled regarding the impact of chemo/radiation treatment on their fertility potential. Forman et al. performed a nationwide survey regarding the approach of medical and gynecological oncologists working in academic facilities to fertility preservation in patients with cancer [39]. Of the 249 oncologists surveyed, 95% responded that they routinely discuss fertility matters with their patients prior to cancer treatment. It was interesting to note that although 82% would refer their patients to an infertility specialist, less than half would actually present to their appointments. Gynecologic oncologists were more likely to consider fertility compared with other oncologists (93% vs. 60%). Most oncologists (86%) would be willing to sacrifice less than a 5% reduction in disease-free survival if a regimen offered better fertility outcomes; 36% felt patients would be willing to sacrifice >5%. Pediatric oncologists were less likely to discuss fertility issues with their patients and parents [40].

It would be helpful to know the proportion of patients with cancer that actually qualify and furthermore who are willing to undergo fertility preservation. In a survey of 304 women with breast cancer among whom 248 were still alive, 107 (43.1%) were counseled regarding fertility preservation prior to treatment and 39 (12.8%) wanted a child before the diagnosis [41]. Eighteen patients have become pregnant, four with more than one pregnancy. One hundred and seven patients were specifically counseled about fertility prior to breast cancer treatment. The mortality due to breast cancer was 10% in nonpregnant patients and 6% in patients who became pregnant after breast cancer.

The optimal approach to fertility preservation depends on the type of cancer, the type of treatment (e.g., radiation and/or chemotherapy), and time available until onset of treatment, patient's age, and whether the patient has a partner. Ovarian transposition remains the standard of care for women undergoing pelvic radiation, although it has been suggested that it may be combined with ovarian tissue cryopreservation. The algorithm of fertility preservation strategies in the female is illustrated in Fig. 7.1.

For patients about to receive chemotherapy or whole body radiation, in vitro fertilization (IVF) with embryo cryopreservation is a well-established treatment with a good success rate. However, it requires delaying cancer treatment for 2–4 weeks and a partner or willingness to use donor sperm. When these criteria cannot be met, more experimental options including oocyte cryopreservation for later IVF and ovarian tissue cryopreservation should be considered [42].

The age of the patient, the type, dose and intensity of chemotherapy and/or radiotherapy are the main factors determining the magnitude of the damage in the ovary. Patients younger than age 40 are more likely to retain or regain menstrual function than those older than age 40 (22–56% vs. 11%) [43]. Older patients have lower ovarian

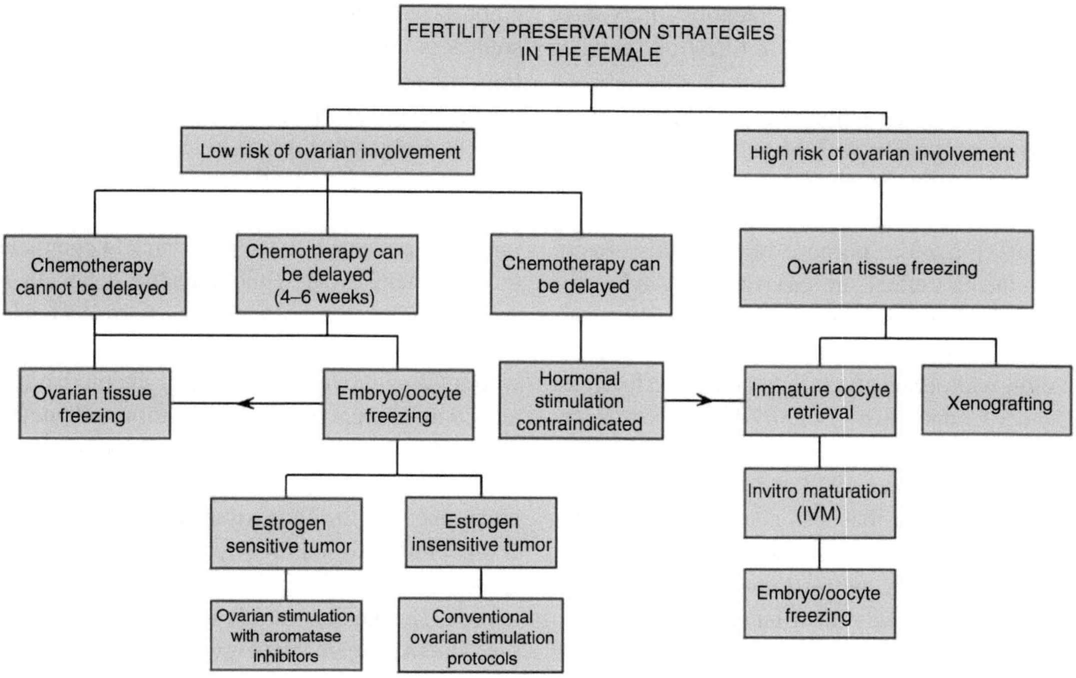

Fig. 7.1 The algorithm of fertility preservation strategies in women

reserve compared to younger ones; therefore have higher risk for ovarian failure during or after chemotherapy or radiation [44]. Furthermore, as noted in the adult survivors of childhood cancers, there are some other adverse extragonadal effects such as abnormalities in the regulation of growth and endocrine functions, and other poor reproductive outcomes that appear later in life such as preterm births and miscarriages [33].

Emergency IVF followed by embryo cryopreservation for future use is a promising technique for enabling preservation of fertility for newly diagnosed cancer patients who are married or have an established relationship. Application of this option is dependent upon the patient's age and previous fertility history, and its availability must be brought to the attention of women before chemo/radiotherapy is begun [45]. There are several issues that should be considered before embarking upon this treatment. Being relatively new in the treatment armamentarium, a cautious approach should be adopted and the patient should not be led into false hopes and undue optimism. One should also bear in mind that there are several social implications of embryo cryopreservation particularly in cancer patients. These include the death of the treated partner, unclaimed embryos due to change in marital status, and factors that affect couples decision about embryo disposition [4]. Decisions made under stressful conditions are more likely to change once stability is achieved.

Creation of Embryos with the Intent to Cryopreserve in Cancer Patients

There has been an intense debate regarding the optimal way of creating embryos for later use in women planned to undergo gonadotoxic chemotherapy for cancer. The concerns can be listed as follows:

Does Delaying Chemotherapy Adversely Affect the Course of the Disease?

It should be emphasized that there are no prospective randomized studies comparing immediate chemotherapy with chemotherapy following harvesting oocytes with the intention of cryopreserving prior to or after fertilization. Studies assessing the effect of the length of time between surgery and the initiation of chemotherapy on the survival of women with breast cancer report no detrimental effect of a delay in treatment if chemotherapy is started within 12 weeks after surgery [46]. The purpose of this study was to examine the effect on survival of delaying the start of adjuvant chemotherapy for early breast cancer for up to 3 months after surgery. In the nation-wide clinical trials of the Danish Breast Cancer Cooperative Group, 7,501 breast cancer patients received chemotherapy within 3 months of surgery between 1977 and 1999. The authors divided the time that was elapsed between surgery and the initiation of chemotherapy into four strata (1–3, 4, 5, and 6–13 weeks). Prognostic factors were evenly distributed in the groups. There appeared to be no evidence for a survival benefit due to early initiation of adjuvant chemotherapy within the first 2–3 months after surgery. A total of 2,782 patients with Stage I-III breast cancer were analyzed in a similar Spanish study [47]. The time interval between surgery and initiation of chemotherapy, and dates of relapse, second primary breast tumor and death were recorded. Patients were divided into four groups (<3 weeks, 3–6 weeks, 6–9 weeks, and >9 weeks) according to the surgery–chemotherapy interval. The authors found no difference in disease-free and 5-year overall survival rates in women receiving early vs. late chemotherapy. Lohrisch et al. studied 2,594 women with stage I and II breast cancer that were registered in the British Columbia Cancer Agency [48]. Relapse-free survival and overall survival were compared among patients grouped by time from definitive curative surgery to start of adjuvant chemotherapy (≤4, >4–8, >8–12, and >12–24 weeks). The authors concluded that chemotherapy started up to 12 weeks after surgery did not affect relapse-free or overall survival rates. Inferior outcomes were recorded in women receiving chemotherapy beyond 12 weeks.

It appears that delaying chemotherapy up to 3 months post surgery does not adversely affect disease-free or overall survival rates in women with breast cancer. However, it must be emphasized that the external validity of these studies is limited to their inclusion criteria, and the potential effect of any delay in oncologic treatment due to fertility preservation procedures must be evaluated on a case-by-case basis together with the treating oncology team [3]. Furthermore, the IVF team should be able to rapidly incorporate these patients into the program and complete the treatment cycle without undue delay. In a recent study, Baynosa et al. retrospectively analyzed the files of 82 breast cancer patients of whom 62 underwent post surgery chemotherapy and 19 underwent ovarian stimulation and oocyte retrieval for fertility preservation and subsequent chemotherapy [49]. Ovarian stimulation was undertaken with Tamoxifen as described by Oktay et al. [50]. Estrogen receptor was positive in 14 of 19 women who underwent ovarian stimulation. The median time from initial diagnosis to reproductive endocrinology consultation was 30.1 days (range 4–133 days) and from referral to oocyte retrieval was 32 days (range 13–66 days). The median times from initial diagnosis to chemotherapy in stimulation/retrieval vs. immediate chemotherapy groups were 71 days (range 45–161 days) and 67 days (range 27–144 days), respectively ($p=0.27$). The median time interval from definitive operation to chemotherapy was similar in the two groups: 30 days (oocyte retrieval; range 14–100 days) and 29 days (chemotherapy; range 12–120 days) ($p=0.79$).

What is the Optimal Way of Harvesting Oocytes?

Oocytes can be harvested in a natural cycle or following ovarian stimulation. Collection of oocytes in a natural cycle rarely results in the generation of more than one embryo. The gains therefore do not appear to outweigh the risks particularly regarding the delay in the initiation of chemotherapy. It is thus generally accepted that some form of ovarian stimulation is necessary to optimize the results. However, oncologists are also concerned with ovarian stimulation in the setting of estrogen-sensitive cancers, as supraphysiologic estrogen levels are reached during the treatment. Evidence indicates that exposure to estrogen is an important determinant of the risk of breast cancer. The mechanisms of carcinogenesis in the breast caused by estrogen include the metabolism of estrogen to genotoxic, mutagenic metabolites, and the stimulation of tissue growth. Together, these processes cause initiation, promotion, and progression of carcinogenesis [51]. Breast cancer risk is increased in women receiving hormone replacement therapy in menopause and in women who use birth control pills for a prolonged period of time. Breast cancer growth is suppressed following oophorectomy that effectively induces hypoestrogenemia. Paradoxically pharmacological doses of estrogen can also induce apoptotic pathways and inhibit the growth of breast tumors. This effect is most pronounced after estrogen deprivation [52]. Treatment with low dose estrogen has been shown to reverse tamoxifen resistance and alternating treatment cycling antiestrogen with estrogen has been suggested [53].

It is questionable today whether transient hyperestrogenemia associated with ovarian stimulation affects tumor growth in women operated for breast cancer and who are awaiting chemotherapy. It is further unknown whether delay in chemotherapy has a differential effect according to stage, tumor cell differentiation, nodal status, and receptor positivity. Estrogen may promote growth even in estrogen receptor negative breast cancer thus receptor status does not render the tumor immune to potential negative effects of estrogen [54, 55]. Moreover, increased E2 levels can be relevant for patients undergoing fertility preservation treatment due to other oncologic or non-oncologic diseases considered to be estrogen sensitive, such as desmoid tumors, systemic lupus erythematosus, or severe endometriosis [3].

Due to concerns regarding the effects of supraphysiological estrogen levels on survival in women with breast cancer, mild ovarian stimulation regimens have been proposed in patients desiring fertility preservation. One novel strategy in breast cancer patients involves the combination of an aromatase inhibitor with low dose gonadotropins [56, 57]. In this protocol an aromatase inhibitor (letrozole) is combined with low-dose gonadotropins for stimulation of the ovaries. Letrozole is started on the second day of the menstrual cycle and gonadotropins were added 2 days later. An antagonist was used to prevent the premature surge of LH. Human chorionic gonadotropin (HCG) was administered when at least two follicles reached19 mm in diameter. Letrozole was reinitiated on the day of oocyte collection in order to prevent a rebound increase in E2 level. The mean peak E2 level was 406 pg/ml (range 58–1,166 pg/ml) in 79 women with breast cancer undergoing ovarian stimulation for embryo or oocyte cryopreservation. An average of 10.3 ± 7.75 oocytes was retrieved and 5.97 ± 4.97 embryos/oocytes cryopreserved per patient. Compared to 136 women who opted against ovarian stimulation, the recurrence and relapse-free survival rates were similar after a median follow up of 23.4 months after definitive surgery. However, it is interesting to note that in the same center 63.3% of breast cancer patients referred for REI consultation declined ovarian stimulation and IVF due to concerns about delay of chemotherapy, effect of ovarian stimulation on cancer, or costs associated with treatment.

Michaan later reported that among a cohort of 22 patients with cancer of whom 12 had breast cancer ovarian response when stimulated with letrozole and gonadotropins was similar to a control group of tubal factor patients [45].

In women with hematological and other malignancies, concerns regarding ovarian stimulation are less compared to women with breast cancer. However, ovarian stimulation and IVF can cause significant delays in initiation of chemotherapy that may adversely affect the course of the primary disease. A joint decision should be reached that involves the affected couple, the oncologist, and the REI specialist. Depending on the urgency to start chemotherapy, ovarian stimulation with gonadotropins or collection of oocytes in natural cycle and in vitro maturation may be considered. Elizur et al. reported the outcome of COS+IVF or IVM in seven women with SLE or other systemic autoimmune rheumatic diseases (SARDs) who were planned to receive cyclophosphamide [58]. No complications were associated with this treatment and chemotherapy was started as planned in all patients.

Legal Aspects

When embryos are cryopreserved in a fertility preservation program, the patient/couple should make an advance decision on the fate of these embryos in the event that they are not transferred for any reason including the patient's failure to survive cancer. Furthermore, a proportion of couples, unfortunately, get divorced when one of the partners is diagnosed with cancer. In such a case, the former male partner will have rights over the embryos, with all possible legal and ethical implications. It should be documented whether the remaining partner is entitled to use the embryos for his own reproductive wishes or whether they are to be donated to a third party and used for research or discarded. Considering these issues, making such decisions can be particularly difficult for a patient who has been recently diagnosed with a life-threatening disease and is facing a demanding treatment period. Therefore, patients should be given appropriate counseling using a multidisciplinary approach involving a psychologist and a legal advisor [3].

In Vitro Maturation

Oocytes may also be harvested in an unstimulated cycle and fertilized following in vitro maturation (IVM), and the resulting embryos can be cryopreserved for future use. This appears to be more acceptable for the patient and the oncologist who is involved in the treatment and can present as a viable option provided that pregnancy rates are comparable to standard IVF.

IVM, initially advanced as a treatment for women with polycystic ovaries who are prone to severe hyperstimulation with its resulting consequences, is gradually becoming an effective and safe alternative treatment option to classical stimulated IVF in almost all infertile couples [59]. IVM presents several advantages above and over classical embryo freezing following ovarian stimulation. These include the absence of transient hyperestrogenism and the waiting period associated with ovarian stimulation. Compared to 2–5 weeks required for a stimulated IVF cycle, immature oocyte retrieval can be done within 2–10 days, depending on the patient's menstrual status [60, 61]. Moreover, immature oocytes can be collected even in the luteal phase [62]. Immature oocytes can also be harvested from ovarian biopsy specimens and can be fertilized or vitrified following IVM [63]. This may provide an important source of gametes especially in women undergoing oophorectomy for non-germ cell tumors. It also carries the potential for harvesting gametes in women who present with recurrent severe ovarian endometriosis and face the risk of gamete depletion following surgical treatment. In summary, IVM combined with embryo or oocyte vitrification provides previously unavailable options for some patients and improves the services provided by a fertility preservation program [3]. Furthermore, the safety of IVM has also been documented albeit in small series of patients [64]. Currently, IVM is not only a recognized treatment alternative for couples who need assisted reproductive technologies, but also is considered as an innovative fertility preservation method which extends the options for patients with various diseases that preclude treatment with conventional methods. However, compared to widespread application of IVF, IVM is still practiced by relatively few clinics worldwide. This unfortunately precludes its use worldwide for fertility preservation.

When Should Cryopreserved Embryos Be Transferred?

In common clinical practice, most clinicians advise women treated for early breast cancer to wait at least 2 years from diagnosis before attempting conception, to allow early recurrences to manifest [65]. However, for women with localized breast cancer and good prognosis, conception 6 months after diagnosis is unlikely to compromise survival. As discussed below, to avoid adverse effect of adjuvant treatment modalities on birth outcome, it also seems advisable to wait 6–12 months after treatment completion to plan pregnancy. According to others, as younger women have significantly lower survival rates and higher local and distant relapse rates than older women, those under 33 years of age might be better advised to delay pregnancy for at least 3 years to reduce the risk of relapse, while patients with lymph node involvement should consider deferring pregnancy for at least 5 years after treatment and those with distant metastases should not consider conception at all because of the intensity of treatment and the poor prognosis [66]. Other factors like residual fertility and age at eventual conception should also be taken into consideration when counseling young patients about interval span between diagnosis and pregnancy [67].

Recently there has been much debate regarding the effect of subsequent pregnancy on the survival outcome of breast cancer. Registry based, cohort and descriptive population-based studies indicate a survival benefit of pregnancy in women with treated breast cancer [68–70]. Few data have been published regarding the influence of subsequent pregnancy after treatment for breast cancer on local and distant recurrences. In the limited number of studies there seemed to be no adverse effect of subsequent pregnancy on the probability of disease recurrences, when compared with women treated for breast cancer that did not become pregnant [71–73]. In fact, in a retrospective Swedish cohort study, the RR of distant metastases for women who became pregnant during the first 5 years after diagnosis of breast cancer in comparison with women without a subsequent pregnancy was 0.48 ($P=0.14$), which suggested a possible decreased risk of distant dissemination. In conclusion, women treated for breast cancer who wish to become pregnant should be informed that it is unlikely that pregnancy would have an adverse effect on their prognosis. On the contrary, there might even be a beneficial effect of subsequent pregnancy in these women. However, they should be made aware that the evidence remains rather scarce and has some limitations [65].

Conclusions

Fertility preservation with embryo cryopreservation is a safe and effective option in women with cancer and other disorders requiring gonadotoxic treatment. Most patients presenting have surgically treated breast cancer who are about to undergo chemotherapy. Other less commonly encountered indications include hematological malignancies and autoimmune diseases such as severe SLE with organ involvement and ITP. Women who are undergoing surgery for endometriosis as well as women with genetic disorders such as Turner syndrome and fragile-X premutation who face similar risks further contribute to the population of women who need fertility preservation procedures. In women with an established partner, fertility preservation entails ovarian stimulation, egg collection, fertilization, and embryo cryopreservation. Several concerns have been voiced over ovarian stimulation and resulting hyperestrogenism in women with surgically treated breast cancer. Alternative ovarian stimulation protocols using aromatase inhibitors combined with gonadotropins have been developed with the aim to obviate some of the potential adverse effects. Collection of immature oocytes followed by in vitro maturation and subsequent fertilization appears to be a viable and effective method that may prevent most of the drawbacks of classical ovarian stimulation.

See Appendix A

References

1. Trounson A, Mohr L. Human pregnancy following cryopreservation, thawing and transfer of an eight-cell embryo. Nature. 1983;305:707–9.
2. Varghese AC, Nagy ZP, Agarwal A. Current trends, biological foundations and future prospects of oocyte and embryo cryopreservation. Reprod Biomed Online. 2009;19:126–40.
3. Ata B, Chian RC, Tan SL, et al. Cryopreservation of oocytes and embryos for fertility preservation for female cancer patients. Best Pract Res Clin Obstet Gynaecol. 2010;24:101–12.
4. Bankowski BJ, Lyerly AD, Faden RR, et al. The social implications of embryo cryopreservation. Fertil Steril. 2005;84:823–32.
5. de Jong D, Eijkemans MJ, Beckers NG, et al. The added value of embryo cryopreservation to cumulative ongoing pregnancy rates per IVF treatment: is

cryopreservation worth the effort? J Assist Reprod Genet. 2002;19:561–8.
6. Le Lannou D, Griveau JF, Laurent MC, et al. Contribution of embryo cryopreservation to elective single embryo transfer in IVF-ICSI. Reprod Biomed Online. 2006;13:368–75.
7. Tiitinen A, Halttunen M, Harkki P, et al. Elective single embryo transfer: the value of cryopreservation. Hum Reprod. 2001;16:1140–4.
8. Wiener-Megnazi Z, Lahav-Baratz S, Rothschild E, et al. Impact of cryopreservation and subsequent embryo transfer on the outcome of in vitro fertilization in patients at high risk for ovarian hyperstimulation syndrome. Fertil Steril. 2002;78:201–3.
9. Gerris J, De Neubourg D, De Sutter P, et al. Cryopreservation as a tool to reduce multiple birth. Reprod Biomed Online. 2003;7:286–94.
10. Gerris J, De Neubourg D, Mangelschots K, et al. Elective single day 3 embryo transfer halves the twinning rate without decrease in the ongoing pregnancy rate of an IVF/ICSI programme. Hum Reprod. 2002;17:2626–31.
11. Henman M, Catt JW, Wood T, et al. Elective transfer of single fresh blastocysts and later transfer of cryostored blastocysts reduces the twin pregnancy rate and can improve the in vitro fertilization live birth rate in younger women. Fertil Steril. 2005;84:1620–7.
12. Society for Assisted Reproductive Technology 2007. Thawed embryo transfers. [cited 16th November 2009] https://www.sartcorsonline.com/rptCSR_Public MultYear.aspx?ClinicPKID=0.
13. Nyboe Andersen A, Goossens V, Bhattacharya S, et al. Assisted reproductive technology and intrauterine inseminations in Europe, 2005: results generated from European registers by ESHRE: ESHRE. The European IVF Monitoring Programme (EIM), for the European Society of Human Reproduction and Embryology (ESHRE). Hum Reprod. 2009;24:1267–87.
14. Balaban B, Yakin K, Isiklar A, et al. Developmental ability of cryo-thawed embryos derived from dysmorphic oocytes. Hum Reprod. 2006;21(Suppl):i38.
15. Karlstrom PO, Bergh C, Fosrberg A, et al. Prognostic factors for the success rate of embryo freezing. Hum Reprod. 1997;12:1263–6.
16. Mandelbaum J, Junca AM, Plachot M, et al. Human embryo cryopreservation, extrinsic and intrinsic parameters of success. Hum Reprod. 1987;2:709–15.
17. Oehninger S, Mayer J, Muasher S. Impact of different clinical variables on pregnancy outcome following embryo cryopreservation. Mol Cell Endocrinol. 2000; 169:73–7.
18. Schalkoff ME, Oskowitz SP, Powers RD. A multifactorial analysis of pregnancy outcome in a successful embryo cryopreservation program. Fertil Steril. 1993; 59:1070–4.
19. Shoukir Y, Chardonnens D, Campana A, et al. The rate of development and time of transfer play different roles in influencing the viabliity of human blastocysts. Hum Reprod. 1998;13:676–81.
20. Toner JP, Veeck L, Muasher S. Basal follicle stimulating hormone level and age affect the chance for and outcome of pre-embryo cryopreservation. Fertil Steril. 1993;59:664–7.
21. Urman B, Balaban B, Yakin K. Impact of fresh-cycle variables on the implantation potential of cryopreserved-thawed human embryos. Fertil Steril. 2007;87:310–5.
22. Kolibianakis EM, Venetis CA, Tarlatzis BC. Cryopreservation of human embryos by vitrification or slow freezing: which one is better? Curr Opin Obstet Gynecol. 2009;21:270–4.
23. Loutradi K, Kolibianakis E, Venetis C, et al. Cryopreservation of human embryos by vitrification or slow freezing: a systematic review and metaanalysis. Fertil Steril. 2008;90:186–93.
24. AbdelHafez F, Desai N, Abou-Setta A, et al. Slow-freezing, vitrification and ultrarapid freezing of human embryos: a systematic review and meta-analysis. RBM Online. 2010;20:209–22.
25. Pinborg A, Loft A, Aaris Henningsen AK, et al. Infant outcome of 957 singletons born after frozen embryo replacement: The Danish National Cohort Study 1995-2006. Fertil Steril. 2010;94:1320–7.
26. Wennerholm UB, Soderstrom-Anttila V, Bergh C, et al. Children born after cryopreservation of embryos or oocytes: a systematic review of outcome data. Hum Reprod. 2009;24:2158–72.
27. The American Cancer Society. Cancer Facts and Figures 2009 http://wwwcancerorg/docroot/STT/STT_0asp.
28. Jemal A, Siegel R, Ward E, et al. Cancer statistics, 2009. CA Cancer J Clin. 2009;59:225–49.
29. Wu X, Groves FD, McLaughlin CC, et al. Cancer incidence patterns among adolescents and young adults in the United States. Cancer Causes Control. 2005;16:309–20.
30. Jemal A, Center MM, Ward E, et al. Cancer occurrence. Methods Mol Biol. 2009;471:3–29.
31. Jemal A, Thun MJ, Ries LA, et al. Annual report to the nation on the status of cancer, 1975–2005, featuring trends in lung cancer, tobacco use, and tobacco control. J Natl Cancer Inst. 2008;100:1672–94.
32. Bleyer W. The impact of childhood cancer on the United States and the world. Cancer J Clin. 1990;40: 355–67.
33. Oktay K, Oktem O. Fertility preservation medicine: a new field in the care of young cancer survivors. Pediatr Blood Cancer. 2009;53:267–73.
34. Oktem O, Oktay K. Fertility preservation for breast cancer patients. Semin Reprod Med. 2009;27: 486–92.
35. Elizur S, Chian R, Holzer H, et al. Cryopreservation of oocytes in a young woman with severe and symptomatic endometriosis: a new indication for fertility preservation. Fertil Steril. 2009;91:293.
36. Huang J, Tulandi T, Holzer H, et al. Cryopreservation of ovarian tissue and in vitro matured oocytes in a female with mosaic Turner syndrome: case report. Hum Reprod. 2008;23:336.

37. Lau N, Huang J, MacDonald S, et al. Feasibility of fertility preservation in young females with Turner syndrome. RBM Online. 2009;18:290–5.
38. Lee SJ, Schover LR, Partridge AH, et al. American Society of Clinical Oncology recommendations on fertility preservation in cancer patients. J Clin Oncol. 2006;24:2917–31.
39. Forman EJ, Anders CK, Behera MA. A nationwide survey of oncologists regarding treatment-related infertility and fertility preservation in female cancer patients. Fertil Steril. 2010;94:1652–6.
40. Vadaparampil S, Quinn G, King L, et al. Barriers to fertility preservation among pediatric oncologists. Patient Educ Couns. 2008;72:402–10.
41. Rippy EE, Karat IF, Kissin MW. Pregnancy after breast cancer: the importance of active counselling and planning. Breast. 2009;18:345–50.
42. Georgescu ES, Goldberg JM, du Plessis SS, et al. Present and future fertility preservation strategies for female cancer patients. Obstet Gynecol Surv. 2008; 63:725–32.
43. Bines J, Oleske DM, Cobleigh MA. Ovarian function in premenopausal women treated with adjuvant chemotherapy for breast cancer. J Clin Oncol. 1996;14: 1718–29.
44. Oktem O, Oktay K. Quantitative assessment of the impact of chemotherapy on ovarian follicle reserve and stromal function. Cancer. 2007;110:2222–9.
45. Michaan N, Ben-David G, Ben-Yosef D, et al. Ovarian stimulation and emergency in vitro fertilization for fertility preservation in cancer patients. Eur J Obstet Gynecol Reprod Biol. 2010;149:175–7.
46. Cold S, During M, Ewertz M, et al. Does timing of adjuvant chemotherapy influence the prognosis after early breast cancer? Results of the Danish Breast Cancer Cooperative Group (DBCG). Br J Cancer. 2005;93:627–32.
47. Jara Sanchez C, Ruiz A, Martin M, et al. Influence of timing of initiation of adjuvant chemotherapy over survival in breast cancer: a negative outcome study by the Spanish Breast Cancer Research Group (GEICAM). Breast Cancer Res Treat. 2007;101:215–23.
48. Lohrisch C, Paltiel C, Gelmon K, et al. Impact on survival of time from definitive surgery to initiation of adjuvant chemotherapy for early-stage breast cancer. J Clin Oncol. 2006;24:4888–94.
49. Baynosa J, Westphal LM, Madrigrano A, et al. Timing of breast cancer treatments with oocyte retrieval and embryo cryopreservation. J Am Coll Surg. 2009;209: 603–7.
50. Oktay K, Buyuk E, Libertella N, et al. Fertility preservation in breast cancer patients: a prospective controlled comparison of ovarian stimulation with tamoxifen and letrozole for embryo cryopreservation. J Clin Oncol. 2005;23:4347–53.
51. Yager JD, Davidson NE. Estrogen carcinogenesis in breast cancer. N Engl J Med. 2006;354:270–82.
52. Lewis JS, Osipo C, Meeke K, et al. Estrogen-induced apoptosis in a breast cancer model resistant to long-term estrogen withdrawal. J Steroid Biochem Mol Biol. 2005;94:131–41.
53. Osipo C, Gajdos C, Cheng D, et al. Reversal of tamoxifen resistant breast cancer by low dose estrogen therapy. J Steroid Biochem Mol Biol. 2005;93:249–56.
54. Gupta PB, Kuperwasser C. Contributions of estrogen to ER-negative breast tumor growth. J Steroid Biochem Mol Biol. 2006;102:71–8.
55. Gupta PB, Proia D, Cingoz O, et al. Systemic stromal effects of estrogen promote the growth of estrogen receptor-negative cancers. Cancer Res. 2007;67:2062–71.
56. Azim A, Costantini-Ferrando M, Oktay K. Safety of fertility preservation by ovarian stimulation with letrozole and gonadotropins in patients with breast cancer: a prospective controlled study. J Clin Oncol. 2008;26:2630–5.
57. Azim A, Oktay K. Letrozole for ovulation induction and fertility preservation by embryo cryopreservation in young women with endometrial carcinoma. Fertil Steril. 2007;88:657–64.
58. Elizur SE, Chian RC, Pineau CA, et al. Fertility preservation treatment for young women with autoimmune diseases facing treatment with gonadotoxic agents. Rheumatology (Oxford). 2008;47:1506–9.
59. Chian RC, Buckett WM, Tan SL. In-vitro maturation of human oocytes. Reprod Biomed Online. 2004;8: 148–66.
60. Lim JH, Yang SH, Chian RC. New alternative to infertility treatment for women without ovarian stimulation. Reprod Biomed Online. 2007;14:547–9.
61. Rao GD, Chian RC, Son WS, et al. Fertility preservation in women undergoing cancer treatment. Lancet. 2004;363:1829–30.
62. Demirtas E, Elizur S, Holzer H, et al. Immature oocyte retrieval in the luteal phase to preserve fertility in cancer patients. Reprod Biomed Online. 2008;17:520–3.
63. Huang JY, Tulandi T, Holzer H, et al. Combining ovarian tissue cryobanking with retrieval of immature oocytes followed by in vitro maturation and vitrification: an additional strategy of fertility preservation. Fertil Steril. 2008;89:567–72.
64. Chian RC, Huang JY, Tan SL, et al. Obstetric and perinatal outcome in 200 infants conceived from vitrified oocytes. Reprod Biomed Online. 2008;16:608–10.
65. de Bree E, Makrigiannakis A, Askoxylakis J, et al. Pregnancy after breast cancer. A comprehensive review. J Surg Oncol. 2010;101:534–42.
66. Averette HE, Mirhashemi R, Moffat FL. Pregnancy after breast carcinoma: the ultimate medical challenge. Cancer. 1999;85:2301–4.
67. Peccatori F, Cinieri S, Orlando L, et al. Subsequent pregnancy after breast cancer. Recent Results Cancer Res. 2008;178:57–67.
68. Blakely LJ, Buzdar AU, Lozada JA, et al. Effects of pregnancy after treatment for breast carcinoma on survival and risk of recurrence. Cancer. 2004;100:465–9.
69. Ives A, Saunders C, Bulsara M, et al. Pregnancy after breast cancer: population based study. BMJ. 2007; 334:194.

70. Mueller BA, Simon MS, Deapen D, et al. Childbearing and survival after breast carcinoma in young women. Cancer. 2003;98:1131–40.
71. Hortobagyi GN, Hung MC, Buzdar AU. Recent developments in breast cancer therapy. Semin Oncol. 1999;26:11–20.
72. Malamos NA, Stathopoulos GP, Keramopoulos A, et al. Pregnancy and offspring after the appearance of breast cancer. Oncology. 1996;53:471–5.
73. von Schoultz E, Johansson H, Wilking N, et al. Influence of prior and subsequent pregnancy on breast cancer prognosis. J Clin Oncol. 1995;13:430–4.

8 Oocyte Cryopreservation

Andrea Borini and Veronica Bianchi

Abstract

Since the 1940s, the cryopreservation field has been deeply studied in order to increase the number of options available for human reproductive technologies. Initially, sperm cells were the only type to be preserved due to their size and high number available; these characteristics made the task relatively easy. Later on, with the advances made in the laboratories, embryo freezing became an available option in in vitro fertilization (IVF) clinics. In 1983, the first pregnancy was obtained from a cryopreserved embryo [1] and provided an important new option, influencing daily practice in the field of reproductive medicine. Consequently, the number of multiple pregnancies was significantly reduced and in the event that pregnancy was not achieved with a fresh cycle, cryopreserved embryos could be used in subsequent cycles without the need to undergo a new stimulation cycle. The overall efficiency was definitely improved. This new success pushed the clinicians further and the investigation on oocyte cryopreservation started. Initially, outcomes were not very encouraging, mainly because of the particular features of the oocyte compared to sperm or embryos. The oocyte, especially in the human, has unique composition in term of water content, membrane stability and intracellular structures. All of these characteristics made it very difficult to cryopreserve oocytes and for many years clinical applications have not been routinely adopted. The low survival rates and objectionable developmental competence of oocytes initially obtained after cryopreservation suggest that this technique could not be safely applied to patients. Moreover, the good results obtained with embryo cryopreservation were satisfying enough to limit research in the area of oocyte freezing for many years.

A. Borini (✉)
Tecnobios Procreazione Centre for Reproductive Health,
Via Dante 15, 40125 Bologna, Italy
e-mail: borini@tecnobiosprocreazione.it

E. Seli and A. Agarwal (eds.), *Fertility Preservation: Emerging Technologies and Clinical Applications*,
DOI 10.1007/978-1-4419-1783-6_8, © Springer Science+Business Media, LLC 2012

Keywords

Cryopreservation • Oocyte cryopreservation • Fertility • In vitro fertilization • Vitrification in oocyte preservation

Since the 1940s, the cryopreservation field has been deeply studied in order to increase the number of options available for human reproductive technologies. Initially, sperm cells were the only type to be preserved due to their size and high number available; these characteristics made the task relatively easy. Later on, with the advances made in the laboratories, embryo freezing became an available option in in vitro fertilization (IVF) clinics. In 1983, the first pregnancy was obtained from a cryopreserved embryo [1] and provided an important new option, influencing daily practice in the field of reproductive medicine. Consequently, the number of multiple pregnancies was significantly reduced and in the event that pregnancy was not achieved with a fresh cycle, cryopreserved embryos could be used in subsequent cycles without the need to undergo a new stimulation cycle. The overall efficiency was definitely improved.

This new success pushed the clinicians further and the investigation on oocyte cryopreservation started. Initially, outcomes were not very encouraging, mainly because of the particular features of the oocyte compared to sperm or embryos. The oocyte, especially in the human, has unique composition in terms of water content, membrane stability and intracellular structures. All of these characteristics made it very difficult to cryopreserve oocytes and for many years clinical applications have not been routinely adopted. The low survival rates and objectionable developmental competence of oocytes initially obtained after cryopreservation suggest that this technique could not be safely applied to patients. Moreover the good results obtained with embryo cryopreservation were satisfying enough to limit research in the area of oocyte freezing for many years.

Several indications lead to a re-evaluation of the importance of this technique but, in Italy the situation has changed since 2004. The Italian Parliament issued a law (40/2004) which limited the options in the IVF clinics, the most striking change was that embryo freezing was no longer allowed. As a consequence, a new wave of interest towards oocyte freezing developed in order to optimize the stimulation cycles and improve the cumulative pregnancy rates.

The first studies dated back to the 1980s when Chen [2] and Al Hasani et al. [3] each achieved a pregnancy using frozen oocytes. No other important works were published in for a long period of time; and it took a decade to reproduce these results [4].

Indication for Oocyte Cryopreservation

Oocyte cryopreservation undoubtedly offers several advantages in IVF including the ability to preserve fertility in cancer patients. Every year approximately 650,000 women are diagnosed with cancer in the United States; many of these women have not yet had a child. Moreover, the survival rate in cancer patients has been dramatically improved in the last 2 decades thanks to new therapies and disease prevention. Consequently, offering the option of fertility preservation to these women helps improve their quality of life.

The pharmacological treatments used for most types of cancers can lead to a partial or total loss of ovarian function; in the worse case scenario even the reproductive organs may need to be removed. Therefore, it is essential to act in a short window of time to provide these women with a chance of preserving their fertility, and at the same time not to interfere with the oncological therapies.

In patients with a partner, embryo cryopreservation is a valuable option since it has been applied worldwide successfully; but, in women who do not have a partner, oocyte cryopreservation represents the best option available.

Besides the benefits to cancer patients, oocyte cryopreservation is of great importance for those women who decide to delay childbearing. it is well known that fertility in women dramatically decreases with age and thus it has been proposed that it would be beneficial to freeze "young" oocytes for use later in life. The fertility potential of these gametes would be preserved thereby improving the outcomes for older patients. This would also minimize the need for oocyte donation, which is also forbidden in Italy and many other countries.

The development of infertility treatments has driven the need to improve cryopreservation strategies. As previously discussed, embryo cryopreservation was first adopted as a consequence of the limited number of embryos transferred, in order to maximize the use of harvested oocytes following ovarian hyperstimulation.

The collection of multiple oocytes has allowed cryopreservation of embryos. More recently, the improvements made in cryopreservation protocols and embryo culture led to the transfer of fewer embryos and better survival rate improving the pregnancy rate per transfer. At this point, the production of a great number of supernumerary embryos raised ethical concerns for certain couples and, in certain countries it is considered ethically inappropriate due to religious concerns.

Another crucial point emerged in cases of divorce; since the embryos are owned by both partners their fate after the separation of the couple is still unclear. Oocyte cryopreservation may offer a way to avoid the above-mentioned complications during treatments.

Another indication for oocyte freezing is the event of unforeseen non-production of sperm the day of oocyte retrieval, giving the male partner another chance to try in a less stressful moment without compromising the final outcome.

Beyond these indications, oocyte cryopreservation provides a unique and safe opportunity to optimize donor oocyte cryo-banking, since it allows safe quarantine, cost reduction and better synchronization between donor and recipient.

Patient Selection for Oocyte Freezing

A marked age-related decline in fertility has been seen in women over the age of 35 years predominantly due to reduced oocyte quality. Consequently, the best patients for oocyte freezing should be younger than 36 years of age. The most frequent indications for cryopreservation of oocytes include treatments and diseases that may impair ovarian function such as radiation and chemotherapy for cancer treatment, surgical intervention requiring removal of the ovaries and ovary function-related problems (premature ovarian failure, anovulation, polycystic ovarian syndrome). Other infertility indications include absent or blocked fallopian tubes, failed salpingoplastic, oligo-ovulation and unexplained infertility. Women who desire to delay motherhood and/or do not have a partner may choose to cryopreserved their oocytes for later use, in addition to donating their oocytes for recipients who cannot use their own oocytes (in countries where this is allowed).

Stimulation Protocols

Ovarian stimulation is a key factor in assisted reproductive technology (ART) procedures, aiming at the maturation of a high number of good quality oocytes, in order to enhance the chance of success of the treatment. It is very important to tailor the protocol according to the patients' characteristics in term of age, basal FSH level, endometriosis, number of antral follicles, body mass index or previous attempts.

Gonadotropin Releasing Hormone Agonists

The long gonadotropin releasing hormone (GnRH) agonist protocol is currently the most widely used in IVF, since it has greatly increased pregnancy rates through an enhancement of follicular recruitment, allowing for the recovery of a larger number of oocytes [5]. Moreover, it improved the routine patient treatment schedule, but it needs a long treatment period until desensitisation occurs. The costs for this protocol are relatively high due to the higher amount of gonadotropin required and the increased need for hormonal and ultrasonographic measurements.

Two kinds of gonadotropins are currently available: the urinary derived and the recombinants, which are more bioactive and have greater batch to batch consistency, with less variability in FSH activity [6].

The starting dose in patients undergoing their first treatment and younger than 40 years usually ranges from 100 to 250 IU/day, but there is not a universal agreement on the most advantageous initial FSH dose. The chance of pregnancy is not influenced by gonadotropin doses; higher doses can be more effective in terms of oocyte yield and reduced cycle cancellation, but are associated with a higher incidence of side-effects, especially a greater number of ovarian hyperstimulation syndrome (OHSS) cases requiring hospitalization [7, 8].

The actual role of LH in controlled ovarian stimulation is still a matter of debate. A therapeutic "window" of LH concentrations, below which estradiol production is not adequate and above which LH may be detrimental to follicular development has been described [9] with GnRH agonist depot (leuprorelin or triptorelin), half dosage (1.87 mg) i.m. on day 21 of the cycle [10]. At the onset of menses, patients begin gonadotropin stimulation as previously reported [11] with 300 IU per day of r-hFSH (follitropin α) for 2 days and 150 IU per day for 4 days. In women aged 40 years or more the FSH daily dose can be increased by 75–150 IU according to basal FSH levels. The dose is then adjusted according to individual response as assessed by 17β-oestradiol (E_2) assay and ultrasound scanning performed every other day. Recombinant HCG, 250 μg, is administered when at least three follicles reach a maximum diameter of 20 mm, of which at least one is >23 mm. Transvaginal oocyte retrieval is performed under ultrasound guidance 36 h after HCG administration.

GnRH Antagonist

A more recent alternative to prevent premature LH surge during IVF cycles is the use of GnRH antagonists. The mechanism by which they suppress the secretion of gonadotropins is totally different: gonadotropin secretion decreases in a few hours, pituitary reserve and synthesis of follicle stimulating hormone (FSH) and LH are not affected; as a result, the recovery of pituitary function is rapid.

A major advantage of GnRH antagonists over the agonists is that their administration can be limited to the few days of the stimulation preceding human chorionic gonadotropin (HCG) injection, when the risk of premature LH surge is higher. Compared to a long luteal agonist protocol, the treatment is shorter and requires a lower amount of gonadotropins. Pregnancy rate seems to be slightly lower [12], but recent meta-analyses suggest a similar likelihood of live births [13]. Moreover, a decrease in the incidence of severe OHSS is reported by several studies. Published studies show a trend towards higher pregnancy rates with fixed antagonist initiation on day 6 as compared to flexible initiation based on a follicle size of 14–16 mm [14].

Aged or poor ovarian reserve women may take advantage from a GnRH antagonist protocol. This treatment may also be useful when there is not enough time to organize a long luteal agonist protocol. Our group usually employs a flexible multiple dose protocol. Oral contraceptives may be used to program the cycles; they have also been proposed to improve synchronization of the recruitable cohort of follicles.

Gonadotropin stimulation begins on the second day of the menstrual cycle. Highly purified human menopausal gonadotropin or r-hFSH are administered at a daily dose between 150 and 450UI (depending on age or ovarian reserve) for 5 days. Monitoring by 17β-oestradiol assay and ultrasound scanning begins on day 6 of stimulation. The dose of gonadotropin is then adjusted according to individual response. Antagonist injections (Cetrorelix or Ganirelix at a daily dose of 0.25 mg) start as soon as the follicles reach a size of 14–15 mm and continue until HCG administration.

Oocyte Quality Before Cryopreservation

An important aspect of oocyte quality affecting outcome of cryopreserved oocytes is their morphology prior to freezing. Survival rate is probably the most important parameter to consider when an embryologist chooses a cryopreservation protocol. However, if the oocyte quality is sub-optimal, the number of oocytes recovered will be lower.

It has been well demonstrated that less than half of mature oocytes are capable of being fertilized and produce viable embryos [15]. Therefore, a careful evaluation of oocyte appearance is very important in order to choose the cohort of oocytes to freeze. In this way it might be possible for embryologists to predict the implantation potential of cryopreserved oocytes leading to better outcomes in terms of post-thaw survival, fertilization rate and good embryo morphology and also a possible cost reduction per cycle.

Although the relationship between morphology and IVF outcome is well accepted in conventional IVF [16] it is still controversial in ICSI. This discrepancy may be due to the fact that more accurate grading is possible with denuded oocytes [17, 18].

Generally speaking the gross morphology of oocytes at metaphase II includes a certain degree of variability in the shape, granularity, colour and homogeneity of the cytoplasm, size and characteristics of the perivitelline space, vacuolization, inclusions and abnormalities of the first polar body or the zona pellucida. Some oocytes show one or more abnormalities which makes it difficult to define the relationship between their appearance and their development potential.

Morphological analyses occupy an extensive proportion of the literature and among all the features analyzed, the first polar body is of specific interest. Studies have shown that human metaphase II oocytes with atypical characteristics of the first polar body at collection (fragmented, very small, or overly large) demonstrate slight decreases in normal fertilization along with increases in abnormal fertilization [19]. Indeed, Xia et al. observed a significant correlation between fertilization rate, embryo quality and human oocyte morphology (three factors were taken into account: first polar body morphology, size of perivitelline space and cytoplasmic inclusions). Oocytes without inclusions, with an intact first polar body and normal perivitelline space had significantly higher rates of fertilization and produced better quality embryos compared to the other groups. Moreover, the same group evidenced that, the proportion of mature, immature and degenerated oocytes was strongly related to the total number of oocytes retrieved and the estradiol concentration on the day of HCG administration [20].

Similarly, Ebner et al. [21] confirmed a significant correlation between polar body morphology, fertilization rate and embryo quality, while cytoplasmic criteria were not associated with the outcome. Later the same group [22] showed that fragmentation on day 2 was increased in embryos derived from oocytes with a fragmented first polar body compared to those coming from oocytes with an intact one. The same difference was detected in blastocyst development and implantation rate which were significantly higher in the oocytes with an intact polar body.

Another important aspect is the evaluation of oocyte maturity. Although the treatment with gonadotropin-releasing hormone analogues allows a more synchronous cohort of follicles to be recruited, some of the retrieved oocytes may be from slower developing follicles. Consequently, the oocyte cytoplasm would be at a different maturational stage when exposed to HCG for the resumption of meiosis in stimulated cycles. Therefore, asynchrony of nuclear and cytoplasmic maturation could occur in oocytes at the metaphase II stage.

All the data about oocyte morphological evaluation underline the importance of an early selection which may be helpful in identifying embryos with a good prognosis outcome. Therefore, prior to freezing it is important to track all the morphological features and make a strict selection in order to improve the pregnancy rate post thawing.

As recently described by Patrizio et al. [23], in addition to the established methods described above, more advanced methods of comparative genomic hybridization and molecular genetic technologies may soon be utilized to improve the understanding of oocyte genetics, identification of viability markers and proteomic profiles, to enhance oocyte and embryo selection for improved outcomes.

Available Cryopreservation Protocols

The difficulties encountered in achieving success with human mature oocytes mainly arose from the unique characteristics of this cell in terms of membrane permeability and amount of water in the cytoplasm. Another issue is the susceptibility of the mammalian spindle to damage induced by freezing. Such problems were highlighted by initial reports which indicated increased rates of post thaw aneuploidy in cryopreserved mouse oocytes [24].

In order to survive the freeze/thaw process, the cells must maintain their structural integrity throughout the cryopreservation procedure. To accomplish this, it is important to minimize the damage caused by intracellular ice formation. This is usually achieved by dehydrating the cells before or during the cooling procedure. If dehydration is inadequate, large intracellular ice crystals may form, damaging the oocyte irreversibly.

In general, the cryopreservation of gametes and embryos involves an initial exposure to cryoprotectants, cooling to subzero temperatures and the final storage in liquid nitrogen. At this very low temperature, the metabolic activities in the cell are almost undetectable and thus the storage can be prolonged indefinitely. When the samples are thawed, cryoprotectants are removed by stepwise dilution with final return to a physiologic environment.

Major factors affecting survival of cryopreserved cells include the species, developmental stage, cryoprotectants and method of cryopreservation. The three major freezing protocols analyzed in the literature are: slow cooling, rapid/ultra rapid freezing and vitrification. They are characterized first by different cooling temperature curves and also by the use of different cryoprotectant concentrations.

Cryopreservation procedures used for mammalian oocytes and embryos have advanced since the very first attempts in 1970s. Slow cooling protocols have been the most widely analyzed and a substantial amount of data are available in the literature. Rapid freezing protocols require the presence of higher concentrations of cryoprotectants than the slow cooling methods and they allow embryos to be plunged directly into liquid nitrogen or liquid nitrogen vapour from temperatures of 0°C. Due to poor results, this methodology was abandoned after the first experiments [25, 26]. Vitrification is a rapidly growing technique that uses very high concentrations of cryoprotectant to make the solution and its contents vitrify (forming a glass-like solid rather than ice crystals during the cooling phase) and to remain this way during warming [27].

Slow Freezing

Over the last 20 years, the slow freezing technique has been the only one available and undoubtedly the most studied. It was initially used for embryo freezing and, due to the positive results, it was directly applied to oocyte cryopreservation. However the first experiments were not encouraging, after sporadic pregnancies [2, 3] the inability to recover high rates of stored material after thawing lead the clinics to put this procedure aside. A renewed interest came up during the last decade due to the need to preserve fertility mostly in cancer patients but also, in certain countries, due to legal restrictions on embryo cryopreservation imposed by the government. In the meantime, the protocols have advanced and, as a consequence the results in terms of survival rate and pregnancy outcomes have significantly improved.

From a more technical point, freezing cells at very low temperature allows their maintenance for an indefinite time since the biological processes are essentially suspended. Under those conditions, water is in a solid state and thus the only possible alteration to DNA may be due to background reactions. This does not alter in any way the survival of the oocytes and their development potential.

The main problems encountered with oocyte freezing are instead related to its unique size, content of water and membrane permeability characteristics. It is important to design a specific protocol for oocyte cryopreservation that takes into account several factors that might alter oocyte physiology. For example, it has been

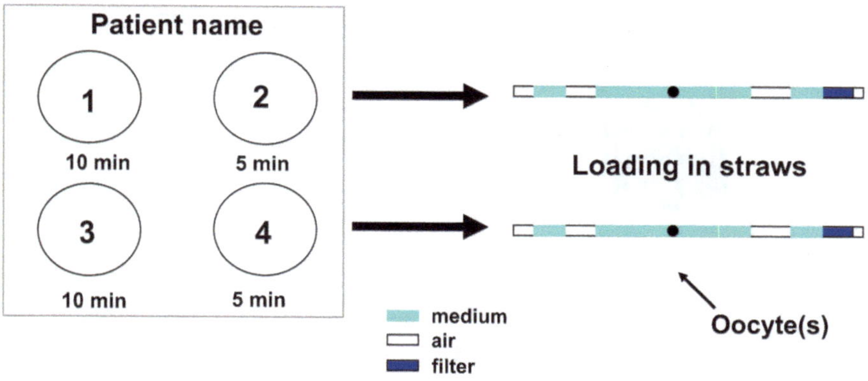

Fig. 8.1 Slow freezing. Wells 1 and 3: 1.5 M PROH+20% albumin. Wells 2 and 4: 1.5 M PROH+0.2 M Sucrose+20% albumin. Each row correspond to one straw

shown that oocyte exposure to conventional permeating cryoprotectants, such as propanediol, ethylene glycol and DMSO, lead to an increase in oocyte intracellular calcium concentration that, in turn, is responsible for oocyte activation and consequent zona hardening.

Standard slow freezing protocols have been designed to minimize cell damage by avoiding ice crystal formation derived from the solidification of intracellular water. The goal was reached by using a mixture of different cryoprotectants in the solution in order to sufficiently dehydrate the oocyte. The cryoprotectant agents (CPAs) can be classified in two groups:

- Permeating agents which enter the cell (glycerol, dimethyl sulphoxide (DMSO), ethylene glycol, and 1,2-propanediol (PROH) are generally used).
- Non-permeating agents which do not enter the cell but withdraw the water from the cell cytoplasm. They include large sugar molecules such as sucrose, ficoll, and raffinose, as well as proteins and lipoproteins.

The slow cooling protocols usually are based on a mixture of PROH or DMSO in a 1.5 M concentration plus sucrose or trehalose in variable concentrations (Fig. 8.1). The freezing curve is characterized by a slow decrease of the temperature that allows for sufficient and progressive dehydration of the oocyte. The thawing is instead based on a rapid dilution of the cryoprotactants (Fig. 8.2). Briefly, the oocytes are maintained at room temperature for 10 min in a 1.5-M PROH solution with 20% protein supplement for equilibration phase (Fig. 8.1). During this time, the PROH enters the cell as water exits the oocyte. During the second step, the oocytes are placed in a 1.5-M PROH solution supplemented with sucrose (at different concentration according to the protocols) and 20% protein. The exposure time for the loading solution is about 5 min and this increases the oocyte dehydration proportionally to the amount of sucrose contained in the solution. The oocytes are subsequently loaded in straws and placed in an automated Kryo 10 series biological vertical freezer (Planer Kryo GB). Freezing solutions are cooled from 20 to −7°C at a rate of 2°C/min. Manual seeding of oocytes within straws is performed at near −7°C and this temperature is maintained for 10 min in order to allow uniform ice propagation. The temperature is then decreased to −30°C at a rate of 0.3°C/min and then rapidly lowered to −150°C at a rate of 50°C/min. Straws are then directly plunged into liquid nitrogen at −196°C and stored.

Thawing consists of rapid re-warming (air for 30 s then 40 s in a 30°C water bath) and subsequent stepwise dilution of the cryoprotectants (Fig. 8.2); first in 1.0 M and then in 0.5 M PROH solutions supplemented with sucrose (depending on the sucrose concentration used during freezing procedure) for 5 min each and then in a sucrose solution for 10 min and in PBS solution for an additional 10 min. Finally, the oocytes are returned in culture media at 37°C to support recovery.

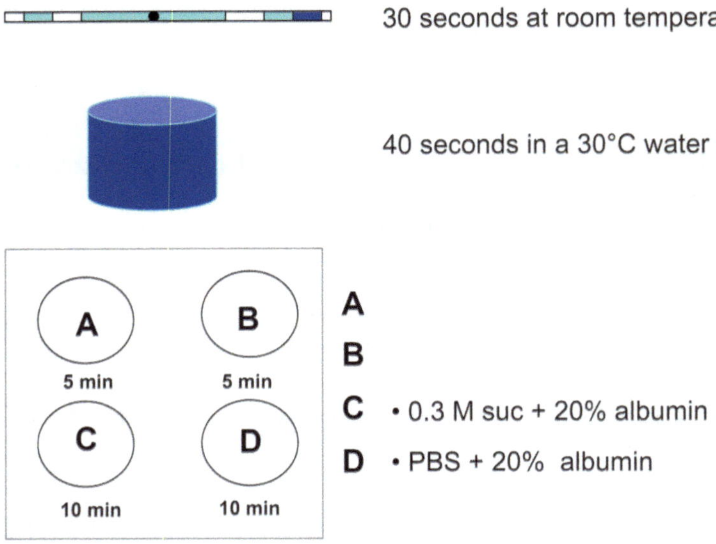

Fig. 8.2 Thaw protocol after slow freezing

Vitrification

Vitrification is an alternative cryopreservation protocol that involves ultra rapid freezing and warming. Reports in the literature have demonstrated improved survival rates for oocytes and ovarian tissue and suggest that this might become the method of choice for cryopreservation. The process of vitrification involves the use of high concentrations of cryoprotectants and rapid cooling to form a highly viscous glass-like solution without the formation of ice crystals. Although mechanical damage may be prevented by the use of these higher concentrations of cryoprotectants, they may cause cytotoxicity or osmotic shock. Nevertheless more recent studies using metabolome and proteome quantification of the oocyte have highlighted that, whereas slow freezing seems to have a dramatic effect on cell physiology, vitrification appears to cause limited alterations [28]. Despite the high cryoprotectant concentrations, the limited exposure is probably better for preserving the oocyte.

Generally speaking, the primary considerations when designing a vitrification protocol include the concentration of cryoprotectants, sample volume, cooling and warming rates. A lower concentration of CPAs (or combinations of CPAs) can be used when vitrification is performed with minimal volume and rapid freezing rates, leading to less potential cryodamage. In contrast, the likelihood of ice nucleation is higher if larger volumes are used, which limit maximal rates of cooling and can cause the entire specimen to freeze instantaneously. The probability of vitrification is directly related to cooling/warming rate and viscosity, and inversely related to volume [29].

It is well known that CPAs at high concentrations might have toxic consequences, which is one of the main challenges with vitrification procedures. In order to improve the outcome it is best to increase the cooling rate, minimize the sample volume and time of exposure to the high concentrations of permeating cryoprotectants. The cooling rate can be increased using liquid nitrogen slush at −210°C while the loading volume can be reduced to less than 1 μL using newly developed devices (Fig. 8.3).

The cryoprotectants in the vitrification procedure serve two main functions: to remove the water from the cell and, at the same time, permeate into the cell to form the amorphous state in the cytoplasm, preventing the cell from low temperature damage.

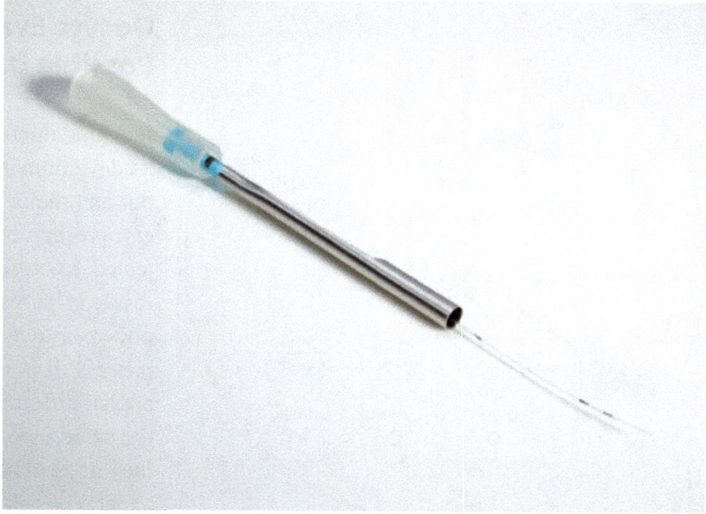

Fig. 8.3 CryoTip®. Pulled straw. Inner diameter ≈200 μm, volume ~1 μL, closed system, heat sealed at both ends

Initially just penetrating agents were used in vitrification mixtures but, more recently the solutions have been changed to include both penetrating and non-penetrating agents. This increased survival rates and made significant advancements in the success of the procedure.

The use of vitrification in embryology was first reported with mouse embryos in 1985 [27], followed by successful vitrification of oocytes in 1991 [30], yet the general application of vitrification in assisted reproduction has been rather limited until recently. The use of vitrification has been described in the literature for several mammalian species, including humans, with varying degrees of success depending upon the wide variety of tools and procedures applied [31, 32]. Numerous recent publications have shown outstanding results for survival and clinical outcomes using vitrification compared to slow cooling [33]. Vitrification methods have been modified over the years to optimize results in humans, by using minimal volumes and very rapid cooling rates, allowing for lower concentrations of cryoprotectants to reduce injuries related to chemical toxicity, osmotic shock, chilling sensitivity and ice nucleation [32, 33].

The general methodology involves a two-step sequential exposure of oocytes to vitrification solutions containing one or more cryoprotectants in increasing concentrations up to 40% (v/v), loading the oocyte in a minimal volume (<1 μL) of solution onto a carrier device (open or closed system), and very rapid cooling by plunging directly into liquid nitrogen. The time and temperature of exposure to such solutions are critical to avoid toxicity. Moreover it is a very fast procedure that does not require any electronic equipment and allows freezing of specimens in a very short time. Conversely, warming rates must also be rapid to prevent ice nucleation during the warming process in order to achieve optimal results. After warming, the oocytes are then moved through at least three solutions with decreasing concentrations of sucrose to effectively remove the permeating cryoprotectants and rehydrate the oocytes.

The most widely used vitrification protocol (Fig. 8.4) involves gradual exposure of oocytes to the equilibration solution (7.5% ethylene glycol, 7.5% DMSO, 20% Serum Substitute Supplement (SSS) in HEPES buffered Medium 199 [M199 H]) for approximately 8 min and then vitrification solution (15% ethylene glycol, 15% DMSO, 0.5 M sucrose and 20% SSS in M199-H) up to 110 s. Samples are then loaded onto a carrier device and plunged into liquid nitrogen. The thawing solutions (Fig. 8.5) are based on a series of solutions with decreasing sucrose concentration (1.0, 0.5 and 0 M) with 20% of SSS in M199-H.

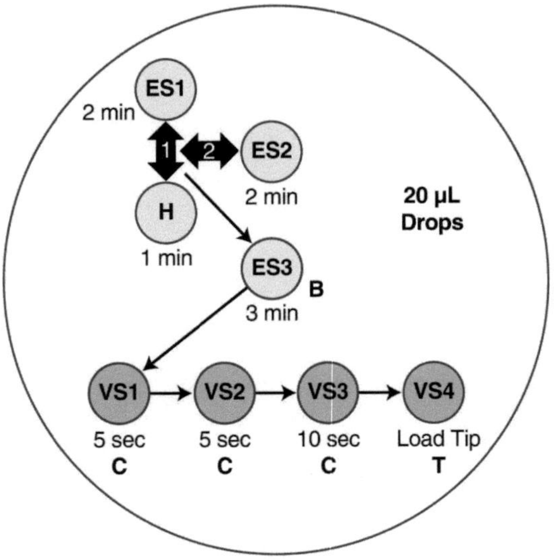

Fig. 8.4 Oocyte vitrification protocol. Performed at room temperature (22–27°C) in ~10 min. Set up Vitrification dish with drops of H and ES solution prior to beginning procedure. Add VS drops during ES exposure time (as in diagram). *H* HEPES buffered culture medium with protein (e.g. mHTF+12 mg/mL HSA); *ES* equilibration solution (3 drops); VS vitrification solution (4 drops); *B* bottom; *C* centre; *T* top. Timing of exposure to VS is critical (110 s maximum)

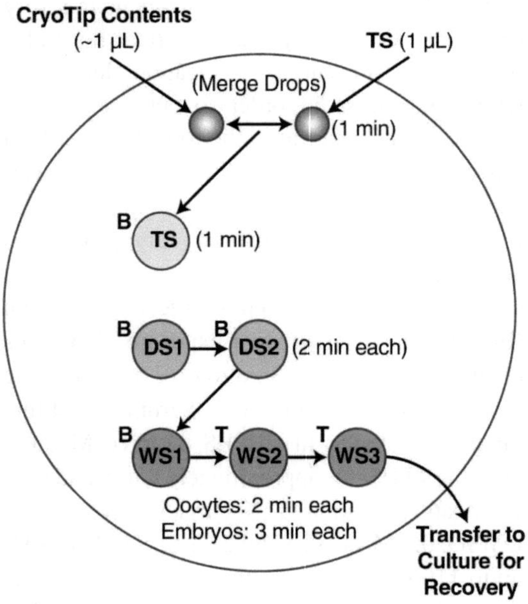

Fig. 8.5 Thaw protocol after vitrification. Performed at room temperature (22–27°C). Set up thawing dish with drops of TS and DS solutions prior to beginning procedure (as in diagram). Add WS during DS exposure time. *TS* thawing solution (1 drop); *DS* dilution solution (2 drops); *WS* washing solution (3 drops); *B* bottom; *T* top

Oocyte Evaluation After Thawing: Possible Damages

It is fundamental that the oocyte needs to maintain its developmental competence after thawing without any major alteration. Nevertheless, during the cryopreservation process, oocytes may be subjected to different kinds of damage that might result in partial alteration of the structure of the oocyte or to its total degeneration after the thawing process. It is important to evaluate if oocytes that appear different from their fresh counterparts have the same theoretical developmental potential or if they need to be discarded prior to insemination.

The possible alterations that may be observed in the oocyte are mainly in the cytoplasm, where, it is possible to visualize cortical granularity, changes in the cytoplasm colour or alterations in the shape of the oocyte.

Moreover, refined studies showed that damage may also include alterations in the intracellular calcium concentration, disruption–depolarization of the meiotic spindle structure, reduction in the amount of mitochondria, or modification of the appearance of the Golgi reticulum [34].

It has been described that inappropriately timed evaluation of survival may lead to erroneous conclusions; in fact, some oocytes may look perfectly normal just after thawing but may degenerate within a couple of hours. This was well described in one of the first studies about oocyte freezing [35] where the survival rate decrease from 51 to 41% after 3 h culture.

Ultrastructural Damages

Freezing procedures may induce critical damage to the delicate structures of the human metaphase II oocyte. This may happen because at this stage the chromosomes are aligned and in strict relation with the meiotic spindle which, in turn, is responsible for the segregation of the chromatids which follows the resumption of meiosis. This whole substructure is very sensitive to cryodamages and its alteration might lead to subsequent embryo aneuploidy.

There are several other structures involved in the maintenance of oocyte stability after thawing which have been analyzed. One of primary consideration in the past has been the zona pellucida which plays a key role in normal fertilization. In fact, when the sperm binds specific glycoproteins on its surface the so-called cortical reaction takes place leading to hardening of the zona. This event, in turn, prevents other sperm from entering the cell, thereby avoiding abnormal fertilization. It has been demonstrated that freezing procedures may induce a premature release of the cortical granules by increasing the intracellular calcium concentration; as a consequence fertilization by standard IVF did not ensure a high number of normal zygote. This issue was overcome by the introduction of intracytoplasmic sperm injection (ICSI) as the elective methodology to inseminate thawed oocytes.

Other structures can undergo a change during cryopreservation; mithocondria or smooth endoplasmatic reticulum (SER) are very sensitive to low temperature but they were not deeply investigated initially since they did not seem to correlate with the clinical outcome. Nevertheless, in the last 5 years, several studies focusing on this topic were published. Gualtieri et al. [34] analyzed mitochondrial features and showed that fresh samples had a regular shape with few short cristae mithocondria while the frozen/thawed group displayed a high percentage of oocytes (72%) with decreased electron density mithocondria.

Another interesting manuscript has been published by Nottola et al. [36]. The author analyzed fresh and frozen/thawed oocytes by slow cooling using PROH plus sucrose (0.1 or 0.3 mol/L) as cryoprotectants. The oocytes were analyzed using electronic microscopy and while in the fresh group they mostly displayed a homogeneous cytoplasm, the cryopreserved oocytes showed some area of microvacuolization. The number of mitochondria was normal and the zona pellucida intact.

Meiotic Spindle Evaluation

At the MII maturational stage, the chromosomes are aligned on the metaphase plate in strict relation to the meiotic spindle. This is a delicate structure responsible for the correct chromosome segregation during the maturational process that ends with extrusion of the second polar body. The meiotic spindle is a highly dynamic bipolar structure made of microtubules that have the ability to undergo disassembly and reassembly under particular conditions. They are dimers composed of α and β tubulin, more dispersed at the microtubule organizing centre (MTOC) and more compacted at both ends in contact with the kinetochore; the spindle has a typical barrel-shaped structure with the chromosomes suspended within it.

Several studies in the literature correlate temperature decrease with spindle alteration. It has been shown that the tubulin depolymerise in response to a decreae in temperature causing a transient loss of the structure but when returned to 37°C, the spindle reappears in the cytoplasm [37, 38].

It is evident that if the spindle undergoes permanent alteration, the fate of the related oocyte is uncertain. Nevertheless, chromosome alteration in embryos derived from frozen thawed oocytes do not seem to be statistically higher compared to fresh embryos [39].

The main problem related to the meiotic spindle observation is the maintenance of cell viability. Most of the techniques available to date are based on irreversible fixation or staining which result in the death of the oocyte. Thus, it is possible to collect data using, for example, electronic or confocal microscopy but this would not be useful in the routine analyses of the viable oocytes used for ongoing treatment. In order to use a methodology for spindle evaluation, it must allow maintenance of cell viability.

The Poloscope® is a polarized light-based microscopy device which allows multiple observations over time thanks to an optical system that can visualize highly ordered structures such as the spindle. Beside viewing its location in the oocyte cytoplasm, the poloscope can measure microtubule density as a correlation to spindle retardance and birefringence [40]. The poloscope is a useful tool to evaluate possible modifications after cryopreservation and provides an ideal non-invasive method for routine visualization of the spindle since it does not compromise oocyte viability.

There is general agreement in the literature regarding the correlation of the spindle to fertilization rate [41, 42] or embryo development [43] during IVF. Additionally, oocytes without a visible meiotic spindle show a high degree of aneuploidy [44, 45]. Consequently, spindle recovery after cryopreservation of oocytes has been evaluated using different protocols to select the thawed oocytes with the best potential for fertilization and ongoing development.

Studies using the Poloscope were conducted by several authors [46, 47] who agreed that, during slow cooling cryopreservation the meiotic spindle transiently disappeared and does not reappear until some time after thawing. This phenomenon was consistent despite the protocol and sugar concentration in the freezing solution. Similar findings were reported using confocal microscopy [48] in fresh and frozen thawed oocytes fixed at different times after thawing. The spindle which was visualized in all the fresh oocytes was not immediately recovered after rewarming of frozen oocyte, but, after a period of culture this recovery rate was significantly increased. These data and the Poloscope observations confirmed that a minimum of 2 h culture at 37°C is recommended prior to insemination. What emerges from other confocal microscopy analysis [49] is a correlation between spindle recovery and sucrose concentration in the freezing mixture. The use of a higher amount of sugar (0.3 M instead of 0.1 M) not only improves the survival rate but also better preserves the meiotic spindle. This is probably correlated to the higher shrinkage of the cell which results in a lower intracellular ice crystal formation.

Spindle recovery rate has been investigated in vitrified oocyte as well although with contrasting outcomes. Ciotti et al. [50] recorded a significantly faster recovery rate in the vitrification compared to the slow freezing group, postulating that the high cryoprotectant concentration may have a protecting effect on the meiotic spindle. Nevertheless, while the spindle was detectable during the entire incubation in the slow freezing it was not detectable during vitrification due to the high shrinkage of the cell. However, an early spindle recovery would allow for an earlier insemination window after thawing which seems to be positively correlated with the outcome [51].

A different result was obtained by Noyes et al. [52], who found no difference in spindle recovery rate in oocytes cryopreserved with either slow cooling or vitrification (93% vs. 88% respectively). It was confirmed, instead, that a higher percentage of embryos suitable for transfer on day 3 derived from oocytes who displayed the spindle before cryopreservation. After thawing, spindle-positive oocytes showed a higher rate of blastocyst formation and embryo quality suitable for day-5 transfer.

Clinical Outcome: Slow Cooling

The slow cooling approach was the first method used for oocyte cryopreservation; originally designed for embryo freezing by Lassale et al. [53], it was applied to oocyte freezing, keeping the same cryoprotectant concentrations, exposure times and decreasing temperature curve. Unfortunately, the results were not comparable to the embryo outcomes.

In 2004, Borini et al. [54] tracked clinical data on 68 patients. The survival rate was very low (37%) as was the fertilization rate (45.4%); on the contrary the cleavage rate (86.3%) and the pregnancy outcome rate (22% per patient) were acceptable. After 2001, the increase in sucrose concentration introduced by Fabbri et al. [55] led to a significant change in the field. The post-thaw viability drastically improved and was proportional to the sugar concentration in the freezing/thawing solutions (34% with 0.1 M, 60% with 0.2 M and 82% with 0.3 M); this increase in the survival rates has been related to a higher extent of oocyte dehydration, thereby reducing the amount of intracellular water and consequent ice formation.

The initial optimism from these results was extinguished when additional reports on clinical data became available. Fabbri et al. [55] in fact, just reported biological evaluations, but no pregnancy outcome. Several following clinical studies [56, 57] demonstrated that the implantation potential of these oocytes were not satisfying (5.2% and 3.5% respectively).

In a more recent study, Bianchi et al. [58] were able to significantly increase the implantation rate (13.4%) while maintaining high survival rates (76%) by using a reduced amount of sucrose in the freezing mixture (0.2 M) and a higher sugar concentration in the thawing solutions.

Another important multicenter study in Italy was reported by Borini et al. [59] after the application of the Italian law when just three oocytes could be inseminated. The protocol was based on a 2-step PROH – sucrose based solution with the same sucrose concentration previously described by Bianchi et al. [58]. Despite the good results the implantation per oocytes was lower (10.1%) than before.

More recently Noyes et al. [60] published a comprehensive summary of almost all the data available in the literature on oocyte freezing. Although slow cooling was started in 1986 while the first report on vitrification was in 1999, the latter has since achieved essentially the same number of reported outcomes to date. A total of 58 reports (43 using slow freeze, 12 vitrification and three both methods) from 1986 to 2008 were reviewed. The number of babies born was distinguished according to the technique used: 308 from slow freezing, 289 from vitrification and 12 from a combination of both protocols. The rate of single pregnancy was 81% compared to 19% for multiples. Moreover, the overall rate of birth anomalies has been reported showing eight anomalies: three ventricular septal cardiac defects, one choanal atresia, one biliary atresia, one Rubinstein–Taybi syndrome, one clubfoot and one skin haemangioma. Three of these anomalies resulted from slow freezing while five from vitrification, demonstrating a birth defect rate that was no different than that seen for natural conception.

Clinical Outcomes: Vitrification

The literature and available data for vitrification is more limited than that for slow cooling since it is a relatively new technique. The first human live birth with vitrification, was reported by Kuleshova et al. [61] using ethylene glycol and sucrose as CPAs and an open-pulled straw device. In the following few years, other sporadic pregnancies were reported, Yoon et al. [62] (2003) achieved three pregnancies in seven patients using vitrified oocytes and, in the same year Katayama et al. [63] (2003) obtained other two pregnancies out of six transfers. Nevertheless, it was not until 2005 that a higher number of pregnancy was achieved. Kuwayama et al. [32] published a 91% survival rate after thawing and a 50% blastocyst stage development. Twenty-nine embryo transfers resulted in 12 pregnancies.

A meta analysis by Oktay et al. [64] reported four papers on vitrification regarding the outcome of 503 thawed oocytes . A total of 10 live births were achieved with a rate of live birth per thawed oocyte of 2.0%.

Similar results were more recently reported by Antinori et al. [65] using the cryotop vitrification method. In this study, 330 oocytes were thawed with a survival rate of 99.4%. The fertilization, pregnancy and implantation were 92.9, 32.5 and 13.2%, respectively.

Cobo et al. [66] compared the outcomes achieved with fresh and cryopreserved donor oocytes vitrified by the Cryotop method. The vitrified group included 231 MII oocytes while the fresh group had 219. The survival rate after thawing was very high at 96.9%. The fertilization rates (82.2% vs. 76.3%) and cleavage rates on day 3 (84.6% vs. 77.6%) were higher in the fresh group compared to the vitrified group, but the differences were not significant. The blastocyst rates was almost the same for fresh (47.5%) and frozen/thawed (48.7%) specimens. The pregnancy and implantation rates for the frozen/thawed group was very high (65.2% and 40.8% respectively).

More recently, Chian et al. [67] reported the obstetric and perinatal outcomes in 165 pregnancies and 200 infants conceived following oocyte vitrification in three different clinics. The results indicate that the mean birth weight and the incidence of congenital anomalies are comparable to that of spontaneous conceptions in fertile women or infertile women undergoing in vitro fertilization treatment. These data confirmed the previous described by Noyes et al. [60].

Smith et al. [68] compared slow cooling and vitrification outcomes; it emerged that oocyte survival was significantly higher following vitrification compared with slow cooling (67% vs. 81% respectively). Although similar numbers of embryos were transferred the ones coming from vitrified oocytes showed higher clinical pregnancy rates compared with embryos resulting from slow cryopreserved (38% vs. 13% respectively). Even though from these data vitrification seems to be the most efficient technique the protocol used for the slow freezing was not the one with better outcome as previously demonstrated [58].

Another recent analysis from Grifo and Noyes [69] reported IVF outcome from fresh and frozen thawed oocytes using both slow cooling and vitrification as freezing methodologies. The survival rates were not different (88% for slow cooling and 95% for vitrification). The same happened for fertilization and blastocyst formation. Of the 23 transfers, 14 (61%) resulted in clinical pregnancy. These results add a new paragraph in the comparison between different protocols showing that either can be successful. At this point, a randomized clinical trial with a higher number of patients and, using the best protocols available for both procedures is strongly required.

Ubaldi et al. [70] just published an interesting analyses on the cumulative pregnancy rate obtained using the cryotop vitrification method. Out of 182 ICSI treatments, the overall pregnancy rate reported per stimulation cycle was 53.3%. This rate perfectly matches the data published by Borini et al. in 2009 [71] using the slow freezing protocol previously described [58]. The overall pregnancy rate achieved was, in fact, 57% in a 2-years analyse (2007–2009). Moreover, both authors reported clinical outcomes according to the female's age and, even in this case no difference were obtained using either protocols. A common trend observed by the two groups was related to a significant decrease in the pregnancy rate with the increasing maternal age.

Cryopreservation of Immature Oocytes: A Possible Alternative

Although the refinements in cryopreservation procedures have improved the success of mature oocyte vitrification outcomes, the option of immature oocyte cryopreservation remains an important area of investigation. The conditions under which oocytes are retrieved are being investigated in depth. In general, the cryopreservation of oocytes has focused on the use of mature MII oocytes obtained following ovarian stimulation protocols. This is not feasible for women faced with chemotherapy intervention or other therapies that preclude hormonal stimulation. The only option for such women is to cryopreserve immature oocytes retrieved during the follicular phase, which must then be matured in vitro after thawing/warming. Alternatively, the immature oocytes can be matured in vitro prior to cryopreservation.

After the first studies on the human immature oocyte outcome by Toth et al. [72, 73], the field was not pursued due to the progress with ovarian stimulation cycles which gave improved outcomes.

In recent years, a renewed interest in immature oocyte and in vitro maturation protocols have been reported mostly to preserve fertility in patients for whom ovarian stimulation was not feasible. The use of in vitro immatured oocytes coupled with cryopreservation was described for three patients in a clinical report by Demirtas [74]. Their ages were 21, 30 and 40 years seeking to preserve their fertility following cancer diagnosis and without male partners. The oocytes retrieved were successfully matured and then vitrified suggesting that this technique can be used in the event there is not sufficient time for conventional stimulated IVF cycle.

The most comprehensive reports in the literature are by the group at McGill University. In 2009, Chian et al. [75] reported obstetric outcomes following vitrification of in vitro and in vivo matured oocytes. They observed that the oocyte survival and fertilization rates along with the cumulative embryo score were superior in the stimulated cycles compared to IVM cycles. Nevertheless, the differences in clinical pregnancy/started cycle, implantation and live birth rates were not statistically significant in the two groups. These data represent a step forward in the use of IVM as a novel approach for cryopreservation of oocytes.

Laboratory Requirements and Safety

The availability and routine use of liquid nitrogen for cryopreservation has advanced the field across a broad range of applications. However, care must be taken with regards to safety and potential cross-contamination. The use of liquid nitrogen presents obvious dangers regarding possible explosions of containers or vials due to rapid expansion upon warming, as well as skin burns that can result from touching extremely cooled surfaces when handling specimens. To prevent injuries, it is advisable to wear eye protection and safety gloves. It is also important to work in well-ventilated areas to prevent the potential of localized oxygen deprivation in the event of a spill and vaporization when using large volumes of liquid nitrogen. For sample storage, it is important to consider the device, and preferential use of closed containers that retain a barrier from external liquid nitrogen in cryotanks, and/or storage in vapour phase to prevent potential cross-contamination of the specimens. Finally, in order to assure an accurate identification of patient specimens it is fundamental to have a detailed and permanent labelling using a standardize cryo-resistent method.

See Appendix B

References

1. Trounson A, Mohr L. Human pregnancy following cryopreservation, thawing and transfer of an eight-cell embryo. Nature. 1983;305(5936):707–9.
2. Chen C. Pregnancy after human oocyte cryopreservation. Lancet. 1986;1:884–6.
3. Al-Hasani DK, van der Ven H, Reinecke A, et al. Cryopreservation of human oocytes Hum. Reproduction. 1987;2:695–700.
4. Porcu E, Fabbri R, Seracchioli R, et al. Birth of a healthy female after intracytoplasmic sperm injection of cryopreserved human oocytes. Fertil Steril. 1997;68(4):724–6.
5. Liu H, Lai Y, Davis O, et al. Improved pregnancy outcome with gonadotropin releasing hormone agonist stimulation is due to the improvement in oocyte quantity rather than quality. J Assist Reprod Genet. 1992;9:338–42.
6. Hugues JN, Barlow DH, Rosenwaks Z, et al. Improvement in consistency of response to ovarian stimulation with recombinant human follicle stimulating hormone resulting from a new method for calibrating the therapeutic preparation. Reprod BioMed Online. 2003;6:185–90.
7. Out HJ, David I, Ron-El R, et al. A randomized, double-blind clinical trial using fixed daily doses of 100 or 200 IU of recombinant FSH in ICSI cycles. Hum Reprod. 2001;16:1104–9.
8. Yong PY, Brett S, Baird DT, et al. A prospective randomized clinical trial comparing 150 IU and 225 IU of recombinant follicle-stimulating hormone (Gonal-F*) in a fixed-dose regimen for controlled ovarian stimulation in in vitro fertilization treatment. Fertil Steril. 2003;79:308–15.
9. Tesarik J, Mendoza C. Effects of exogenous LH administration during ovarian stimulation of pituitary down-regulated young oocyte donors on oocyte yield and developmental competence. Hum Reprod. 2002; 17:3129–37.
10. Dal Prato L, Borini A, Coticchio G, et al. Half-dose depot triptorelin in pituitary suppression for multiple ovarian stimulation in assisted reproduction technology: a randomized study. Hum Reprod. 2004;19: 2200–5.
11. Dal Prato L, Borini A, Travisi MR, et al. Effect of reduced dose of triptorelin at the start of ovarian stimulation on the outcome of IVF: a randomized study. Hum Reprod. 2001;16(7):1409–14.
12. Al-Inany HG, Abou-Setta AM, Aboulghar M. Gonadotropin-releasing hormone antagonists for assisted conception: a Cochrane review. Reprod Biomed Online. 2007;14(5):640–9.
13. Kolibianakis EM, Collins J, Tarlatzis BC, et al. Among patients treated for IVF with gonadotropins and GnRH analogues, is the probability of live birth dependent on the type of analogue used? A systematic review and meta-analysis. Hum Reprod Update. 2006;12(6): 651–71.
14. Al-Inany H, Aboulghar MA, Mansour RT, et al. Optimizing GnRH antagonist administration: meta-analysis of fixed versus flexible protocol. Reprod Biomed Online. 2005;10(5):567–70.
15. Patrizio P, Bianchi V, Lalioti M, et al. High rate of biological loss in assisted reproduction: it is in the seed, not in the soil. Reprod Biomd Online. 2007;14(1): 92–5.
16. Bedford JM, Kim HH. Sperm/oocyte binding patterns and oocyte cytology in retrospective analysis of fertilization failure in vitro. Hum Reprod. 1993;8:453–63.
17. Balaban B, Urman B, Sertac A, et al. Oocyte morphology does not affect fertilization rate, embryo

quality and implantation rate after intracytoplasmic sperm injection. Hum Reprod. 1998;13:3431–3.
18. Serhal PF, Ranieri DM, Kinis M, et al. Oocyte morphology predicts outcome of intracytoplasmic sperm injection. Hum Reprod. 1997;12:1267–70.
19. Veeck LL. Oocyte assessment and biological performance. Ann NY Acad Sci. 1988;541:259–74.
20. Xia P. Intrcytoplasmic sperm injection: correlation of oocyte grade based on polar body, perivitelline space and cytoplasmic inclusions with fertilization rate and embryo quality. Human Reprod. 1997;12:1750–5.
21. Ebner T, Yaman C, Moser M, et al. Prognostic value of first polar body morphology on fertilization rate and embryo quality in intracytoplasmic sperm injection. Hum Reprod. 2000;15:427–30.
22. Ebner T, Moser M, Yaman C, et al. Prospective hatching of embryos developed from oocytes exhibiting difficult oolemma penetration during ICSI. Hum Reprod. 2002;17:1317–20.
23. Patrizio P, Fragouli E, Bianchi V, et al. Molecular methods for selection of the ideal oocyte. Reprod Biomed Online. 2007;15(3):346–53.
24. Kola I, Kirby C, Show J, et al. Vitrification of mouse oocytes results in aneuploid zygotes and malformed fetuses. Teratology. 1988;38:467–74.
25. Biery KA, Seidel Jr GE, Elsden RP. Cryopreservation of mouse embryos by direct plunging into liquid nitrogen. Theriogenology. 1986;25:140.
26. Shaw JM, Diotallevi L, Trounson AO. A simple rapid dimethyl suiphoxide freezing technique for the cryopreservation of one-cell to blastocyst stage preimplantation mouse embryos. Reprod Fertil Dev. 1991;3:621.
27. Rall WF, Fahy GM. Ice-free cryopreservation of mouse embryos at - 196°C vitrification. Nature. 1985; 313:573.
28. Gardner DK, Sheehan CB, Rienzi L, et al. Analysis of oocyte physiology to improve cryopreservation procedures. Theriogenology. 2007;67:64–72.
29. Yavin S, Arav A. Measurement of essential physical properties of vitrification solutions. Theriogenology. 2007;67(1):81–9.
30. Kono T, Kwon OY, Nakahara T. Development of vitrified mouse oocytes after in vitro fertilization. Cryobiology. 1991;28(1):50–4.
31. Liebermann J, Nawroth F, Isachenko V, et al. Potential importance of vitrification in reproductive medicine. Biol Reprod. 2002;67:1671–80.
32. Kuwayama M, Vajta G, Kato O, et al. Highly efficient vitrification method for cryopreservation of human oocytes. Reprod Biomed Online. 2005;11:300–8.
33. Vajta G, Nagy ZP. Are programmable freezers still needed in the embryo laboratory? Review on vitrification. Reprod Biomed Online. 2006;12(6):779–96.
34. Gualtieri R, Iaccarino M, Mollo V, et al. Slow cooling of human oocytes: ultrastructural injuries and apoptotic status. Fertil Steril. 2009;91:1023–34.
35. Gook DA, Osborn SM, Johnston WI. Cryopreservation of mouse and human oocytes using 1,2-propanediol and the configuration of the meiotic spindle. Hum Reprod. 1993;8:1101–9.
36. Nottola SA, Coticchio G, De Santis L, et al. Ultrastructural of human mature oocytes after slow cooling cryopreservation using different sucrose concentrations. Hum Reprod. 2007;22:1123–33.
37. Pickering SJ, Johnson MH. The influence of cooling on the organization of the meiotic spindle of the mouse oocyte. Hum Reprod. 1987;2:207–16.
38. Almeida PA, Bolton VN. The effect of temperature fluctuation on the cytoskeletal organisation and chromosomal constitution of the human oocyte. Zygote. 1995;3:357–65.
39. Cobo A, Rubio C, Gerli S, et al. Use of fluorescence in situ hybridization to assess the chromosomal status of embryos obtained from cryopreserved oocytes. Fertil Steril. 2001;75:354–60.
40. Sato H, Ellis GW, Inouè S. Microtubular origin of mitotic spindle from birefringence: demonstration of the applicability of Wiener's equation. J Cell Biol. 1975;67:501–17.
41. Rienzi L, Ubaldi F, Martinez F, et al. Relationship between meiotic spindle location with regard to polar body position and oocyte developmental potential after ICSI. Hum Reprod. 2003;18:1289–93.
42. De Santis L, Cino I, Rabelotti E, et al. Polar body morphology and spindle imaging as predictors of oocyte quality. Reprod Biomed Online. 2005;11:36–42.
43. Madaschi C, Carvalho de Souza Bonetti T, Paes de Almeida Ferreira Braga D, et al. Spindle imaging: a marker for embryo development and implantation. Fertil Steril. 2008;90:194–8.
44. Wang WH, Meng L, Hackett RJ, et al. Developmental ability of human oocytes with or without birefringent spindles imaged by PoloScope before insemination. Hum Reprod. 2001;16:1464–8.
45. Wang WH, Meng L, Hackett RJ, et al. The spindle observation and its relationship with fertilization after intracytoplasmic sperm injection in living human oocytes. Fertil Steril. 2001;75:348–53.
46. Rienzi L, Martinez F, Ubaldi F, et al. Poloscope analysis of meiotic spindle changes in living metaphase II human oocytes during the freezing and thawing procedures. Hum Reprod. 2004;19:655–9.
47. Bianchi V, Coticchio G, Fava L, et al. Meiotic spindle imaging in human oocytes frozen with a slow freezing procedure involving high sucrose concentration. Hum Reprod. 2005;20(4):1078–83.
48. Bromfield JJ, Coticchio G, Hutt K, et al. Meiotic spindle dynamics in human oocytes following slow-cooling cryopreservation. Hum Reprod. 2009;24(9): 2114–23.
49. Coticchio G, De Santis L, Rossi G, et al. Sucrose concentration influences the rate of human oocytes with normal spindle and chromosome configuration after slow cooling cryopreservation. Hum Reprod. 2006; 21(7):1771–6.
50. Ciotti P, Porcu E, Notarangelo L, et al. Meiotic spindle recovery is faster in vitrification of human oocytes

compared to slow freezing. Fertil Steril. 2009;91(6):2399–407.

51. Parmigiani L, Cognigni GE, Bernardi S, et al. Freezing within 2 h from oocytes retrieval increases the efficiency of human oocyte cryopreservation when using slow freezing/rapid protocol with high sucrose concentration. Hum Reprod. 2008;23(8):1771–7.
52. Noyes N, Knopman J, Labella P, et al. Oocyte cryopreservation outcomes including pre-cryopreservation and post-thaw meiotic spindle evaluation following slow cooling and vitrification of human oocytes. Fertil Steril. 2010;94:2076–82.
53. Lassalle B, Testart J, Renard JP. Human embryo features that influence the success of cryopreservation with the use of 1,2 propanediol. Fertil Steril. 1985;44:645–51.
54. Borini A, Bonu MA, Coticchio G, et al. Pregnancies and births after oocyte cryopreservation. Fertil Steril. 2004;82:601–5.
55. Fabbri R, Porcu E, Marsella T, et al. Human oocyte cryopreservation: new perspectives regarding oocyte survival. Hum Reprod. 2001;16:411–6.
56. Borini A, Sciajno R, Bianchi V, et al. Clinical outcome of oocyte cryopreservation after slow cooling with a protocol utilizing a high sucrose concentration. Hum Reprod. 2006;21:512–7.
57. Levi Setti PE, Albani E, Novara PV, et al. Cryopreservation of supernumerary oocytes in IVF/ICSI cycles. Hum Reprod. 2006;21:370–5.
58. Bianchi V, Coticchio G, Distratis V, et al. Differential sucrose concentration during dehydration (0.2 mol/L) and rehydration (0.3 mol/L) increases the implantation rate of frozen human oocytes. Reprod Biomed Online. 2007;14:64–71.
59. Borini A, Levi Setti PE, Anserini P, et al. Multicenter observational study on slow-cooling oocyte cryopreservation: clinical outcome. Fertil Steril. 2010;94:1662–8.
60. Noyes N, Porcu E, Borini A. Over 900 oocyte cryopreservation babies born with no appearent increase in congenital anomalies. Reprod Biomed Online. 2009;18(6):769–76.
61. Kuleshova L, Gianaroli L, Magli C, et al. Birth following vitrification of a small number of human oocytes. Hum Reprod. 1999;14(12):3077–9.
62. Yoon T, Kim T, Park S, et al. Live births after vitrification of oocytes in a stimulated in vitro fertilization-embryo transfer program. Fertil Steril. 1993;79:1323–6.
63. Katayama KP, Stehlik J, Kuwayama M, et al. High survival rate of vitrified human oocytes results in clinical pregnancy. Fertil Steril. 2003;80:223–4.
64. Oktay K, Cil PA, Bang H. Efficiency of oocyte cryopreservation: a meta-analysis. Fertil Steril. 2006;86:70–80.
65. Antinori S, Licata E, Dani G, et al. Cryotop vitrification of human oocytes results in high survival rate and healthy deliveries. Reprod Biomed Online. 2007;14:72–9.
66. Cobo A, Kuwayama M, Perez S, et al. Comparison of concomitant outcome achieved with fresh and cryopreserved donor oocytes vitrified by the Cryotop method. Fertil Steril. 2008;89:1657–64.
67. Chian R-C, Huang J, Tan S, et al. Obstetric and perinatal outcome in 200 infants conceived from vitrified oocytes. Reprod BioMed Online. 2008;16:608–10.
68. Smith G, Serafini C, Fioravanti J, et al. Prospective randomized comparison of human oocyte cryopreservation with slow-rate freezing or vitrification. Fertil Steril. 2010;94:2088–95.
69. Grifo J, Noyes N. Delivery rate using cryopreserved oocytes is comparable to conventional in vitro fertilization using fresh oocytes: potential fertility preservation for female cancer patients. Fertil Steril. 2010;93(2):391–6.
70. Ubaldi F, Anniballo R, Romano S, et al. Cumulative ongoing pregnancy rate achieved with oocyte vitrification and cleavage stage transfer without embryo selection in a standard infertility program. Hum Reprod. 2010;25(5):1199–205.
71. Borini A, Bonu MA. Success rates from oocyte cryopreservation. In: Borini A, Coticchio G, editors. Preservation of human oocytes. London: Informa Healthcare; 2009. p. 235–45.
72. Toth TL, Baka SG, Veeck LL, et al. Fertilization and in vitro development of cryopreserved human prophase I oocytes. Fertil Steril. 1994;61:891–4.
73. Toth TL, Hassen WA, Lanzendorf SE, et al. Cryopreservation of human prophase I oocytes collected from unstimulated follicles. Fertil Steril. 1994;61:1077–82.
74. Demirtas E. Immature oocyte retrieval in the luteal phase to preserve fertility in cancer patients. Reprod Biomed Online. 2008;17(4):520–3.
75. Chian C, Huang J, Gilbert L, et al. Obstetric outcomes following vitrification of in vitro and in vivo matured oocytes. Fertil Steril. 2009;91(6):2391–8.

Ovarian Tissue Cryopreservation and Autotransplantation

9

Mohamed A. Bedaiwy, Gihan M. Bareh,
Katherine J. Rodewald, and William W. Hurd

Abstract

This chapter focuses on an approach to protecting ovarian tissue: the surgical removal and cryopreservation of ovarian tissue (ovarian cortical fragments or the intact ovary) prior to cancer therapy, followed by later autotransplantation. The obvious advantage of this currently experimental approach is that ovarian tissue is completely protected from the deleterious effects of chemotherapeutic agents or irradiation. Challenges of this approach include those related to the surgical techniques, limiting ovarian damage from both cryopreservation and hypoxia, and adapting in vitro fertilization/embryo transfer (IVF/ET) techniques when required. Despite these challenges, this approach has been demonstrated to be feasible and shows great promise for preservation of ovarian function and fertility in women treated for malignancy.

Keywords

Ovarian tissue cryopreservation • Ovarian tissue autotransplantation • Autotransplantation of ovarian tissue • In vitro fertilization • Fertility in women treated for malignancy

Cancer is the leading cause of death among women of reproductive age. Fortunately, advances in chemotherapy and radiotherapy have increased the long-term survival of these patients. According to the National Cancer Institute, currently there are >10 million cancer survivors in the United States [1]. It is has been estimated that childhood cancer survivors will soon comprise 1 in 250 young adults in the industrialized world [2].

Unfortunately, many cancer survivors will confront gonadal failure and infertility after successful treatment of their malignancies, since gonadal damage is a common consequence of chemotherapy and radiotherapy [3]. Ovarian exposure to both chemotherapy and radiotherapy results in depletion or complete destruction of the primordial follicle reserve [4]. A common result is decreased ovarian reserve and premature ovarian failure (POF). Since normal ovarian function and fertility are critical quality-of-life components for

W.W. Hurd (✉)
Department of Obstetrics and Gynecology,
University Hospitals Case Medical Center, Case Western Reserve University, 11100 Euclid Avenue, Cleveland, OH 44106, USA

E. Seli and A. Agarwal (eds.), *Fertility Preservation: Emerging Technologies and Clinical Applications*,
DOI 10.1007/978-1-4419-1783-6_9, © Springer Science+Business Media, LLC 2012

most women, developing strategies to preserve ovarian function has become a priority [5].

During the last two decades, a number of approaches have been developed in an attempt to minimize or prevent the toxic effects of chemotherapy and radiotherapy on ovarian tissue in an effort to preserve subsequent ovarian function and/or fertility [6–10]. The most successful method for preserving fertility in adult women appears to be the removal of oocytes prior to chemotherapy and/or radiotherapy. Removal of oocytes followed by in vitro fertilization (IVF) and cryopreservation of embryos for later embryo transfer (ET) has become an accepted strategy with a reasonable pregnancy rate [11, 12]. Cryopreservation of oocytes prior to fertilization has been more difficult, but recent technical advances have resulted in successful pregnancies using this approach as well. However, neither approach preserves ovarian hormonal function.

A second approach is to protect ovarian tissue from the detrimental effects of chemotherapy and radiotherapy. Surgical ovarian transposition away from the planned field of radiation is often effective in preserving subsequent hormonal function in women who do not undergo radiotherapy, but is less effective in preserving fertility [4]. Administration of gonadotropin-releasing hormone (GnRH) agonists during chemotherapy has been found in some studies to decrease ovarian damage and preserve ovarian function, but it remains uncertain how much this approach improves subsequent fertility [9].

This chapter focuses on a third approach to protecting ovarian tissue: the surgical removal and cryopreservation of ovarian tissue (ovarian cortical fragments or the intact ovary) prior to cancer therapy, followed by later autotransplantation. The obvious advantage of this currently experimental approach is that ovarian tissue is completely protected from the deleterious effects of chemotherapeutic agents or irradiation. Challenges of this approach include those related to the surgical techniques, limiting ovarian damage from both cryopreservation and hypoxia, and adapting IVF/ET techniques when required. Despite these challenges, this approach has been demonstrated to be feasible and shows great promise for preservation of ovarian function and fertility in women treated for malignancy.

Rationale

Techniques continue to be developed for ovarian tissue cryopreservation and autoimplantation because many females with malignancies are not candidates for standard IVF and embryo cryopreservation for later attempts at pregnancy (Table 9.1). Children are not candidates for this approach, and adult women without male partners often do not desire the use of anonymous sperm to create embryos. Women with estrogen-sensitive

Table 9.1 Comparison of the different techniques available for fertility preservation

Aspect	Oocytes cryopreservation	Embryo cryopreservation	Ovarian tissue cryopreservation
Can be offered to prepubertal children	No	No	Yes
Requires delay in chemotherapy	Yes	Yes	No
Might stimulate estrogen-sensitive malignancies	Yes	Yes	No
Requires partner or sperm donation	No	Yes	No
Requires surgery	No	No	Yes
Protects endocrine function	No	No	Yes
Risk of reseeding cancer	No	No	Yes
Live births reported	Few	Many	Few

breast cancer are another group who are not candidates for standard IVF and embryo cryopreservation because there is reason to believe that ovarian stimulation prior to cancer treatment might worsen their prognosis [13]. Finally, many women with malignancies do not want to delay their cancer therapy for the 2–4 weeks minimally required to perform an IVF cycle.

Definitions

Transplantation of human tissue and organs has developed over decades in multiple areas of medicine and thus has attained its own nomenclature. By convention, the following terms are used to describe the specific method for ovarian tissue transplantation in women with malignancies.

Autotransplantation vs. Allotransplantation

The most common methods used to treat women with malignancies is surgical removal of ovarian tissue and transplantation of this tissue after cancer treatment back into the same individual, which is referred to as autotransplantation. This is in contrast to allotransplantation, where ovarian tissue removed from one individual and transplanted into a different individual. Since allotransplantation is prone to immunity-related tissue rejection issues, its use to date in the present context has been limited to transplantation of ovarian cortex fragments from one identical twin to another, as described below [14, 15].

Orthotopic vs. Heterotopic

Once frozen and thawed, the ovarian tissue can be transplanted into a number of locations. Orthotopic transplantation refers to transplantation of ovarian tissue back to ovary or the pelvic ovarian fossa [16]. This is in contrast to heterotopic transplantation, where ovarian tissue is transplanted into an alternative area in the body, usually subcutaneous or subperitoneal [17, 18].

History

Heterotopic autotransplantation of autologous organs and tissues from one location to another in the same individual is a well-established technique in several fields. During total thyroidectomy, it has become common to preserve parathyroid gland function by surgical removal and selective autotransplantation of fresh or frozen-thawed parathyroid tissue [19]. Autotransplantation of bone and skin has become standard for many orthopedic and plastic procedures [20]. Similar techniques were first described in humans over 30 years ago [21, 22].

Heterotopic autotransplantation of frozen-thawed ovarian tissue resulting in a subsequent pregnancy was first described in the sheep model almost 2 decades ago by Gosden et al. [23]. Improvements in cryopreservation techniques have resulted in numerous additional successful pregnancies in a number of animal models [23, 24].

In 1996, Newton et al. reported cryopreservation of human ovarian tissue with the intent of later autotransplantation in women treated for malignancies [25]. The first successful pregnancy after orthotopic autotransplantation of frozen-thawed ovarian tissue in a woman after treatment of a malignancy was reported in 2004 by Donnez et al. [16]. To date, a variety of techniques have led to 14 human pregnancies, of which 10 have resulted in the birth of a child, and thus this approach is still considered experimental (Table 9.2) [26].

Ovarian Cortex Fragments vs. Intact Ovaries

Ovarian Cortex Fragments

Autotransplantation of frozen-thawed ovarian cortex fragments has become the preferred approach in this setting. Methods have been developed for the successful cryopreservation of ovarian cortex fragments, as discussed below. Since no blood supply is conserved with these fragments, expertise in vascular anastomosis is not required.

Table 9.2 Human pregnancies after autotransplantation of ovarian cortex fragments

References	Year of report	Patient age at time of ovarian cryopreservation	Ovarian transplantation method	Assisted reproduction?	Outcome
Donnez et al. [16]	2004	25	Orthotopic	Spontaneous	Live birth
Meirow et al. [41]	2005	28	Orthotopic	IVF	Live birth
Demeestere et al. [81]	2006	24	Orthotopic/heterotopic	Spontaneous	Spontaneous abotion
Demeestere et al. [69, 81]	2007	29	Orthotopic	Spontaneous	Live birth
Oktay [18]	2006	28	Heterotopic	Spontaneous	Live birth
Rosendahl et al. [70]	2006	28	Heterotopic	IVF/ICSI	Biochemical
Demeestere et al. [69]	2007	31	Orthotopic and heterotopic	Spontaneous	Spontaneous abortion; live birth
Andersen et al. [67]	2008	25	Orthotopic and heterotopic	IVF	Spontaneous abortion
	2008	26	Orthotopic and heterotopic	IVF	Live birth
		27	Orthotopic	IVF	Live birth
Silber et al. [29]	2008	14	Orthotopic	Spontaneous	Live birth
Roux et al. [31]	2010	20	Orthotopic	Spontaneous	Live birth
Ernst et al. [26]	2010	27	Orthotopic	Spontaneous	Live birth

However, transplantation of ovarian cortex fragments has some disadvantages. These fragments experience ischemic damage both prior to cryopreservation and after transplantation during the time needed to establish a vascular supply [27, 28]. This results in significant follicular loss and subsequent compromise in ovarian function. The functional duration of ovarian cortex fragment transplants has thus far been limited to a matter of months to years [29].

Intact Ovaries

Autotransplantation of intact ovaries has not been reported for women with malignancies for two reasons. First, cryopreservation is required prior to thawing and autotransplantation, and difficulties encountered when attempting to freeze intact ovaries without significant damage have yet to be resolved (see below). Second, reproductive surgeons rarely have expertise in the microvascular anastomosis techniques required for this approach.

Surgical Removal of Ovarian Tissue

The goal is to obtain a reasonable amount of ovarian cortex fragments prior to chemotherapy and/or radiation therapy. The superficial cortex layer of the ovary is rich in oocytes. Therefore, cryopreservation methods are more successful with these small pieces. Although retrieval of ovarian tissue was originally performed via laparotomy, most recent reports have utilized a less invasive laparoscopic approach [17, 30].

Obtaining ovarian cortex fragments can be done by one of three surgical methods: ovarian biopsy, ovarian wedge resection, or oophorectomy. Performing ovarian biopsies is the least invasive method for obtaining ovarian cortex fragments and has minimal effect on ovarian function [31]. However, it is difficult to obtain enough ovarian cortex, and thus this approach is not commonly used.

Others have reported performing an ovarian wedge resection [32, 33]. This approach obtains a greater amount of tissue while still retaining ovarian function, even in women with only a single ovary. This is somewhat more technically challenging and the amount of ovarian cortex obtained is limited by the size of the resection.

The most commonly used method to obtain ovarian cortex for subsequent autotransplantation is the removal of an entire ovary followed by dissection of ovarian cortex fragments in the laboratory [30, 34]. This approach obtains the maximal amount of ovarian cortex and does not appreciably alter ovarian function in women with two functioning ovaries.

Techniques for Oophorectomy

A standard laparoscopic approach for oophorectomy has been adapted so as to minimize the time of ovarian ischemia, referred to as the ischemic interval [30]. In brief, the ovary is dissected from the utero-ovarian ligament, mesovarium, and finally the infundibulopelvic ligament using electrosurgery or other standard power source for hemostasis. The ischemic interval can be minimized by severing the infundibulopelvic ligament as the final step in the procedure [35]. Once freed, the ovary is removed from the abdomen with the aid of an endoscopic specimen bag and immediately processed for cryopreservation.

Another method used to reduce the ischemic interval is perfusion of the ovary with heparinized cryoprotectant solution prior to removal [30]. For this method, the utero-ovarian ligaments are occluded at the beginning of the procedure with clips and a heparinized cryoprotectant solution is injected into the ovarian artery.

In order to minimize the ischemic interval further, the time between ovarian removal and cryopreservation should be minimized [30, 36]. When the ovarian specimen is removed from the abdomen, it is placed in cold normal saline and handed off the operating field for immediate processing. The most superficial layer of ovarian cortex is dissected into 5×5 mm fragments that are 1–2 mm thick, and placed in individual tubes of cryoprotectant as described below.

Cryopreservation and Thawing

Slow Freezing vs. Vitrification

Cryopreservation of ovarian cortex fragments is performed using one of two methods: slow freezing or vitrification. With both methods, the challenges are to avoid (1) the formation of potentially lethal intracellular ice during cooling, (2) osmotic injury as water moves across the cell membrane during freezing and thawing, and (3) the toxic effects of cryoprotectants.

It remains controversial as to which method is most effective [28, 32, 37]. One comparative study found that both techniques are effective in preservation of follicles, but vitrification resulted in significantly better morphological integrity of ovarian stroma as compared with slow freezing [38].

Slow Freezing

Ovarian cortex fragments are placed for 30 min in individual tubes containing a combination of permeating and nonpermeating cryoprotectants [16, 39–41]. Permeating cryoprotectants, which enter the cell cytoplasm, include dimethylsulphoxide, 1–2 propanediol, dimethylsulphoxide, and/or ethylene glycol. Commonly used nonpermeating cryoprotectants include sucrose and human serum albumin.

Ovarian cortex fragments are then placed into 0.25 ml straws and slowly cooled in a stepwise manner with the aid of a computer-controlled freezing machine. Temperature is reduced at rates ranging from 2 to 0.3°C/min until the tissue reaches a temperature of −30°C. The straws are then immersed in liquid nitrogen (−70°C) without the risk of internal crystallization.

Vitrification

Vitrification is a promising new cryopreservation method that involves exposure of ovarian cortex fragments to higher concentrations of permeating cryoprotectants for shorter period of time (1–5 min) followed by plunging the tissue directly into −70°C liquid nitrogen [37]. This induces a glass-like (vitrified) intracellular state and avoids the formation of destructive ice crystals.

Compared to slow freezing, vitrification is less complicated, less time-consuming and less expensive [42, 43]. A disadvantage of vitrification is that some techniques involve direct contact of the tissue with liquid nitrogen, which is a potential source of microbial contamination [44, 45].

Intact Ovary Cryopreservation

Methods for cryopreservation of intact ovaries has been developed in animal models [46–49]. The first case of restoration of fertility after whole frozen-thawed ovary transplantation was described in the rat by Wang et al. in 2002 [50]. However, most species studied have ovaries much smaller than humans, measuring in mm rather than cm, making cryopreservation much easier.

Cryopreservation of large intact ovaries has several technical challenges. Chief among these are difficulties in assuring adequate diffusion of cryoprotectants, uneven cooling deep within a large specimen, and vascular injury caused by intravascular ice formation [51]. For intact ovaries, the most effective methods for exposing tissue to cryoprotectants is perfusion of the ovary via the ovarian artery [46, 52, 53].

The cryopreservaton methods for large intact ovaries, such as those of sheep and humans, must be modified to prevent tissue damage. Conventional slow freezing devices operate on the principle of multidirectional heat transfer by convection, where tissue temperature changes are dependent upon the thermal conductivity and the geometric shape of the container. For large intact ovaries, this method does not uniformly control the temperature between the periphery of the container and the ovarian core [54, 55]. A multithermal gradient freezer has been developed in an effort to achieve accurate, uniform cooling rate through the entire tissue [46, 56–58].

Thawing

Thawing both ovarian cortex fragments and intact ovaries is relatively simple. Ovarian cortex fragments are thawed rapidly by bathing them in a decreased sucrose gradient at 37°C [39]. Intact ovaries also require concomitant vascular perfusion with 10 ml HEPES-Talp medium supplemented with a single step 0.5 mol/l sucrose and 10 IU/ml heparin while thawing [59]. For both cortex fragments and intact ovaries, these thawing approaches result in a remarkable 75% follicular survival rate with minimal post-thawing apoptosis [60, 61].

Ovarian Cortex Fragment Autotransplantation

At least 1 year after completion of cancer therapy, thawed ovarian cortex fragments are transplanted back into the women from whom they were removed. Reported successful techniques included orthotopic transplantation into the ovarian fossa or onto an intact ovary, heterotopic transplantation into the subcutaneous location of the forearm and abdominal wall, or a combination of both these techniques [62–66].

Orthotopic Autotransplantation

The first pregnancy in this field was reported by Donnez in 2004 after orthotopic autotransplantation of ovarian cortex fragments into the ovarian fossa using a two-step transplantation procedure [16]. During the initial laparoscopy, a window was created in the peritoneum of the ovarian fossa to stimulate angiogenesis and neovascularization. One week later, a second laparoscopy was performed to reimplant frozen–thawed ovarian cortex fragments in this area. This patient conceived spontaneously approximately 11 months after the transplant.

Using a similar technique, Andersen et al. reported two pregnancies among 6 women after frozen-thawed ovarian cortex fragments were transplanted into peritoneal “pockets” created near the ovary [67]. Two of these women also underwent heterotopic transplantation of cortex fragments to two subcutaneous pockets. These pregnancies required IVF/ET.

Others have reported pregnancies with the aid of IVF/ET after orthotopic transplantation where strips of frozen-thawed ovarian cortex were sutured onto the ovary [29, 31, 68]. Others have reported using a combination orthotopic and heterotopic autotransplantation, where ovarian tissue was placed in both subcutaneous and subperitoneal sites [69].

Allotransplantation of fresh ovarian cortex fragments between 8 sets of identical twin after one of the twins had POF using similar methods [14, 15]. In each case, fresh ovarian cortex fragments were dissected from an ovary removed from the twin with functioning ovaries. The fragments were immediately sewn to the surface of the nonfunctioning ovaries of the other twin via minilaparotomy. To date ovarian function has been restored in eight twins with POF, and six spontaneous pregnancies have occurred. Although this approach did not require cryopreservation of the ovarian cortex fragments, it does demonstrate the potential efficacy of the transplantation technique.

Heterotopic Autotransplantation

The first live birth resulting from heterotopic autotransplantation of frozen-thawed ovarian tissue was reported by Oktay in 2006 [18]. This group transplanted ovarian cortex fragments subcutaneously in the suprapubic region. After spontaneous ovulation, an oocyte was retrieved for IVF/ET. Another grouped established a biochemical pregnancy by using IVF/ET after transplanting ovarian cortex fragments in a midline subperitoneal pocket on the lower abdominal wall [70].

Intact Ovary Autotransplantation

Autotransplantation of a frozen-thawed intact ovary has not yet been reported in humans, primarily because of the challenges related to cryopreservation described above. However, in the sheep model, autotransplantation of a fresh intact ovary along with its vascular pedicle has been shown to improve the quality of oocytes and the longevity of the transplant apparently by minimizing hypoxic stress [71].

Orthotopic vs. Heterotopic

When cryopreservation difficulties are solved for intact ovaries, orthotopic transplantation of the intact ovary to the ovarian fossa might ultimately prove to be the ideal approach. However, this approach has not yet been reported in cancer survivors, primarily because laparotomy is required for microvascular anatomoses.

For this reason, heterotopic transplantation of an intact ovary to other locations has been considered based on vascular anatomy, the relative ease of subcutaneous transplantation, cosmetic factors, and the ease of accessibility of the transplant for oocytes retrieval for IVF/ET. Proposed locations have included the neck (carotid vessels), pectoral region (internal mammary vessels), antecubital fossa (bracheal vessels), lower part of the anterior abdominal wall (inferior epigastric vessels, Fig. 9.1), and the inguinal region (femoral vessels) [35]. Based on these considerations, the best location appears to be the lower part of the anterior abdominal wall with vascular anastomoses to the inferior epigastric vessels.

Vascular Anastomosis Techniques

The surgical challenge of intact ovarian transplantation is microsurgical anastomosis of small blood vessels, particularly when the vessels to be anastomosed are of different diameters. A dramatic difference between vessel diameters at the site of anastomosis results in turbulent blood flow which can lead to platelet aggregation and eventually ovarian vessel thrombosis [72].

In animal models, this problem has been addressed by using a section of aorta, although this approach is not applicable to humans [71, 73, 74]. In humans, several surgical techniques have been developed, depending on the degree of

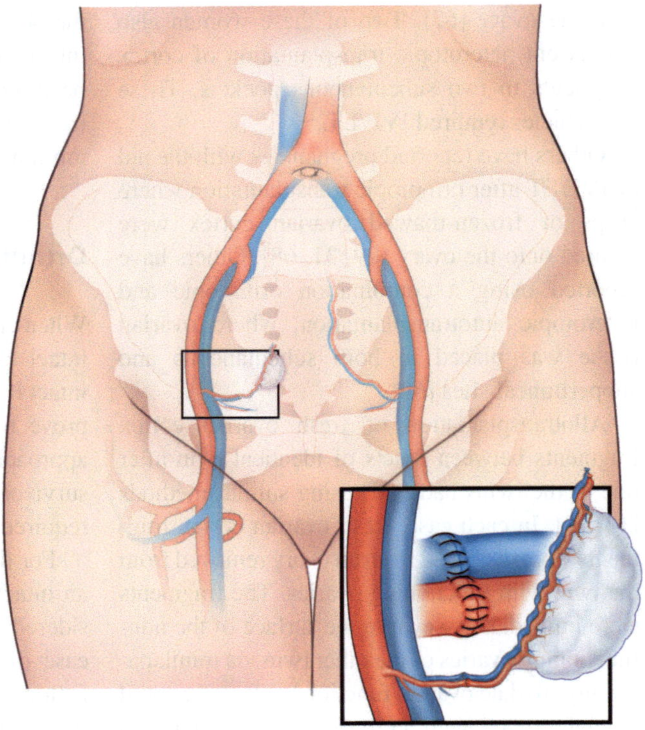

Fig. 9.1 Transplantation of the intact ovary to the anterior abdominal wall with vascular anastomoses to the inferior epigastric vessels

discrepancy of the vessel diameters [35]. When the diameter ratios of the vessels to be anastomosed is ≤1:1.5, the smaller vessel can be dilated with a jeweler's forceps prior to anastomosis [72]. When vessel diameter ratio is >1:1.5, recommended techniques include oblique cut, fish mouth cut, or end-to-side anastomosis [35].

Microscopic anastomoses using interrupted 9–0 or 10–0 nonabsorbable suture has been reported in the sheep model to have a subsequent patency rate >60% [52]. Future advances in the field of vascular surgery are likely to increase patency rates further, such as vascular clips, the use of glues or adhesives, and laser-assisted anastomosis [30, 75].

Outcome

Animal Models

A great deal of research has been done in animals, particularly in the sheep model since ovarian size is similar to humans. In this model, intact ovary cryopreservation and autotransplantation has been used to restore ovarian function and fertility [52, 53, 76, 77]. However, long-term ovarian function after intact ovary cryopreservation and autotransplantation remains a problem, and the longest documented function thus far has been 3 years [46].

Humans

Restoration of Ovarian Function

At least temporary restoration of ovarian function has been demonstrated in humans after heterotopic and orthotopic autotransplantation of frozen-thawed ovarian cortex fragments. Autotransplantation of frozen-thawed intact ovaries has not been reported in humans.

Ovarian cortex fragments appear to take 6 weeks to 9 months after transplantation to resume hormonal functioning [63, 78, 79]. In one carefully studied patient, early signs of restored ovarian function (decreased FSH levels and increased AMH) were detectable 9 weeks after the transplant, but spontaneous ovulation could not be documented until 4 months [31].

Restoration of Fertility

To date, ten successful human pregnancies have been reported after autotransplantation of frozen-thawed ovarian cortex fragments to both orthotopic (ovarian fossa peritoneum or ovarian surface) and heterotopic locations (subcutaneous or subperitoneal locations in the anterior abdominal wall) [80]. In the 28.5% of the orthotopic and 100% of the heterotopic cases, IVF/ET was required.

The first pregnancy resulting in a live birth after autotransplantation of frozen-thawed ovarian cortex fragments was reported by Donnez et al. in 2004 [16]. Prior to chemotherapy and radiotherapy for Hodgkin's lymphoma, one ovary was surgically removed from a 25-years-old patient and ovarian cortex fragments were cryopreserved. Six years later, these fragments were transplanted onto the perioneum of the ovarian fossa. Spontaneous ovulation was documented within 5 months, and spontaneous pregnancy occurred 11 months after transplantation.

Since this initial accomplishment, nine more successful pregnancies and three nonviable pregnancies have been reported (Table 9.2). In five cases, spontaneous pregnancy occurred after orthotopic transplantation [16, 29, 81] . In the remaining 7 cases (2 orthotopic, 1 heterotopic, and 4 both orthotopic and heterotopic combination), IVF/ET was required for conception [67, 68, 81]. The overall miscarriage rate in the pregnancies reported to date is 23%, which appears to be in the range of pregnancies conceived in healthy women.

Risks

Surgical Risks

Although not negligible, the risk of surgery and/or IVF related to autotransplantation of ovarian tissue is small and acceptable to most patients. Once ovarian tissue is removed, heterotopic transplantation to a subcutaneous space minimized the surgical risk, but appears to shorten the functional longevity of the transplants [82].

Ovarian Metastasis

Whenever ovarian tissue is removed from a woman with a malignancy for later autotransplantation, there is a risk of re-introducing tumor cells from the original malignancy back into the patient. This is more than just a theoretical concern. A case has reported of a women treated for Burkett's lymphoma where malignant cells were detected in ovarian fragments, and autotransplantation was averted, as discussed below [83].

In light of this risk, it is recommended that certain precautions be taken [65, 84, 85]. For example, harvested tissue should be assessed histologically for micrometastases before cryopreservation. In some types of cancers, molecular markers can be used to identify a very small amount of metastatic cells in the harvested tissue. With proper counseling, patients may be given the option of harvesting ovarian tissue after the first cycle of chemotherapy or radiation in those with metastases or cancers with high risk of ovarian involvement.

Patients with lymph node metastases are poor candidates for this approach, particularly breast cancer with its propensity for ovarian metastases [86]. In addition, patients with blood-borne malignancies, such as leukemia, Burkitt's lymphoma, and neuroblastoma are not candidates for cryopreservation and autotransplantation. However, some authors believe it is appropriate to use this approach even in high-risk patients after appropriate counseling [87].

For all patients considering ovarian autotransplantation, gross ovarian metastasis should be excluded prior to surgery using transvaginal ultrasonography. Once ovarian tissue is surgically removed, sections should be examined histologically to exclude malignancy, as well as to document sufficient follicle numbers [87].

Experimental methods that might be more sensitive for detecting subclinical tumor cells in ovarian specimens have been reported. Immunohistchemistry and reverse transcription polymerase chain reaction (RT-PCR) were used to detect malignant cells in the ovarian sample from the patient with Burkett's lymphoma mentioned about [83]. Xenotransplantation of resected ovarian tissue into immunodeficient mice is another promising method for detecting the presence of cancer cells in tissue prior to autotransplantation [87, 88]. These experimental methods for detecting subclinical amounts of cancer cells might someday become routine prior to autotransfusion of ovarian tissue in women treated for malignancies.

Pregnancy Complications

Many candidates for the techniques described in this chapter will undergo both chemotherapy and radiotherapy. Pelvic radiotherapy in women increases the risk of subsequent pregnancy complications, presumably by damaging both uterus musculature and blood supply. Radiation-related endometrial insufficiency has been shown to decrease the implantation capacity of embryos [89–91]. Women who achieve pregnancy after pelvic irradiation are at increased risk for spontaneous miscarriage and intrauterine growth retardation, especially if conception occurred less than a year after radiation exposure [92, 93]. For this reason, it is recommended that attempts at pregnancy be delayed for at least 1 year after the completion of pelvic radiotherapy.

See Appendix C

References

1. American Cancer Society. Cancer facts and figures. 2002. Available at: http://www.cancer.org. Last accessed: 13 May 2007.
2. Blatt J. Pregnancy outcome in long-term survivors of childhood cancer. Med Pediatr Oncol. 1999;33: 29–33.
3. Putowski L, Kuczynski W. Strategies for fertility preservation after anti-cancer therapy. Ginekol Pol. 2003;74(8):638–45.
4. Falcone T, Attaran M, Bedaiwy MA, Goldberg JM. Ovarian function preservation in the cancer patient. Fertil Steril. 2004;81(2):243–57.
5. Rodriguez-Wallberg KA, Oktay K. Fertility preservation medicine: options for young adults and children with cancer. J Pediatr Hematol Oncol. 2010; 32(5):390–6.
6. Jeruss JS, Woodruff TK. Preservation of fertility in patients with cancer. N Engl J Med. 2009;360(9): 902–11.
7. Lobo RA. Potential options for preservation of fertility in women. N Engl J Med. 2005;353(1):64–73.
8. Blumenfeld Z, Avivi I, Ritter M, Rowe JM. Preservation of fertility and ovarian function and minimizing chemotherapy-induced gonadotoxicity in young women. J Soc Gynecol Investig. 1999;6(5): 229–39.
9. Glode LM, Robinson J, Gould SF. Protection from cyclophosphamide-induced testicular damage with an analogue of gonadotropin-releasing hormone. Lancet. 1981;1(8230):1132–4.
10. Nicholson HS, Byrne J. Fertility and pregnancy after treatment for cancer during childhood or adolescence. Cancer. 1993;71(10 Suppl):3392–9.
11. Sonmezer M, Oktay K. Assisted reproduction and fertility preservation techniques in cancer patients. Curr Opin Endocrinol Diabetes Obes. 2008;15(6):514–22.
12. The American Society of Reproductive Medicine-The Ethics Committee. Fertility preservation and reproduction in cancer patients. Fertil Steril. 2005;83:1622–8.
13. Lawrenz B, Neunhoeffer E, Henes M, Lessmann-Bechle S, Kramer B, Fehm T. Management of fertility preservation in young breast cancer patients in a large breast cancer centre. Arch Gynecol Obstet. 2010; 282:547–51.
14. Silber SJ, Lenahan KM, Levine DJ, et al. Ovarian transplantation between monozygotic twins discordant for premature ovarian failure. N Engl J Med. 2005;353:58–63.
15. Silber SJ, Gosden RG. Ovarian transplantation in a series of monozygotic twins discordant for ovarian failure. N Engl J Med. 2007;356(13):1382–4.
16. Donnez J, Dolmans MM, Demylle D, et al. Livebirth after orthotopic transplantation of cryopreserved ovarian tissue. Lancet. 2004;364(9443):1405–10.
17. Sonmezer M, Shamonki MI, Oktay K. Ovarian tissue cryopreservation: benefits and risks. Cell Tissue Res. 2005;322(1):125–32.
18. Oktay K. Spontaneous conceptions and live birth after heterotopic ovarian transplantation: is there a germline stem cell connection? Hum Reprod. 2006;21(6): 1345–8.
19. Kuriloff DB, Kizhner V. Parathyroid gland preservation and selective autotransplantation utilizing topical lidocaine in total thyroidectomy. Laryngoscope. 2010;120(7):1342–4.
20. Siemionow MW, Zor F, Gordon CR. Face, upper extremity, and concomitant transplantation: potential concerns and challenges ahead. Plast Reconstr Surg. 2010;126(1):308–15.

21. Wells Jr SA, Gunnells JC, Leslie JB, Schneider AS, Sherwood LM, Gutman RA. Transplantation of the parathyroid glands in man. Transplant Proc. 1977; 9(1):241–3.
22. Wells Jr SA, Stirman Jr JA, Bolman 3rd RM, Gunnells JC. Transplantation of the parathyroid glands. Clinical and experimental results. Surg Clin North Am. 1978;58(2):391–402.
23. Gosden RG, Baird DT, Wade JC, Webb R. Restoration of fertility to oophorectomized sheep by ovarian autografts stored at −196 degrees C. Hum Reprod. 1994;9(4):597–603.
24. Bordes A, Lornage J, Demirci B, et al. Normal gestations and live births after orthotopic autograft of vitrified-warmed hemi-ovaries into ewes. Hum Reprod. 2005;20(10):2745–8.
25. Newton H, Aubard Y, Rutherford A, Sharma V, Gosden R. Low temperature storage and grafting of human ovarian tissue. Hum Reprod. 1996;11:1487–91.
26. Ernst E, Bergholdt S, Jorgensen JS, Andersen CY. The first woman to give birth to two children following transplantation of frozen/thawed ovarian tissue. Hum Reprod. 2010;25(5):1280–1.
27. Baird DT, Campbell B, de Souza C, Telfer E. Long-term ovarian function in sheep after ovariectomy and autotransplantation of cryopreserved cortical strips. Eur J Obstet Gynecol Reprod Biol. 2004;113:S55–9.
28. Rahimi G, Isachenko E, Isachenko V, et al. Comparison of necrosis in human ovarian tissue after conventional slow freezing or vitrification and transplantation in ovariectomized SCID mice. Reprod Biomed Online. 2004;9(2):187–93.
29. Silber SJ, DeRosa M, Pineda J, et al. A series of monozygotic twins discordant for ovarian failure: ovary transplantation (cortical versus microvascular) and cryopreservation. Hum Reprod. 2008;23(7):1531–7.
30. Jadoul P, Donnez J, Dolmans MM, Squifflet J, Lengele B, Martinez-Madrid B. Laparoscopic ovariectomy for whole human ovary cryopreservation: technical aspects. Fertil Steril. 2007;87(4):971–5.
31. Roux C, Amiot C, Agnani G, Aubard Y, Rohrlich PS, Piver P. Live birth after ovarian tissue autograft in a patient with sickle cell disease treated by allogeneic bone marrow transplantation. Fertil Steril. 2010;93(7):2413 e2415–9.
32. Huang L, Mo Y, Wang W, Li Y, Zhang Q, Yang D. Cryopreservation of human ovarian tissue by solid-surface vitrification. Eur J Obstet Gynecol Reprod Biol. 2008;139(2):193–8.
33. Elizur SE, Tulandi T, Meterissian S, Huang JY, Levin D, Tan SL. Fertility preservation for young women with rectal cancer–a combinded approach from one referral center. J Gastrointest Surg. 2009;13(6): 1111–5.
34. Donnez J, Dolmans MM, Demylle D, et al. Restoration of ovarian function after orthotopic (intraovarian and periovarian) transplantation of cryopreserved ovarian tissue in a woman treated by bone marrow transplantation for sickle cell anaemia: case report. Hum Reprod. 2006;21(1):183–8.
35. Bedaiwy MA, Falcone T. Harvesting and autotransplantation of vascularized ovarian grafts: approaches and techniques. Reprod Biomed Online. 2007;14: 360–71.
36. Lee SJ, Schover LR, Partridge AH, et al. American Society of Clinical Oncology recommendations on fertility preservation in cancer patients. J Clin Oncol. 2006;24(18):2917–31.
37. Gandolfi F, Paffoni A, Papasso-Brambilla E, Bonetti S, Brevini TA, Ragni G. Efficiency of equilibrium cooling and vitrification procedures for the cryopreservation of ovarian tissue: comparative analysis between human and animal models. Fertil Steril. 2006;85 Suppl 1:1150–6.
38. Keros V, Xella S, Hultenby K, et al. Vitrification versus controlled-rate freezing in cryopreservation of human ovarian tissue. Hum Reprod. 2009;24(7): 1670–83.
39. Hovatta O, Silye R, Krausz T, et al. Cryopreservation of human ovarian tissue using dimethylsulphoxide and propanediol-sucrose as cryoprotectants. Hum Reprod. 1996;11(6):1268–72.
40. Schmidt KL, Ernst E, Byskov AG, Nyboe Andersen A, Yding Andersen C. Survival of primordial follicles following prolonged transportation of ovarian tissue prior to cryopreservation. Hum Reprod. 2003;18(12): 2654–9.
41. Meirow D, Levron J, Eldar-Geva T, et al. Pregnancy after transplantation of cryopreserved ovarian tissue in a patient with ovarian failure after chemotherapy. N Engl J Med. 2005;353(3):318–21.
42. Wang Y, Xiao Z, Li L, Fan W, Li S. Novel needle immersed vitrification: a practical and convenient method with potential advantages in mouse and human ovarian tissue cryopreservation. Hum Reprod. 2009;7(24):1767–8.
43. Li Y, Zhou C, Yang G, Wang Q, Dong Y. Modified vitrification method for cryopreservation of human ovarian tissues. Chin Med J. 2007;120(2):110–4.
44. Bielanski A, Bergeron H, Lau PC, Devenish J. Microbial contamination of embryos and semen during long term banking in liquid nitrogen. Cryobiology. 2003;46(2):146–52.
45. Tedder RS, Zuckerman MA, Goldstone AH, et al. Hepatitis B transmission from contaminated cryopreservation tank. Lancet. 1995;346(8968):137–40.
46. Arav A, Revel A, Nathan Y, et al. Oocyte recovery, embryo development and ovarian function after cryopreservation and transplantation of whole sheep ovary. Hum Reprod. 2005;20:3554–9.
47. Ishijima T, Kobayashi Y, Lee DS, et al. Cryopreservation of canine ovaries by vitrification. J Reprod Dev. 2006;52(2):293–9.
48. Chen CH, Chen SG, Wu GJ, Wang J, Yu CP, Liu JY. Autologous heterotopic transplantation of intact rabbit ovary after frozen banking at −196 degrees C. Fertil Steril. 2006;86 Suppl 4:1059–66.
49. Yin H, Wang X, Kim SS, Chen H, Tan SL, Gosden RG. Transplantation of intact rat gonads using vascular anastomosis: effects of cryopreservation,

ischaemia and genotype. Hum Reprod. 2003;18(6): 1165–72.
50. Wang X, Chen H, Yin H, Kim SS, Lin Tan S, Gosden RG. Fertility after intact ovary transplantation. Nature. 2002;415:385.
51. Donnez J, Jadoul P, Squifflet J, et al. Ovarian tissue cryopreservation and transplantation in cancer patients. Best Pract Res Clin Obstet Gynaecol. 2010;24(1):87–100.
52. Bedaiwy MA, Jeremias E, Gurunluoglu R, et al. Restoration of ovarian function after autotransplantation of intact frozen-thawed sheep ovaries with microvascular anastomosis. Fertil Steril. 2003;79(3): 594–602.
53. Imhof M, Bergmeister H, Lipovac M, Rudas M, Hofstetter G, Huber J. Orthotopic microvascular reanastomosis of whole cryopreserved ovine ovaries resulting in pregnancy and life birth. Fertil Steril. 2006;85:1208–15.
54. Armitage WJ. Cryopreservation of animal cells. Symp Soc Exp Biol. 1987;41:379–93.
55. Karlsson JO, Toner M. Long-term storage of tissues by cryopreservation: critical issues. Biomaterials. 1996;17(3):243–56.
56. Covens AL, van der Putten HW, Fyles AW, et al. Laparoscopic ovarian transposition. Eur J Gynaecol Oncol. 1996;17(3):177–82.
57. Saragusty J, Gacitua H, Pettit MT, Arav A. Directional freezing of equine semen in large volumes. Reprod Domest Anim. 2007;42(6):610–5.
58. Elami A, Gavish Z, Korach A, et al. Successful restoration of function of frozen and thawed isolated rat hearts. J Thorac Cardiovasc Surg. 2008;135(3): 666–72. 672 e661.
59. Arav A, Gavish Z, Elami A, et al. Ovarian function 6 years after cryopreservation and transplantation of whole sheep ovaries. Reprod BioMed. 2010;20(1): 48–52.
60. Martinez-Madrid B, Donnez J. Technical challenges in freeze–thawing of human ovary: reply of the authors. Fertil Steril. 2005;83(4):1069–70.
61. Martinez-Madrid B, Donnez J. Cryopreservation of intact human ovary with its vascular pedicle–or cryopreservation of hemiovaries? Hum Reprod. 2007; 22(6):1795–6.
62. Donnez J, Squifflet J, Van Eyck AS, et al. Restoration of ovarian function in orthotopically transplanted cryopreserved ovarian tissue: a pilot experience. Reprod Biomed Online. 2008;16(5):694–704.
63. Donnez J, Dolmans MM, Demylle D, Jadoul P, Pirard C, Squifflet J, et al. Restoration of ovarian function after orthotopic (intraovarian and periovarian) transplantation of cryopreserved ovarian tissue in a woman treated by bone marrow transplantation for sickle cell anaemia: case report. Hum Reprod. 2006;21:183–8.
64. Oktay K, Newton H, Aubard Y, Salha O, Gosden RG. Cryopreservation of immature human oocytes and ovarian tissue: an emerging technology? Fertil Steril. 1998;69(1):1–7.
65. Meirow D, Ben Yehuda D, Prus D, et al. Ovarian tissue banking in patients with Hodgkin's disease: is it safe? Fertil Steril. 1998;69(6):996–8.
66. Gosden RG. Prospects for oocyte banking and in vitro maturation. J Natl Cancer Inst Monogr. 2005;34: 60–3.
67. Andersen CY, Rosendahl M, Byskov AG, et al. Two successful pregnancies following autotransplantation of frozen/thawed ovarian tissue. Hum Reprod. 2008; 23(10):2266–72.
68. Meirow D, Levron J, Eldar-Geva T, Ben Yehuda D, Dor J. Ovarian tissue storing for fertility preservation in clinical practice: successful pregnancy, technology, and risk assessment. Fertil Steril. 2005;84:4.
69. Demeestere I, Simon P, Emiliani S, Delbaere A, Englert Y. Fertility preservation: successful transplantation of cryopreserved ovarian tissue in a young patient previously treated for Hodgkin's disease. Oncologist. 2007;12(12):1437–42.
70. Rosendahl M, Loft A, Byskov AG, et al. Biochemical pregnancy after fertilization of an oocyte aspirated from a heterotopic autotransplant of cryopreserved ovarian tissue: case report. Hum Reprod. 2006;21(8): 2006–9.
71. Jeremias E, Bedaiwy MA, Gurunluoglu R, Biscotti CV, Siemionow M, Falcone T. Heterotopic autotransplantation of the ovary with microvascular anastomosis: a novel surgical technique. Fertil Steril. 2002; 77(6):1278–82.
72. Cakir B, Akan M, Akoz T. The management of size discrepancies in microvascular anastomoses. Acta Orthop Traumatol Turc. 2003;37(5):379–85.
73. Goding JR, McCracken JA, Baird DT. The study of ovarian function in the ewe by means of a vascular autotransplantation technique. J Endocrinol. 1967;39: 37–52.
74. Scott JR, Keye WR, Poulson AM, Reynolds WA. Microsurgical ovarian transplantation in the primate. Fertil Steril. 1981;36(4):512–5.
75. Lopez-Monjardin H, de la Pena-Salcedo JA. Techniques for management of size discrepancies in microvascular anastomosis. Microsurgery. 2000; 20(4):162–6.
76. Revel A, Elami A, Bor A, Yavin S, Natan Y, Arav A. Whole sheep ovary cryopreservation and transplantation. Fertil Steril. 2004;82(6):1714–5.
77. Courbiere B, Massardier J, Salle B, Mazoyer C, Guerin JF, Lornage J. Follicular viability and histological assessment after cryopreservation of whole sheep ovaries with vascular pedicle by vitrification. Fertil Steril. 2005;84 Suppl 2:1065–71.
78. Schmidt KL, Andersen CY, Loft A, Byskov AG, Ernst E, Andersen AN. Follow-up of ovarian function postchemotherapy following ovarian cryopreservation and transplantation. Hum Reprod. 2005;20(12):3539–46.
79. Donnez J, Martinez-Madrid B, Jadoul P, Van Langendonckt A, Demylle D, Dolmans MM. Ovarian tissue cryopreservation and transplantation: a review. Hum Reprod Update. 2006;12(5):519–35.

80. Sonmezer M, Oktay K. Orthotopic and heterotopic ovarian tissue transplantation. Clin Obstet Gynaecol. 2010;24(1):113–26.
81. Demeestere I, Simon P, Buxant F, et al. Ovarian function and spontaneous pregnancy after combined heterotopic and orthotopic cryopreserved ovarian tissue transplantation in a patient previously treated with bone marrow transplantation: case report. Hum Reprod. 2006;21(8):2010–4.
82. Callejo J, Salvador C, Miralles A, Vilaseca S, Lailla JM, Balasch J. Long-term ovarian function evaluation after autografting by implantation with fresh and frozen-thawed human ovarian tissue. J Clin Endocrinol Metab. 2001;86:4489–94.
83. Meirow D, Hardan I, Dor J, et al. Searching for evidence of disease and malignant cell contamination in ovarian tissue stored from hematologic cancer patients. Hum Reprod. 2008;23(5):1007–13.
84. Wallace WH, Anderson RA, Irvine DS. Fertility preservation for young patients with cancer: who is at risk and what can be offered? Lancet Oncol. 2005;6(4):209–18.
85. Seshadri T, Gook D, Lade S, et al. Lack of evidence of disease contamination in ovarian tissue harvested for cryopreservation from patients with Hodgkin lymphoma and analysis of factors predictive of oocyte yield. Br J Cancer. 2006;94(7):1007–10.
86. Perrotin F, Marret H, Bouquin R, Fischer-Perrotin N, Lansac J, Body G. Incidence, diagnosis and prognosis of ovarian metastasis in breast cancer. Gynécol Obstét Fertil. 2001;29(4):308–15.
87. von Wolff M, Donnez J, Hovatta O, et al. Cryopreservation and autotransplantation of human ovarian tissue prior to cytotoxic therapy–a technique in its infancy but already successful in fertility preservation. Eur J Cancer. 2009;45(9):1547–53.
88. Kim SS, Radford J, Harris M, et al. Ovarian tissue harvested from lymphoma patients to preserve fertility may be safe for autotransplantation. Hum Reprod. 2001;16(10):2056–60.
89. Critchley HO, Wallace WH, Shalet SM, Mamtora H, Higginson J, Anderson DC. Abdominal irradiation in childhood; the potential for pregnancy. Br J Obstet Gynaecol. 1992;99:392–4.
90. Critchley HO, Wallace WH. Impact of cancer treatment on uterine function. J Natl Cancer Inst Monogr. 2005;34:64–8.
91. Bath LE, Critchley HO, Chambers SE, Anderson RA, Kelnar CJ, Wallace WH. Ovarian and uterine characteristics after total body irradiation in childhood and adolescence: response to sex steroid replacement. Br J Obstet Gynaecol. 1999;106(12):1265–72.
92. Wallace WH, Anderson R, Baird D. Preservation of fertility in young women treated for cancer. Lancet Oncol. 2004;5(5):269–70.
93. Fenig E, Mishaeli M, Kalish Y, Lishner M. Pregnancy and radiation. Cancer Treat Rev. 2001;27(1):1–7.

Oocyte In Vitro Maturation: Formidable Obstacles on the Road to Fertility Preservation

10

David F. Albertini

Abstract

Oocyte maturation is a complex process that is initiated in the ovarian follicle during pre-antral development and culminates with the release of the mature ovum at the time of ovulation. As fertility preservation strategies emerge, it has become increasingly apparent that supporting oocyte maturation in as physiological a context as is possible is an important goal. This chapter summarizes a body of evidence consistent with the view that enabling and sustaining the metabolic interplay between the oocyte and cumulus cells will be essential to realizing a mature state for the ovum capable of producing viable embryos. Two key parameters are addressed: (1) the changing demands and regulatory mechanisms that distinguish communication between oocyte and granulosa during the growth and maturative phases of oogenesis and (2) the introduction of a paradigm for dissecting periovulatory maturation into several components that extend beyond the traditional nuclear and cytoplasmic properties that have been previously adopted. This framework provides a novel range of attributes in the cumulus-oocyte complex that could be assayed as biomarkers of oocyte quality.

Keywords

Oocyte in vitro maturation • Fertility preservation • In vitro fertilization • Oocyte maturation

Modern medical science is obsessed with the concept of ameliorating disease and suffering through the use of emerging technologies. Fertility preservation is no exception to this trend. Recognizing the growing need for protecting the reproductive capacity of humans unfortunate enough to have had their reproductive organs compromised as a result of trauma, lifestyle, or age provides impetus for exploring ways to manipulate our fecundity. Whereas ex vivo systems for the maintenance and control of cell and tissue level functions of the soma is a central tenet of contemporary medical research, tailoring technologies that would sustain the function of the male and female germ line has only more

D.F. Albertini (✉)
University of Kansas Cancer Center,
3901 Rainbow Boulevard, Kansas City, KS 66103, USA
e-mail: dalbertini@kumc.edu

E. Seli and A. Agarwal (eds.), *Fertility Preservation: Emerging Technologies and Clinical Applications*,
DOI 10.1007/978-1-4419-1783-6_10,

recently assumed momentum for treatments of infertility. It is in this spirit that the present chapter will consider a major tenet of infertility strategies that bears on the problem of oocyte maturity. Advances in fertility preservation will require conducting oocyte maturation in vitro in a safe and efficacious manner. That we have established a robust and reliable technology for achieving oocyte maturity in vitro is much less certain [1].

Even though the use of the term in vitro is a misnomer, it is often categorically assigned to the use of diluted culture solutions designed to meet the metabolic and environmental requirements for cells or communities of cells. In the case of ovarian function, many studies beginning in the 1930s sought to isolate and maintain the various physiological compartments that underscore its two-principle secretory activities: secretion of hormones and secretion of cells otherwise known as oocytes. Given that the oocyte in most mammals undergoes a protracted multistep transformation from a primordial stage to a fully developed and fertilization-ready state, it is said to have "matured" along its journey to ovulation [2]. Importantly, it should be appreciated that the pathway to maturity is anything but autonomous. Rather, oocyte development invokes a parallel and equally sophisticated program that drives, pauses, and activates the multicellular hormone factory known as the ovarian follicle [1]. Thus, it is not unreasonable to state at the outset that the maturation of the follicle is one and the same with the maturation of the oocyte. In fact it is owing to the use of in vitro systems to culture oocytes and/or follicles that the functional integration of these cell types has been most fully appreciated [3]. It is also likely that the utility of in vitro maturation in the context of preserving ovarian function will summarily require definition, modeling and engineering aimed at a deeper understanding of the integration of the germ and somatic cell types of the follicle [4]. What follows then is a survey of this field that unlike most contemporary works builds on strategic and technical foundations formulated many years ago. Revitalizing these principles will add to efforts aimed at optimizing the use of in vitro maturation to obtain healthy and developmentally poised oocytes for embryo production.

Defining Oocyte Maturation

Oocyte maturation refers to the final stages of oogenesis when the oocyte acquires the properties required to initiate and sustain embryonic development [5, 6]. Under in vivo circumstances, this process is elicited by the LH surge at ovulation and in humans it is estimated that the entire process occurs over a time period of 36–38 h [7, 8]. Central to an understanding of the problem of oocyte maturity is the subdivision of the maturation process into both spatial and temporal parameters that undergo significant changes in vivo in response to the LH surge. These features include alterations in cell cycle state [9], modifications in oocyte metabolism, and cytoplasmic organization that are collectively referred to as "cytoplasmic maturation" [10], and the reductive divisions of chromosomes during meiosis 1 and 2 often referred to as "nuclear maturation" [11, 12]. While the signals that elicit this series of changes in the oocyte, and link the response of follicular somatic cells to the oocyte, are being uncovered through studies in rodent animal models [2], they remain ill-defined in the human and will not be the subject of further consideration. Rather, a detailed examination of the maturation process will be reviewed from a historical perspective that incorporates both old and new concepts into a contemporary paradigm aimed at implementing this process in ARTs.

The mammalian oocyte is arrested at the prophase 1 or diplotene stage of meiosis within the developing follicle. In this state, the chromatin is organized into a partially condensed form and the nucleus, known as the germinal vesicle, contains one or more prominent nucleoli [11, 13]. From a cell cycle perspective, the oocyte is best described as being in G2 having completed the growth phase of oogenesis and primed to enter the M-phase of the cell cycle. What keeps the oocyte from entering M-phase precociously is an inhibitory signal established within the follicle that maintains the arrested prophase state. Upon reception of LH or as a result of physical removal from the follicle, the oocyte will entrain a series of events through not one, but two sequential M-phases without an intervening S-phase [6, 14, 15]. These events entail chromosome condensation, spindle assembly, alignment, and segregation of chromosomal bivalents with extrusion of the first polar body and subsequent arrest of the oocyte at metaphase of meiosis 2 [7]. The biochemical machinery that drives these events is well characterized and highly conserved and may be involved in the many errors in chromosome segregation that distinguish the oocytes of humans relative to other mammalian species [15]. In the end, it is this cascade of CDK1/cyclin kinases that ensures the timing of nuclear maturation and the reduction of chromosome number to a haploid complement.

What happens in the cytoplasm during oocyte maturation is another matter altogether. It is here that a number of proteins are synthesized and localized into positions that will protect them during the processes of fertilization and egg activation, events during which degradation of mRNAs and proteins are under exacting control [5, 11, 16, 17]. Sequestration of calcium into storage vesicles and positioning of these vesicles and cortical granules into the oocyte cortex engenders aspects of cytoplasmic maturation that will play directly into the process of fertilization. And remodeling of the oocyte cortex through changes in the state of actin assembly has recently emerged as a key component of maturation that appears to be controlled by a distinct set of tyrosine kinases [9].

Together, the picture that is now in hand is one invoking a complicated rearrangement of the cytoskeleton in oocytes that will impact both early and somewhat later events during the course of preimplantation development [18]. Precisely defining these events from a regulatory point of view will be required to best design systems for oocyte in vitro maturation that meet the rigorous standards expected for human ARTs.

Historical Perspective of Oocyte Maturation

As long ago as the 1930s, Pincus was investigating the maturation and cleavage division of rabbit oocytes using simple media formulations based on Locke's medium (see Albertini and Akkoyunlu [1]). At that time, in vitro culture systems were being adapted from the work of Warren Lewis at Johns Hopkins University and already the importance of maintaining cultures in high concentrations of serum were recognized (cited in Pincus [19]). The remarkable fact that oocytes retrieved from follicles enclosed in their cumulus cell investment would initiate and complete maturation in vitro set the tempo for all later studies. The implications were profound. Since the oocyte remained arrested at the dictyate stage of meiosis in the follicle, something about disturbing this environment was necessary and sufficient to trigger maturation, although the extent to which this process was normal was yet to be clarified. It took some 30 years, including experimentation with a variety of mammalian systems before Edwards fully appreciated this phenomenon is his quest to understand the behavior of chromosomes during the meiotic divisions of the oocyte. Further refinements in culture technique, which included lowering the serum concentrations led to the first detailed characteriziation of the kinetics of oocyte maturation that would set the stage for later attempts to accomplish in vitro fertilization [13]. The second major breakthrough appeared in 1971 when Cho, Sterns, and Biggers demonstrated that a cyclc AMP analog (DcAMP) reversibly arrested the onset of maturation. This discovery implicated the Protein Kinase A pathway as the first of what was to become an array of signaling pathways subsequently invoked to explain the regulation of oocyte maturation in vitro. Foundational papers from the Eppig laboratory and many others refined these procedures until it became almost commonplace to produce viable embryos and pups in the mouse model [2]. The recent demonstration that IVM mouse oocytes can produce healthy young has buttressed efforts to extend this work in the human and subhuman primate [20].

One of the principal advantages of the mouse model is the ability to manipulate IVM conditions in accordance with mechanistic insights gleaned from various genetic alterations known to produce defects in the processes underlying either nuclear or cytoplasmic maturation. Among these is the recognition that EGF-like factors, rather than LH itself, are important in stimulating intrafollicular changes in gene expression within the cumulus that would control the onset and extent of meiotic progression in cumulus-enclosed oocytes. Interestingly, the studies of Eppig et al.,[20] included EGF in IVM medium and with exception to the exclusion of amino acids, these conditions supported development to term at a somewhat lower rate than in vivo matured oocytes. This work in the mouse provides a baseline for comparison of health outcomes in the human or other mammalian models for IVM.

Factors Influencing In Vitro Maturation

As mentioned above, with exception to the mouse model, there is little or no consensus as to the optimal culture conditions for conducting IVM even though in systems such as the bovine, current media formulations are known to be sufficient to support term development of in vitro produced embryos [5]. An examination of the acceptable formulations for the bovine has identified the key components as gonadotropins (both FSH and LH), serum, pyruvate, and lactate as energy sources and standard antibiotic combinations [7, 10, 21–25]. The importance of LH and FSH has been recognized for many years as in the absence of these factors, meiotic progression is halted at metaphase-1 due to a failure of somatic cell signaling via transzonal projections (TZPs) [21]. Of course, the persistence of gonadotropins in IVM medium is hardly a reflection of the pulsatile nature of delivery to the follicle that would be expected under physiological conditions but despite this, it is assumed that signaling of the cumulus proceeds unabated to support full maturation and the acquisition of developmental competence. Recent insights into the metabolic demands imposed by the oocyte during maturation have been validated by examination of various supplements with respect to embryo quality [26].

Classic studies from DeMatos et al., [27] first revealed a beneficial effect of *N*-acetyl cystamine (NAC) on oocyte quality during IVM in the bovine. This work reinforced earlier studies in pointing to the extreme dependence that the oocyte has on the provision of metabolites and substrates through TZPs and gap junctions that originate in the cumulus cells [6, 28]. In the case of NAC, it was clearly shown that conversion to reduced glutathione (GSH) and subsequent uptake of GSH in the oocyte result in what we have called "metabolic loading" to indicate the infusion of compounds that would otherwise not be available to the oocyte during or after meiotic maturation [27, 29–31]. In addition to an essential role in regulating redox metabolism in the embryo, GSH plays a critical role in the reduction of sperm protamines that is permissive and required for the formation of the male pronucleus. It is interesting to note that as a paradigm for the design of IVM media, this component and likely others have not yet been recognized as an important constituent for improving oocyte quality. Moreover, the essential role played by cumulus cells is also often ignored despite the fact that IVM generally leads to the loss of TZPs as a result of COC isolation from the follicle. Recent systematic studies in the mouse have reinforced this concept [9] and are illustrated in Fig. 10.1.

A series of investigations from our laboratory have demonstrated the existence of discrete differences in the properties of mouse oocytes matured under in vivo or in vitro conditions. Besides conferring a developmental advantage on the resulting embryos, this body of work also demonstrates the importance of medium supplements in attaining partial levels of developmental competence when undertaking IVM. Our inquiries into the fundamental differences between oocytes matured in vivo or in vitro have revealed a pattern of changes in cytoplasmic and spindle organization suggestive of both the importance of cumulus contact in acquiring cytoplasmic and nuclear maturation (Fig. 10.1). For example, the size of the meiotic spindles is carefully regulated and minimized under in vivo conditions implying that the utilization of vital resources for spindle assembly may be conserved as the oocyte completes the two meiotic M-phases prior to fertilization and cleavage. These resources appear to include the global

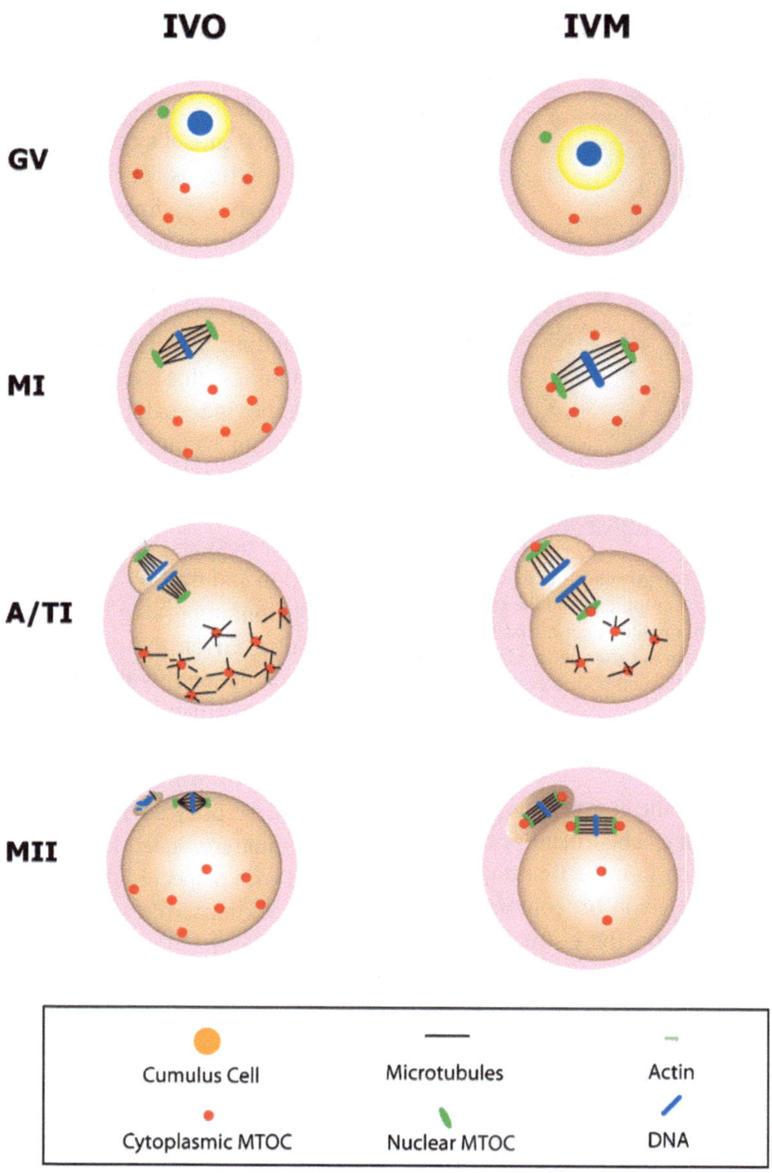

Fig. 10.1 Schematic of mouse oocytes at various stages of meiotic maturation (*top*, *bottom*) that compares in vivo matured oocytes on the *left* with comparable stages for in vitro matured oocytes on the *right*. Note the variations in germinal vesicle (GV) location, meiotic spindle size, and polar body size (Courtesy of Stan Fernald)

microtubule subunit of tubulin, the variant known as gamma-tubulin, and components of the microtubule organizing centers (MTOCs) that direct both spindle assembly and cell-cycle progression (e.g., pericentrin). Besides conserving on the utilization of maternal proteins that participate in embryonic cell cycles, several examples of macromolecular localization are distinct when one compares oocytes that have matured in vivo or in vitro. First, we have consistently observed an elaboration of a cortical actin meshwork under in vivo conditions that is rarely as developed in in vitro matured oocytes. Second, the germinal vesicle location in mouse oocytes is cortical during in vivo maturation whereas after isolation and culture, it migrates to the oocyte center and only

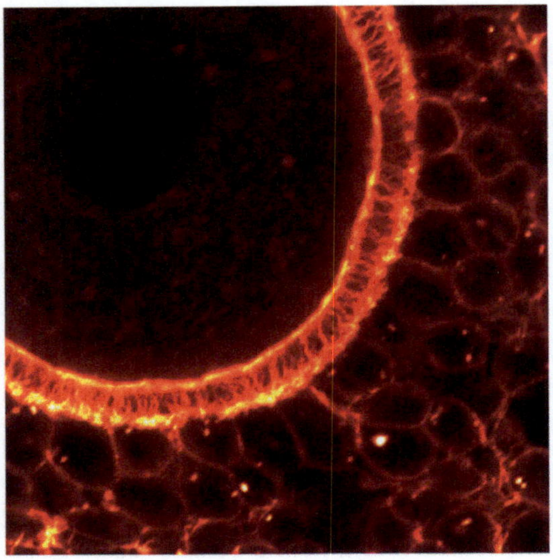

Fig. 10.2 Three-dimensional projection of confocal z-stack of an intact cumulus oocyte complex from the mouse. The *red labeling* represents filamentous actin that is concentrated at the oocyte cortex, in the zona pellucida as a series of striations, and outlining the cell bodies of the surrounding cumulus cells

later repositions itself at the oocyte cortex. Finally, MTOCs are retained in a subcortical location and excluded from inclusion into the first polar body during maturation in vivo. In contrast, when oocytes are matured in vitro, these structures leave their cortical position and become incorporated into the developing spindle only to be excluded upon formation and abscission of the first polar body. Interestingly, oocytes that have matured with minimal MTOCs remaining at metaphase-2 will exhibit a limited number of embryonic cell cycles following fertilization when compared to their in vivo matured counterparts. Collectively, these findings prompt speculation as to the means by which resources are conserved during the process of meiotic maturation and warrant further investigation into the differences between in vivo and in vitro matured human oocytes that may be amenable to manipulation under optimal culture conditions. The status of TZPs looms heavily in these kinds of studies given their essential role in supporting oocyte metabolism (Fig. 10.2).

Directions for the Future

There are clear advantages and disadvantages in using animal model systems for the optimization of IVM conditions for human oocytes. At present, the most experimentally tractable and relevant model is that of the bovine. But even here, the advanced state of human IVM in an applied sense has yet to catch up with the state of our understanding of the molecular and cellular underpinnings that regulate oocyte quality in even the bovine model. Thus, the application of new medium formulations in conjunction with, as an example, characterization of cumulus cell gene expression profiles will be needed to identify human oocytes that are well suited for embryo production and transfer. In the context of fertility preservation, improving human IVM becomes an imperative. Given the predispositions presented by cancer patients with respect to age, marital status, and treatment strategies, there is mounting pressure to identify and define the best resources that can be drawn upon to utilize what in most cases will be oocytes that are retrieved or stored in an immature state. How these materials are handled with respect to the specific needs of each patient seeking fertility preservation will depend on the development of multiple options for oocyte cryopreservation and subsequent maturation in vitro in order to avail viable embryos to this subpopulation of patients. Undoubtedly, once such approaches are firmly validated, the impact and utility of human IVM on the general practice of human ARTs will be materially extended for the benefit of all patients.

Acknowledgments This work has received sustained support from the Hall family Foundation and the ESHE Fund. The author's fascination with oocyte maturation was cultivated by a series of inspiring mentors of the likes of Drs. Everett Anderson, John Biggers, and the late Arthur T. Hertig, and it is an honor to recognize their influence on this work. Figure 10.1 was produced by Stan Fernald and a special thanks to the wonderful graduate students who have directed this line of work for the past 20 years. I am indebted especially to the contributions of Catherine Combelles, Susan Barrett, and Lynda McGinnis.

References

1. Albertini DF, Akkoyunlu G. Ovarian follicle culture systems for mammals. Methods Enzymol. 2010;476:107–21.
2. Su Y-Q, Suguira Y, Eppig JJ. Mouse oocyte control of granulosa cell development and function: paracrine regulation of cumulus cell metabolism. Semin Reprod Med. 2010;27:32–42.
3. McLaughlin M, Bromfield JJ, Albertini DF, Telfer EE. Activin promotes follicular integrity and oogenesis in cultured pre-antral bovine follicles. Mol Hum Reprod. 2010;16:644–53.
4. Atef A, Francois P, Christian V, Marc-Andre S. The potential role of gap junction communication between cumulus cells and bovine oocytes during in vitro maturation. Mol Reprod Dev. 2005;71:358–67.
5. Sirard MA, Coenen K. In vitro maturation and embryo production in cattle. Meth Mol Biol. 2006;348:35–42.
6. Thomas RE, Armstrong DT, Gilchrist RB. Bovine cumulus cell-oocyte gap junctional communication during in vitro maturation in response to manipulation of cell-specific cyclic adenosine 3′,5′-monophosophate levels. Biol Reprod. 2004;70:548–56.
7. Dominko T, First NL. Timing of meiotic progression in bovine oocytes and its effect on early embryo development. Mol Reprod Dev. 1997;47:456–67.
8. Schramm RD, Tennier MT, Boatman DE, Bavister BD. Chromatin configurations and meiotic competence of oocytes are related to follicular diameter in nonstimulated rhesus monkeys. Biology Reprod. 1993;41:349–56.
9. Barrett SL, Albertini DF. Cumulus cell contact during oocyte maturation in mice regulates meiotic spindle positioning and enhances developmental competence. J Assist Reprod Genet. 2010;27:29–39.
10. Levesque JT, Sirard MA. Effects of different kinases and phosphatases on nuclear and cytoplasmic maturation of bovine oocytes. Mol Reprod Dev. 1995;42:114–21.
11. Memili E, Peddinti D, Shack LA, McCarthy F, Sagirkaya H, Burgess SC. Bovine germinal vesicle oocyte and cumulus cell proteomics. Reproduction. 2007;133:1107–20.
12. Tatemoto H, Muto N, Sunagawa I, Shinjo A, Nakada T. Protection of porcine oocytes against cell damage caused by oxidative stress during in vitro maturation: role of superoxide dismutase activity in porcine follicular fluid. Biol Reprod. 2004;71:1150–7.
13. Smitz J, Cortvrindt R. Oocyte in-vitro maturation and follicle culture: current clinical achievements and future directions. Hum Reprod. 1999;14 Suppl 1:145–61.
14. Fair T, Hyttel P, Greve T. Bovine oocyte diameter in relation to maturational competence and transcriptional activity. Mol Reprod Dev. 1995;42:437–42.
15. Pan H, O'Brien MJ, Wigglesworth K, Eppig JJ, Schultz RM. Transcript profiling during mouse oocyte development and the effect of gonadotropin priming and development in vitro. Dev Biol. 2005;286:493–506.
16. Lonergan P, Gutierrez-Adan A, Rizos D, Pintado B, de la Fuente J, Boland MP. Relative messenger RNA abundance in bovine oocytes collected in vitro or in vivo before and 20 hr after the preovulatory luteinizing hormone surge. Mol Reprod Dev. 2003;66: 297–305.
17. Mourot M, Dufort I, Gravel C, Algriany O, Dieleman S, Sirard MA. The influence of follicle size, FSH-enriched maturation medium, and early cleavage on bovine oocyte maternal mRNA levels. Mol Reprod Dev. 2006;73:1367–79.
18. Hutt KJ, Shi Z, Petroff BK, Albertini DF. The environmental toxicant 2,3,7,8-tetrachlorodibenzo-p-dioxin disturbs the establishment and maintenance of cell polarity in preimplantation rat embryos. Biol Reprod. 2010;82:914–20.
19. Pincus G, Enzmann EV. The comparative behavior of mammalian eggs in vivo and in vitro: I. The activation of ovarian eggs. J Exp Med. 1935;62(5):665–75.
20. Eppig JJ, O'Brien MJ, Wigglesworth K, Nicholson A, Zhang W, King BA. Effect of in vitro maturation on the health and lifespan of adult offspring. Human Reprod. 2009;24:922–8.
21. Allworth AE, Albertini DF. Meiotic maturation in cultured bovine oocytes is accompanied by remodeling of the cumulus cell cytoskeleton. Dev Biol. 1993;158:101–12.
22. Senbon S, Fukumi Y, Hamawaki A, Yoshikawa M, Miyano T. Bovine oocytes grown in serum-free medium acquire fertilization competence. J Reprod Dev. 2004;50:541–7.
23. Sutton-McDowall ML, Gilchrist RB, Thompson JG. Cumulus expansion and glucose utilisation by bovine cumulus-oocyte complexes during in vitro maturation: the influence of glucosamine and follicle-stimulating hormone. Reproduction. 2004;128: 313–9.
24. Sutton-McDowall ML, Gilchrist RB, Thompson JG. Effect of hexoses and gonadotrophin supplementation on bovine oocyte nuclear maturation during in vitro maturation in a synthetic follicle fluid medium. Reprod Fertil Dev. 2005;17:407–15.
25. Roth Z, Hansen PJ. Sphingosine 1-phosphate protects bovine oocytes from heat shock during maturation. Biol Reprod. 2004;71:2072–8.
26. Fouladi-Nashta AA, Campbell KHS. Dissociation of oocyte nuclear and cytoplasmic maturation by the addition of insulin in cultured bovine antral follicles. Reproduction. 2006;131:449–60.
27. de Matos DG, Furnus CC, Moses DF. Glutathione synthesis during in vitro maturation of bovine oocytes: role of cumulus cells. Biol Reprod. 1997;57:1420–5.
28. Tseng JK, Chen CH, Chou PC, Yeh SP, Ju JC. Influences of follicular size on parthenogenetic activation and in vitro heat shock on the cytoskeleton in cattle oocytes. Reprod Domest Anim. 2004;39:146–53.
29. Luciano AM, Lodde V, Beretta MS, Colleoni S, Lauria A, Modina S. Developmental capability of denuded bovine oocyte in a co-culture system with intact cumulus-oocyte complexes: role of cumulus cells, cyclic adenosine 3′,5′-monophosphate, and glutathione. Mol Reprod Dev. 2005;71:389–97.

30. Luciano AM, Goudet G, Perazzoli F, Lahuec C, Gerard N. Glutathione content and glutathione peroxidase expression in in vivo and in vitro matured equine oocytes. Mol Reprod Dev. 2006;73: 658–66.
31. Mertens MJ, Lopez-Bejar M, Paramio MT. Intracytoplasmic glutathione levels in heifer oocytes cultured in different maturation media and its effect on embryo development. Reprod Domest Anim. 2005;40:126–30.

Whole Ovary Cryopreservation

11

Hulusi Bulent Zeyneloglu, Gogsen Onalan, and Cengiz Karakaya

Abstract

Surgical, technological, and medical advances in cancer treatment have improved the life expectancy, making quality of life issues in cancer survivors more important than ever before. Within this context, fertility loss due to the side effects of the chemotherapeutic agents or involvement of the gonads has become an active area of investigation. Many patients demand restoration of their fertility after a successful struggle with cancer and ask to discuss fertility preservation options. Currently, the most recommended procedure is to discuss fertility preservation options with the patient prior to the cancer therapy and as a multidisciplinary team. Whole ovary cryopreservation, although still controversial, may soon become an integral part of this discussion.

Keywords

Whole ovary cryopreservation • Cryopreservation of the ovary • Cancer and fertility • Fertility and cancer

Current cancer treatments comprise conservative surgeries, radiotherapy, chemotherapy, and allogeneic bone marrow transplantation. These strategies have raised patients' expectations of a better quality of life, among which fertility preservation represents a forthcoming desire [1]. At present, available modalities to preserve fertility in women are still experimental and *do not guarantee the recovery of fertility*. The only recognized methods are *sperm cryopreservation* in men and in vitro *fertilization (IVF) with embryo cryopreservation* in women, as suggested by The American Society of Reproductive Medicine [2].

Advances in Gamete and Tissue Cryopreservation

Among the experimental approaches to fertility preservation, the most promising options are oocyte and ovarian tissue cryobanking before cancer therapy [3]. At present, proposed strategies to preserve fertility in women with cancer

H.B. Zeyneloglu (✉)
Department of Obstetrics and Gynecology, Baskent University, Kubilay Sok 36, Maltepe, Ankara 06570, Turkey
e-mail: hulusi.zeyneloglu@gmail.com

E. Seli and A. Agarwal (eds.), *Fertility Preservation: Emerging Technologies and Clinical Applications*, DOI 10.1007/978-1-4419-1783-6_11,

include: (1) storage of frozen embryos, (2) storage of frozen oocytes for future fertilization and embryo transfer, (3) storage of frozen ovarian tissue or the whole ovary for future transplantation, (4) frozen storage of ovarian tissue or isolated follicles for in vitro growth and maturation, (5) ovarian transposition before radiotherapy, (6) hormonal protection with GnRH analogs, and (7) pharmacologic protection with antiapoptotic agents (e.g., sphingosine-1-phospate) and (8) in patients with an absent or nonfunctional uterus – a uterine transplant. Only option 1 is considered nonexperimental and could be offered to willing patients as an established fertility preservation strategy [4].

Embryo cryopreservation has been a proven method to protect fertility. However, in nearly all cancers, with the possible exception of breast cancer, chemotherapy is initiated soon after diagnosis. Because preparation and stimulation for oocyte retrieval usually requires 2–3 weeks or longer, it is generally not feasible to freeze embryos from an adult female cancer patient for potential future use. Additionally, not all patients have partners with whom they can create embryos to cryopreserve. Therefore, most female cancer patients of reproductive age do not have the option of utilizing established, assisted reproductive technologies to safeguard their fertility so far [4].

Oocyte cryopreservation is an attractive strategy to preserve female fertility as it does not require surgery, and well-tested stimulation protocols for IVF can be used. However, in most reports, live birth rate per oocyte frozen is very low, discouraging the clinician to offer this strategy as a promising method.

Freezing of ovarian tissue for preservation of fertility among young girls and women facing chemo- or radiotherapy has become a widely accepted procedure since the first successful results of cryopreservation of human ovarian tissue were published [5].

Human ovarian tissue can be successfully cryopreserved as cortical strips, as an intact-whole ovary with its vascular pedicle or as isolated follicles [6] demonstrating significant survival and function after thawing. Until now, ovarian cryopreservation and transplantation procedures have been almost solely limited to avascular cortical fragments in both experimental and clinical studies [7–9] and, to date, more than ten healthy babies have been born worldwide after grafting frozen–thawed ovarian tissue in cancer patients [10–15].

The main problem with ovarian tissue transplantation is that the graft is completely dependent on the establishment of neovascularization and, as a result, a significant proportion of follicles are lost due to ischemic damage by the time neovascularization is achieved. Thus, the major loss seems to occur during the warm ischemic period, probably extending over several days until neovascularization has restored the blood flow to the grafted tissue [16–20].

At least 25% of the primordial follicles are lost as a consequence of cryopreserved xenografts of human ovarian tissue into mice [16, 18]. Some studies estimated that ischemic damage during autograft processes causes the depletion of 60–95% of the follicular reserve, including the loss of virtually the entire population of growing follicles [7, 19, 21]. This phenomenon is associated with a drastical reduction of the graft volume and a noteworthy fibrosis in most grafts. This follicular reduction due to the warm ischemic period observed after ovarian tissue transplantation is a major concern, especially in humans and large animal species that have a dense ovarian cortex, as it may impinge on follicular growth, hormonal environment, and the overall fertility [22].

The time required to accomplish a sufficient perfusion of the transplanted tissue is decisive for the follicular survival and the functional longevity of the graft. In mice, primary perfusion of the autograft is found 3 days posttransplantation [23]. The neovascularization is detected within 48 h in autologous immature transplanted rat ovaries and the tissue is revascularized and functional after 1 week [24]. Using MRI and histology, functional vessels have been observed within ectopic xenotransplanted rat ovarian tissue after only 7 days [25]. In humans, the neovascularization process was detected after only 3 days following ovarian tissue transplantation onto a chick chorioallantoic membrane [26]. Primordial follicles can endure ischemia for at least 4 h during tissue transport [27], whereas stromal cells surrounding the follicles appeared to be more sensitive to ischemia compared with primordial follicles [28].

Various attempts have been reported to shorten the ischemic period and increase the viability and fertility potential after ovarian graft: such as antioxidants, growth factors, or mechanical factors inducing neovascularization by creating a peritoneal pocket at the transplantation sites 1 week before thine transplantation procedure [11].

Whole Ovarian Cryopreservation

Minimizing the ischemic interval between transplantation and revascularization is vital to maintaining the follicular reserve and extending the life span and function of the graft. In theory, this can be most effectively accomplished by transplantation of intact ovary with vascular anastomosis, allowing immediate revascularization of the transplant. Ovarian vascular transplantation has already been successfully performed using intact fresh ovaries in the mouse [29], rat [30, 31], rabbit [32], dog [33], monkey [34], sheep [35, 36], and human [37–39]. Moreover, fertility in these animal models, with delivery of offspring, was accomplished in the rat [30] and the ewe [40]. In human subjects, Silber reported a series of pregnancies after ovary transplantations between monozygotic twin women in whom one twin had lost ovarian function [9].

For a patient who wishes long-term reproductive and endocrine function after retransplantation, reducing the ischemic interval between transplantation and revascularization is critical for maintaining the viability and functional lifetime of the graft. For this purpose, cryopreservation of the whole ovary with an intact pedicle and vascular supply can potentially overcome this problem because reperfusion occurs immediately upon retransplantation and anastomosis [41–44].

Martinez-Madrid et al. described a cryopreservation protocol for intact human ovary with its vascular pedicle. Ovarian perfusion with cryoprotective solution and slow freezing with a cryofreezing container were performed. Rapid thawing of the ovaries was performed by perfusion and bathing with a decreased sucrose gradient. High survival rates of follicles (75.1%), small vessels, and stroma, as well as a normal histologic structure, were documented in all the ovarian components after thawing [42]. Postthaw induction of apoptosis, assessed by both the terminal deoxynucleotidyl transferase biotin-dUTP nick-end labeling method and immunohistochemistry for active caspase-3, was not observed in any cell type [43]. They confirmed that the majority (96.7%) of primordial follicles were intact and that their endothelial cells had a completely normal ultrastructure after cryopreservation confirmed by using transmission electron microscopy (TEM) [45]. The percentage of active caspase-3-positive endothelial cells was <1% [43].

Challenges Associated with Whole Ovary Cryopreservation

Challenges Related to Whole Organ Cryopreservation

At present, whole ovary cryopreservation is associated with several challenges. Firstly, inadequate or uneven diffusion or perfusion of cryoprotectant into all tissue mass is a major obstacle, as the whole organ (ovary) is composed of multicellular systems, such as follicles formed by the oocytes, granulosa cells, and theca cells. In addition, blood vessels and nerves are important for the function after transplantation. Primordial ovarian follicles which comprise 70–90% of all ovarian follicles in human ovary are the main target of cryoperservation approaches because they are small, less differentiated, contain a smaller number of organelles and no zona pellucida, and are relatively metabolically quiescent resulting in less sensitivity to cryopreservation damage [46]. In contrast; secondary, preantral, and antral follicles do not tolerate freezing and thawing and probably do not survive cryopreservation because of improper penetration of cryoprotectant into all of these cells in a timely manner.

Second challenge is the heat transfer and the nonhomogenous rate of cooling between core and periphery of the organ [47]. Cooling and heating rates are directly related by the dimensions and geometry of the organ. Uniform cooling rate through the entire organ is very problematic. Multithermal gradient freezing methodology has been suggested a solution for this technical limitation [48].

Another issue in multicellular systems is extracellular ice crystal formation, especially intravascular ice formation, potentially leading to vascular injury. Cryoinjury to the vascular compartment could result in organ failure after transplantation. Ice crystal formation has to be minimized by choosing optimal freezing and thawing rates. The choice of a cryoprotectant with optimal permeation with lowest toxicity and least ice crystal formation is specific for each cell and tissue type [49]. On the basis of current knowledge, due to chemical toxicity and osmotic shock following exposure to very high (≥50%) concentrations of cryoprotectant solution with vitrification procedure, the standard method for human whole ovary cryopreservation is slow-programmed freezing using human serum albumin-containing medium and propanediol, dimethylsulfoxide (DMSO), or ethylene glycol as a cryoprotectant, combined or not with sucrose [5]. The use of increasing concentration of sucrose as nonpenetrating cryoprotectant, instead of sodium with choline, has improved survival rates by 80% [50].

Challenges Related to Surgical Technique

While laparatomy can help reduce ischemic period during the harvesting of human ovaries with vascular pedicles, a laparoscopic approach is considered superior for the purposes of cryopreservation and successive transplantation [51]. Indeed, a minimally invasive approach can facilitate the dissection of the ovarian pedicle up to the pelvic brim, allowing the ovarian artery and veins to be dissected in sufficing length, which, in turn, facilitates their subsequent anastomosis to vessels of similar diameter during transplantation [52]. In addition, performing laparoscopy in cancer patients ensures superior wound healing, which is consistent with patients' desires to begin their chemotherapeutic treatment cycles without delay [53]. During the procedure, sharp dissection and suturing are preferred to electrocoagulation, and longer and wider caliber vessels are obtained through ligating the ovarian vessels as proximal to the origin as possible. An endobag is placed through the 10-mm trocar to transport the ovary with its vascular pedicle outside the peritoneal cavity. For avoiding the crushing of the ovary and the blood vessels against the narrow port site, an extended port incision should be used.

Discrepancy between the diameters of ovarian vessels and the recipient vessels is a significant challenge that needs to be overcome to ensure the patency of the anastomosis. The abrupt alteration of caliber between the cut end diameters of the vessels encountered in approximately one-third of anastomoses may result in turbulence, thus predisposing the vessel to platelet aggregation [54]. The surgical approach to reanastomosis should be determined based on the degree of discrepancy between vascular diameters and range from dilatation using a jeweler's forceps [54] in case of simple discrepancies (<1:1.5) to using the oblique cut, fish-mouth cut, or end-to-side anastomosis when discrepancies exceed 1:1.5 [54]. Sleeve anastomosis is performed when discrepancies are larger and when the upstream donor vessel is smaller than the recipient one [55].

Microvascular thrombosis after transplantation may cause a reduced longevity of the graft [56]. Because of the thrombotic events in the reanastomosed vascular pedicle, cryopreservation of the contralateral ovarian cortical tissue has been suggested until the efficiency of whole ovary cryopreservation is validated with adequate attention to the postthaw function of the ovarian vascular system [53]. Others proposed to leave one of the ovaries in situ to ensure the availability of an intact pedicle (for exchanging the sterilized organ with the frozen and thawed ones), once the patient is ready for autotransplantation [30]. The cryoperfusion of ovary with a cryoprotective medium via ovarian artery was proposed by some authors [40, 56, 57] while others opposed the idea as it could cause toxic injury leading to loss of tissue viability [4].

Reseeding of Cancer

Cryopreservation of whole ovaries in women with cancer, followed by transplantation, carries a risk of reseeding cancer. The extent of this risk would depend on the cancer type and stage among other factors. Incidence of ovarian metastasis has been reported for many cancer types and this information is reviewed in Chap. 12.

Conclusions

There have been significant advances in whole ovary cryopreservation and transplantaiton; however, challenges remain. Optimization of cryopreservation and transplantation techniques, as well as better understanding of the extent of risks associated with cancer reseeding, is necessary before this approach finds its place within routine discussion of fertility preservation.

References

1. Jemal A et al. Cancer statistics, 2009. CA Cancer J Clin. 2009;59(4):225–49.
2. Ethics Committee of the American Society for Reproductive Medicine. Fertility preservation and reproduction in cancer patients. Fertil Steril. 2005; 83(6):1622–8.
3. Kim SS. Fertility preservation in female cancer patients: current developments and future directions. Fertil Steril. 2006;85(1):1–11.
4. Bedaiwy MA et al. Reproductive outcome after transplantation of ovarian tissue: a systematic review. Hum Reprod. 2008;23(12):2709–17.
5. Hovatta O. Methods for cryopreservation of human ovarian tissue. Reprod Biomed Online. 2005;10(6): 729–34.
6. Donnez J et al. The role of cryopreservation for women prior to treatment of malignancy. Curr Opin Obstet Gynecol. 2005;17(4):333–8.
7. Baird DT et al. Long-term ovarian function in sheep after ovariectomy and transplantation of autografts stored at −196 C. Endocrinology. 1999;140(1):462–71.
8. Oktay K et al. Embryo development after heterotopic transplantation of cryopreserved ovarian tissue. Lancet. 2004;363(9412):837–40.
9. Silber SJ et al. Ovarian transplantation between monozygotic twins discordant for premature ovarian failure. N Engl J Med. 2005;353(1):58–63.
10. Andersen CY et al. Two successful pregnancies following autotransplantation of frozen/thawed ovarian tissue. Hum Reprod. 2008;23(10):2266–72.
11. Demeestere I et al. Orthotopic and heterotopic ovarian tissue transplantation. Hum Reprod Update. 2009;15(6):649–65.
12. Donnez J et al. Livebirth after orthotopic transplantation of cryopreserved ovarian tissue. Lancet. 2004;364(9443):1405–10.
13. Meirow D et al. Pregnancy after transplantation of cryopreserved ovarian tissue in a patient with ovarian failure after chemotherapy. N Engl J Med. 2005; 353(3):318–21.
14. Sanchez-Serrano M et al. Twins born after transplantation of ovarian cortical tissue and oocyte vitrification. Fertil Steril. 2010;93(1):268 e11–3.
15. Ernst E et al. The first woman to give birth to two children following transplantation of frozen/thawed ovarian tissue. Hum Reprod. 2010;25(5):1280–1.
16. Nisolle M et al. Histologic and ultrastructural evaluation of fresh and frozen-thawed human ovarian xenografts in nude mice. Fertil Steril. 2000;74(1):122–9.
17. Liu J et al. Early massive follicle loss and apoptosis in heterotopically grafted newborn mouse ovaries. Hum Reprod. 2002;17(3):605–11.
18. Newton H et al. Low temperature storage and grafting of human ovarian tissue. Hum Reprod. 1996;11(7): 1487–91.
19. Candy CJ, Wood MJ, Whittingham DG. Effect of cryoprotectants on the survival of follicles in frozen mouse ovaries. J Reprod Fertil. 1997;110(1):11–9.
20. Gunasena KT et al. Live births after autologous transplant of cryopreserved mouse ovaries. Hum Reprod. 1997;12(1):101–6.
21. Aubard Y. Ovarian tissue xenografting. Eur J Obstet Gynecol Reprod Biol. 2003;108(1):14–8.
22. Kim SS. Time to re-think: ovarian tissue transplantation versus whole ovary transplantation. Reprod Biomed Online. 2010;20(2):171–4.
23. Nugent D et al. Protective effect of vitamin E on ischaemia-reperfusion injury in ovarian grafts. J Reprod Fertil. 1998;114(2):341–6.
24. Dissen GA et al. Immature rat ovaries become revascularized rapidly after autotransplantation and show a gonadotropin-dependent increase in angiogenic factor gene expression. Endocrinology. 1994;134(3):1146–54.
25. Israely T et al. Angiogenesis in ectopic ovarian xenotransplantation: multiparameter characterization of the neovasculature by dynamic contrast-enhanced MRI. Magn Reson Med. 2004;52(4):741–50.
26. Martinez-Madrid B et al. Chick embryo chorioallantoic membrane (CAM) model: a useful tool to study short-term transplantation of cryopreserved human ovarian tissue. Fertil Steril. 2009;91(1):285–92.
27. Schmidt KL et al. Survival of primordial follicles following prolonged transportation of ovarian tissue prior to cryopreservation. Hum Reprod. 2003;18(12): 2654–9.
28. Kim SS et al. Quantitative assessment of ischemic tissue damage in ovarian cortical tissue with or without antioxidant (ascorbic acid) treatment. Fertil Steril. 2004;82(3):679–85.
29. Migishima F et al. Successful cryopreservation of mouse ovaries by vitrification. Biol Reprod. 2003;68(3):881–7.
30. Wang X et al. Fertility after intact ovary transplantation. Nature. 2002;415(6870):385.
31. Yin H et al. Transplantation of intact rat gonads using vascular anastomosis: effects of cryopreservation, ischaemia and genotype. Hum Reprod. 2003;18(6): 1165–72.
32. Winston RM, Browne JC. Pregnancy following autograft transplantation of Fallopian tube and ovary in the rabbit. Lancet. 1974;2(7879):494–5.
33. Paldi E et al. *Genital organs.* Auto and homotransplantation in forty dogs. Int J Fertil. 1975;20(1): 5–12.

34. Scott JR et al. Microsurgical ovarian transplantation in the primate. Fertil Steril. 1981;36(4):512–5.
35. Jeremias E et al. Heterotopic autotransplantation of the ovary with microvascular anastomosis: a novel surgical technique. Fertil Steril. 2002;77(6): 1278–82.
36. Goding JR, McCracken JA, Baird DT. The study of ovarian function in the ewe by means of a vascular autotransplantation technique. J Endocrinol. 1967; 39(1):37–52.
37. Leporrier M et al. *A new technique to protect ovarian function before pelvic irradiation.* Heterotopic ovarian autotransplantation. Cancer. 1987;60(9):2201–4.
38. Mhatre P, Mhatre J, Magotra R. Ovarian transplant: a new frontier. Transplant Proc. 2005;37(2):1396–8.
39. Hilders CG et al. Successful human ovarian autotransplantation to the upper arm. Cancer. 2004; 101(12):2771–8.
40. Imhof M et al. Orthotopic microvascular reanastomosis of whole cryopreserved ovine ovaries resulting in pregnancy and live birth. Fertil Steril. 2006;85 Suppl 1:1208–15.
41. Bromer JG, Patrizio P. Fertility preservation: the rationale for cryopreservation of the whole ovary. Semin Reprod Med. 2009;27(6):465–71.
42. Martinez-Madrid B et al. Freeze-thawing intact human ovary with its vascular pedicle with a passive cooling device. Fertil Steril. 2004;82(5):1390–4.
43. Martinez-Madrid B et al. Apoptosis and ultrastructural assessment after cryopreservation of whole human ovaries with their vascular pedicle. Fertil Steril. 2007;87(5):1153–65.
44. Falcone T et al. Ovarian function preservation in the cancer patient. Fertil Steril. 2004;81(2):243–57.
45. Nottola SA et al. Cryopreservation and xenotransplantation of human ovarian tissue: an ultrastructural study. Fertil Steril. 2008;90(1):23–32.
46. Nugent D et al. Transplantation in reproductive medicine: previous experience, present knowledge and future prospects. Hum Reprod Update. 1997;3(3): 267–80.
47. Balasubramanian SK, Coger RN. Heat and mass transfer during the cryopreservation of a bioartificial liver device: a computational model. ASAIO J. 2005;51(3):184–93.
48. Arav A et al. Ovarian function 6years after cryopreservation and transplantation of whole sheep ovaries. Reprod Biomed Online. 2010;20(1):48–52.
49. Fuller B, Paynter S. Fundamentals of cryobiology in reproductive medicine. Reprod Biomed Online. 2004;9(6):680–91.
50. Coticchio G et al. Criteria to assess human oocyte quality after cryopreservation. Reprod Biomed Online. 2005;11(4):421–7.
51. Bedaiwy MA, Falcone T. Harvesting and autotransplantation of vascularized ovarian grafts: approaches and techniques. Reprod Biomed Online. 2007;14(3): 360–71.
52. Bisharah M, Tulandi T. Laparoscopic preservation of ovarian function: an underused procedure. Am J Obstet Gynecol. 2003;188(2):367–70.
53. Jadoul P et al. Laparoscopic ovariectomy for whole human ovary cryopreservation: technical aspects. Fertil Steril. 2007;87(4):971–5.
54. Cakir B, Akan M, Akoz T. The management of size discrepancies in microvascular anastomoses. Acta Orthop Traumatol Turc. 2003;37(5):379–85.
55. de la Pena-Salcedo JA, Cuesy C, Lopez-Monjardin H. Experimental microvascular sleeve anastomosis in size discrepancy vessels. Microsurgery. 2000;20(4):173–5.
56. Bedaiwy MA et al. Restoration of ovarian function after autotransplantation of intact frozen-thawed sheep ovaries with microvascular anastomosis. Fertil Steril. 2003;79(3):594–602.
57. Arav A et al. Oocyte recovery, embryo development and ovarian function after cryopreservation and transplantation of whole sheep ovary. Hum Reprod. 2005;20(12):3554–9.

Risk of Transplanting Cryopreserved Ovarian Tissue in Women with Malignancies

12

Javier Domingo del Pozo, María Sánchez-Serrano, and Antonio Pellicer

Abstract

The restoration of the ovarian function after chemotherapy or radiotherapy has two main goals: to improve quality of life and restore reproductive function. Breast cancer is the most frequent diagnosis of patients demanding fertility preservation treatments, followed by hematological cancers. When ovarian tissue cryopreservation is considered, it becomes necessary to assess the possibility of reintroducing metastatic cells within the reimplant, an issue that has not been addressed systematically. Current data indicates that ovarian cortical grafting for fertility preservation is contraindicated in patients with leukemia owing to the risk of reintroducing malignant cells, but there seems to be no risk for women with Hodgkin's disease or those in the early stages of breast cancer.

Keywords

Ovarian tissue • Breast cancer • Leukemia • Neoplastic cells • Transplantation of cryopreserved ovarian tissue

Ovarian tissue cryopreservation is a fast-developing strategy for preservation of fertility in cancer patients, but the possibility of reseeding tumor cells into cured patients limits its clinical application.

Consequences in both oncological and nononcological patients receiving chemo- or radiotherapy are now considered of great relevance. Menopause and infertility are two of the main causes for concern to those patients who survive cancer. Furthermore, other problems related to the acquired menopause, such as vasomotor, skeletal, or cardiovascular alterations are causes for concern. The main objective of ovarian tissue cryopreservation, similar to other fertility preservation techniques, is to elude or prevent these consequences.

The most frequent cancers diagnosed in young women aged 15–24 years include Hodgkin's lymphoma, malignant melanoma, and leukemia [1]. Breast cancer is the most frequent malignancy in reproductive age, representing a third of all cancers diagnosed in young women [2]. More than 15% of all breast cancers appear in women

A. Pellicer (✉)
Department of Obstetrics and Gynecology, Hospital Universtario Doctor Peset, Valencia, Spain

Department of Obstetrics and Gynecology, Valencia University, Valencia, Spain

E. Seli and A. Agarwal (eds.), *Fertility Preservation: Emerging Technologies and Clinical Applications*,
DOI 10.1007/978-1-4419-1783-6_12,

under 40 years, with more than 8,000 new cases per year and about 600 new cases per year in women under 30 in the USA [2].

Most of these patients are treated with chemotherapy, including highly gonadotoxic agents as cyclophosphamide, which has the potential to induce ovarian failure and infertility (in 30–40% of women aged 30–39 years and 65% of women aged 40–45%) [3]. In addition to cyclophosphamide, chlorambucil, procarbazine, cisplatin, melphalan, carboplatin, and busulfan are chemotherapeutic agents with gonadotoxic effects that are used in women diagnosed with cancer [4].

As the age of childbearing has been substantially delayed in our society, the consequences chemotherapy has on human reproduction have become more important, and thus the percentage of women requesting fertility preservation procedures has increased, especially for breast cancer, where the demand is higher than initially expected.

Nowadays, fertility preservation procedures have become an important part of cancer treatment. At IVI Fertility Preservation Programme, the number of patients demanding oocyte cryopreservation prior to chemotherapy has increased in recent years, with breast cancer as the most frequent diagnosis in people undergoing this treatment, followed by Hodgkin's lymphoma (Fig. 12.1). The Valencia Programme for Fertility Preservation was established in 2005 and offers the cryopreservation and later implantation of ovarian cortex in women with high risk of premature ovarian failure. Breast cancer is also the most frequent diagnosis (Fig. 12.1) [5].

Indeed, breast cancer is the most common malignancy during the reproductive age and women with breast cancer are those who most frequently demand fertility preservation techniques, including ovarian tissue cryopreservation. The main concern with ovarian tissue cryopreservation is the possibility of reintroducing metastatic cells within the implant, an issue that has not been addressed systematically and whose risk must be discarded.

Since fertility preservation procedures have been shown to be of benefit for patients treated with gonadotoxic agents, these procedures must not be limited to preserving and restoring fertility in patients undergoing cancer therapies, but also be applied to any situations where the reproductive function is threatened.

An Overview of Ovarian Tissue Cryopreservation Within the Context of Fertility Preservation: Pros and Cons

Gonadotoxicity depends on several factors, such as age, previous statement of ovaries, type of agent used, and the cumulative doses [6]. Ovaries have a limited capability of recovering follicles after being damaged. With ovarian cryobanking, abundant primordial follicles containing small and less-differentiated oocytes are efficiently stored, as immature oocytes in primordial follicles are less sensitive to cryopreservation damage.

Some cancer patients treated with chemotherapy or low doses of radiotherapy recover their ovarian function and even get pregnant spontaneously. But often, oocyte quality may be suboptimal, and consequently the possibility of achieving a pregnancy might be diminished. In addition to the reduction in the number of germ cells, a loss of steroid hormones and the possibility of mutation or teratogenic consequences are also described [7]. Thus, in case of pregnancy, higher incidence of miscarriage, intrauterine growth restriction, or prematurity is reported [8].

Different strategies have been proposed to protect or preserve the ovarian function in patients who are going to be treated with gonadotoxic agents. Some have demonstrated their efficiency while others are under evaluation and still need to be improved.

Trend on fertility preservation possibly is directed to a combination of some of these techniques: ovarian tissue cryopreservation and further retrieval of immature oocytes followed by in vitro maturation and vitrification [9]. But immature egg retrieval for further in vitro oocyte maturation (IVM) still does not offer feasible options to the patient, so the two main techniques we can consider nowadays with better possibilities are oocytes or embryo vitrification and ovarian tissue cryopreservation.

At the time of considering a fertility preservation procedure, it is important to analyze first the

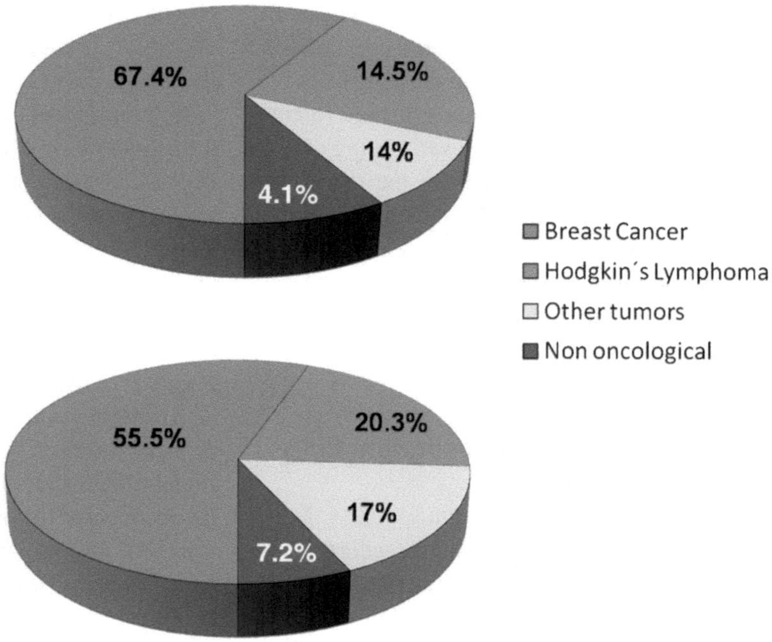

Fig. 12.1 Oncological and nononcological request for oocyte vitrification at IVI Fertility Preservation Programme and ovarian cortex cryopreservation at Valencia Programme for Fertility Preservation. IVI Clinics. Hospital Universitario Dr. Peset, Valencia, Spain

patient's risk of infertility and the prognosis of the cancer depending on the nature of the malignancy. Then, if cryopreservation of oocytes or embryos is planned, the consequences of delaying the initiation of chemotherapy a few weeks to complete the ovarian stimulation should be considered and compared to the risk of reimplanting tumoral cells with the cryopreserved ovarian tissue [10]. Fertility preservation procedures that do not involve transplantation of cryopreserved ovarian tissue to the patient, such as embryo or oocyte cryopreservation, or gonadotropin-releasing hormone agonist cotreatment have been studied extensively [10–22] and are discussed in separate chapters (Chaps. 7, 8, and 13). Similarly, ovarian tissue cryopreservation (Chap. 9) and whole ovary cryopreservation (Chap. 11) are also discussed in detail elsewhere. However, a brief overview is presented below as an introduction to our overall discussion.

Ovarian Tissue Cryopreservation

Orthotopic transplantation of the frozen/thawed ovarian cortex can restore the ovarian function and would allow natural fertility with unlimited cycles while the graft is active. If not, in vitro fertilization would still remain as an option. This approach has so far been limited to avascular cortical fragments, but recently transplantation of intact ovary with vascular anastomosis has been proposed [23]. One of the advantages of whole transplantation is that the ischemic interval between transplantation and revascularization might be reduced.

Heterotopic transplantation of ovarian cortical tissue has been attempted. However, it is not clear which is the best place for implantation. Furthermore, follicle size reaches only 15 mm, and no pregnancy has been obtained yet.

A disadvantage of ovarian tissue cryopreservation is that it requires surgery – laparoscopy – to obtain the ovarian cortex or the whole ovary. Ischemic damage and reduced follicular pool usually appear after transplantation. The active life of the transplanted tissue depends on neoangiogenesis and new vascularization, and it may have a short life.

Currently, frozen ovarian cortical tissue is solely used for transplantation. Potentially, frozen/thawed ovarian tissue could also serve as a source of follicle for IVM. However, IVM from primordial human follicles is still experimental and has not yet resulted in a fertilized oocyte or a live birth.

Ovarian tissue cryopreservation might be the only option in case of prepubertal girls and the first choice when ovarian stimulation is not recommended or there is not enough time to complete it in adult women. However, storing ovarian tissue in cancer patients carries the risk of the presence of malignant cells that could lead to recurrence of cancer after the reimplant [24]. This chapter summarizes current data on the risks of reseeding cancer in women considering reimplantation of cryopreserved ovarian tissue.

Risk of Reintroducing Malignant Cells with the Implant

The risk of cryopreserving and reseeding malignant cells should be tested separately for each disease according to the risk of ovarian metastasis and the ability to detect single malignant cells.

When considering the ovarian tissue cryopreservation process, it is convenient to discard affected ovaries in order to avoid unnecessary surgery and storage of tissue. Thus, to know the natural history of the tumor and preoperative imaging techniques before excision of the ovarian cortex is of great relevance.

Specific and sensitive methods to detect minimal residual disease have been developed over the past years. Currently available methods include histology and immunohistochemistry, PCR amplification of antigen-receptor genes or fusion transcripts, flow cytometric detection of abnormal immunophenotypes, or rearranged immunoglobulin.

If there is any consistent doubt about the possibility of reintroducing malignant cells with the cryopreserved-thawed ovarian tissue, other possibilities for fertility preservation should be considered, including oocyte or embryo cryopreservation. Although transplantation is not recommended for some pathologies at this time, it is possible to keep the tissue frozen for the future expecting the development of new methods of diagnoses for residual disease.

The risk of reintroducing malignant cells with the implant for the most frequent diagnosis of patients asking for fertility preservation is as follows.

Hematological Diseases

Hodgkin's disease in the ovary is extremely rare [25]. It may result from lymphoid tissue located at the hilium of the ovary, but not in ovarian cortex. In patients with non-Hodgkin's lymphoma, ovarian involvement is also rare, with the exception of Burkitt's type in which usually both ovaries are affected [26]. Nevertheless, in non-Hodgkin's lymphomas diagnosed during the leukemic phase, the ovaries may be involved, together with neighboring and distant organs [27].

Sometimes, it is difficult to differentiate primary ovarian lymphoma from a localized initial manifestation of a generalized disease, as involvement of the ovary by malignant lymphoma is a well-known late manifestation of disseminated nodal disease. Although non-Hodgkin's lymphoma may involve the female genital tract, particularly the ovaries, primary ovarian lymphoma is rare with only a few cases of primary lymphoblastic lymphoma of the ovary reported [28].

To date, there are no molecular markers for Hodgkin's lymphoma while T- and B-cell rearrangement can be detected using PCR techniques. This method can serve for the detection of minimal residual lymphoma in tissues.

In leukemias, ovarian tissue cryopreservation procedure is considered of high risk because cancer cells are in the bloodstream and there is high probability of blastic cells reaching different organs, including the cryopreserved ovary [29]. Leukemic infiltration of the ovaries may be submicroscopic and undetectable.

In the same way, intratesticular transplantation of testicular cells from leukemic rats has been shown to cause transmission of leukemia [30]. Decontamination of leukemic cells and enrichment of germ cells from testicular samples in rats were attempted by flow cytometric sorting, but it was unable to decontaminate testicular samples effectively. The poor specificity of spermatogonial surface markers and aggregation of germ and leukemic cells limited the positive selection of germ cells while immunophenotypic variation among lymphoblastic leukemia cells prevented adequate deletion of leukemic cells.

When germ cell selection was performed in combination with leukemic cell deletion, it prevented leukemia transmission in association with intratesticular injection of the sorted cells, but only 0.23% of the original testicular cells were recovered. Therefore, flow cytometric purification of germ cells seems ineffective and/or unsafe for clinical use [31, 32].

Meirow et al. analyzed a series of 58 young patients with hematological tumors that requested ovarian tissue cryopreservation for fertility preservation before or after chemotherapy treatment [33]. After excluding ovarian pathology or pelvic metastases with preoperative imaging, ovarian cortex was harvested from one of the ovaries. A small part of this tissue was cryopreserved separately for further investigation in the moment of the reimplant to detect malignant cells if possible. The excised ovarian cortex was histologically studied in all cases to confirm the presence of primordial follicles and the absence of malignant cells. Ovarian tissue was stored only in patients with normal laparoscopy and absence of malignant cells in the histology.

After ovarian thawing, histological evaluation and immunohistochemical staining to detect Reed–Sternberg cells in Hodgkin's lymphoma and molecular markers with PCR, real-time PCR (RT-PCR), or quantitative RT-PCR in non-Hodgkin's lymphoma and in chronic myelogenous leukemia were performed. Thawed ovarian tissue was evaluated for minimal residual disease in five patients. In *Hodgkin's lymphomas*, no minimal residual disease was found, even in patients with stage IV disease. The patient transplanted showed no evidence of disease 2 years later. Similarly, other authors have retrospectively examined subclinical involvement of harvested ovarian tissue from Hodgkin's lymphoma using morphology–immunohistochemistry, and no evidence of involvement was detected [34]. In *non-Hodgkin's lymphomas*, previous histological evaluation for lymphoma cells in the ovarian samples examined was negative as were the two thawed ovarian tissues, where the T-cell receptor PCR was negative. Thus, in Hodgkin's lymphoma and in T- and B-cell lymphomas, we can consider ovarian cryopreservation and later transplantation a safe procedure regarding the possibility of reseeding tumor metastatic cells [33, 35].

In the same study, cryopreserved ovarian tissues from two patients with chronic myeloid leukemia (CML) were thawed and analyzed. Using quantitative RT-PCR, results were positive in one patient. Other leukemic patients (acute myelocytic leukemia, chronic lymphoid leukemia, acute lymphoblastic leukemia) had not yet requested ovarian transplantation. But these preliminary results advise that the risks associated with this procedure have to be discussed in detail and that further investigation is needed.

Ewing Sarcoma

Ewing sarcoma (EWS) is a highly metastatic tumor that affects young patients. Ovarian cryopreservation is the elective option for fertility preservation in prepubertal girls, but ovarian involvement in EWS may exist.

This was elucidated in a study in which ovarian samples from eight patients between 13 and 20 years old were fixed for light microscopy and frozen in liquid nitrogen for RNA extraction followed by RT-PCR. Immunostaining for the adhesion receptor CD99 and sensitive molecular methods were used to detect both histopathological features and translocations causing the formation of tumor-specific EWS-Friend leukemia virus integration site 1 fusion gene (EWS-FLI1) [36]. In one of eight patients, the RT-PCR showed the EWS translocation, although without pathological confirmation. Thus, in young girls with EWS who want to preserve their fertility by cryopreserving ovarian tissue, ovarian tissue should be examined for traces of malignancy at both the pathological and molecular levels previous to the grafting to exclude the risk of reimplanting affected tissue.

Breast Cancer

One of the main concerns in breast cancer is the possibility of reintroducing metastatic cells within the reimplant. Different authors have suggested that in the absence of clinical and radiological evidence for distant metastasis, it is extremely unlikely to find ovarian involvement in breast cancer patients [29] with early stage tumors while 11% of the autopsies of women with stage IV breast cancer show ovarian metastases [37].

Detection of Metastases in Breast Cancer

Although breast cancer is one of the most important indications for ovarian cortex cryopreservation in western countries, currently, there is only one published study that assessed the presence of minimal disease in ovarian tissue from breast cancer patients undergoing ovarian cortex cryopreservation [38]. In this section, we refer to that study and data from Valencia Programme for Fertility Preservation regarding this important issue.

Since 2005, ovarian cortex has been cryopreserved in patients who were going to be treated with chemotherapy or had just received it due to oncological or nononcological diseases in the Valencia Programme for Fertility Preservation; seven reimplantations have been performed.

Thus, 100 ovarian cortex biopsies were analyzed from 63 women (average age 32.8 years (range 19–39)) diagnosed of infiltrating ductal breast carcinoma, stages I–IIIa. Of these, 11 patients had chemotherapy prior to the ovarian cortex extraction. None of these women carried mutations for BRCA-1, BRCA-2, or HER2-neu. Lymph node involvement was absent in 31 cases (49.2%) and present in 32 (50.8%). Ovarian macroscopic affection was discarded previously with a vaginal ultrasound in all patients.

In addition, six entire ovarian cortex pieces (four from women with stage IV breast cancer in whom a pregnancy was not recommended and two from women who did not consider pregnancy due to social reasons) were donated for research after cryopreservation. The entire ovarian cortex was processed after thawing in five to seven blocks per case, in the same way as the 100 biopsies from 63 women described above.

In our program, most of the right ovarian cortex is removed for cryopreservation leaving the left ovary in its place. For the grafting, if the patient has ovarian failure once she is cured, the left cortex is removed and the right thawed cortical pieces are implanted onto the left ovarian medulla in an attempt to reduce ischemia and provide an anatomical site for the transplantation what would allow natural gestation [39]. A biopsy from the surgically removed right ovary is always done before cryopreservation. Additionally, in this study, a biopsy of the contralateral ovary was obtained in 37 patients before cryopreservation.

The methodology used to search for cancer cells was based on the concept of the sentinel lymph node, as advances in this field could be applied to the search of micrometastases in ovarian cortex [40]. It consisted in serial sections per block that are stained with hematoxylin-eosin intercalated with some immunohistochemical markers.

The antibody panel included Gross Cystic Disease Fluid protein-I5 (GCDFPI5), Mammaglobin 1 (MGBI), Cytokeratin CAM 5.2, and Wilms tumor antigen-I (WTI).

GCDFPI5 and MGB1 are specific for breast cancer, but are absent from the ovarian tissue. *GCDFPI5* is a marker of apocrine differentiation present in 77% of breast carcinomas. When positive, it confers 99% specificity for breast origin. When negative, it does not exclude a breast origin. *MGB1* is also a marker of breast cancer origin. GCDFPI5 or MGB1 positive reaction is consistent with a breast carcinoma metastasis, but a negative reaction is not informative.

WT1 is a specific marker of ovarian tissue that is absent in breast cancer. It can identify the surface epithelium of the ovary and serves to differentiate metastatic from primary ovarian tumor cells. It is expressed in the nuclei of surface ovarian epithelium and in 100% of serous ovarian or extraovarian carcinomas and 80% of ovarian transitional carcinomas. Breast carcinomas are negative for WTI (100%). Thus, a negative result for breast markers (GCDFPI5 or MGB1) combined with a positive result for WT1 provides almost 100% specificity and sensitivity for ovarian tissue origin, therefore excluding breast cancer.

Cytokeratin CAM 5.2 is a universal marker of epithelia with high sensitivity to detect ovarian surface epithelium, ovarian follicular cells, and breast cancer cells.

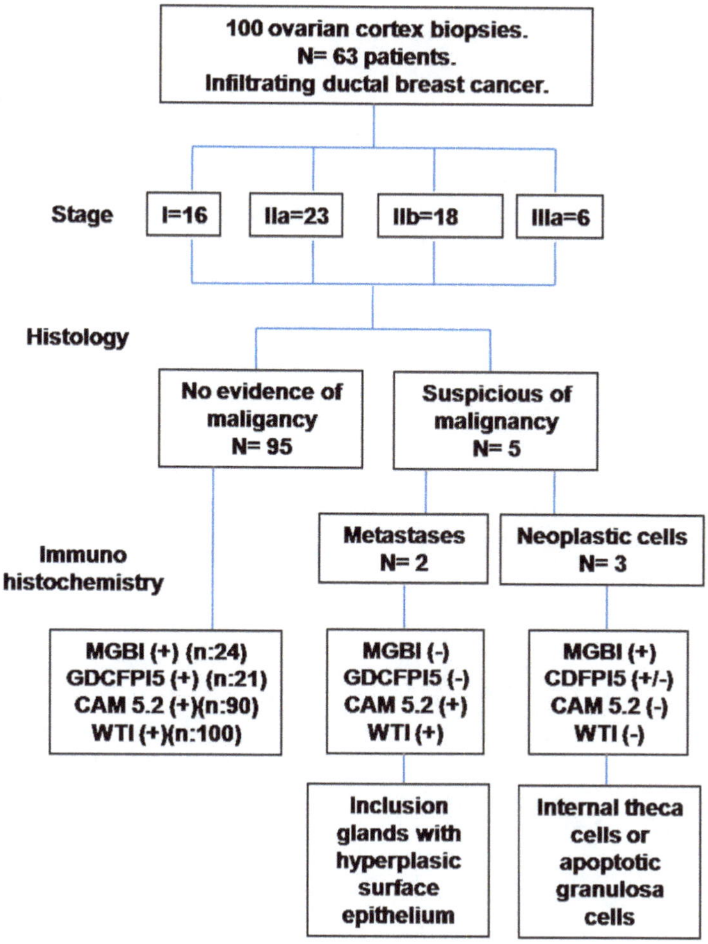

Fig. 12.2 Results after histological and immunohistochemical analysis of 100 ovarian cortex biopsies from 63 patients diagnosed of infiltrating ductal breast cancer

Micrometastases were defined as a neoplastic focus of 0.2–2 mm, and isolated tumor cells were defined as isolated neoplastic cells or metastatic foci of 0.2 mm detected through immunohistochemistry and morphology. Results were considered positive for metastases when samples showed histological findings and the immunohistochemical panel showed positivity for CAM 5.2, GCDFPI5, or MGBI with WTI negative. Samples were considered suspicious when morphologic findings were seen but not confirmed immunohistochemically.

Primordial follicles were seen in only 88 biopsies while in the other 12 the cortical tissue was fibrous stroma. In the six entire cortex analyzed, primordial follicles were seen in all of them.

Employing only morphologic criteria, suspicious neoplastic cells were detected in only five cases, but none in the six entire cortex analyzed. Two biopsies presented elongated ducts with an atypical epithelial lining, which could be considered as metastases from ductal carcinoma or atypical inclusion glands. The three biopsies left showed nests of small clear cells with atypical active nuclei near atretic follicles. They were considered as suspicious neoplastic cells or irregular apoptotic granulose cells.

These five cases were reclassified after the immunohistochemical study (Fig. 12.2). Two cases were diagnosed as hyperplasic surface epithelium or inclusion cysts (MGBI (–), GDCFPI5 (–), CAM 5.2 (+), WTI (+)), and the other three as internal theca cells or apoptotic granulose cells (MGBI (+), CAM 5.2 (–), GCDFPI5 (±), WTI (–)).The immunohistochemical characteristics of the other 95 biopsies are shown in Figs. 12.2 and 12.3.

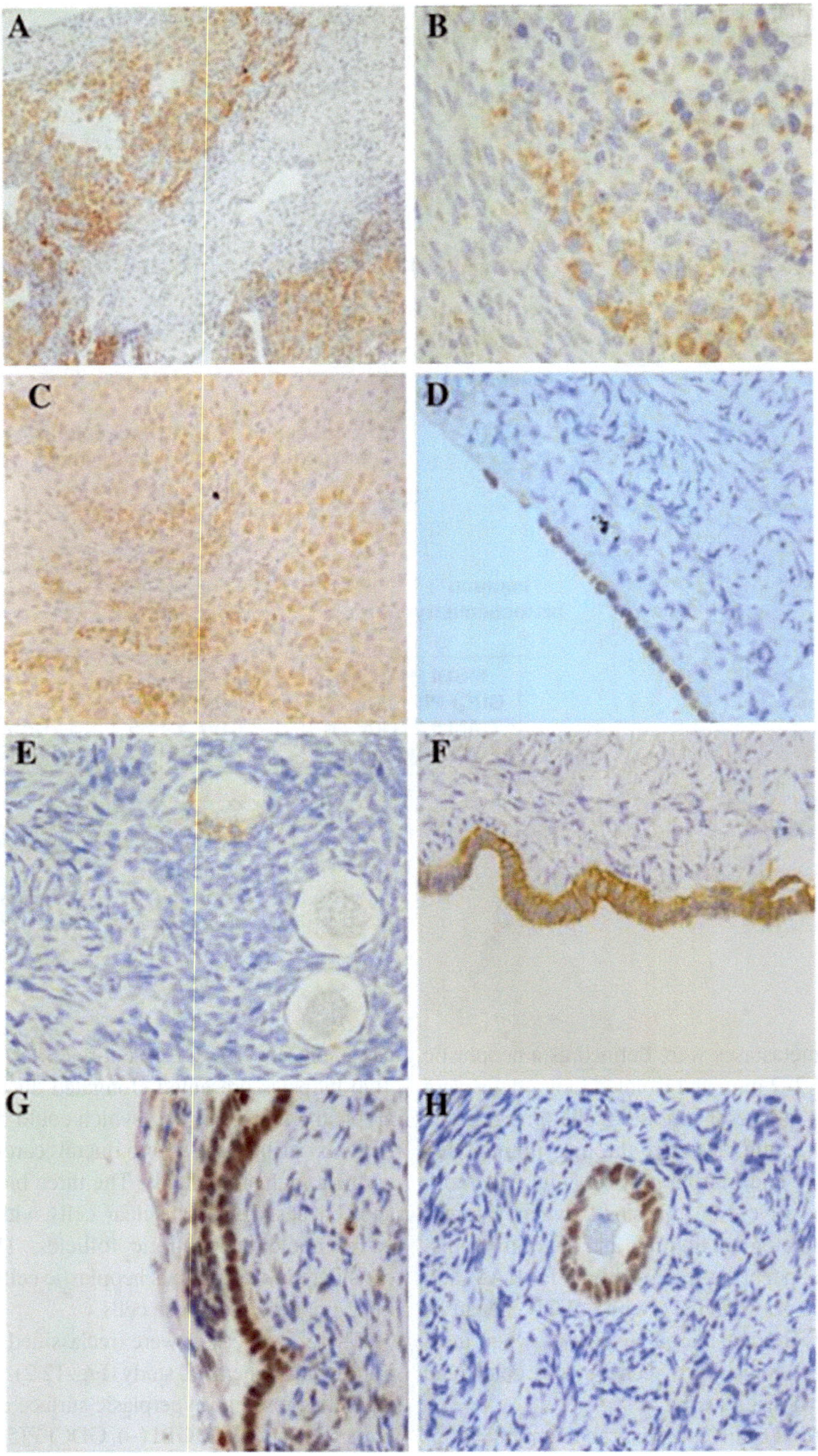

Fig. 12.3 Mammaglobin staining in luteal cells (**a**) and theca interna cells as intracytoplasmic granules (**b**). GDFP-15 positive staining in luteal cells (**c**) and light positivity in surface epithelium (**d**). CAM 5.2 staining with a diffuse intracytoplasmic pattern in granulosa cells (**e**) and strong staining in surface epithelium (**f**). WTI positive staining in surface epithelium (**g**) and granulosa cells (**h**)

One of these patients underwent reimplantation of the ovarian cortex 2 years later, once she was considered free of disease and after obtaining the consent of her oncologist. She recovered her ovarian function and menstruated regularly, although she had received chemotherapy prior to the cryopreservation of the ovarian cortex.

In another patient in whom an orthotopic transplantation was performed and as a natural pregnancy was not achieved, an IVF cycle was planned. Due to the patient's low response, vitrification of the retrieved oocytes was carried on after several stimulation cycles to accumulate them for an IVF cycle. Thus, 16 mature oocytes were obtained after four ovarian stimulations. They all survived after thaw. Fecundation rate was 77%, and two-day 3 embryos were transferred. The strategy of the combination of cryostorage, grafting, and further ovarian stimulation and vitrification resulted in a twin pregnancy [40].

As a conclusion, based on our experience, ovarian cortex storage and subsequent orthotransplantation seem to be a safe procedure that can be recommended to women with breast cancer requesting fertility preservation.

However, future studies are needed to provide more accurate methods of cancer detection to confirm the safety of ovarian cortex cryopreservation in these patients. It would be preferable if these methods were also applicable for ovarian cortex already cryopreserved and achieved high sensitivity in the detection of such micrometastases.

Controversies Regarding Ovarian Tissue Cryopreservation in Women with Breast Cancer

Patchy distribution of malignant cells is common in solid tumors, and negative evaluation of the ovarian biopsies may not exclude the presence of malignant cells in the ovarian tissue that is going to be cryopreserved. Although we analyze histologically random sections of the ovarian tissue, we are only studying a very small piece of the tissue, not the totality, as most of the tissues are frozen to preserve their fertility and are never examined.

It is noteworthy that in six cases, where the whole ovarian cortices were analyzed because they were donated for research (four due to the appearance of distant metastases), absence of ovarian metastasis was confirmed.

Finally, it should be noticed that our experience deals with breast cancers in early stages, and thus the possibility of advanced breast cancer presenting metastases in the ovary cannot entirely be excluded. In these situations, where ovarian metastases cannot be fully excluded, other strategies of fertility preservation that do not carry the possible risk of transmitting cancer cells can be considered, such as embryo or oocyte cryopreservation. Thus, caution recommends multiple biopsies when cryopreservation and grafting are chosen as fertility preservation methods, and a careful screening of the patient before the graft, although any distant metastases that could appear at a later stage would contraindicate a pregnancy and consequently the graft.

References

1. CancerStats report: breast cancer – UK. Cancer Research UK; 2009. http://info.cancerresearchuk.org/cancerstats/index.htm
2. Surveillance, Epidemiology, and End Results Program, 1975–2003, Division of Cancer Control and Population Sciences, National Cancer Institute; 2006.
3. Bines J, Oleske DM, Cobleigh MA. Ovarian function in premenopausal women treated with adjuvant chemotherapy for breast cancer. J Clin Oncol. 1996;14:1718–29.
4. Sonmezer M, Oktay K. Fertility preservation in young women undergoing breast cancer therapy. Oncologist. 2006;11:422–34.
5. Sánchez M, Novella-Maestre E, Teruel J, et al. The valencia programme for fertility preservation. Clin Transl Oncol. 2008;10:433–8.
6. Poniatowski BC, Grimm P, Cohen G. Chemotherapy-induced menopause: a literature review. Cancer Invest. 2001;19:641–8.
7. Meirow D, Epstein M, Lewis H, et al. Administration of cyclophosphamide at different stages of follicular maturation in mice: effects on reproductive performance and fetal malformations. Hum Reprod. 2001;16:632–7.
8. Ganz PA, Greendale GA, Petersen L, et al. Breast cancer in younger women: reproductive and late health effects of treatment. J Clin Oncol. 2003;21(22):4184–93.
9. Huang JY, Tulandi T, Holzer H, et al. Combining ovarian tissue cryobanking with retrieval of immature oocytes followed by in vitro maturation and

vitrification: an additional strategy of fertility preservation. Fertil Steril. 2008;89:567–72.
10. The Ethics Committee of the American Society for Reproductive Medicine. Fertility preservation and reproduction in cancer patients. Fertil Steril. 2005;83: 1622–8.
11. Kuwayama M, Vajta G, Kato O, et al. Highly efficient vitrification method for cryopreservation of human oocytes. Reprod Biomed Online. 2005;11:300–8.
12. Tao T, del Valle A. Human oocyte and ovarian tissue cryopreservation and its application. J Assist Reprod Genet. 2008;25:287–96.
13. Cobo A, Kuwayama M, Pérez S, et al. Comparison of concomitant outcome achieved with fresh and cryopreserved oocytes vitrified by the Cryotop method. Fertil Steril. 2008;89:1657–64.
14. Chian RC, Huang JY, Tan SL, et al. Obstetric and perinatal outcome in 200 infants conceived from vitrified oocytes. Reprod Biomed Online. 2008;16:608–10.
15. Venn A, Watson L, Bruinsma F, et al. Risk of cancer after use of fertility drugs with in vitro fertilization. Lancet. 1999;354:1586–90.
16. Platet N, Cathiard AM, Gleizes M, et al. Estrogens and their receptors in breast cancer progression: a dual role in cancer proliferation and invasion. Crit Rev Oncol Hematol. 2004;51:55–67.
17. Oktay K, Buyuk E, Libertella N, et al. Fertility preservation in breast cancer patients: a prospective controlled comparison of ovarian stimulation with tamoxifen and letrozole for embryo cryopreservation. J Clin Oncol. 2005;23:4347–53.
18. Requena A, Herrero J, Landeras J, et al. Use of letrozole in assisted reproduction: a systematic review and meta-analysis. Hum Reprod Update. 2008;14(6): 571–82.
19. Oktay K, Sönmezer M, Oktem O, et al. Absence of conclusive evidence for the safety and efficacy of gonadotropin-releasing hormone analogue treatment in protecting against chemotherapy-induced gonadal injury. Oncologist. 2007;12:1055–66.
20. Badawy A, Elnashar A, El-Ashry M, et al. Gonadotropin-releasing hormone agonists for prevention of chemotherapy-induced ovarian damage: prospective randomized study. Fertil Steril. 2009;91:694–7.
21. Emous G, Grndker C, Ganthert AR, et al. GnRH antagonists in the treatment of gynaecological and breast cancer. Endocrinol Relat Cancer. 2003;10: 291–9.
22. Gonfloni S, Di Tella L, Caldarola S, et al. Inhibition of the c-Abl–TAp63 pathway protects mouse oocytes from chemotherapy-induced death. Nat Med. 2009;15: 1179–85.
23. Andersen CY, Rosendahl M, Byskov AG, et al. Successful autotransplantation of frozen/thawed ovarian tissue. Hum Reprod. 2008;23:2266–72.
24. Dolmans M, Donnez J, Camboni A, et al. IVF outcome in patients with orthotopically transplanted ovarian tissue. Hum Reprod. 2009;24:2778–87.
25. Khan MA, Dahill SW, Stewart KS. Primary Hodgkin's disease of the ovary. Br J Obstet Gynaecol. 1986;93: 1300–1.
26. Eldar AH, Futerman B, Abrahami G, et al. Burkitt lymphoma in children: the Israeli experience. J Pediatr Hematol Oncol. 2009;31(6):428–36.
27. Butrón K, Ramírez M, Germes F, et al. Systemic lymphoma cells with T precursor condition of extreme female genital tract. A case report and literature review. Gynecol Obstet Mex. 2009;77(6):291–9.
28. Iyengar P, Ismiil N, Deodhare S. Precursor B-cell lymphoblastic lymphoma of the ovaries: an immunohistochemical study and review of the literature. Int J Gynecol Pathol. 2004;23(2):193–7.
29. Oktay K, Buyuk E. Ovarian transplantation in humans: indications, techniques and the risk of reseeding cancer. Eur J Obstet Gynecol Reprod Biol. 2004;113: S45–7.
30. Jahnukainen K, Hou M, Petersen C, et al. Intratesticular transplantation of testicular cells from leukemic rats causes transmission of leukemia. Cancer Res. 2001;61(2):706–10.
31. Hou M, Andersson M, Zheng C, et al. Decontamination of leukemic cells and enrichment of germ cells from testicular samples from rats with Roser's T-cell leukemia by flow cytometric sorting. Reproduction. 2007; 134(6):767–79.
32. Yokota S, Okamoto T. The minimal residual disease in hematological malignancies. Gan To Kagaku Ryoho. 2001;28(6):762–8.
33. Meirow D, Hardan I, Dor J, et al. Searching for evidence of disease and malignant cell contamination in ovarian tissue stored from hematologic cancer patients. Hum Reprod. 2008;23(5):1007–13.
34. Seshadri T, Gook D, Lade S, et al. Lack of evidence of disease contamination in ovarian tissue harvested for cryopreservation from patients with Hodgkin lymphoma and analysis of factors predictive of oocyte yield. Br J Cancer. 2006;94:1007–10.
35. Demeestere I, Simon P, Emiliani S, et al. Fertility preservation: successful transplantation of cryopreserved ovarian tissue in a young patient previously treated for Hodgkin's disease. Oncologist. 2007;12: 1437–42.
36. Abir R, Feinmesser M, Yaniv I, et al. Occasional involvement of the ovary in Ewing sarcoma. Hum Reprod. 2010;25(7):1708–12.
37. Lee MS, Chang KS, Cabanillas F, et al. Detection of minimal residual cells carrying the t(14;18) by DNA sequence amplification. Science. 1987;237:175–8.
38. Sánchez-Serrano M, Novella E, Roselló E, et al. Malignant cells are not found in ovarian cortex from breast cancer patients undergoing ovarian cortex cryopreservation. Hum Reprod. 2009;24(9):2238–43.
39. Sánchez M, Alamá P, Gadea B, et al. Fresh human orthotopic ovarian cortex transplantation: long-term results. Hum Reprod. 2007;22:786–91.
40. Branagan G, Hughes D, Jeffrey M, et al. Detection of micrometastases in lymph nodes from patients with breast cancer. Br J Surg. 2002;89:86–9.
41. Sánchez-Serrano M, Crespo J, Mirabet V, et al. Twins born after transplantation of ovarian cortical tissue and oocyte vitrification. Fertil Steril. 2010;93: 268–71.

Gonadotropin-Releasing Hormone Agonists in Fertility Preservation

13

Hakan Cakmak and Emre Seli

Abstract

Fertility preservation in females diagnosed with cancer has become an important area of investigation due to increasing cancer survival rates combined with delayed childbearing. Initial studies using gonadotropin-releasing hormone (GnRH) agonist cotreatment with chemotherapy demonstrated promising results for fertility preservation. If this protective effect of GnRH agonists is confirmed by the ongoing prospective randomized clinical trials, GnRH agonist cotreatment may become valuable option for fertility preservation in reproductive-age women undergoing chemotherapy. Thus, ovarian protection may enable the preservation of future fertility in survivors and, in addition, prevent other adverse effects of premature menopause, such as bone density loss, sexual dysfunction, and vasomotor symptoms.

Keywords

Gonadotropin-releasing hormone agonist • Fertility preservation • Cancer and fertility • GnRH in fertility • GnRH treatment with chemotherapy • Chemotherapy and GnRH treatment

Cancer is not uncommon among younger women of reproductive age. More than 710,000 new female cancer cases were estimated to be diagnosed in 2009 in the USA [1]. Approximately 10% of female cancer cases occur under the age of 40 years [2]. Over the past three decades, there has been a remarkable improvement in the survival rates due to progress in cancer treatment. With improvements in treatment outcomes, 82% of women younger than 45 years diagnosed with cancer survived between 1999 and 2006 [2]. Moreover, it was previously estimated that by 2010, one in 250 adult women will be a cancer survivor [3]. As a consequence of the increase in the number of patients surviving cancer, greater attention has been focused on the effects of cancer treatment on the quality of life of the survivor.

The treatment for most of the common cancer types in younger women involves either removal

E. Seli (✉)
Department of Obstetrics, Gynecology and Reproductive Science, Yale University School of Medicine, 333 Cedar Street, New Haven, CT 06510, USA
e-mail: emre.seli@yale.edu

E. Seli and A. Agarwal (eds.), *Fertility Preservation: Emerging Technologies and Clinical Applications*, DOI 10.1007/978-1-4419-1783-6_13, © Springer Science+Business Media, LLC 2012

of reproductive organs or cytotoxic treatment (chemotherapy and/or radiotherapy) that could partially or definitively affect reproductive function. Early loss of ovarian function not only puts the patients at risk for menopause-related complications at a very young age, but is also associated with loss of fertility.

In addition, women in the USA have been delaying initiation of childbearing to later in life [4] (see Chap. 1 for detailed review). In other words, more women in their 30s to early 40s are attempting to get pregnant for the first time than ever before. Since the incidence of most cancers increases with age, delayed childbearing results in more female cancer survivors interested in fertility preservation.

Multiple strategies have emerged aiming to preserve fertility in females with different types of malignancies. These include embryo and oocyte cryopreservation, cortical and whole ovary cryopreservation, ovarian transplantation, ovarian transposition, and gonadotropin-releasing hormone (GnRH) agonist protection [5]. In this chapter, we discuss the use of GnRH agonist cotreatment with chemotherapy for the protection of the ovarian reserve.

Distribution and Roles of the GnRH/GnRH Receptor System

GnRH

GnRH is a decapeptide (Fig. 13.1) synthesized by specific neurons located in the arcuate nucleus and in the preoptic area of the hypothalamus, and is released into the portal blood in a pulsatile fashion. On the surface of gonadotroph cells in the anterior pituitary, GnRH binds with high affinity to specific G protein-coupled receptors and induces biosynthesis and secretion of follicle-stimulating hormone (FSH) and luteinizing hormone (LH). These hormones act directly on the ovary, stimulating steroidogenesis and gametogenesis (Fig. 13.2).

Twenty-three different isoforms of GnRH have been identified in various vertebrate species [6]. All of these peptides consist of ten amino acids and have a similar structure with at least 50% sequence identity [7]; they have been named according to the species from which they were initially isolated [7, 8].

In mammals, the hypophysiotropic GnRH that stimulates the hypophysiotropic gonadotropin release was first isolated in pigs [9] and is designated as GnRH-I (mammalian GnRH) [10] while the early evolved and highly conserved new isoform that was discovered in the chicken is designated as GnRH-II (chicken GnRH) [11, 12]. In addition, in mammals, a third isoform, the salmon GnRH named GnRH-III, was reported [13]. In the human genome, only the GnRH-I and GnRH-II have been found.

In humans, in addition to the hypothalamus, the expression of GnRH-I mRNA has been demonstrated in somatic and gonadal tissues, such as the placenta, ovary, endometrium, trophoblast, and the fallopian tubes [14]. In support of these observations, GnRH-I immunoreactivity has been found mainly in the median eminence of the brain, in all endometrial cell types with the most intense staining during the secretory phase, and

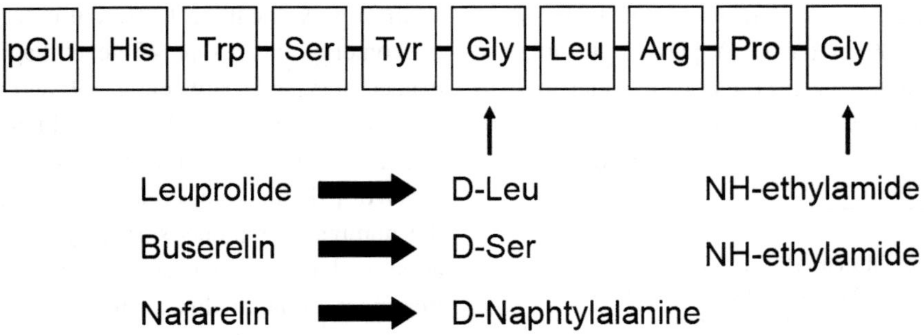

Fig. 13.1 Amino acid sequence of GnRH and GnRH agonists

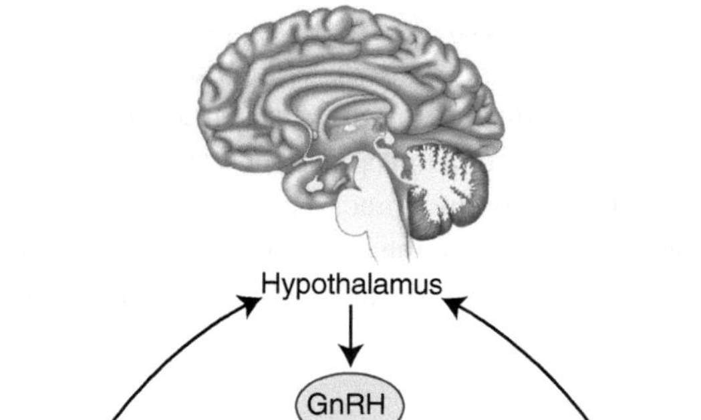

Fig. 13.2 Hypothalamic-hypophyseal-gonadal axis. Pulsatile hypothalamic GnRH secretion induces pituitary FSH excretion, which stimulates follicular growth and estrogen production in the follicular phase. As estrogen level rises later in the follicular phase, there is a positive feedback on the release of LH, resulting in the LH surge and ovulation. During the luteal phase, FSH, LH, and GnRH are significantly suppressed through the negative feedback effect of elevated circulating estrogen and progesterone. This inhibition persists until estrogen and progesterone levels decline near the end of luteal phase as a result of corpus luteum regression

in the extravillous cytotrophoblast of the placenta [14]. Similarly, the GnRH-II peptide in humans is widely distributed in the central nervous system and also in many peripheral tissues. In the brain, GnRH-II is present mainly in the preoptic and mediobasal hypothalamic area (where GnRH-I is found), but it is particularly expressed in the midbrain, in such areas as the hippocampus, caudate nucleus, and amygdale [15]. mRNA encoding for GnRH-II has also been detected in kidney, bone marrow, endometrium, ovary, trophoblast, placenta, breast cells, prostate, and ovarian cancer cells [15].

GnRH Receptor

The GnRH receptor (GnRHR) belongs to serpent-like membrane receptors, which consist of seven hydrophobic transmembrane chains connected to each other with extracellular and intracellular loops. Transmembrane chains participate in receptor activation and the transmission of signals, and intracellular loops are involved in the interaction with G proteins and also other proteins participating in the intracellular signal transmission. A common feature of the GnRHR in mammals is the absence of a carboxyl-terminal tail that exists in all other receptors that belong to the G protein-coupled receptor family [16, 17].

Two homologous GnRHR have been identified: type I and type II. The absence of the carboxyl-terminal tail results in slow internalization of type I GnRHR and prevents rapid desensitization of the receptor [16, 17]. In humans, the conventional type I GnRHR is expressed in pituitary cells, placenta, breast, ovary, uterus, prostate, and the corresponding cancer cells.

The type II GnRHR has been cloned in marmoset as well as in both African green and rhesus monkey [18, 19]. This receptor also has the characteristic carboxyl-terminal tail, which allows its rapid desensitization [20]. However, a full-length type II GnRHR mRNA is absent in humans, as the open reading frame from the putative human type II GnRHR gene is disrupted by a frame shift, resulting in a premature stop codon [21].

GnRH and GnRHR Expression in the Ovary

The human GnRHR was cloned in 1992 [22]. Since then, GnRHR mRNA has been detected in extrapituitary tissues, including the ovary. In human ovaries, GnRHR mRNA expression was initially demonstrated in granulosa-luteal cells aspirated from preovulatory follicles of women undergoing infertility treatment with in vitro fertilization (IVF) [23]. Later, other researchers confirmed the presence of mRNA encoding for GnRHR in human granulosa-luteal cells and corpus luteum [24, 25]. It is noteworthy that the levels of GnRHR mRNA in the ovary are almost 200-fold lower than in the pituitary [26].

GnRH-I, GnRH-II, and GnRHR protein expressions were not detected by immunostaining in the follicles from the primordial to the early antral stage [27]. However, in preovulatory follicles, both forms of GnRH and their common receptor were immunolocalized, predominantly to the granulosa cell layer, whereas the theca interna layer was weakly positive. In the corpus luteum, significant levels of GnRH-I, GnRH-II, and GnRHR were observed in granulosa-luteal cells, but not in theca-luteal cells. Both GnRH isoforms and the type I GnRHR were also immunolocalized to the ovarian surface epithelium [27].

Direct Effect of GnRH in the Ovary

Consistent with its expression in the ovary, where its receptor is also detected, a direct effect of GnRH on ovarian cells has been demonstrated in animal models and human subjects. GnRH-I was found to inhibit DNA synthesis in vitro and to induce apoptosis in rat granulosa cells [28, 29]. In humans, GnRH-I was reported to exert inhibitory action on ovarian steroidogenesis, decreasing progesterone production. However, other studies reported either a stimulatory effect or an absence of any effect [30]. GnRH-II, similar to GnRH-I, inhibits progesterone secretion in human granulosa-luteal cells. GnRH-II also downregulates the receptors of FSH and LH [31]. Apart from its effects on ovarian steroidogenesis, GnRH is also implicated in downregulation of cell proliferation and induction of apoptosis. GnRH-I has been suggested as a luteolytic factor, increasing the number of apoptotic luteinized granulose cells [32]. GnRH-I induced an increase in the number of apoptotic human granulosa cells obtained during oocyte retrieval for IVF [32]. Both GnRH-I and GnRH-II act as negative autocrine/paracrine regulators of cell proliferation in ovarian epithelial cells [33]. Therefore, in addition to its essential function of stimulating gonadotropin synthesis and secretion, GnRH seems to act as an autocrine and/or paracrine factor in the ovary, where it downregulates steroidogenesis and cell proliferation and promotes apoptosis.

Gonadotropins and Their Receptors

FSH and LH

FSH and LH are glycoproteins, and their protein dimer contains two polypeptide units: alpha and beta subunits. Although the alpha subunits of FSH and LH are identical, their beta subunits are specific to the gonadotropin. Beta subunits of gonadotropins confer their specific biologic action and are responsible for interaction with their receptors.

Both LH and FSH are secreted by the gonadotrope cells localized in anterior pituitary gland in response to the pulsatile stimulation of GnRH. Binding of GnRH to its receptor in the pituitary results in release of gonadotropins from their secretory granules.

FSH and LH Receptors

FSH and LH function through cell surface G protein-coupled receptors with seven transmembrane domains [34]. Both receptors are unique in having a large extracellular domain which is involved in hormone recognition and binding [35]. The FSH and LH receptor genes consist of 10 and 11 exons, respectively: the first 9 and 10 exons encode the extracellular domains while exon 10 (FSH receptor) and 11 (LH receptor) encode the transmembrane and intracellular domains [36].

In human ovary, FSH receptor expression pattern is striking with no expression in primordial and primary follicles (Fig. 13.3). However, its expression is uniformly present in as early as 3–4 granulosa layer preantral follicles [37]. This reflects the expression of FSH receptor in growing follicles rather than in the passive primordial reserve. Granulosa cells are the only ovarian cell type to express FSH receptor [38]. Ovarian LH

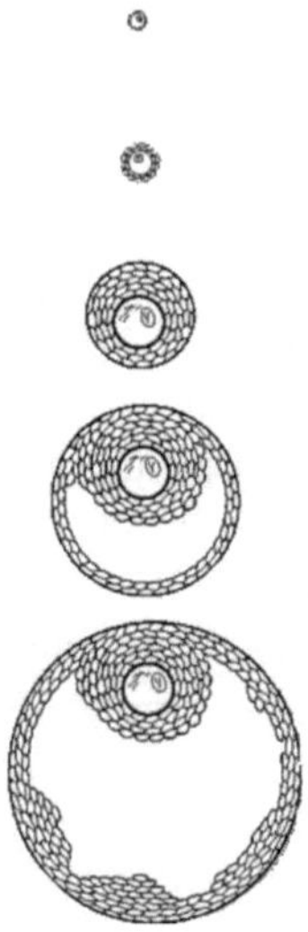

	FSHR	LHR	GnRHR
Primordial Follicle	-	-	-
Primary Follicle	-	-	-
Preantral Follicle	+	+	-
Antral Follicle	+	+	-
Preovulatory Follicle	+	+	+

Fig. 13.3 GnRH, FSH, and LH receptor expressions of ovarian follicle at different developmental stages. *GnRHR* GnRH receptor; *FSHR* FSH receptor; *LHR* LH receptor

receptor expression is limited to the theca interna in antral follicles, but it is also found in granulosa cells of Graafian follicles [39].

GnRH Agonists

GnRH agonists are decapeptides with similar structure to native GnRH and a great affinity to the GnRHRs (see Fig. 13.1). After their binding to the receptors, they initially cause gonadotropin release (flare-up effect), and after several days of continuous administration they result in a dramatic drop of the circulating concentrations of FSH and LH through a desensitization mechanism. GnRH agonists have greater affinity for the GnRHR than native GnRH; they also have greater resistance to enzymatic breakdown and a prolonged half-life compared to native GnRH (i.e., native GnRH has a half-life of about 2–4 min compared to 3 h for GnRH agonist, leuprolide) [40]. The internalization of receptors and the slow dissociation of the agonist–receptor complex decrease the total number of functional GnRHRs, leading to a prolonged desensitization [41]. During treatment with GnRH agonists, high concentrations of GnRH agonists circulate and bind all the available GnRHR, both in central and peripheral tissues.

The studies on the effects of GnRH agonists on ovarian steroidogenesis have come to contradictory conclusions. Some studies have shown a lack of direct effects [42–45] while others have suggested direct effects, reporting either inhibition or stimulation of the production of estrogen, progesterone, or both [46–53]. These controversial results may be at least partly explained by differences in the experimental settings. It is possible that not all GnRH agonists have the same effect on ovarian steroidogenesis [54].

Animal Studies Using GnRh Analogs to Achieve Fertility Preservation

Based on the debated role of gonadal suppression in men in preserving testicular function against chemotherapy and the belief that prepubertal girls are not affected by gonadotoxic cancer treatment, the effect of GnRH agonists in preserving fertility by creating a prepubertal hormonal milieu has been investigated.

Animal studies demonstrated a protective role of GnRH agonist treatment against chemotherapy-induced gonadal damage [55, 56]. Primordial follicle loss associated with cyclophosphamide treatment was significantly lower in rhesus monkeys receiving GnRH agonist treatment compared with untreated animals (65 vs. 29%, respectively) [57]. Interestingly, using the same model, protective effect of GnRH agonist treatment was not observed after radiation therapy [58]. Germinal epithelium is extremely sensitive to irradiation, and it seems unlikely that radiotherapy-induced gonadal damage can be prevented by gonadal suppression [58, 59].

In contrast to the protective effects of GnRH agonists, GnRH antagonists do not seem to protect the ovary from chemotherapeutic destruction. In a murine model, administration of GnRH antagonists not only did not prevent chemotherapeutic destruction of primordial follicles, but also depleted primordial follicles possibly through a direct effect on the ovary [60]. However, in another murine model, pretreatment with a GnRH antagonist significantly decreased the extent of primordial follicle damage induced by chemotherapy [61]. The discrepancy between these two studies may be due to the different treatment regimens or different strains of mice that were used.

Clinical Applications of Using GnRh Agonist to Achieve Fertility Preservation

Following encouraging findings in animal models, nonrandomized studies with short-term follow-up suggested a protective role for GnRH agonist cotreatment in women undergoing gonadotoxic chemotherapy (Table 13.1) [62–69]. These studies were criticized for their lack of randomization, different follow-up periods for treatment and control groups, and the use of ovarian failure as end point which may not reflect the decrease in primordial follicle count in response to chemotherapy in young women [59].

One of the largest studies including 111 patients with Hodgkin's lymphoma was published by Blumenfeld et al. [70]. In that study, a monthly depot injection of GnRH analog was administered starting before chemotherapy for up to 6 months, in parallel to, and until the end of chemotherapeutic treatment. Patients receiving GnRH agonist were compared with a control group of patients of comparable age, who were similarly treated with chemotherapy but without the GnRH agonist. The cumulative doses of each chemotherapeutic agent and the mean or median radiotherapy exposure did not differ between the groups. The main significant difference was the rate of premature ovarian failure (POF), which was 3% (2/65) in the GnRH agonist cotreatment group vs. 37% (17/46) in the control group. In the GnRH agonist treated group, 48 pregnancies occurred in 34 patients, who were 18–33 years old at chemotherapy administration, compared with 22 pregnancies in 16 patients in the control group who were 16–26 years old at chemotherapy administration [70].

Huser et al. [66] have reported similar results in patients with Hodgkin's lymphoma in a case–control study with historical controls. Both the experimental and control groups were of similar age and received the same protocols. Significantly lower rate of POF was found in the study group (20.8%) compared to that in control group (71.1%) [66].

Similarly, Castelo-Branco et al. [64] have prospectively treated 30 Hodgkin's disease patients with GnRH analog in addition to chemotherapy and compared them with 26 controls of similar age (14–45 years) and treatment and who did not receive the agonist. The control group was composed of patients who were treated for Hodgkin's disease during the same period and who had identical disease stages and chemotherapy schedules, but did not wish to wait for the effect of GnRH analog on ovarian function before starting chemotherapy. Whereas 20 of the 26 (76.9%) controls suffered POF, only 3 out of the 30 (10.0%) patients in the study group developed POF [64].

Recchia et al. [71] investigated the protective effects on ovarian function and the efficacy and tolerability of goserelin, a GnRH analog, added to adjuvant chemotherapy for early breast cancer. After a median follow-up of 55 months, 86% of the patients had resumed normal menses, 84% of patients were disease-free, and 94% were alive.

Table 13.1 Rate of premature ovarian failure following GnRH analog cotreatment during gonadotoxic chemotherapy

References	Disease	Study design	GnRH protocol	GnRH group	Control	Protective effect
Waxman et al. [69]	Lymphoma	Prospective, controlled	Buserelin 200 μg: Intranasally three times a day started 1 week before and continued throughout chemotherapy	50.0% (4/8)	66.7% (6/9)	No
Blumenfeld et al. [62]	SLE	Prospective (historic controls)	Decapeptyl C.R 3.75 mg i.m.: Monthly with chemotherapy for up to 6 months	0% (0/8)	55.6% (5/9)	Yes
Pereyra Pacheco et al. [67]	Lymphoma, leukemia	Prospective (historic controls)	Leuprolide acetate 3.75 mg i.m.: 1 week before and then every 4 weeks until 30 days after completion of chemotherapy	0% (0/12)	100% (4/4)	Yes
Dann et al. [65]	Non-Hodgkin's lymphoma	Prospective, controlled	Decapeptyl C.R. 3.75 mg i.m.: Monthly during chemotherapy	0% (0/7)	17% (1/6)	No effect of chemotherapy on ovarian function
Somers et al. [68]	SLE	Prospective, controlled	Leuprolide acetate 3.75 mg i.m.: At least 10 days before and then every 4 weeks for the duration of chemotherapy	5.0% (1/20)	30.0% (6/20)	Yes
Castelo-Branco et al. [64]	Hodgkin's lymphoma	Prospective, controlled	Triptorelin 3.75 mg i.m.: 1–2 weeks before and then every 4 weeks for the duration of chemotherapy. Tibolone was administered daily as add-back	10.0% (3/30)	76.9% (20/26)	Yes
Blumenfeld et al. [70]	Hodgkin's lymphoma/ non-Hodgkin's lymphoma	Prospective nonrandomized study with concurrent and historical controls	Decapeptyl C.R. 3.75 mg i.m.: 10–14 days before and then every 4 weeks for the duration of chemotherapy	3.1% (2/65)	63.0% (29/46)	Yes
Huser et al. [66]	Hodgkin's lymphoma	Prospective (historic controls)	Triptorelin SR 3 mg i.m.: 1 week before and then every 4 weeks for the duration of chemotherapy	20.8% (15/72)	71.1% (32/45)	Yes

The 1-, 3-, and 5-year projected recurrence-free survival rates were 100, 81, and 75%, respectively. Five years after treatment, one patient had a pregnancy that ended with a normal birth. These data show that the addition of goserelin to adjuvant therapy of premenopausal patients with early breast cancer is well-tolerated and may protect long-term ovarian function [71].

In another study, Recchia et al. [72] retrospectively evaluated 100 consecutive premenopausal women with high-risk early breast carcinoma who received a GnRH agonist as ovarian protection during adjuvant chemotherapy. After a median follow-up of 75 months, normal menses were resumed in all patients younger than 40 years and by 56% of patients older than 40 years. Three pregnancies were observed that resulted in two normal deliveries and one elective termination of pregnancy.

Similarly, Del Mastro et al. [73] have treated 30 patients with breast cancer of a median age of 38 years (range 29–47) with chemotherapy and GnRH agonist. Sixteen out of 17 patients (94%) younger than 40 years resumed cyclic ovarian function. Although this study did not include a control group, it revealed low rate of premature menopause despite chemotherapy [73].

In contrast, Waxman et al. [69] reported that GnRH agonist treatment to be ineffective in preserving fertility in patients receiving chemotherapy for Hodgkin's disease. In that study, 18 women were randomly allocated to receive GnRH agonist prior to, and for the duration of, cytotoxic chemotherapy. Eight women received buserelin. After 3 years of follow-up, four of eight in the treatment and six of nine female controls became amenorrheic. However, the power of the study was too low to show a difference [69].

The GnRH agonist cotreatment may also be applied to young women receiving cytotoxic chemotherapy for noncancerous, benign diseases. Somers et al. [68] have also demonstrated that the treatment with GnRH agonist in parallel to cyclophosphamide therapy was associated with a significant reduction of POF in young women with severe systemic lupus erythematosus (SLE). In their study, POF developed in 1 of 20 women treated with GnRH agonist (5%) compared with that in 6 of 20 controls (30%) matched by age and cumulative cyclophosphamide dose [68].

The protective effect observed in previous preliminary studies of GnRH agonists and chemotherapy on future ovarian function is now being studied in larger and prospective randomized studies that are currently ongoing (Table 13.2). Clinical trial investigating incidence of chemotherapy-induced early menopause in premenopausal women undergoing adjuvant chemotherapy in combination with vs. without triptorelin for previously resected stage I–III breast cancer has been completed, but the results have not been released yet [74]. Another phase III clinical trial that is sponsored by Southwest Oncology Group comparing the rate of POF in women with stage I–IIIA hormone receptor-negative breast cancer treated with chemotherapy with vs. without goserelin is still recruiting participants [75]. Zoladex Rescue of Ovarian function (ZORO) study (sponsored by German Breast Group), which is a prospective randomized multicenter study investigating the effect of GnRH agonist goserelin in the prevention of chemotherapy-induced ovarian failure in young hormone-insensitive breast cancer patients receiving anthracycline-containing (neo-)adjuvant chemotherapy, is also ongoing [76]. Similarly, two other phase III randomized studies (with sponsorships of M.D. Anderson Cancer Center and AstraZeneca [PROOF study]) studying a goserelin acetate for preservation of ovarian function in patients with primary breast cancer are in progress [77, 78].

Manger et al. tested the concentrations of FSH and LH before, during, and after cyclophosphamide treatment in 63 premenopausal women with SLE without ovarian protection concluding that most of them suffered POF, and they therefore initiated the ongoing Prospective Randomized study on protEction against GOnadal toxicity (PREGO)-Study investigating the effect of monthly injection of depot GnRH agonist in preservation of ovarian function [79].

Suggested Mechanisms of Gonadotoxic Protection by GnRH Agonists

In the adult ovary, ovarian reserve is primarily made up of primordial follicles that constitute about 90% of the total follicle pool [80]. These follicles are at resting stage and contain an oocyte arrested at the prophase of first meiotic division. Primordial follicles initiate follicle growth through an unknown mechanism which is FSH-independent [37, 81]. FSH receptors are

Table 13.2 Ongoing prospective randomized clinical trials investigating the use of GnRH analogs for fertility preservation in women receiving gonadotoxic therapies

Title	Sponsor	Disease	GnRH	Protocol
Prevention of chemotherapy-induced menopause by temporary ovarian suppression with triptorelin. A randomized phase III multicenter study (PROMISE) [74]	Gruppo Italiano Mammella (GIM)	Previously resected stage I–III breast cancer	Triptorelin	Triptorelin i.m.: 1 week before and then every 4 weeks for the duration of chemotherapy
Phase III trial of LHRH analog administration during chemotherapy to reduce ovarian failure following chemotherapy [75]	Southwest Oncology Group	Stage I–IIIA hormone receptor-negative breast cancer	Goserelin	Goserelin s.c.: 1 week before and then every 4 weeks for the duration of cyclophosphamide-containing chemotherapy
Prospective randomized multicenter study to prevent chemotherapy-induced ovarian failure with the GnRH agonist goserelin (ZORO) [76]	German Breast Group	Hormone-insensitive breast cancer	Goserelin	Goserelin 3.6 mg s.c.: At least 2 weeks before and then every 4 weeks for the duration of anthracycline-containing (neo-)adjuvant chemotherapy
Phase III randomized study of a goserelin acetate for preservation of ovarian function [77]	M.D. Anderson Cancer Center	Stage I–III breast cancer	Goserelin	Goserelin 3.6 mg s.c.: 1 week before and then every 4 weeks until 3 weeks after the last chemotherapy dose
Protection of ovarian function with goserelin acetate in premenopausal early breast cancer: an open-label, randomized, multicenter, phase IIIb study (PROOF) [78]	Astra Zeneca	Stage I–III breast cancer	Goserelin	Goserelin 3.6 mg s.c. along with adjuvant chemotherapy
Prospective randomized study on protection against gonadal toxicity (PREGO study) [79]	Astra Zeneca	Systemic lupus erythematosus	Goserelin	Monthly goserelin s.c. before starting and throughout cyclophosphamide treatment

not expressed in primordial or primary follicles, but the expression is uniformly present in as early as 3–4 granulosa layer preantral follicles (Fig. 13.3).

It has been hypothesized that the gonadotoxic chemotherapy induces an accelerated rate of follicular demise due to low estrogen and inhibin levels resulting in supraphysiological FSH [63]. According to this theory, high FSH levels accelerate the rate of preantral follicles' recruitment to enter the unidirectional process of maturation, resulting in their further exposure to gonadotoxic effects of chemotherapy and in accelerated follicular demise. The administration of GnRH agonists could interrupt this vicious cycle by inducing pituitary desensitization, preventing the increase in FSH concentrations despite low inhibin and estrogen levels [63]. However, as discussed above (Fig. 13.3), primordial follicles constitute the great majority for follicle reserve in adult ovary and do not express FSH receptors. Therefore, there is no molecular basis for this theory.

Another potential mechanism proposed to explain the beneficial effect of GnRH agonist on decreasing chemotherapy-associated gonadotoxicity is the decrease in utero-ovarian perfusion resulting from the hypoestrogenic state generated by pituitary–gonadal desensitization [82, 83]. The decreased utero-ovarian perfusion may result in a lower total cumulative exposure of the ovaries to the chemotherapeutic agents, in turn resulting in decreased gonadotoxic effect [63].

Sphingosine-1-phosphate (S-1-P) is a pleiotropic lipid mediator of cell survival, and may be involved in prevention of chemotherapy-induced oocyte apoptosis [84]. It was hypothesized that GnRH agonist may upregulate the ovarian S-1-P [85]. S-1-P molecule may prevent doxorubicin-induced oocyte death in vitro [86]. Moreover, administration of S-1-P prevented the radiation-induced destruction of the ovarian follicles [87]. However, this concept of GnRH agonist acting on S-1-P metabolism is still speculative and needs further evaluation.

The effects of GnRH analog may also be explained through its direct actions in the ovary. GnRHR activation by its ligands may decrease cellular apoptosis [88]. More recently, Imai et al. have demonstrated that GnRH analog may decrease the in vitro gonadotoxic effect of chemotherapy, independently of the hypogonadotropic milieu [89].

Overall, current experimental data fails to explain how GnRH agonist could protect against gonadotoxic effects of chemotherapeutic agents. However, promising evidence from clinical trials warrants investigation of possible mechanistic pathways.

Conclusions

Fertility preservation in females diagnosed with cancer has become an important area of investigation due to increasing cancer survival rates combined with delayed childbearing. Initial studies using GnRH agonist cotreatment with chemotherapy demonstrated promising results for fertility preservation. If this protective effect of GnRH agonists is confirmed by the ongoing prospective randomized clinical trials, GnRH agonist cotreatment may become valuable option for fertility preservation in reproductive-age women undergoing chemotherapy. Thus, ovarian protection may enable the preservation of future fertility in survivors and, in addition, prevent other adverse effects of premature menopause, such as bone density loss, sexual dysfunction, and vasomotor symptoms.

References

1. American Cancer Society. Cancer Facts & Figures 2009. Atlanta: American Cancer Society. 2009;1–8.
2. Altekruse SF, Krapcho M, Neyman N, Aminou R, Waldron W, Ruhl J, et al., editors. SEER cancer statistics review, 1975–2007. Bethesda: National Cancer Institute; 2010.
3. Bleyer WA. The impact of childhood cancer on the United States and the world. CA Cancer J Clin. 1990;40:355–67.
4. Hamilton BE Martin J, Ventura SJ. National vital statistics reports, births: preliminary data for 2007, Volume 57, Number 12.
5. Seli E, Tangir J. Fertility preservation options for female patients with malignancies. Curr Opin Obstet Gynecol. 2005;17:299–308.
6. Millar RP, Lu ZL, Pawson AJ, Flanagan CA, Morgan K, Maudsley SR. Gonadotropin-releasing hormone receptors. Endocr Rev. 2004;25:235–75.
7. Dubois EA, Zandbergen MA, Peute J, Goos HJ. Evolutionary development of three gonadotropin-releasing hormone (GnRH) systems in vertebrates. Brain Res Bull. 2002;57:413–8.
8. King JA, Millar RP. Evolutionary aspects of gonadotropin-releasing hormone and its receptor. Cell Mol Neurobiol. 1995;15:5–23.
9. Matsuo H, Baba Y, Nair RM, Arimura A, Schally AV. Structure of the porcine LH- and FSH-releasing hormone. I. The proposed amino acid sequence. Biochem Biophys Res Commun. 1971;43:1334–9.
10. Sealfon SC, Weinstein H, Millar RP. Molecular mechanisms of ligand interaction with the gonadotropin-releasing hormone receptor. Endocr Rev. 1997;18: 180–205.
11. Miyamoto K, Hasegawa Y, Nomura M, Igarashi M, Kangawa K, Matsuo H. Identification of the second gonadotropin-releasing hormone in chicken hypothalamus: evidence that gonadotropin secretion is probably controlled by two distinct gonadotropin-releasing hormones in avian species. Proc Natl Acad Sci USA. 1984;81:3874–8.
12. White SA, Bond CT, Francis RC, Kasten TL, Fernald RD, Adelman JP. A second gene for gonadotropin-releasing hormone: cDNA and expression pattern in the brain. Proc Natl Acad Sci USA. 1994;91:1423–7.
13. Montaner AD, Somoza GM, King JA, Bianchini JJ, Bolis CG, Affanni JM. Chromatographic and immunological identification of GnRH (gonadotropin-releasing hormone) variants. Occurrence of mammalian and a salmon-like GnRH in the forebrain of an eutherian mammal: hydrochaeris hydrochaeris (Mammalia, Rodentia). Regul Pept. 1998;73: 197–204.
14. Cheng CK, Leung PC. Molecular biology of gonadotropin-releasing hormone (GnRH)-I, GnRH-II, and their receptors in humans. Endocr Rev. 2005;26: 283–306.

15. Millar RP. GnRHs and GnRH receptors. Anim Reprod Sci. 2005;88:5–28.
16. Blomenrohr M, Heding A, Sellar R, Leurs R, Bogerd J, Eidne KA, et al. Pivotal role for the cytoplasmic carboxyl-terminal tail of a nonmammalian gonadotropin-releasing hormone receptor in cell surface expression, ligand binding, and receptor phosphorylation and internalization. Mol Pharmacol. 1999;56: 1229–37.
17. Vrecl M, Heding A, Hanyaloglu A, Taylor PL, Eidne KA. Internalization kinetics of the gonadotropin-releasing hormone (GnRH) receptor. Pflugers Arch. 2000;439:R19–20.
18. Millar R, Lowe S, Conklin D, Pawson A, Maudsley S, Troskie B, et al. A novel mammalian receptor for the evolutionarily conserved type II GnRH. Proc Natl Acad Sci USA. 2001;98:9636–41.
19. Neill JD, Duck LW, Sellers JC, Musgrove LC. A gonadotropin-releasing hormone (GnRH) receptor specific for GnRH II in primates. Biochem Biophys Res Commun. 2001;282:1012–8.
20. Millar RP. GnRH II and type II GnRH receptors. Trends Endocrinol Metab. 2003;14:35–43.
21. Morgan K, Conklin D, Pawson AJ, Sellar R, Ott TR, Millar RP. A transcriptionally active human type II gonadotropin-releasing hormone receptor gene homolog overlaps two genes in the antisense orientation on chromosome 1q.12. Endocrinology. 2003;144: 423–36.
22. Kakar SS, Musgrove LC, Devor DC, Sellers JC, Neill JD. Cloning, sequencing, and expression of human gonadotropin releasing hormone (GnRH) receptor. Biochem Biophys Res Commun. 1992;189:289–95.
23. Peng C, Fan NC, Ligier M, Vaananen J, Leung PC. Expression and regulation of gonadotropin-releasing hormone (GnRH) and GnRH receptor messenger ribonucleic acids in human granulosa-luteal cells. Endocrinology. 1994;135:1740–6.
24. Nathwani PS, Kang SK, Cheng KW, Choi KC, Leung PC. Regulation of gonadotropin-releasing hormone and its receptor gene expression by 17beta-estradiol in cultured human granulosa-luteal cells. Endocrinology. 2000;141:1754–63.
25. Olofsson JI, Conti CC, Leung PC. Homologous and heterologous regulation of gonadotropin-releasing hormone receptor gene expression in preovulatory rat granulosa cells. Endocrinology. 1995;136:974–80.
26. Minaretzis D, Jakubowski M, Mortola JF, Pavlou SN. Gonadotropin-releasing hormone receptor gene expression in human ovary and granulosa-lutein cells. J Clin Endocrinol Metab. 1995;80:430–4.
27. Choi JH, Choi KC, Auersperg N, Leung PC. Differential regulation of two forms of gonadotropin-releasing hormone messenger ribonucleic acid by gonadotropins in human immortalized ovarian surface epithelium and ovarian cancer cells. Endocr Relat Cancer. 2006;13:641–51.
28. Billig H, Furuta I, Hsueh AJ. Gonadotropin-releasing hormone directly induces apoptotic cell death in the rat ovary: biochemical and in situ detection of deoxyribonucleic acid fragmentation in granulosa cells. Endocrinology. 1994;134:245–52.
29. Saragueta PE, Lanuza GM, Baranao JL. Inhibitory effect of gonadotrophin-releasing hormone (GnRH) on rat granulosa cell deoxyribonucleic acid synthesis. Mol Reprod Dev. 1997;47:170–4.
30. Olsson JH, Akesson I, Hillensjo T. Effects of a gonadotropin-releasing hormone agonist on progesterone formation in cultured human granulosa cells. Acta Endocrinol (Copenh). 1990;122:427–31.
31. Kang SK, Tai CJ, Nathwani PS, Leung PC. Differential regulation of two forms of gonadotropin-releasing hormone messenger ribonucleic acid in human granulosa-luteal cells. Endocrinology. 2001;142:182–92.
32. Zhao S, Saito H, Wang X, Saito T, Kaneko T, Hiroi M. Effects of gonadotropin-releasing hormone agonist on the incidence of apoptosis in porcine and human granulosa cells. Gynecol Obstet Invest. 2000;49:52–6.
33. Radovick S, Wondisford FE, Nakayama Y, Yamada M, Cutler Jr GB, Weintraub BD. Isolation and characterization of the human gonadotropin-releasing hormone gene in the hypothalamus and placenta. Mol Endocrinol. 1990;4:476–80.
34. Minegishi T, Nakamura K, Takakura Y, Ibuki Y, Igarashi M, Minegish T. Cloning and sequencing of human FSH receptor cDNA. Biochem Biophys Res Commun. 1991;175:1125–30.
35. Tilly JL, Aihara T, Nishimori K, Jia XC, Billig H, Kowalski KI, et al. Expression of recombinant human follicle-stimulating hormone receptor: species-specific ligand binding, signal transduction, and identification of multiple ovarian messenger ribonucleic acid transcripts. Endocrinology. 1992;131:799–806.
36. Simoni M, Gromoll J, Nieschlag E. The follicle-stimulating hormone receptor: biochemistry, molecular biology, physiology, and pathophysiology. Endocr Rev. 1997;18:739–73.
37. Oktay K, Briggs D, Gosden RG. Ontogeny of follicle-stimulating hormone receptor gene expression in isolated human ovarian follicles. J Clin Endocrinol Metab. 1997;82:3748–51.
38. Zheng W, Magid MS, Kramer EE, Chen YT. Follicle-stimulating hormone receptor is expressed in human ovarian surface epithelium and fallopian tube. Am J Pathol. 1996;148:47–53.
39. Takao Y, Honda T, Ueda M, Hattori N, Yamada S, Maeda M, et al. Immunohistochemical localization of the LH/HCG receptor in human ovary: HCG enhances cell surface expression of LH/HCG receptor on luteinizing granulosa cells in vitro. Mol Hum Reprod. 1997;3:569–78.
40. Chillik C, Acosta A. The role of LHRH agonists and antagonists. Reprod Biomed Online. 2001;2:120–8.
41. Ortmann O, Weiss JM, Diedrich K. Gonadotrophin-releasing hormone (GnRH) and GnRH agonists: mechanisms of action. Reprod Biomed Online. 2002;5 Suppl 1:1–7.
42. Asimakopoulos B, Nikolettos N, Nehls B, Diedrich K, Al-Hasani S, Metzen E. Gonadotropin-releasing hormone antagonists do not influence the secretion of

steroid hormones but affect the secretion of vascular endothelial growth factor from human granulosa luteinized cell cultures. Fertil Steril. 2006;86: 636–41.
43. Casper RF, Erickson GF, Rebar RW, Yen SS. The effect of luteinizing hormone-releasing factor and its agonist on cultured human granulosa cells. Fertil Steril. 1982;37:406–9.
44. Lanzone A, Panetta V, Di Simone N, Arno E, Fulghesu AM, Caruso A, et al. Effect of gonadotrophin-releasing hormone and related analogue on human luteal cell function in vitro. Hum Reprod. 1989;4:906–9.
45. Weiss JM, Oltmanns K, Gurke EM, Polack S, Eick F, Felberbaum R, et al. Actions of gonadotropin-releasing hormone antagonists on steroidogenesis in human granulosa lutein cells. Eur J Endocrinol. 2001;144:677–85.
46. Bussenot I, Azoulay-Barjonet C, Parinaud J. Modulation of the steroidogenesis of cultured human granulosa-lutein cells by gonadotropin-releasing hormone analogs. J Clin Endocrinol Metab. 1993;76: 1376–9.
47. Dor J, Bider D, Shulman A, Levron JL, Shine S, Mashiach S, et al. Effects of gonadotrophin-releasing hormone agonists on human ovarian steroid secretion in vivo and in vitro-results of a prospective, randomized in-vitro fertilization study. Hum Reprod. 2000;15: 1225–30.
48. Gaetje R. Influence of gonadotrophin releasing hormone (GnRH) and a GnRH-agonist on granulosa cell steroidogenesis. Clin Exp Obstet Gynecol. 1994;21: 164–9.
49. Guerrero HE, Stein P, Asch RH, de Fried EP, Tesone M. Effect of a gonadotropin-releasing hormone agonist on luteinizing hormone receptors and steroidogenesis in ovarian cells. Fertil Steril. 1993;59:803–8.
50. Miro F, Sampaio MC, Tarin JJ, Pellicer A. Steroidogenesis in vitro of human granulosa-luteal cells pretreated in vivo with two gonadotropin releasing hormone analogs employing different protocols. Gynecol Endocrinol. 1992;6:77–84.
51. Parinaud J, Beaur A, Bourreau E, Vieitez G, Pontonnier G. Effect of a luteinizing hormone-releasing hormone agonist (Buserelin) on steroidogenesis of cultured human preovulatory granulosa cells. Fertil Steril. 1988;50:597–602.
52. Pellicer A, Miro F. Steroidogenesis in vitro of human granulosa-luteal cells pretreated in vivo with gonadotropin-releasing hormone analogs. Fertil Steril. 1990;54:590–6.
53. Uemura T, Namiki T, Kimura A, Yanagisawa T, Minaguchi H. Direct effects of gonadotropin-releasing hormone on the ovary in rats and humans. Horm Res. 1994;41 Suppl 1:7–13.
54. Ortmann O, Weiss JM, Diedrich K. Embryo implantation and GnRH antagonists: ovarian actions of GnRH antagonists. Hum Reprod. 2001;16:608–11.
55. Ataya K, Moghissi K. Chemotherapy-induced premature ovarian failure: mechanisms and prevention. Steroids. 1989;54:607–26.
56. Bokser L, Szende B, Schally AV. Protective effects of D-Trp6-luteinising hormone-releasing hormone microcapsules against cyclophosphamide-induced gonadotoxicity in female rats. Br J Cancer. 1990;61: 861–5.
57. Ataya K, Rao LV, Lawrence E, Kimmel R. Luteinizing hormone-releasing hormone agonist inhibits cyclophosphamide-induced ovarian follicular depletion in rhesus monkeys. Biol Reprod. 1995; 52:365–72.
58. Ataya K, Pydyn E, Ramahi-Ataya A, Orton CG. Is radiation-induced ovarian failure in rhesus monkeys preventable by luteinizing hormone-releasing hormone agonists? Preliminary observations. J Clin Endocrinol Metab. 1995;80:790–5.
59. Sonmezer M, Oktay K. Fertility preservation in female patients. Hum Reprod Update. 2004;10:251–66.
60. Danforth DR, Arbogast LK, Friedman CI. Acute depletion of murine primordial follicle reserve by gonadotropin-releasing hormone antagonists. Fertil Steril. 2005;83:1333–8.
61. Meirow D, Assad G, Dor J, Rabinovici J. The GnRH antagonist cetrorelix reduces cyclophosphamide-induced ovarian follicular destruction in mice. Hum Reprod. 2004;19:1294–9.
62. Blumenfeld Z, Shapiro D, Shteinberg M, Avivi I, Nahir M. Preservation of fertility and ovarian function and minimizing gonadotoxicity in young women with systemic lupus erythematosus treated by chemotherapy. Lupus. 2000;9:401–5.
63. Blumenfeld Z, von Wolff M. GnRH-analogues and oral contraceptives for fertility preservation in women during chemotherapy. Hum Reprod Update. 2008;14: 543–52.
64. Castelo-Branco C, Nomdedeu B, Camus A, Mercadal S, Martinez de Osaba MJ, Balasch J. Use of gonadotropin-releasing hormone agonists in patients with Hodgkin's disease for preservation of ovarian function and reduction of gonadotoxicity related to chemotherapy. Fertil Steril. 2007;87:702–5.
65. Dann EJ, Epelbaum R, Avivi I, Ben Shahar M, Haim N, Rowe JM, et al. Fertility and ovarian function are preserved in women treated with an intensified regimen of cyclophosphamide, adriamycin, vincristine and prednisone (Mega-CHOP) for non-Hodgkin lymphoma. Hum Reprod. 2005;20:2247–9.
66. Huser M, Crha I, Ventruba P, Hudecek R, Zakova J, Smardova L, et al. Prevention of ovarian function damage by a GnRH analogue during chemotherapy in Hodgkin lymphoma patients. Hum Reprod. 2008; 23:863–8.
67. Pereyra Pacheco B, Mendez Ribas JM, Milone G, Fernandez I, Kvicala R, Mila T, et al. Use of GnRH analogs for functional protection of the ovary and preservation of fertility during cancer treatment in adolescents: a preliminary report. Gynecol Oncol. 2001;81:391–7.
68. Somers EC, Marder W, Christman GM, Ognenovski V, McCune WJ. Use of a gonadotropin-releasing hormone analog for protection against premature ovarian failure during cyclophosphamide therapy in women with severe lupus. Arthritis Rheum. 2005;52:2761–7.

69. Waxman JH, Ahmed R, Smith D, Wrigley PF, Gregory W, Shalet S, et al. Failure to preserve fertility in patients with Hodgkin's disease. Cancer Chemother Pharmacol. 1987;19:159–62.
70. Blumenfeld Z, Avivi I, Eckman A, Epelbaum R, Rowe JM, Dann EJ. Gonadotropin-releasing hormone agonist decreases chemotherapy-induced gonadotoxicity and premature ovarian failure in young female patients with Hodgkin lymphoma. Fertil Steril. 2008;89: 166–73.
71. Recchia F, Sica G, De Filippis S, Saggio G, Rosselli M, Rea S. Goserelin as ovarian protection in the adjuvant treatment of premenopausal breast cancer: a phase II pilot study. Anticancer Drugs. 2002;13:417–24.
72. Recchia F, Saggio G, Amiconi G, Di Blasio A, Cesta A, Candeloro G, et al. Gonadotropin-releasing hormone analogues added to adjuvant chemotherapy protect ovarian function and improve clinical outcomes in young women with early breast carcinoma. Cancer. 2006;106:514–23.
73. Del Mastro L, Catzeddu T, Boni L, Bell C, Sertoli MR, Bighin C, et al. Prevention of chemotherapy-induced menopause by temporary ovarian suppression with goserelin in young, early breast cancer patients. Ann Oncol. 2006;17:74–8.
74. National Cancer Institute. Triptorelin in preventing early menopause in premenopausal women who are receiving chemotherapy for stage I stage II or stage III breast cancer that has been removed by surgery. http://www.clinicaltrials.gov/ct/show/NCT00311636. Accessed 23 May 2010.
75. National Cancer Institute. Goserelin in preventing ovarian failure in women receiving chemotherapy for breast cancer. http://clinicaltrials.gov/ct/show/NCT00068601. Accessed 23 May 2010.
76. National Cancer Institute. Prospective randomized multicenter study to prevent chemotherapy induced ovarian failure with the GnRh-agonist goserelin in young hormone insensitive breast cancer patients receiving anthracycline containing adjuvant chemotherapy. http://clinicaltrials.gov/ct/show/NCT00196846. Accessed 23 May 2010.
77. National Cancer Institute. Phase III randomized study of a goserelin acetate for preservation of ovarian function in patients with primary breast cancer. http://clinicaltrials.gov/ct/show/NCT00429403. Accessed 23 May 2010.
78. National Cancer Institute. Protection of ovarian function with goserelin acetate in premenopausal early breast cancer patients with chemotherapy (PROOF). http://clinicaltrials.gov/ct/show/NCT00888082. Accessed 23 May 2010.
79. Manger K, Wildt L, Kalden JR, Manger B. Prevention of gonadal toxicity and preservation of gonadal function and fertility in young women with systemic lupus erythematosus treated by cyclophosphamide: the PREGO-study. Autoimmun Rev. 2006;5:269–72.
80. Oktay K, Nugent D, Newton H, Salha O, Chatterjee P, Gosden RG. Isolation and characterization of primordial follicles from fresh and cryopreserved human ovarian tissue. Fertil Steril. 1997;67:481–6.
81. McNatty KP, Heath DA, Lundy T, Fidler AE, Quirke L, O'Connell A, et al. Control of early ovarian follicular development. J Reprod Fertil Suppl. 1999;54: 3–16.
82. Kitajima Y, Endo T, Nagasawa K, Manase K, Honnma H, Baba T, et al. Hyperstimulation and a gonadotropin-releasing hormone agonist modulate ovarian vascular permeability by altering expression of the tight junction protein claudin-5. Endocrinology. 2006;147: 694–9.
83. Saitta A, Altavilla D, Cucinotta D, Morabito N, Frisina N, Corrado F, et al. Randomized, double-blind, placebo-controlled study on effects of raloxifene and hormone replacement therapy on plasma no concentrations, endothelin-1 levels, and endothelium-dependent vasodilation in postmenopausal women. Arterioscler Thromb Vasc Biol. 2001;21:1512–9.
84. Tilly JL. Commuting the death sentence: how oocytes strive to survive. Nat Rev Mol Cell Biol. 2001;2: 838–48.
85. Blumenfeld Z. How to preserve fertility in young women exposed to chemotherapy? The role of GnRH agonist cotreatment in addition to cryopreservation of embryo, oocytes, or ovaries. Oncologist. 2007;12: 1044–54.
86. Morita Y, Perez GI, Paris F, Miranda SR, Ehleiter D, Haimovitz-Friedman A, et al. Oocyte apoptosis is suppressed by disruption of the acid sphingomyelinase gene or by sphingosine-1-phosphate therapy. Nat Med. 2000;6:1109–14.
87. Paris F, Perez GI, Fuks Z, Haimovitz-Friedman A, Nguyen H, Bose M, et al. Sphingosine 1-phosphate preserves fertility in irradiated female mice without propagating genomic damage in offspring. Nat Med. 2002;8:901–2.
88. Grundker C, Emons G. Role of gonadotropin-releasing hormone (GnRH) in ovarian cancer. Reprod Biol Endocrinol. 2003;1:65.
89. Imai A, Sugiyama M, Furui T, Tamaya T, Ohno T. Direct protection by a gonadotropin-releasing hormone analog from doxorubicin-induced granulosa cell damage. Gynecol Obstet Invest. 2007;63: 102–6.

Fertility Preservation in Gynecologic Malignancies

14

Christine E. Richter and Peter E. Schwartz

Abstract

Twenty-one percent of gynecologic cancers affect women in the reproductive-age group. Since an increasing number of women are delaying childbearing and overall survival of many gynecologic malignancies has improved, fertility preservation is an important issue for these young patients. The standard treatment of gynecologic cancers often includes the removal of the ovaries and/or the uterus. Conservative surgery consists of the preservation of at least the uterine corpus and part of one ovary in order to maintain the reproductive potential. While fertility-preserving options are promising, the majority of these procedures are not considered standard of care. The selection of appropriate patients and counseling of interested patients are challenging. Close collaborations between the gynecologic oncologists, reproductive endocrinologists, geneticists, and maternal-fetal medicine specialists are necessary. This chapter presents the current standard of care for each type of gynecologic malignancy and discusses the options of fertility preservation for each type of cancer, focusing on the indications, techniques, and reproductive and oncological outcomes of fertility-preserving surgery.

Keywords

Fertility preservation in gynecologic malignancy • Gynecologic malignancy and fertility • Ovarian cancer and fertility • Uterine cancer and fertility • Conservative surgery in gynecologic malignancy • Ovarian preservation • Trachelectomy • Tumorectomy

Although the majority of gynecologic malignancies are diagnosed in postmenopausal women, 21% of these cancers affect women in the reproductive-age group [1, 2]. A diagnosis during the reproductive age is more common in cervical and ovarian cancers (41.3% and 12.2%, respectively) than in cancers of the uterus, vagina, or vulva (≤10%) [3]. Since an increasing number of women are delaying childbearing and overall survival of many gynecologic malignancies has improved, fertility preservation is an important

P.E. Schwartz (✉)
Department of Obstetrics, Gynecology and Reproductive Sciences, Yale University School of Medicine, 333 Cedar Street, New Haven, CT 06510, USA

E. Seli and A. Agarwal (eds.), *Fertility Preservation: Emerging Technologies and Clinical Applications*, DOI 10.1007/978-1-4419-1783-6_14, © Springer Science+Business Media, LLC 2012

issue for young patients [1]. While the significant influence of concerns about infertility on the treatment decisions of young women with cancer has been well-documented, counseling regarding fertility-preserving treatment choices remains poor [4]. The American Society of Clinical Oncology recommends that oncologists "should address the possibility of infertility with patients treated during their reproductive years and be prepared to discuss possible fertility preservation options or refer appropriate and interested patients to reproductive specialists" [5]. Despite these recommendations, only half of the U.S. physicians are following these referral guidelines [6].

The standard treatment of gynecologic cancers often includes the removal of the ovaries and/or the uterus [7]. Conservative surgery consists of the preservation of at least the uterine corpus and part of one ovary in order to maintain the reproductive potential [7]. While fertility-preserving options are promising for young women with gynecologic cancer, the majority of these procedures are not considered standard of care [1]. Selecting the appropriate patients is challenging, and counseling of the patient and her family is crucial [1]. Unfortunately, the data on fertility-preserving procedures for the treatment of gynecologic malignancies is very limited [1]. While this should not deprive a young woman of her hope of future reproductive function, she needs to understand that compliance with a close follow-up regimen is highly important and that the risk of recurrence and death is currently undefined [1]. Along with the gynecologic oncologist, reproductive endocrinologists, perinatologists, geneticists, and maternal-fetal medicine specialists should be part of the counseling and treatment [1].

The goal of this chapter is to present the available fertility-preserving surgical procedures for the different types of gynecologic cancers. It consists of a brief introduction of each carcinoma and the current standard of care. It subsequently introduces options of fertility preservation for each type of cancer addressed. The focus is on the indications, techniques, and reproductive and oncological outcomes for fertility-preserving surgery. The current literature on fertility-preserving surgery is summarized in tables accompanying most sections of this chapter.

Cervical Carcinoma

Cervical carcinoma is the second leading cause of death in women worldwide and has a mortality rate of 50% [8–10]. While the median age at diagnosis is 48 years, approximately 40% of the cervical cancer patients are of reproductive age [8, 10–12]. Twenty-six percent of cervical cancer patients are diagnosed between the age of 35–44, 15% between age 20–34, and <1% are younger than age 20 [3]. Most cervical carcinomas are squamous cell cancers, but there is an increasing incidence of adenocarcinomas in the USA, especially in younger women [4]. Cancers of the cervix are the most common reproductive organ cancers to occur in pregnancy.

Standard of Care

The treatment of cervical cancer depends on the stage. Stage IA1 is treated with a cone biopsy when fertility preservation is important or simple hysterectomy when fertility preservation is not an issue [2, 4, 13]. Treatment options for stage 1A2 to IB1 include radical hysterectomy with pelvic lymphadenectomy or radiation, radiation only, or chemoradiation [2, 4, 13]. Experience utilizing radical trachelectomy and partial vaginectomy to allow fertility preservation is now being reported. Patients presenting with stage IB2 to IV disease are treated with chemoradiation [13].

Fertility Preservation

Conization

Indications

Reproductive-age patients with stage IA1 squamous cell cervical carcinoma without lympho-vascular space invasion (LVSI) who desire fertility preservation and some patients with stage IA2 cervical cancer are candidates for a cervical conization [7, 13].

Technique

Laser cone biopsies, loop electrosurgical excision of the transformation zone (LEETZ) procedures, and cold-knife cone biopsies appear to have similar efficacies in the treatment of stage IA1 cervical cancer [14]. Some authors recommend the performance of a cold-knife cone rather than an LEETZ procedure to avoid fragmentation of the specimen [2]. If the margins on the specimen are positive, additional excision is necessary [7].

Oncological Outcome

The risk of pelvic lymph node metastases in stage IA1 disease is <1% in patients without LVSI, with a depth of invasion ≤3 mm and negative endocervical curettings (ECCs) [2, 15]. The reported risk of residual microinvasive carcinoma after complete excision of the lesion is approximately 5%, and the risk of recurrence after conservative surgery <0.5% [1, 2]. In the largest series of 200 patients with stage IA1 cervical carcinoma without LVSI treated by conization, all patients were alive and recurrence-free at a median follow-up time of 117 months [16]. This is consistent with results by Tseng et al. [17]. It is extremely important that the endocervical margin and curettings are negative since there is a 10% chance of the tumor being stage IA2 with positive margins and curettings [1]. In patients with LVSI or a depth of invasion >3 mm, the risk of pelvic lymph node metastases increases to 8% [1].

For stage IA2 cervical cancer, the risk of lymph node metastases is 5–9% [2]. Depending on the size of the lesion and the depth of invasion, a cervical cone biopsy in combination with pelvic lymph node dissection or a sentinel pelvic lymph node dissection may be possible in cases with negative margins and in the absence of other risk factors [4]. While the literature shows no relapses in cases without LVSI, the number of patients studied is very limited [2].

The data on fertility-preserving surgery in patients with stage IA1 adenocarcinoma is conflicting. Bisseling et al. and McHale et al. noted no recurrences patients with cervical adenocarcinoma who underwent conservative treatment [18, 19]. Pahisa et al., however, reported a recurrence rate of 25% in conservatively treated stage IA1 adenocarcinomas of the cervix [14]. It is important to be aware of the fact that adenocarcinoma in situ (AIS) may be associated with skip lesions. Invasive adenocarcinomas have been found in hysterectomy specimens in which the uterus was removed for the treatment of AIS. Similar findings are possible to occur in women with stage IA1 adenocarcinomas of the cervix treated with a cone biopsy and having negative margins.

Fertility Outcome

Pregnancy rates after cervical conization are 50–60% [7]. Cervical conization is associated with a significantly increased risk of preterm birth<37 weeks, birth weight<2,500 g, and a significantly higher risk of cesarean section [7, 20]. In addition, patients who have undergone large LEETZ procedures are also at increased risk for premature rupture of membranes [20]. Careful counseling of the patient is paramount, and fertility preservation should only be offered to the well-informed patient.

Radical Trachelectomy and Partial Vaginectomy

Indications

It is estimated that 50% of women with cervical cancer diagnosed under age 40 are eligible for a fertility-preserving radical trachelectomy (RT) and partial vaginectomy, the removal of the uterine cervix and adjacent tissue [1, 5, 11, 21]. Dargent reported his first series of patients treated by vaginal radical hysterectomy in 1994 [7, 22]. Today, there are four surgical options for young patients with low-risk cervical cancer who desire fertility preservation: radical vaginal trachelectomy (RVT), radical abdominal trachelectomy (RAT), radical laparoscopic trachelectomy (RLT), and radical robotic trachelectomy (RRT) [4, 23]. The following patients are candidates for a radical trachelectomy: [1, 5, 7, 8, 24–26]

- Age<40 years
- Desire to preserve fertility
- No clinical evidence of impaired fertility
- Stage IA to IB1
- Lesion size<2 cm
- Limited endocervical canal involvement at colposcopy
- No evidence of pelvic node metastases

Some authors further limit this procedure to patients with tumors with a depth of invasion of <10 mm [1]. LVSI and adenocarcinomas are relative contraindications [1, 7]. Patients with small-cell carcinomas or sarcomas are not candidates for RT [7].

Specific indications for RAT are pediatric patients, bulky exophytic tumors, distorted vaginal anatomy, or cervical cancer during the first half of pregnancy [27]. RAT can also be offered to patients who meet the criteria for RT but have limited access to the upper vagina via the perineal route. Of note, patients diagnosed with cervical cancer during the first half of pregnancy may also be offered neoadjuvant chemotherapy and undergo definitive surgery postpartum.

Technique

A thorough history and physical exam, an examination under anesthesia, and an MRI for lymph node assessment are part of the preoperative assessment [4, 5, 7, 9]. The RT is a two-step procedure: first, the pelvic lymph node dissection (with or without para-aortic lymphadenectomy) with frozen section analysis of the lymph nodes is performed [26–28]. This can be done by laparotomy, laparoscopy, or robotically [10, 26, 27]. If the lymph nodes are negative for malignancy, the RT follows [2, 4, 26–29]. It consists of a vaginotomy, paravaginal and cervical dissection, ligation of the descending branch of the uterine artery, ureterolysis, and transsection of the distal cervix [2, 7, 9]. A frozen section of the proximal margin of the cervix is performed [1]. Positive margins require a radical hysterectomy [1]. Some authors recommend the placement of a cerclage, but its benefit for the prevention of preterm delivery is not clear [9, 27].

A time period of 4–6 weeks between a cervical cone biopsy and an RT is recommended to avoid bleeding associated with a dissection through inflammatory tissue [7].

The RAT is usually performed through a lower midline laparotomy and allows for a wider parametrial resection than the RVT [30]. The abdominal approach of this procedure shares greater similarity with the radical hysterectomy than the RVT and may, therefore, be easier for gynecological oncologists who are mainly trained in radical abdominal surgery than the RVT [27, 29–31].

Cervical stenosis, vaginal passage of the cerclage, and dyspareunia are potential postoperative complaints [31].

Oncological Outcome

Compared to patients treated with a radical hysterectomy and RAT, the RVT is associated with less intraoperative blood loss and fewer transfusions, a shorter hospital stay, shorter time to normal urine residual with a similar operating time, and intraoperative complications [30, 32–34].

Successful RVT has been described in 968 cases (Table 14.1) [1, 9, 21, 28, 30, 32–40]. The procedure is aborted in 11–12% of cases most commonly because of lymph node metastases, positive cervical margins, or more extensive disease than expected [1]. The recurrence rate of 4% is comparable to that of a radical hysterectomy, and tumor size is the best predictor of recurrence [1, 2, 7, 12]. Patients with tumors >2 cm in size have a recurrence rate of up to 29%, most commonly in the parametria, para-aortic lymph nodes, and the pelvic sidewall [2, 30]. Use of the sentinel node in cervical cancer for the assessment of parametrial spread has shown some promising results, but the number of patients are small and further research is necessary [2].

The radical abdominal approach yields more than a 50% greater parametrial specimen than the radical vaginal approach [21]. Within the 145 successful RATs with information on recurrences, 7 recurrences (4.8%) and no deaths have been described (Table 14.1) [26, 29–31, 41–43]. Advantages of the RAT are the greater radicality of the parametrial resection and the possibility to adjust the radicality of the parametrial resection according to prognostic factors [27]. The disadvantage is the necessity of laparotomy [27].

The laparoscopic/robotic approaches do not appear to compromise lymph node counts or parametrial length, but patient numbers are very limited [23, 28, 44]. Their potential advantages include that they allow for a more radical parametrial resection than the vaginal procedures while avoiding blood loss, abdominal incision, and a longer postoperative recovery period [45].

Fertility Outcome

By 2010, 1,168 radical abdominal and vaginal trachelectomies have been reported and 308 postoperative pregnancies with 195 live births have been reported (Table 14.1) [9, 11]. Pregnancy rates after RVT are 50–60% [1, 2, 7]. The reported

Table 14.1 Cervical cancer 2000–2010

First author	Year	Patients	Procedure	Pregnancies	Attempted pregnancies	Live births	Preterm deliveries <36 w GA	Term deliveries	Recurrences	Deaths
RVT[a]										
Beiner	2008	90	RVT	NR[b]	NR	NR	NR	NR	5	3
Diaz	2008	40	RVT+RAT	9	5	4	1	3	1	1
Einstein	2008	26	RVT	NR	NR	NR	NR	NR	NR	NR
Shepherd	2008	158	RVT	88	NR	44	25	19	4	4
Sonoda	2008	41	RVT	11	14	4	0	4	1	0
Marchiole	2007	118	RVT	NR	NR	NR	NR	NR	7	5
Hertel	2006	100	RVT	18	NR	12	NR	NR	4	2
Plante	2005	72	RVT	50	39	36	NR	NR	NR	2
Ungár	2005	30	RAT	3	5	2	0	2	0	0
Bernardini	2003	80	RVT	22	39	18	6	12	1	NR
Burnett	2003	18	RVT	4	NR	3	2	1	0	0
Covens	2003	81	RVT	22	37	18	6	12	5	NR
Mathevet	2003	95	RVT	56	42	34	5	29	4	1
Schlaerth	2003	10	RVT	4	NR	2	0	2	0	0
Total RVT		959		287	181	177	45	84	32	18
RAT[c], *RLT*[d], *RRT*[e]										
Cibula	2009	20	RAT+RLT	6	6	5	2	3	1	NR
Kim	2009	27	RLT	3	6	1	0	1	1	1
Nishio	2009	61	RAT	4	29	4	2	2	6	0
Olawaiye	2009	10	RAT	3	3	4	0	4	0	0
Ramirez	2009	4	RRT	NR	NR	NR	NR	NR	0	NR
Abu-Rustum	2008	19	RAT	NR	NR	NR	NR	NR	0	NR
Einstein	2008	13	RAT	NR	NR	NR	NR	NR	NR	NR
Bader	2005	1	RAT	NR	NR	NR	NR	NR	1	0
Ungár	2005	30	RAT	3	5	2	0	2	0	NR
Palfalvi	2003	21	RAT	2	5	1	0	1	0	NR
Rodriguez	2001	3	RAT	NR	1	1	0	1	0	0
Total RAT, RLT, RRT		209		21	55	18	4	14	9	1
Total all RT[f]		1,168		308	236	195			41	19

[a]RVT = radical vaginal trachelectomy
[b]NR = not recorded
[c]RAT = radical abdominal trachelectomy
[d]RLT = radical laparoscopic trachelectomy
[e]RRT = radical robotic trachelectomy
[f]RT = radical trachelectomy

pregnancy rates for patients undergoing RAT are lower (range from 14 to 70%) [11, 31]. The majority of the RAT patients required fertility treatments to conceive [11]. This may in part be caused by the selection of patients, with patients requiring a more extensive procedure being selected for the RAT rather than RVT [27].

The risk of prematurity (<36 weeks) after RT has varied from 10 to 33%, most likely associated with an infectious etiology [5, 8, 9, 11, 26]. There is also an increased risk of midtrimester losses (12%) [5, 8, 9, 11, 26]. Placement of a cerclage for the prevention of prematurity is controversial and can lead to cerclage-related complications [11]. The mode of delivery after an RT is cesarean section [1].

Thirty percent of women after RT have fertility-related issues, with the majority being attributed to cervical issues [11, 26]. Cervical or isthmic stenosis may lead to menstrual disorders [11, 26]. Return of menstrual function is seen in the majority of patients (>90%) [26, 30].

It is important to keep in mind that there is still a need for further standardization of patient selection, for longer follow-up and training for this challenging procedure [12].

Sarcoma Botryoides

Sarcoma botryoides is a variant of an embryonal rhabdomyosarcoma of the female genital tract [46, 47]. It is very rare, accounting for 4–6% of all malignancies in childhood and adolescence [48–50]. Sarcoma botryoides is characterized by a grape-like appearance caused by spindle cells pushing up beneath the mucosal layer [50]. Sarcoma botryoides typically presents in childhood and adolescence [46–52]. The childhood variant is most commonly localized in the vagina and usually presents before age 4. In adolescence, the sarcoma is typically localized in the cervix and uterine corpus [46–52]. Cervical sarcoma botryoides tends to behave less aggressively than the vaginal or uterine type, with a 60–80% survival for cervical and 96% survival for vaginal sarcomas [47, 48, 50, 51]. The extent of the disease at diagnosis and its location are the most important prognostic factors [50].

Standard of Care

Since the 1960s, a combination of surgery (i.e., simple hysterectomy or local excision) and combination chemotherapy has been the standard of care for sarcoma botryoides [50, 51, 53]. The most commonly used chemotherapeutic regimen consists of vincristine, dactinomycin, and cyclophosphamide (VAC) [50]. The role of radiotherapy in the treatment of sarcoma botryoides is controversial [52].

Fertility Preservation

Tumorectomy

Indications

Patients with localized disease and favorable prognostic factors (lesions confined to a single polyp, absence of deep myometrial invasion) who want to preserve their fertility are potential candidates for conservative surgical treatment with chemotherapy [47, 49]. The presence of extensive extrauterine disease or metastases is a contraindication for fertility-sparing surgery [47].

Technique

Wide local excision of the tumor after preoperative chemotherapy may be an option for patients who desire to preserve their fertility [50]. After removal of the tumor, negative margins need to be confirmed pathologically [51]. If the margins are positive, more aggressive surgery is warranted. The addition of pelvic radiation is controversial, particularly since there is concern of radiation-induced ovarian damage in patients who are interested in fertility preservation [50, 51].

Oncological Outcome

In localized disease with favorable prognostic parameters, the outcome of patients treated with fertility-preserving surgery appears to be similar to patients treated with radical hysterectomy or pelvic exenteration [47, 49, 51]. To date, the number of patients treated with fertility-preserving surgery is small and consists primarily of case reports. Table 14.2 summarizes the available data: among the 61 patients treated conservatively, the documented recurrence rate is 23%

Table 14.2 Sarcoma botryoides 1985–2010

First author	Year	Patients	Procedure	Fertility data	Recurrences	Deaths
Bernal	2004	1 (cvx[a])	Local excision, VA[b]×6 cycles	NR[c]	0	0
Gruessner	2004	1 (cvx)	Polypectomy, cervical conization, AVI[d]	NR	0	0
Behtasth	2003	2 (cvx)	Local excision	NR	1	0
Porzio	2000	1 (cvx)	Polypectomy, epidoxorubicin×6 cycles	NR	0	0
Fawole	1999	1 (vag[e])	VAC[f]×12 cycles	NR	0	0
Martelli	1999	38 (vulva, vagina, uterus)	AVI Local excision only if no response to chemotherapy	NR	4	3
Lin	1995	1 (cvx)	Cervicectomy, VAC×9 cycles	NR	1	0
Gordon	1990	1 (cvx)	Local excision, VCN[g]×3 cycles	NR	1	0
Zanetta	1990	3 (cvx)	Local excision, DI[h]×3–4 cycles	NR	1	0
Daya	1988	6 (cvx)	Polypectomy/cervicectomy	NR	1	1
Hays	1988	6 (cvx)	Polypectomy/transvaginal biopsy/partial excision, VAC/VCN	NR	5	2
Total		61			14	6

[a] Cvx = cervix
[b] VA = vincristine, actinomycin
[c] NR = not recorded
[d] AVI = actinomycin, vincristine, ifosfamide
[e] Vag = vagina
[f] VAC = vincristine, dactinomycin, cyclophosphamide
[g] VCN = vincristine
[h] DI = doxorubicin, ifosfamide

with a mortality of 10%. More than half of these patients were initially treated with neoadjuvant chemotherapy only [53, 54]. In this subgroup of patients, the recurrence risk was 10% with a mortality of 8% (Table 14.2) [53, 54].

Fertility Outcome

Data on the fertility outcome of patients after fertility-preserving treatment of the sarcoma botryoides are scarce. Isolated reports state that the vast majority of patients resume normal ovarian function after the completion of their treatment, but larger studies are needed to evaluate the treatment effect [47]. Chemotherapeutic agents may impair fertility, depending on the agent, dose, and age of the patient [55]. This is particularly true for cyclophosphamide, an alkylating agent that is part of the routine VAC treatment regimen for sarcoma botryoides [55]. It is gonadotoxic and has been shown to cause ovarian failure with an odds ratio of 4.0 [55].

To date, there is no data on congenital malformations in the offspring of women treated for sarcoma botryoides. However, large studies evaluating the risk of congenital malformations in the offspring of female childhood cancer survivors have failed to show a significant increase in malformations compared to the general population [55].

Ovarian Carcinoma

Borderline Ovarian Carcinoma

Ten to twenty percent of all ovarian carcinomas are borderline ovarian carcinomas, i.e., ovarian tumors of low malignant potential [2, 4, 56–59]. They share the pathological features of invasive carcinomas (cellular proliferation, nuclear atypia, mitotic activity), but show no evidence of ovarian stromal invasion [57, 60]. Unlike invasive epithelial carcinoma, borderline ovarian carcinoma primarily affects reproductive-age women [4, 12, 15]. The median age at diagnosis is 39 years, so fertility preservation is of particular interest to many of these patients [4, 12]. The majority (70%) of borderline ovarian tumors is diagnosed in stage I and has an excellent prognosis [61]. Stage I disease patients have a 5-year survival >95% for stage I disease [4, 56, 57, 62].

Histologically, the most common form of a borderline ovarian cancer is serous and represents approximately two-thirds of all ovarian borderline carcinomas. Mucinous tumors represent approximately one-third of borderline ovarian carcinomas [62]. Rupture of the latter tumors preoperatively or at the time of surgery may be associated with the long-term complication of pseudomyxoma peritonei. Rarely, one may find borderline ovarian tumors of endometrioid or clear cell histology.

Ovarian tumors of borderline malignant potential may be associated with two forms of metastases, noninvasive or invasive. The latter have been associated with a poor prognosis in some series, but not in others. Burks et al. have suggested dividing serous ovarian tumors or low malignant potential into two forms, one a proliferative form that is benign and the other, which expresses a micropapillary pattern similar to that of a well-differentiated invasive serous cancer [63].

Standard of Care

The standard surgical management for women with borderline ovarian cancers is similar to invasive ovarian carcinoma [57, 64]. When fertility is not an issue, it consists of a surgical staging procedure, including a hysterectomy, bilateral salpingo-oophorectomy (BSO), peritoneal washings, omentectomy, removal of all macroscopic lesions, and peritoneal biopsies [57, 64]. The role of pelvic and para-aortic lymphadenectomies and postoperative chemotherapy is controversial [56, 57]. Advanced-stage disease is usually treated with surgery [57]. It has recently been recognized that the micropapillary form of serous borderline ovarian cancers have a worse prognosis because these tumors do not appear to respond to chemotherapy [63].

Fertility Preservation

In fertility-sparing surgery, the uterus and ovarian tissue in one or both ovaries are preserved [57]. In general, fertility preservation is recommended for young patients with borderline ovarian carcinoma confined to a single ovary [57].

Cystectomy or Unilateral Salpingo-Oophorectomy with Preservation of the Uterus and Contralateral Ovary

Indications

Because of the high recurrence rate associated with this procedure, ovarian cystectomy as the treatment of borderline ovarian carcinoma is highly controversial [58]. Patients who desire fertility preservation and already had one ovary removed or patients with bilateral ovarian tumors may be candidates for this controversial procedure [58]. Anecdotal experience at Yale-New Haven Hospital has shown that a cystectomy may leave invasive cancer in the residual ovary. Bell et al. have proposed histologic criteria as to when it is appropriate to remove the residual ovary [65].

There are no clear guidelines regarding the selection of candidates for the performance of a unilateral salpingo-oophorectomy (USO) with preservation of the uterus and contralateral ovary for the treatment of borderline ovarian carcinoma [64]. Generally, it is offered to young, nulliparous patients with early-stage disease [59, 64].

Technique

Ovarian cystectomy or USO can be performed by laparotomy or laparoscopy [59]. A thorough inspection of the peritoneal cavity and pelvic washings prior to the ovarian cystectomy or the USO is important to rule out extensiveness of potentially invasive disease [59].

While some authors have recommended not biopsying the contralateral ovary, the authors have routinely done this with no obvious untoward effects. An abnormal-appearing contralateral ovary should be biopsied [57, 59].

Oncological Outcome

Ovarian cystectomy is the fertility-preserving procedure for the treatment of borderline ovarian cancer that is associated with a recurrence rate of up to 30% [58]. This rate is significantly higher than the recurrence rate with USO (2–6%) and is particularly high with intraoperative cyst rupture or advanced-stage disease [58, 59, 66, 67]. The majority of the recurrences occur in the ipsilateral ovary and are borderline lesions [56, 58, 68]. Most of the patients with recurrent disease can be salvaged by more radical surgery, and overall survival does not appear to be negatively affected by disease recurrence [56, 58, 68]. There is a low risk of progression to invasive carcinoma (<5%) and disease-related death [56, 62].

Conservative surgery for the treatment of borderline ovarian carcinoma may, therefore, be an acceptable option to the well-counseled patient who desires fertility preservation. The majority of studies reporting fertility preservation include patients with stage I disease only [58].

Fertility Outcome

After fertility-preserving treatment of the ovarian borderline carcinoma, the successful pregnancy rate is ≥50% (Table 14.3) [62, 66, 68]. Pregnancy does not appear to adversely affect the outcome of conservatively treated borderline ovarian carcinoma [62, 66, 68]. The majority of pregnancies are delivered at term and vaginally (see Table 14.3) [64, 66, 67, 69–74]. No major fetal malformations in pregnancies delivered after conservative treatment of borderline ovarian carcinoma have been reported [64].

Interestingly, from the time ovarian tumors of low malignant potential were first recognized as a clinical entity by Taylor in 1928 ("semimalignant ovarian cancers"), it has been noted that they were associated with infertility [62, 74]. The recently reported pretreatment incidence of infertility among patients with borderline ovarian tumors is 10–35% [74, 75]. Only age at diagnosis appears to predict fertility outcome in patients with borderline ovarian carcinoma who underwent fertility-preserving treatment [64]. Infertility drugs may be used safely in patients with infertility after conservative treatment of early-stage borderline ovarian carcinoma [75].

Invasive Eptithelial Ovarian Carcinoma

Epithelial ovarian cancer is the leading cause of death from gynecologic malignancies [76, 77]. It is estimated that 21,550 women were diagnosed with epithelial ovarian cancer in 2009, and there were 14,600 deaths from ovarian cancer [78]. While the incidence of epithelial ovarian cancer increases with age, 3–17% of all epithelial ovarian cancers are diagnosed in women of reproductive age [2, 7, 79–82]. Seven to eight percent of all stage I epithelial ovarian cancers are diagnosed in women≤35 years of age [80]. Young women tend to present at an earlier stage, with low-grade disease and tend to have a better overall prognosis than postmenopausal women [83].

The prognosis of ovarian cancer depends on the stage at diagnosis. Early-stage disease has an excellent prognosis, with a 5-year survival of 94% for stage IA, 92% for stage IB, and 84% for stage IC disease [83]. Mucinous and endometrioid ovarian tumors have the best prognosis in stage I disease [15]. Despite this, the mortality is still 30–40% among women with early-stage disease, and potential candidates for fertility-preserving surgery must be carefully selected and counseled [78].

Standard of Care

The standard treatment of invasive epithelial ovarian carcinoma consists of surgery and adjuvant chemotherapy [78, 81, 84, 85]. Surgical staging includes a total abdominal hysterectomy, BSO, tumor debulking, omentectomy, pelvic and para-aortic lymphadenectomy, peritoneal biopsies, and pelvic and abdominal washings [81, 84]. Adjuvant combination chemotherapy with carboplatin and paclitaxel is part of the standard treatment in all patients, except those with well-differentiated stage IA or IB disease [86].

Fertility Preservation

USO with Staging

Indications

Fertility-preserving surgery is only an option for patients with disease that is confined to the ovaries with favorable histology [84, 87, 88]. There is some controversy regarding the eligibility of patients with moderately or poorly differentiated tumors or with stage IC disease for fertility-sparing surgery. While some offer conservative surgery to patients with well-differentiated stage IA disease only, others also consider fertility-sparing surgery in combination with adjuvant chemotherapy for patients with stage IA ovarian cancer with clear cell histology or with stage IC disease [88, 89]. 12 Fertility-preserving surgery is not

Table 14.3 Borderline ovarian carcinoma 2000–2010

First author	Year	Patients	Procedure	Pregnancies	Attempted pregnancies	Live births	Preterm deliveries	Term deliveries	Recurrences	Deaths
Laurent	2009	9	USO[a]	6	NR[b]	3	0	3	5	0
		1	Cystectomy							
Park	2009	4	Cystectomy	33	31/130 (no follow-up data on rest)	34	0	34	1	0
		168	USO						8	1
Poncelet	2006	33	Cystectomy	NR	NR	NR	NR	NR	10	0
		100	USO						11	0
Romagnolo	2006	21	Cystectomy	8	12	7	0	7	6	0
		32	USO						7	1
Boran	2005	15	Cystectomy	13	25	5	0	5	1	0
		47	USO			5	0	5	3	0
Fauvet	2005	20	Cystectomy	5	30	18	0	17 (twins × 1)	27	0
		45	USO	16						0
Rao	2005	5	Cystectomy	4	NR	4	0	4	0	0
		33	USO						6	0
Olaszewska	2004	6	Cystectomy	13	NR	13	NR	NR	2	0
		36	USO							0
Chan	2003	1	Cystectomy	5	6	5	0	5	0	0
		24	USO						0	0
Donnez	2003	2	Cystectomy	12	11	12	0	12	0	0
		14	USO						3	0
Camatte	2002	9	Cystectomy	26	29	19	3	16	16	0
		59	USO							0
Demeter	2002	12	USO	6	12	NR	NR	NR	1	0
Morice	2001	11	Cystectomy	17	NR	10	0	10	4	0
		33	USO						5	0
Seracchioli	2001	11	Cystectomy	6	12	6	0	6	1	0
		8	USP						0	0
Zanetta	2001	74	Cystectomy	NR	NR	NR	NR	NR	31	0
		110	USO							
Morris	2000	8	Cystectomy	25	24	16	0	16	4	0
		35	USO						10	1
Total		986		195	192	157	3	140	162	3

[a]USO = unilateral salpingo-oophorectomy
[b]NR = not recorded

recommended for patients with high-grade stage IA or high-grade stage IC disease [88].

Technique

After performance of a USO, thorough staging with peritoneal cytology, omentectomy, pelvic and para-aortic lymphadenectomy, and peritoneal biopsies is recommended [1, 84, 87]. This is important since 10–25% of women with suspected stage I disease have occult lymph node metastases [84]. Routine biopsies of the contralateral ovary are not recommended since the risk of occult metastases in normal-appearing ovaries is small [1, 84, 87, 90]. Careful examination of the contralateral ovary is extremely important, and suspicious areas should be biopsied [2]. A dilation and curettage is recommended by some authors, particularly in the setting of granulosa cell tumors and endometrioid ovarian cancers [1].

Oncological Outcome

At baseline, the prognosis of young women with ovarian cancer is significantly better than for postmenopausal women with ovarian cancer [84]. Patients who are younger than age 30 have a 5-year survival rate of 56% with stage III and IV disease, compared to 22% for patients ≥60 years [84].

Reproductive-age patients with stage IA or IC disease who undergo conservative treatment for epithelial ovarian cancer have 5-year survival rates similar to patients treated with a total abdominal hysterectomy, BSO, and staging [78, 81, 84].

The largest series to date reports a recurrence rate of 9% [88]. Two percent of the conservatively treated patients died as a result of their disease [88].

Smaller retrospective studies report recurrence rates of 4–28% for ovarian cancer patients treated with conservative surgery [84]. Table 14.4 summarizes the results of studies of conservatively treated ovarian cancer patients and reveals a recurrence rate of 11% and a mortality of 5%. Morice et al. documented that all recurrences occurred in patients with incomplete staging, underlining the importance of complete staging [87].

The contralateral adnexa is a common place of recurrences [81, 87]. The question whether the remaining adnexa should be removed after the completion of childbearing must be decided on an individual basis since long-term outcome data is lacking [84].

Fertility Outcome

The reported pregnancy rate ranges from 71 to 79% after conservatively treated ovarian cancer [77, 80, 81]. This is consistent with the data presented in Table 14.4: 75% of the women attempting pregnancy managed to conceive. The incidence of congenital anomalies does not appear to be higher than in the general population [81].

Ovarian cancer is associated with nulliparity and infertility and requires fertility treatments to conceive [91]. In a Medline analysis, Mahdavi et al. detected no detrimental effects of fertility drugs on the outcome of conservatively treated ovarian cancer patients [91]. So far, no larger, prospective studies are available to definitively exclude a potential association [91]. Theoretically, the fertility-preserving staging surgery might cause more adhesions than a simple USO and could potentially cause infertility. However, no difference in reproductive outcomes has been noticed between patients treated with a simple USO and those undergoing staging surgery [85].

Postoperative Chemotherapy

Indications

Adjuvant chemotherapy is indicated for all, except stage IA or low-risk-stage IC patients with epithelial ovarian cancer [86]. The indications for adjuvant chemotherapy are identical for patients undergoing conservative and radical surgery [2]. The majority of patients who are eligible for conservative surgery, therefore, do not need adjuvant chemotherapy [4, 15, 88]. Additionally, some authors recommend adjuvant chemotherapy for patients with stage IA disease with clear cell histology [88].

Technique

Standard first-line chemotherapeutic agents for the adjuvant treatment of epithelial ovarian carcinoma are carboplatin and paclitaxel, usually given for a total of six cycles [1, 86].

Table 14.4 Invasive epithelial ovarian carcinoma 2000–2010

First author	Year	Patients	Procedure	Pregnancies	Attempted pregnancies	Live births	Preterm deliveries	Term deliveries	Recurrences	Deaths
Satoh	2010	211	USO[a]/unilateral ovarian cystectomy ± staging	76	NR[b]	66	NR	NR	18	5
Kwon	2009	21	USO ± staging	5	NR	5	0	5	1	0
Schlaerth	2009	20	USO + staging	9	NR	9	0	9	3	3
Park	2008	62	USO/UO[c] + staging	24	19	22	0	22	11	6
Zanagnolo	2005	14 EOC (+23 LMP 33 MOGCT 5 SCST)	USO + staging	23	20	18	1	17	0	0
Schilder	2002	52	USO + staging	17	24	26	0	26	5	2
Morice	2001	31	USO/unilateral ovarian cystectomy ± staging	4	NR	3	NR	NR	7	3
Total		411		158	63	149	1	79	45	19

[a]USO = unilateral salpingo-oophorectomy
[b]NR = not recorded
[c]UO = unilateral oophorectomy

Oncological Outcome

Adjuvant chemotherapy does not improve the outcome of patients with stage IA grade 1 or 2 disease [85, 88]. After the administration of adjuvant chemotherapy, patients with stage IC or clear cell disease appear to have similar outcomes after conservative surgery compared to patients after radical surgery [85, 88].

Fertility Outcome

Many chemotherapeutic agents are associated with a risk of premature ovarian failure, and this risk has been reported to be as high as 68% [1]. It is influenced by the patient's age and the chemotherapeutic agent used [92]. Data regarding the ovarian toxicity of the standard chemotherapeutic combination used in the treatment of ovarian cancer, carboplatin, and paclitaxel is scarce [85]. Cisplatin analogs have been estimated to cause ovarian failure with an odds ratio of 1.77 [93]. In animal experiments, paclitaxel has been shown to only have minor effects on fertility [94]. No congenital anomalies have been described after the administration of carboplatin and paclitaxel [87, 88].

Malignant Germ Cell Ovarian Tumors

About 25% of all ovarian tumors are germ cell tumors [95]. Only a minority of these tumors (3%) are malignant [95–97]. The current classification divides them into three categories: (1) the primitive germ cell tumors, including the yolk sac tumor (endodermal sinus tumors), dysgerminomas, embryonal carcinomas, choriocarcinomas, polyembryomas, and mixed germ cell tumors; (2) biphasic or triphasic teratomas, including immature teratomas; and (3) monodermal teratomas and somatic-type tumors associated with biphasic and triphasic teratomas. The latter tumors are rare, occur most often in older adults, and are not discussed further [98]. Primitive germ cell tumors and bi- and triphasic teratomas are usually diagnosed at an early stage and in young women [97]. One-fourth are diagnosed in premenarchal girls [96, 97, 99]. Malignant ovarian germ cell tumors (MOGCTs) are predominantly unilateral and are very chemosensitive [81, 96, 100]. Therefore, fertility-preserving surgery is both an important issue and an attractive option for this young patient population [96, 100, 101].

Standard of Care

The current standard of care for the treatment of MOGCTs consists of the removal of the affected ovary while preserving the contralateral ovary and the uterus [95, 99, 101]. Removal of the contralateral adnexa and uterus does not appear to affect survival, even in metastatic or advanced disease [81, 97, 99, 102]. Adjuvant chemotherapy is recommended for all stages and histological grades of primitive, nondysgerminomatous germ cell tumors, advanced-stage dysgerminoma, stages ≥IB immature teratomas, and all grade 3 immature teratomas regardless of stage [95, 96].

Fertility Preservation

USO with Staging

Indications

A USO with staging is recommended for all stages and histological grades of immature teratomas and primitive germ cell tumors [81, 95, 100–104]. Embryonal carcinomas are rare. The current standard of care consists of a USO with staging followed by postoperative chemotherapy [101, 105].

Technique

After performance of the USO, an omentectomy, peritoneal biopsies, and pelvic and abdominal washings are performed [81, 97, 99, 101]. A routine pelvic and para-aortic lymph node biopsy is recommended for all patients who have pure dysgerminomas or mixed tumors containing dysgerminomas as a microscopic finding of metastases. Dysgerminomas selectively spread by the lymphatic route while the remaining primitive germ cell tumors and the immature teratomas spread most often by intraperitoneal dissemination [97, 106]. Routine biopsies of a normal-appearing contralateral ovary are indicated in patients with dysgerminomas or mixed tumors containing dysgerminoma [97, 104]. In the Yale series, 27% of patients with dysgerminomas had contralateral ovarian involvement. Two of the latter patients preserved that ovary and went on to have full-term pregnancies [102].

Oncological Outcome

Multiple studies have shown that the outcome after fertility-preserving surgery is similar to the outcome after more radical procedures, even in advanced-stage disease [81, 95, 96, 101, 104]. This is particularly true in the setting of routine postoperative chemotherapy. Based on the studies summarized in Table 14.5, the risk of recurrence is 8%, with a mortality of 2%. The risk of bilateral ovarian involvement is extremely small, and removal of the contralateral ovary does not affect survival [99, 101, 104].

Fertility Outcome

Since many of the women diagnosed with MOGCTs are premenarchal or adolescent, follow-up obstetrical data is difficult to obtain. As documented in Table 14.5, >95% of the patients who were attempting to conceive delivered a live infant.

Instead, several studies have used the onset of normal menstruation or the resumption of normal menses as a marker for reproductive function [96, 107]. In the Yale series, those patients who experienced menstrual abnormalities were the ones most likely to have difficulty in conceiving [95]. Consistent with prior studies, the data summarized in Table 14.5 reveals that the majority of patients (91%) either underwent normal menarche or resumed normal menstruation [96, 107].

Postoperative Chemotherapy

Indications

Given the high sensitivity of MOGCTs to chemotherapy, adjuvant chemotherapy is part of the standard of care for all patients diagnosed with MOGCTs [95, 96, 100–102, 108]. Adjuvant chemotherapy is routinely recommended for all patients with nondysgerminomatous primitive germ cell tumors, stage IB or greater dysgerminomas, all grade 3 immature teratomas, and immature teratomas stage IB or greater stage.

Technique

A combination regimen consisting of bleomycin, etoposide, and cisplatin (BEP) is currently part of the standard treatment of MOGCTs [96].

Oncological Outcome

With adjuvant chemotherapy, the outcome of patients treated with conservative surgery is similar to the outcome of patients after radical surgery [81, 95, 96, 101].

Fertility Outcome

As with any chemotherapeutic regimen, there is a risk of premature ovarian failure after the treatment with BEP. Tangir et al. and Gershenson et al. reported premature menopause after postoperative chemotherapy in 3% of the patients [95, 107].

There is a mild, but nonsignificant, increase in the rate of congenital malformations in pregnancies of patients treated with postoperative chemotherapy with BEP in some, but not all, studies [95, 99, 109].

Overall, women who undergo fertility-preserving treatment and chemotherapy for ovarian germ cell tumors are very likely to preserve fertility after completion of treatment [95, 100, 101, 107, 108].

Ovarian Sex Cord-Stromal Tumors

Ovarian sex cord-stromal tumors represent approximately 8% of all ovarian neoplasms [110, 111]. Ovarian sex cord-stromal tumors are a group of heterogeneous tumors that originate from the non-germ cell component of the ovary [111, 112]. Tumors of the thecoma-fibroma group (thecomas, fibroma–fibrosarcomas, sclerosing stromal tumors) as well as Sertoli-Leydig cell tumors, granulosa stromal cell tumors, gynandroblastomas, and sex cord tumors with anular tubules are part of this entity [110, 111]. The majority of the malignant sex cord-stromal cell tumors are granulosa cell tumors [111]. Ovarian sex cord-stromal tumors affect all age groups and can have estrogenizing or virilizing effects [110]. Most of the sex cord-stromal tumors (>90%) are unilateral and confined to the ovary at diagnosis [110, 111]. Overall, the prognosis of ovarian sex cord-stromal tumors is excellent, with the long-term overall survival ranging from 75 to 90% for all stages [111].

Table 14.5 Malignant germ cell ovarian tumors 2000–2010

First author	Year	Patients	Procedure	Resumed menstruation/ normal menarche	Pregnancies	Attempted pregnancies	Live births	Preterm deliveries	Term deliveries	Recurrences	Deaths
Yoo	2010	25	USO[a] + staging	23	NR[b]	NR	NR	NR	NR	0	0
De la Motte Rouge	2008	41	USO + staging	39	19	16	15	NR	NR	2	0
Gershenson	2007	71	USO + staging	62	NR	NR	37	NR	NR	NR	NR
Nishio	2006	30	USO + staging	4	8	12	7	NR	NR	0	0
Tangir	2003	64	USO	4 premenarchal	29	38	38	NR	NR	NR	NR
Zanetta	2001	138	USO + staging	128	28	32	40	0	40	16	3
Low	2000	74	USO + staging	3	19	20	14	NR	NR	7	2
Total		443		263	103	118	151			25	5

[a]USO = unilateral salpingo-oophorectomy
[b]NR = not recorded

Standard of Care

The standard treatment of ovarian sex cord-stromal tumors is similar to the treatment of invasive epithelial ovarian cancer [111, 113]. It consists of a hysterectomy, BSO, omentectomy, peritoneal cytology, inspection of the abdominopelvic cavity, and peritoneal biopsies [110, 111, 114]. In retrospective multivariate analyses, complete surgical staging has been shown to be an independent predictor of survival [115]. Cytoreductive surgery is particularly important in tumors that are less chemosensitive, such as granulosa cell tumors [115]. The role of a lymphadenectomy in the treatment of ovarian sex cord-stromal tumors remains controversial [110, 114]. Adjuvant chemotherapy is recommended for patients with high-grade stage I and stage II–IV disease [110, 111, 115]. Some authors also recommend postoperative chemotherapy for patients with large tumors, a high mitotic index, or tumor rupture [111, 112]. Bleomycin, etoposide, and a platinum agent (BEP) or paclitaxel and a platinum agent are the most commonly used chemotherapeutic regimens [110, 114]. The routine use of radiation is not recommended [111].

Fertility Preservation

Indications

Patients with stage IA disease who desire to preserve their fertility are candidates for fertility-preserving surgery [111, 113]. Patients with extraovarian disease should not be offered fertility-preserving surgery [111].

Technique

A USO, along with an omentectomy, peritoneal cytology, inspection of the abdominopelvic cavity, and peritoneal biopsies, is performed [110, 111, 114]. If fertility preservation is desired, endometrial curetting must be performed to rule out a concomitant endometrial pathology [111]. Routine biopsies of a normal-appearing contralateral ovary are not recommended [116]. It remains controversial whether or not the uterus and contralateral ovary should be removed after the completion of childbearing [111, 117]. The indications for adjuvant chemotherapy for patients treated with fertility-preserving surgery are similar to the indications for patients treated with radical surgery [118].

Oncological Outcome

Because of the rarity of the disease, reports on the oncological outcome after fertility-preserving surgery are rare. Evans et al. reported a worse outcome of patients treated with fertility-preserving surgery compared to patients treated with radical surgery [119]. It is important to note, however, that the patients included in this study were mainly diagnosed with high-stage disease. In contrast, an analysis of 11 conservatively treated patients with ovarian sex cord-stromal tumors revealed no statistically different overall survival between the patients undergoing radical or conservative treatment [118]. Larger prospective trials are needed to definitively evaluate the oncological outcome after fertility-preserving surgery.

Fertility outcome

As is true for the oncological outcome, data on the fertility outcome after conservative treatment of ovarian sex cord-stromal tumors is very scarce and often based on case reports [111, 113, 116, 117]. Overall, the pregnancy outcomes for patients with early-stage disease who undergo fertility-preserving treatment are encouraging [111, 113, 116]. In their series of 11 patients, Zanagnolo et al. report a pregnancy rate of 45% of the patients undergoing fertility-preserving treatment. Larger studies are necessary to validate these results [118].

Uterine Carcinoma

Atypical Endometrial Hyperplasia

Prolonged exposure to unopposed estrogens causes excessive endometrial proliferation and can eventually lead to endometrial hyperplasia [120, 121]. Endometrial hyperplasia is primarily a disease of the peri- and postmenopausal woman and often presents with abnormal vaginal bleeding [120]. In reproductive-age women, hyperestrogenic conditions, such as chronic anovulation and obesity or estrogen-secreting ovarian tumors, predispose to the development of endometrial hyperplasia [120]. Endometrial hyperplasia can be divided into three different categories: simple, complex, and atypical [122, 123]. While only 1–3% of non-atypical hyperplasias progresses to carcinoma, about 8–30% of atypical hyperplasias progresses to carcinoma over a period of 4 years [122, 124–126]. Atypical endometrial hyperplasia is, therefore, considered as a precursor of endometrial cancer [122, 123].

Standard of Care

The goal of the treatment of endometrial hyperplasia is to achieve a reduction in the amount of abnormal bleeding and a prevention of endometrial cancer [122]. Currently, the standard treatment for atypical endometrial hyperplasia is a total hysterectomy [122]. Progestin therapy is the standard of care for the treatment of women diagnosed with simple hyperplasia, but there is no consensus regarding the dose, duration of treatment, or route of administration [123, 126, 127].

Fertility Preservation

Progestin Therapy

Indications

Young women diagnosed with atypical hyperplasia who desire fertility preservation are candidates for medical treatment with progestins as long as there are no medical contraindications to progestin treatment, such as a prior history of thromboembolic events [125]. Progestins affect the cell differentiation, inhibit estrogen receptors and endometrial mitoses, and have antiangiogenic effects [128].

Technique

There is no consensus regarding the dose, duration of treatment, and route of administration of progestins for the treatment of atypical endometrial hyperplasia [126, 127]. The most commonly used progestin for oral treatment is medroxyprogesterone acetate (MPA), but there are reports using 17α-hydroxy-progesterone caproate and megestrol [125]. Oral progestins cause severe side effects, including nausea, headaches, and weight gain, along with increasing the risk of thromboembolic events [124]. The advantage of depot progestins, such as a progesterone-releasing intrauterine device, is that they deliver high concentrations of progestins (usually, 20 μg of levonorgestrel per day) locally to the uterus without causing the systemic side effects observed with oral progestin therapy [123, 126].

Follow-up endometrial sampling at regular intervals is recommended in order to exclude the presence of persistent or progressive lesions, but the exact follow-up schedule for patients undergoing progestin treatment for endometrial hyperplasia has not been established [120]. Most authors recommend routine endometrial sampling every 3–6 months [120, 129]. A hysterectomy and a BSO after the completion of childbearing are controversial [121].

Oncological Outcome

The literature reports high response rates of greater than 90% to progestin therapy in patients with non-atypical endometrial hyperplasia [120, 122, 124, 130].

Regression rates for atypical hyperplasia treated with oral progestins are lower, ranging from 67 to 83% [122, 125, 130]. The largest study to date, a prospective trial by Varma et al., reported regression rates of 67% for atypical hyperplasia [122]. The treatment of atypical hyperplasia remains controversial. Horn et al. described a progression to carcinoma in three out of the seven women with atypical hyperplasia who received oral progestin treatment [129]. Overall, failure rates with oral progestin therapy appears to be higher with low-dose therapy (3–10 mg per day) than with high-dose treatment [126]. The mean time to response is 12 months, but recurrences in up to one-third of patients with an initial response have been reported [131].

The compliance of patients with oral progestin therapy is often poor because of the significant side effects [124, 125]. Depot progestins might avoid these side effects while delivering sufficient doses of progestins. Data on the use of progesterone-releasing IUDs in atypical hyperplasia remains controversial. Ørbo et al. compared the response of patients with atypical and non-atypical endometrial hyperplasia to oral progestin treatment to the response to progestin-releasing IUDs [126]. They concluded that the progestin-releasing IUDs were more effective than oral progestins in patients with atypical as well as non-atypical hyperplasia [126]. Some reports, however, showed progression of atypical endometrial hyperplasia to endometrial carcinoma with a levonorgestrel (LNG)-IUD in place [132]. Although larger trials suggest that progestin-releasing IUDs can be safe in the treatment of atypical endometrial hyperplasia with close follow-up, larger studies are needed to validate these results.

A literature review reveals an initial response rate of 81% for patients with atypical hyperplasia treated with LNG-IUD and of 76% for patients treated with oral progestins (Table 14.6).

A Gynecologic Oncology Group study revealed concurrent endometrial carcinoma in the hysterectomy specimens of women diagnosed with atypical endometrial hyperplasia on endometrial biopsy in 43% of the cases, stressing the importance of representative endometrial sampling for the appropriate treatment of endometrial hyperplasia [133].

Fertility Outcome

Data on the fertility outcome of women with endometrial hyperplasia who underwent medical treatment is very limited. Yu et al. describe a pregnancy rate of 40% in patients with atypical hyperplasia treated with oral progesterone [134]. Qi et al. report successful pregnancies in both patients treated with LNG-IUD [135]. A review of the literature reveals an overall pregnancy rate of 60% for patients treated with LNG-IUD and of 40% for patients treated with oral progestins (Table 14.6). Larger studies are needed to validate these results.

Endometrial Adenocarcinoma

Endometrial adenocarcinoma (EAC) mainly affects postmenopausal women [136]. Two to fourteen percent of all cases, however, are diagnosed in women under age 40 [125, 131, 136–139]. Prolonged unopposed estrogen exposure is a risk factor for EAC [140]. EAC in premenopausal women can present with infertility, obesity, irregular bleeding, uterine polyps, or a thickened endometrium [131, 137, 139]. Young women are most commonly diagnosed with early-stage, well-differentiated, and estrogen-dependent EAC [125, 139]. The incidence of myometrial invasion and lymph node metastases is low [139]. The prognosis of these women diagnosed is excellent, with a 5-year survival>93% [125, 131, 136, 141]. Many of the premenopausal women diagnosed with EAC are nulliparous and desire preservation of fertility.

Standard of Care

Hysterectomy, BSO with or without pelvic and para-aortic lymphadenectomy, is the standard treatment for EAC [125, 131, 136, 140]. Depending on the stage and grade of the disease, adjuvant brachytherapy and chemotherapy may be indicated.

Fertility Preservation

Progesterone Therapy (Oral, Intrauterine)

Indications

Patients with grade 1 EAC at presumed stage IA are candidates for progestin-only therapy since these cancers are associated with a low risk of extrauterine spread [131, 138, 139]. Estrogen and progesterone receptor-positive tumors show a better response to progestin treatment than receptor-negative tumors [142]. Patients with myometrial invasion to the outer third of the myometrium have positive pelvic and para-aortic lymph nodes in 25% and 17%, respectively, and are therefore not ideal candidates for oral progestin treatment [125]. Based on data from the Gynecologic Oncology Group, the risk of ovarian metastases in disease apparently confined to the uterus is 5% [143]. Careful counseling for young women with EAC who desire to preserve fertility is crucial. Progestin therapy for the treatment of EAC is only appropriate in patients who are highly motivated to preserve fertility [138].

Technique

There are no standardized protocols regarding the dosing, route, or duration of administration of oral progestins for the treatment of early EAC

Table 14.6 Endometrial hyperplasia 2000–2010

First author	Year	Patients	Treatment	Endometrial response	Pregnancies	Live births	Term deliveries	Recurrences	EAC[a]	Deaths
LNG-IUD[b]										
Ercan	2010	1 atypical	MPA[c] 250 mg/d × 6 months, then megestrol acetate 160 mg/d × 3 months, then LNG-IUD × 9 months	1	1	2	0	1	0	0
Signorelli	2009	10 atypical	Cyclic progesterone 200 mg/d	6	8 (5 women)	NR[d]	NR	0	2	0
Yu	2009	17 atypical	MPA 100–500 mg/d, megestrol acetate, hydroxyprogesterone caproate	14	7	3	3	3	1	0
Kresowik	2008	1 atypical	LNG-IUD × 6 months	0	NR	NR	NR	1	1	0
Ørbo	2008	151 non-atypical and atypical	MPA 10 mg/d × 3–6 months $n=85$, LNG-IUD × 3–108 months $n=66$	46/85 MPA 66/66 LNG-IUD	NR	NR	NR	NR	0	0
Qi	2008	2 atypical	LNG-IUD × 6 months	2	2	2	2	0	0	0
Varma	2008	105: 96 non-atypical, 9 atypical	LNG-IUD × 1 year	94:88/96 non-atypical, 6/9 atypical	NR	NR	NR	NR	2	0
Wildermeersch	2007	20: 12 non-atypical, 8 atypical	LNG-IUD	19	NR	NR	NR	1	0	0
Total IUD		307		248	18	7	5	6	6	0
PO										
Ushijima	2007	17 atypical	MPA 600 mg/d × 26 weeks	17	8	4	NR	6	2	0
Digabel	2006	8 atypical	Po[e] progestin	5	3	2	NR	2	0	0
Rhattana-chaiyanont	2005	135 non-atypical	Cyclic: MPA/norethisterone/medrogestone 10 mg/d or dydrogesterone 20 mg/d × 6 months Continuous: MPA 2.5 mg/d or depot-MPA 150 mg/month × 6 months	134	NR	NR	NR	NR	3	0
Horn	2004	214: 208 complex, 7 atypical	Po progestins × 3–5 months	128:128 complex, 0 atypical	NR	NR	NR		5:2 complex , 3 atypical	0
Goker	2001	1 atypical	Po progestins	1	1	1	1	NR	NR	NR
Kaku	2001	18 atypical	MPA 100–800 mg/d × 1–23 months	15	5	4	4	1	1	0
Total PO		393		300	17	11	5	9	11	0
Total		700		548						

[a]EAC = endometrial adenocarcinoma
[b]LNG-IUD = levonorgestrel-releasing intrauterine device
[c]MPA = medroxyprogesterone acetate
[d]NR = not recorded
[e]Po = oral

[125, 139, 140]. MPA and megestrol acetate (MA) are the most frequently used progestins [125, 141]. The doses range from 200 to 1,500 mg/day for MPA and 160–600 mg/day for MA (Table 14.7). Systemic progestins cause side effects, such as thrombophlebitis, weight gain, and hypertension [142]. They also lead to an increased risk of thromboembolism [142]. The LNG-releasing intrauterine device does not lead to systemic side effects and has been used in case reports for the treatment of EAC [144–147].

It is recommended to perform a uterine dilation and curettage as part of the initial evaluation in patients who are interested in progestin therapy [138]. An MRI is often performed in order to evaluate the depth of the myometrial invasion and to rule out extrauterine disease [131, 138, 148]. Some authors recommend a laparoscopy to evaluate the ovaries for potential macroscopic metastases [145]. The role of hysteroscopy and ultrasound in the evaluation of these patients is unknown [138].

It remains controversial whether a BSO should be performed after the completion of childbearing [140]. There are no established guidelines for follow-up of patients after the completion of the progestin therapy [125]. Regular endometrial sampling by endometrial curettage is necessary to rule out recurrent endometrial carcinoma [125, 138, 149]. Some authors recommend a maintenance regimen after the initial response to progestins, but this also remains controversial [138, 141].

Oncological Outcome

The majority of reports including some prospective studies suggest that patients with low-risk EAC can safely be treated with oral progestins alone and that conservative treatment does not worsen the prognosis [131, 134, 138–141, 148–151]. Response rates range from 62 to 100% and are achieved after various treatment durations (Table 14.7) [125, 130, 138, 140, 151–153]. In most patients, however, a response can be observed after 3 months of treatment [140]. Thigpen et al. described a higher response rate among patients whose tumors were progesterone or estrogen receptor-positive [142].

Treatment of endometrial carcinoma with progestins alone is not risk-free [138]. The recurrence rate ranges from 0 to 44% among initial responders to oral progestin therapy in studies including >1 patient (Table 14.7) [125, 138, 141]. Even if a recurrence is diagnosed, however, extension beyond the uterus is rare [141]. There are case reports of disease progression on oral progestin therapy [154, 155].

Although the data is very limited regarding the treatment of EAC with a LNG-releasing intrauterine device, the available case reports suggest a recurrence rate of 50% (Table 14.7) [144–147].

Fertility Outcome

EAC is associated with a high rate of anovulation and infertility in young women [138]. This may contribute to the relatively low pregnancy rates that have been reported, ranging from 0 to 54% (Table 14.7). Since young patients with EAC often have fertility problems, it is possible that the use of assisted reproductive techniques increases the pregnancy rates in this patient population [138, 139]. Most authors recommend the initiation of fertility treatment after two consecutive normal endometrial samplings [140]. Currently, there is no data on fertility outcomes after treatment with LNG-releasing intrauterine devices.

Hysteroscopic Resection ± Progestins

Indications

As with medical treatment alone, strict guidelines regarding the patient selection for the treatment of EAC with hysteroscopic resection with or without progestins are unavailable [156]. Patients with low-grade, presumed stage IA endometrial carcinoma with a strong wish to preserve fertility are presumed to be the best candidates [156, 157].

Technique

No standard recommendations for the dose, route of administration, or duration of treatment exist, nor are there strict guidelines for follow-up evaluations [156, 157]. The tumor resection is usually performed hysteroscopically with resection of the endometrium and surrounding myometrium [156].

Table 14.7 Endometrial adenocarcinoma 2000–2010

First author	Year	Patients	Treatment	Endom. Response[a]	Pregnancies	Attempted pregnancies	Live births	Term deliveries	Recurrences	Deaths
LNG-IUD[b]										
Vandenput	2009	1	LNG-IUD	1	NR[c]	NR	NR	NR	1	0
Kothari	2008	1	LNG-IUD + MA[d] 40 mg four times/d × 6 m	1	NR	NR	NR	NR	1	0
Dhar	2005	4	LNG-IUD (20 μg/d) × 6–36 m	1	NR	NR	NR	NR	0	0
Giannopoulos	2004	1	LNG-IUD (20 μg/d) + Provera 200 mg twice/d	1	NR	NR	NR	NR	0	0
Total LNG-IUD		7		4					2	0
PO[e]										
Hahn	2009	35	MPA[f] 250–1,500 mg/d/MA 160 mg/d × 3–6 m	23	10	12	8	NR	9	0
Signorelli	2009	11	Natural progesterone 200 mg/d	5	8	NR	NR	NR	NR	NR
Yu	2009	8	MPA 250–500 mg/d × 3–6 m after remission	6	4	NR	0	0	1	0
Hurst	2008	1	MA 160 mg/d × 6 m	1	0	1	0	0	1	0
Ushijima	2007	28	MPA 600 mg/d × 26 w, then estrogen–progestin × 6 m	9	4	NR	3	3	8	0
Yamazawa	2007	9	MPA 400 mg/d × 6 m	9	4	8	3	3	2	0
Le Digabel	2006	4	Progestin × 3–25 m	2	0	NR	0	0	0	0
Park	2006	1	MA 320–600 mg/d × 24 w	1	1	1	1	1	0	0
Ferrandini	2005	1	Dihydrogesterone 20 mg/d × 3 m	1	1	1	1	0	1	1
Gotlieb	2003	13	MA 160 mg/d/MPA 600 mg/d/OH-progesterone[g] 2–3 g/w/norethindrone acetate 5 mg/d × ≥3 m	13	7	NR	9	NR	6	0
Wang	2002	9	MA 160 mg/d + tamoxifen 30 mg/d × 6 m	8	4	NR	3	3	4	0
Kaku	2001	12	MPA 200–800 mg/d × 2–14 m	9	2	NR	1	1	2	0
Total PO		132		87	45	23	29	11	34	1
Resection ± progestins										
Mazzon	2010	6	Hysteroscopic resection + MPA 160 mg/d × 6 m	6	4	6	5	5	0	0
Vilos	2007	16	Hysteroscopic resection	16	NR	NR	NR	NR	0	0
Montz	2002	13	Hysteroscopic resection + LNG-IUD (65 μg/d)	7	NR	NR	NR	NR	1	0
Total resection		35		29	4	6	5	5	0	0
Total LNG-IUD, PO, resection		174		120	49	29	34	16	36	1

[a]Endom. Response = endometrial response
[b]LNG-IUD = levonorgestrel-releasing intrauterine device
[c]NR = not recorded
[d]MA = megestrol acetate
[e]Po = oral
[f]MPA = medroxyprogesterone acetate
[g]OH-progesterone = hydroxyprogesterone caproate

The advantage of the hysteroscopic tumor resection is that it might allow a more accurate pathological evaluation of tumor grade and myometrial invasion than endometrial curettings [156]. Patients are followed by regular endometrial sampling by endometrial curetting postoperatively to rule out cancer recurrence [156, 157]. Hysteroscopic surgery is associated with an overall complication rate of 3% and a rate of uterine perforation of 1% [156]. There is a report of peritoneal dissemination of malignant cells during hysteroscopy, but the number of patients evaluated is small [156].

Oncological Outcome

Overall, the outcome for early-grade, presumed stage IA endometrial cancer patients treated with hysteroscopic resection and progestins appears to be favorable, with response rates of 65–100% and few recurrences (Table 14.7) [156, 157]. Vilos et al. report no change in the 5-year survival in patients with endometrial carcinoma treated by hysteroscopic resection only [158]. Larger studies are needed to confirm the results of these small studies.

Fertility Outcome

Data on the fertility outcome after hysteroscopic resection of a low-risk EAC with or without progestin treatment are extremely limited (Table 14.7). Mazzon et al. report a pregnancy rate of 67% in a group of six patients treated with resection and oral progestins (Table 14.7) [156]. Further studies are necessary to validate these results.

Uterine Stromal Tumors

Uterine stromal sarcomas include leiomyomas, smooth muscle tumors of uncertain malignant potential (STUMP), leiomyosarcomas, endometrial stromal nodules, low-grade endometrial stromal sarcomas (ESS), undifferentiated sarcomas, adenomyxomas and fibromas, adenosarcomas, and carcinosarcomas (Table 14.8). The different types of uterine stromal tumors are often confusing. A logical division is the separation in benign tumors, tumors of indeterminate malignant behavior and cancers (Table 14.8). Malignant uterine stromal tumors consist of leiomyosarcomas, undifferentiated sarcomas, and carcinosarcomas (Table 14.8) [159].

Patients with benign uterine stromal tumors (cc) have an excellent prognosis [159, 160].

Leiomyosarcomas are the most common sarcomas among young patients, followed by low-grade ESS [159, 161, 162]. Leiomyosarcomas are composed of smooth muscle cells and show mitoses, nuclear atypia, and tumor cell necrosis [162]. They can be divided into low-grade and high-grade leiomyosarcomas [161]. Tumors that show worrisome features histologically, but do not meet all the histological criteria of leiomyosarcomas, are classified as STUMP [159, 162]. Leiomyosarcomas are very aggressive, leading to a recurrence rate of 53–71% and a poor prognosis even for patients with early-stage disease [159, 161, 163].

ESS originate from cells that resemble endometrial stromal cells of the proliferative endometrium

Table 14.8 Uterine stromal tumors

	Cell of origin		
Clinical behavior	Smooth muscle	Endometrial stromal cells	Mixed tumors
Benign	Leiomyoma	Nodule	Adenomyxoma Adenofibroma
Indeterminate malignant behavior	STUMP[a]	Low-grade ESS[b]	Adenosarcoma
Cancer	Leiomyosarcoma	Undifferentiated sarcoma	Carcinosarcoma

[a]STUMP = smooth muscle tumors of uncertain malignant potential
[b]ESS = endometrial stromal sarcoma

and are hormonally sensitive and usually diagnosed in women<50 years of age [159, 164]. ESS are divided into two types: low-grade ESS (the most common uterine stromal sarcoma) and undifferentiated endometrial sarcomas [159, 160, 165]. Patients with low-grade ESS have an excellent prognosis while undifferentiated endometrial sarcomas are associated with a poor prognosis with frequent local recurrences and distant metastases [159].

Adenosarcomas are very rare tumors of indeterminate malignant behavior [159]. They most commonly arise from the endometrium and affect primarily postmenopausal women [159]. Twenty-five percent of the patients with adenosarcomas eventually die of their disease [159]. Carcinosarcomas (malignant mixed mullerian tumors, MMMTs) typically occur in postmenopausal women, but have also been described in patients<40 years [159]. These tumors are highly aggressive and have a 5-year survival rate of 30% [159].

Standard of Care Treatment

The benign tumors (leiomyoma, endometrial stromal nodule, adenomyoma, and adenofibroma) may be treated with a local excision or hysterectomy, depending on the patient's desire for fertility preservation [159].

The standard treatment for leiomyosarcomas consists of a total abdominal hysterectomy and tumor debulking [159, 163]. The need for a BSO in all patients with leiomyosarcoma is controversial, and adjuvant therapy has not been shown to change outcome [159, 161].

Patients with low-grade ESS generally also undergo a hysterectomy. Because these tumors are sensitive to hormones, some authors recommend the routine performance of a BSO [165]. At least for stage I patients, ovarian preservation does not seem to adversely affect the recurrence [165]. Adjuvant progestin therapy is recommended for patients with low-grade ESS [165].

Adenosarcomas are usually diagnosed as stage I disease and cure may be accomplished with a hysterectomy. However, adenosarcomas with sarcomatous overgrowth, undifferentiated sarcomas, and carcinosarcomas are aggressive tumors that are treated with a total hysterectomy, BSO, and tumor debulking, possibly with adjuvant radio- or chemotherapy for patients with undifferentiated endometrial sarcoma [159, 160, 164, 166–168].

Fertility Preservation

Ovarian Preservation

Technique

The surgeon should perform a thorough inspection of the peritoneal cavity for evidence of any extra-uterine disease. Subsequently, a total abdominal hysterectomy without BSO is performed [161].

Indications

Patients with leiomyosarcoma and low-grade ESS whose disease is limited to the uterus are potential candidates for ovarian preservation [160, 162].

Oncological Outcome

Ovarian preservation did not have an adverse impact on survival in patients with leiomyosarcoma or low-grade ESS (Table 14.9) [160–162, 165–167].

Fertility Outcome

The performance of a hysterectomy without BSO allows the preservation of the ovarian function with the potential of a surrogate pregnancy after in vitro fertilization in the future. In the recent literature, six-term pregnancies have been reported in patients with malignant uterine stromal tumors undergoing fertility- or ovarian-preserving treatment (Table 14.9).

Myomectomy/Local Excision

Indications

The data on fertility preservation in patients diagnosed with leiomyosarcoma is very limited and must be considered experimental [162]. Young patients who have low-grade leiomyosarcomas<5 cm in size or localized low-grade ESS

Table 14.9 Uterine stromal tumors 1995–2010

First author	Year	Patients	Procedure	Pregnancies	Attempted pregnancies	Live births	Preterm deliveries	Term deliveries	Recurrences	Deaths
Leiomyosarcoma										
Cormio	2009	1 LMS[a]	Chemotherapy	NR[b]	NR	NR	NR	NR	1	1
Salman	2007	1 low-grade LMS	Myomectomy	1	1	1	0	1	0	0
Giuntoli	2003	25 LMS	TAH[c]	NR	NR	NR	NR	NR	NR	0
Kagami	2002	1 myxoid LMS	Myomectomy	1	1	1	0	1	1	0
Friedrich	1998	2 LMS	Myomectomy	2	2	NR	NR	NR	0	0
Lissoni	1998	8 LMS	Myomectomy	3	8	2	0	2	1	1
STUMP[d]										
Guntupalli	2009	10 STUMP	Myomectomy	NR	NR	NR	NR	NR	0	0
Total LMS + STUMP		48		7	12	4	0	4	3	2
Low-grade endometrial stromal sarcoma										
Yan	2010	1 ESS[e]	Tumor resection + chemotherapy	1	1	1	0	1	0	0
Koskas	2009	1 ESS	Tumor resection	1	1	1	0	1	1	0
Li	2005	12 ESS	TAH	NR	NR	NR	NR	NR	4	NR
Chu	2003	8 ESS	TAH	NR	NR	NR	NR	NR	4	1
Spano	2003	1 ESS	TAH	NR	NR	NR	NR	NR	1	0
Total ESS		23		2	2	2	0	2	10	1
Endometrial stromal nodule										
Schilder	1999	1 Endometrial nodule	Tumor resection + hormonal treatment	NR	NR	NR	NR	NR	0	0
Total nodule		1		NR	NR	NR	NR	NR	0	0
Total		72		9	14	6	0	6	13	3

[a] LMS = leiomyosarcoma
[b] NR = not recorded
[c] TAH = total abdominal hysterectomy
[d] STUMP = smooth muscle tumor of unknown malignant potential
[e] ESS = low-grade endometrial stromal sarcoma

seem to be the best candidates for fertility-preserving treatment [162, 164, 169]. Evidence of extrauterine disease is a contraindication for local resection or myomectomy. Patients must be highly motivated to preserve fertility and should understand the risk of recurrence and death with conservative treatment [163]. Patients with adenosarcoma, carcinosarcoma, and undifferentiated sarcoma are not candidates for fertility-preserving surgery unless this is performed within a research protocol.

Technique

The published reports describe the performance of an abdominal myomectomy for the fertility-sparing treatment of patients with leiomyosarcomas [170–172]. Local excisions of low-grade ESS with and without adjuvant chemotherapy have been described [164, 169]. As part of the routine follow-up evaluations of patients with malignant stromal tumors undergoing conservative management, serial MRIs of the uterus and pelvis to rule out a recurrence at the primary site as well as chest radiographs to rule out pulmonary metastases are recommended [162]. The need for a hysterectomy after completion of childbearing is controversial [162, 163].

Oncological Outcome

The available data is based on very small series and case reports (Table 14.9). Bonney reported a series of 632 abdominal myomectomies [173]. Among the 632 patients, one patient was found to have a leiomyosarcoma [173]. She rapidly developed recurrent disease with peritoneal spread and subsequently died [173]. Friedrich et al. presented two patients with leiomyosarcoma who preserved fertility. Both are alive and well 3 and 6 years postoperatively [172]. An Italian series included eight patients with leiomyosarcoma who underwent fertility-preserving treatment [171]. One of the eight patients suffered from a recurrence and died of her disease [171]. Van Dinh and Woodruff described six patients with leiomyosarcoma who underwent conservative treatment, with one recurrence [174]. The remainder of the data is based solely on case reports (Table 14.9) [163, 170, 175]. Since 1998, seven patients have undergone myomectomies for leiomyosarcoma, with one recurrence (14%) and one death (14%) (Table 14.9). Overall, myomectomies for the treatment of leiomyosarcoma is based on anecdotal reports only, and a hysterectomy should remain the standard treatment of patients with leiomyosarcoma.

Guntupalli et al. reported no recurrences in ten patients with STUMP who were treated by myomectomy [176]. There is no data on the oncological outcome of patients with carcinosarcoma, adenosarcoma, or undifferentiated sarcoma treated with fertility-preserving surgery.

Fertility Outcome

The data on pregnancy outcomes of patients treated with myomectomy for leiomyosarcoma is very limited. Friedrich et al. reported two patients with leiomyosarcoma who underwent fertility-preserving treatment and subsequently became pregnant [172]. Kagami et al. published on one patient, van Dinh and Woodruff on three patients, and Lissoni et al. also on three patients with pregnancies after fertility-preserving treatment of leiomyosarcoma (Table 14.9) [170, 171, 174]. Patients should be counseled that this data is largely anecdotal.

Gestational Trophoblastic Disease

Gestational trophoblastic disease (GTD) is rare entity, including hydatidiform moles, invasive moles, choriocarcinomas, placental site trophoblastic tumors (PSTTs), and epithelioid trophoblastic tumors (ETTs) [177, 178]. GTD usually affects reproductive-age women and is associated with a prior gestational event. The incidence of GTD shows geographic variations, with a much higher incidence in Asia than in Europe or North America [179].

Hydatidiform moles are the most common form of GTD [180]. Based on their histology, karyotype, and natural history, they can be complete or partial moles [178]. Partial moles have a triploid genome and result from the fertilization of a normal egg by two spermatozoa or from the fertilization of one spermatozoon with duplication [181]. Complete moles are entirely paternal and arise from the fertilization of an empty egg by one spermatozoon with duplication or by the fertilization of an empty egg by two spermatozoa [181]. Follow-up after the diagnosis of a molar pregnancy is important as it can result in persistent GTD in 3–4% of the patients with a partial and 20% of the patients with a complete mole [178, 180].

Choriocarcinoma originates from the villous trophoblast and secretes human chorionic gonadotropin (hCG) [182]. It is chemotherapy-sensitive and highly curable [183]. Most of these young women can achieve a complete remission while preserving their fertility, even with metastatic disease [177, 178, 180, 183].

PSTT is extremely rare and originates from the intermediate-type trophoblast [184–187]. Ten percent of the patients with PSTT are present with metastatic disease [186]. PSTT secretes human placental lactogen (hPL) and hCG [187]. Age>35, pregnancy interval>24 months, hCG>1,000 IU/L, depth of invasion, and pathologic characteristics of the PSTT, like high mitotic index, necrosis, and clear cytoplasm, are associated with worse survival [188]. The overall survival of patients with advanced PSTT is 35–40% [186]. In tumors confined to the uterus treated with hysterectomy and chemotherapy, survival of up to 100% has been reported [187].

ETT is very rare and is derived from the intermediate trophoblast [182, 188, 189]. Available data are extremely limited, but metastases are reported to occur in 25% and death in 10% of patients diagnosed with ETT [189].

Standard of Care

Patients with hydatidiform moles are treated by uterine evacuation or hysterectomy, depending on their desire to preserve fertility [178]. This is curative in the majority of patients. After treatment of the molar pregnancy, hCG levels should be monitored weekly, and the patient should delay fertility for 6 months once the titers are negative [178–180]. Patients with persistent trophoblastic disease are treated with chemotherapy [178, 180]. For patients with low- or moderate-risk disease, the treatment is usually either methotrexate or actinomycin D [178]. For patients with high-risk disease or resistance to single agent therapy, etoposide, methotrexate, actinomycin D, cyclophosphamide, and vincristin (EMA/CO) are administered [178]. Patients should delay fertility for 12 months after normalization of the hCG levels. Patients with a history of hydatidiform mole need to be counseled that they are at increased risk for a recurrent molar pregnancy [179]. The standard treatment of choriocarcinoma consists of a dilation and curettage if disease is present in the uterus, followed by chemotherapy. Patients with a diagnosis of PSTT and ETT are usually treated with hysterectomy with or without BSO since the tumor is considered chemotherapy-resistant [184–186].

Fertility Preservation

Chemotherapy with/Without Tumor Resection

Indications

Fertility-preserving treatment is the standard of care for patients with complete or partial moles and for choriocarcinoma. For patients with PSTT, conservative treatment can only be considered in a patient without evidence of extrauterine spread [185]. Patients>35 years of age with a pregnancy interval>24 months, hCG>1,000 IU/L, deep myometrial invasion, extensive necrosis, and the presence of cells with clear cytoplasm are not good candidates for successful conservative treatment. There is no data on fertility preservation in patients with ETT.

Technique

The chemotherapeutic regimens most commonly used for the conservative treatment of PSTT and ETT are EMA–CO and EMA–etoposide, cisplatin (EP) [177, 184, 186]. After the treatment completion, serial hCG levels are followed [184].

The most frequently applied approach for fertility preservation in PSTT patients is the excision of the tumor through a hysterotomy [184, 185, 187, 190, 191]. A frozen section is obtained to confirm negative margins [185, 190, 191]. Residual disease in the margins of the specimen should prompt a total abdominal hysterectomy [184]. If the uterus can be retained, the myometrium is reconstructed in layers [185, 187, 190, 191]. Preoperatively, imaging modalities, such as ultrasound and MRI, are used to localize the tumor and assess the depth of invasion [185]. Imaging can also evaluate tumor vascularity, size, and metastatic disease [184]. The role of adjuvant chemotherapy in patients with PSTT undergoing fertility-preserving treatment is controversial [184]. Postoperatively, serial hCG and hPL levels help detect recurrences [191].

Oncological Outcome

The prognosis of patients with choriocarcinoma who undergo conservative treatment is similar to the prognosis of patients undergoing a hysterectomy [192]. The data on fertility-preserving treatment for PSTT is very limited and consists mainly of case reports (Table 14.10). Six patients with PSTT underwent fertility-preserving treatment (chemotherapy, resection, or chemotherapy with resection) (Table 14.10). Since 1996, one case of recurrences has been reported, with no deaths (Table 14.10) [191]. Patients need to be made aware of the fact that this data is based on case reports only. In addition, PSTT can present as multifocal disease and can be missed by local resection [191]. Some authors recommend a total hysterectomy with ovarian preservation for a potential surrogate pregnancy even in patients who desire fertility preservation [191].

One larger study evaluating EP/EMA in patients with metastatic PSTT reported an overall response rate of 50% to EP/EMA [193]. The toxicities associated with the EP/EMA regimen were significant, consisting of grade 3 and 4 hematologic toxicities [193]. There is no data on fertility preservation in patients with ETT.

Fertility Outcome

Patients with a history of a complete or partial molar pregnancy can expect normal future fertility [179]. Patients with persistent GTD have a subsequent live birth rate of 69–74% [183]. Since 1996, two successful pregnancies in patients with PSTT have been reported (Table 14.10) [185, 186]. Goto et al. report a higher incidence of congenital heart anomalies in the offspring of patients with choriocarcinoma treated with chemotherapy compared to the general population [192].

Combination chemotherapy can cause ovarian damage and contribute to early ovarian failure [186]. Patients, therefore, should be counseled that despite the preservation of the reproductive organs, ovarian function might suffer significantly because of the chemotherapy [186].

Patients who underwent uterine tumor resection must be followed closely for signs of uterine rupture during subsequent pregnancies [190]. Because a myometrial resection is often part of the tumor resection, the risk of uterine rupture may be much higher than after prior cesarean sections [190]. Some authors recommend following these patients with serial ultrasounds to assess the uterine thickness in the third trimester [190].

Vaginal Carcinoma

Primary vaginal carcinoma is a rare gynecologic malignancy, accounting for 1–3% of all gynecologic cancers [194–196]. Squamous cell carcinoma is the most frequent histologic type of vaginal cancer [194]. Most women diagnosed with vaginal carcinoma are postmenopausal, but about 20% of women are aged 50 years or less [194]. Vaginal carcinoma most commonly spreads locally and via the lymphatics, and the residual vagina, paravaginal lymph nodes, parametrium, and rectovaginal septum are at greatest risk for recurrence after conservative surgery [194]. The risk of pelvic lymph node metastases is 20–35%, even in early-stage disease [197]. The risk factors associated with a poor prognosis of vaginal carcinoma are a nonsquamous histology, disease in the upper two-third of the vagina, and tumor grade and stage [195].

Standard of Care

Radiation is the standard of care and the primary treatment modality in the treatment of vaginal carcinoma [194, 195, 198].

Table 14.10 Gestational trophoblastic disease 1995–2010

First author	Year	Patients	Procedure	Pregnancies	Attempted pregnancies	Live births	Preterm deliveries	Term deliveries	Recurrences	Deaths
Resection ± chemotherapy										
Behtash	2009	1 CC[a]	Craniotomy + EMA–EP[b] + brain XRT[c]	3	1	1	0	1	0	0
Liszka	2009	1 PSTT[d]	Tumor resection	0	1	0	0	0	0	0
Pfeffer	2007	1 PSTT	Partial hysterectomy + MTX[e]/ EMA–CO[f]/gemcitabine	NR[g]	NR	NR	NR	NR	1	0
Rojas-Espaillat	2007	1 persistent GTD[h]	Tumor resection + EMA–CO	NR	NR	NR	NR	NR	0	0
Machtinger	2005	1 PSTT	Hysteroscopic resection + EMA–CO × 3 courses	NR	NR	NR	NR	NR	0	0
Tsuji	2002	1 PSTT	Tumor resection + EMA–CO	NR	NR	NR	NR	NR	0	0
Case	2001	1 persistent GTD	Tumor resection	2	1	2	0	2	0	0
Leiserowitz	1996	1 PSTT	Tumor resection	3	1	1	0	1	0	0
Total resection ± chemotherapy		5 PSTT, 2 persistent GTD, 1 CC		8	4	4	0	4	1	0
Chemotherapy										
Numnum	2006	1 PSTT	EMA–EP	1	1	1	0	1	0	0
Goto	2004	62 CC	MTX/MTX + actinomycin ± cyclophosphamide/etoposide/ etoposide + actinomycin	43	NR	36	NR	NR	NR	12
Total chemothearpy		1 PSTT, 62 CC		44	1	37	0	1	0	12
Total		6 PSTT, 2 persistent GTD, 63 CC		52	5	41	0	5	1	12

[a] CC = choriocarcinoma
[b] EMA–EP = etoposide, methotrexate, actinomycin–etoposide, cisplatinum
[c] XRT = radiation therapy
[d] PSTT = placental site trophoblastic tumor
[e] MTX = methotrexate
[f] EMA–CO = etoposide, methotrexate, actinomycin–cyclophosphamide, vincristine
[g] NR = not recorded
[h] GTD = gestational trophoblastic disease

Fertility Preservation

Tumorectomy with Pelvic Lymphadenectomy

Indications

Patients with low-grade, localized stage I lesions who are strongly desiring fertility preservation are potential candidates for fertility-preserving treatment [195, 198]. Because of the rarity of the disease and this procedure, the indications are based on case report data only. Patients who are interested in this procedure must be counseled that the available data is based on case reports only and must be aware of the risk of recurrence and death after this procedure.

Technique

A wide excision of the tumor is performed through a vaginal approach, and the margins of the specimen are sent for frozen section [194, 197]. Positive margins require further treatment. The pelvic lymphadenectomy is usually performed laparoscopically after the tumor resection [194].

Oncological Outcome

Data on the oncological outcome of patients treated with tumorectomy and pelvic lymphadenectomy is very limited. Four cases of patients with vaginal carcinoma treated with tumorectomy and pelvic lymphadenectomy have been described in the literature (Table 14.11) [197, 199]. The authors describe no recurrences or death (Table 14.11) [194].

Fertility Outcome

The data on reproductive outcome after fertility-preserving surgery is extremely limited [194]. Fujita et al. describe a preterm delivery after tumorectomy for vaginal carcinoma [197]. The remainder of the case reports do not comment on the reproductive outcome (Table 14.11). Cutillo et al. and Fujita et al. reported excellent functional results (no dyspareunia, vaginal dryness, or loss of vaginal sensation) after radical tumorectomy [194, 197].

Partial Vaginectomy/Radical Trachelectomy with Pelvic Lymphadenectomy

Indications

Patients with early-stage vaginal carcinoma in close proximity to the cervix who desire fertility preservation are potential candidates for a partial vaginectomy and radical trachelectomy with pelvic lymphadenectomy [198]. This fertility-preserving procedure is an option for patients who would otherwise require a radical hysterectomy [198]. It must be stressed, however, that data on this procedure in this particular patient population is extremely scarce.

Technique

A radical abdominal trachelectomy is performed through a midline abdominal incision [194]. After inspection of the abdomen and identification of the ureters, the pelvic lymphadenectomy is performed [194, 198]. The radical trachelectomy consists of a vaginotomy, paravaginal and cervical dissection, ligation of the descending branch of the uterine artery, ureterolysis, and transsection of the distal cervix [2, 7, 9]. A frozen section is performed [194, 198]. A positive result on the frozen section warrants further therapy. The upper vagina is resected with a macroscopically free margin, and the vagina and uterus are subsequently reapproximated with sutures [194, 198].

Oncological Outcome

There is very limited case report-based data on the oncological outcome of patients treated with radical trachelectomy, upper vaginectomy, and pelvic lymphadenectomy for vaginal carcinoma. Matthews et al. and Shepherd et al. present a case of a patient with clear cell carcinoma who remained disease-free after the radical trachelectomy, upper vaginectomy, and pelvic lymphadenectomy [198, 200].

Fertility Outcome

There is minimal data on the reproductive outcome of women with vaginal carcinoma treated with radical trachelectomy, upper vaginectomy,

Table 14.11 Vaginal carcinoma 2000–2010

First author	Year	Patients	Procedure	Pregnancy	Attempted pregnancies	Live births	Preterm deliveries	Term deliveries	Recurrences	Deaths
Shepherd	2010	1	RVT[a] + pelvic LND[b]	NR[c]	NR	NR	NR	NR	0	0
Renaud	2009	1	Radical tumorectomy, omentectomy, sentinel lymph node, pelvic LND	NR	NR	NR	NR	NR	NR	NR
Matthews	2007	1	RAT[d] + upper vaginectomy	NR	NR	NR	NR	NR	0	0
Cutillo	2006	4	Three radical tumorectomy + pelvic LND 1 partial hemivaginectomy + pelvic LND	NR	NR	NR	NR	NR	0	0
Fujita	2005	1	Tumorectomy + pelvic LND + brachytherapy	1[e]	1	1	1	0	0	0
Total		8		1	1	1	1	0	0	0

[a] RVT = radical vaginal trachelectomy
[b] LND = lymph node dissection
[c] NR = not recorded
[d] RAT = radical abdominal trachelectomy
[e] Diagnosed during pregnancy

and pelvic lymphadenectomy. Hudson et al. performed a radical trachelectomy on a patient with clear cell vaginal carcinoma, and the patient delivered an infant at 36 weeks gestation via cesarean section [201]. As with radical trachelectomy for other indications, there is an increased risk of preterm delivery after radical trachelectomy [198].

Vulvar Carcinoma

Vulvar carcinomas represent 3–5% of all gynecologic cancers [202, 203]. Although generally a disease of postmenopausal women, 15% of the vulvar carcinomas occurs in women under the age of 40 [202–206]. In young women, vulvar carcinoma is often multifocal and presents with a background of hyperplasia [202]. The majority of vulvar carcinomas are squamous cell carcinomas [207]. Lymphovascular space invasion, tumor multifocality, and the presence of VIN 2 or 3 are independent predictors of recurrence [208].

Standard of Care

The standard treatment of vulvar carcinoma consists of a radical vulvectomy with bilateral inguinal lymphadenectomy [203, 204, 207]. Pelvic lymphadenectomy is no longer recommended as part of the standard treatment of vulvar carcinoma [206]. Wound breakdown, lymphedema of the lower extremity, and psychological problems related to sexual dysfunction and changes in body image are major consequences of this aggressive surgery [207, 209].

Fertility Preservation

Radical Tumorectomy

Indications

Patients with stage I disease who strongly desire preservation of fertility are candidates for conservative surgery for vulvar carcinoma [210]. Some authors even include patients with superficial stage II disease [210].

Technique

For the radical vulvectomy, the incisions are carried down to the inferior fascia of the urogenital diaphragm and the tumor is excised [207]. The inguinofemoral lymph node dissection is indicated in all patients with a depth of invasion >1 mm and can be performed through separate incisions [207, 210]. Margins of at least 1–2 cm are recommended with a radical vulvectomy [207, 210]. Hacker and Van der Velden recommend groin and pelvic radiation therapy if at least one large lymph node was found to be replaced by tumor or if multiple lymph nodes harbor micrometastases [210]. Recently, sentinel lymph node excision has been used to identify occult lymph node spread in women with early (stage I) vulvar cancer. If the sentinel lymph node is free of metastatic disease, the patient can be spared a formal lymphadenectomy and its potential complications.

Oncological Outcome

Arvas et al. and De Hullu et al. compared conservative to radical surgery for patients with vulvar carcinoma and detected no significant difference in the rate of disease-free survival [204, 207]. De Hullu et al. found significantly higher recurrences in the group of conservatively treated patients [207]. Table 14.12 summarizes the available recent literature. In the patients treated with fertility-preserving surgery, the rate of recurrence was 28%. No deaths were reported (Table 14.12). Since the vulvectomy is a less-aggressive surgery, it is associated with a decrease in wound breakdown, lower extremity edema, and a shorter hospital stay [207].

Fertility Outcome

Three cases of a successful pregnancy after fertility-preserving treatment of a vulvar carcinoma are reported in the recent literature [205, 206]. Dicken et al. describe a patient with a term pregnancy after highly individualized treatment of vulvar carcinoma consisting of a radical hemivulvectomy, bilateral inguinofemoral lymphadenectomy, brachytherapy, and external beam radiation [205]. Palmer and Tidy report one case of a patient treated with radical vulvectomy,

Table 14.12 Vulvar carcinoma

First author	Year	Patients	Procedure	Pregnancy	Attempted pregnancies	Live births	Term deliveries	Recurrences	Deaths
Dicken	2010	1	Radical hemivulvectomy + bilateral inguinofemoral LND[a] + brachytherapy + external beam radiation	1	1	1	1	0	0
Palmer	2009	2	1 radical vulvectomy + bilateral inguinofemoral LND + vulvar skin grafting 1 wide excision + bilateral inguinofemoral LND	2	2	2	1	0	0
Arvas	2005	40	29 local radical tumor resection 11 triple-incision vulvectomy	NR[b]	NR	NR	NR	17	NR
De Hullu	2002	85	Local excision + inguinofemoral LND	NR	NR	NR	NR	17	NR
Serkies	2002	1	Local excision	NR	NR	NR	NR	1	0
Total		127		3	3	3	2	35	0

[a] LND = lymph node dissection
[b] NR = not recorded

inguinofemoral lymphadenectomy, and vulvar graft reconstruction and one patient with radical tumorectomy and inguinofemoral lymphadenectomy [206]. One patient delivered at term, and the other was diagnosed with an intrauterine fetal demise at 29 weeks gestation [206]. Pregnancy does not appear to increase the risk of recurrence of vulvar carcinoma [206].

Conclusions

Fertility preservation is a significant concern of young women with gynecologic malignancies and should be addressed by the providing physicians [2, 5]. Conservative treatment regimens have been reported, and many show promising results. The majority of conservative surgical procedures for the preservation of fertility do not routinely require specific technical skills, but the selection of appropriate patients can be challenging [7]. A multidisciplinary approach and early referral to reproductive specialists are important to provide appropriate counseling and offer all available treatment options [5]. Counseling of the patient is crucial, and compliance with follow-up is very important. Fertility-preserving treatment may need to be highly individualized to allow for the best outcome possible.

References

1. Leitao MM, Chi DS. Fertility-sparing options for patients with gynecologic malignancies. Oncologist. 2005;10:613–22.
2. Liou WS, Yap OWS, Chan JK, Westphal LM. Innovations in fertility preservation for patients with gynecologic cancers. Fertil Steril. 2005;84:1561–73.
3. SEER Cancer Statistics Review, 1975–2006, National CancerInstitute.http://seer.cancer.gov/csr/1975_2006/. Accessed 15 Apr 2010.
4. Maltaris T, Boehm D, Dittrich R, Seufert R, Koelbl H. Reproduction beyond cancer: a message of hope for young women. Gynecol Oncol. 2006;103:1109–21.
5. Lee SJ, Schover LR, Partridge AH, et al. American Society of Clinical Oncology recommendations on fertility preservation in cancer patients. J Clin Oncol. 2006;24:2917–31.
6. Quinn GP, Vadaparampil ST, Lee JH, et al. Physician referral for fertility preservation in oncology patients: a national study of practice behaviors. J Clin Oncol. 2009;27:5952–7.
7. Leblanc E, Narducci F, Ferron G, Querleu D. Indications and teaching of fertility preservation in the surgical management of gynecologic malignancies: European perspective. Gynecol Oncol. 2009;114:S32–6.
8. Schwartz S. Young cervical cancer patients and fertility. Semin Oncol Nurs. 2009;25:259–67.
9. Shepherd JH, Milliken DA. Conservative surgery for carcinoma of the cervix. Clin Oncol. 2008;20:395–400.
10. Ramirez P, Schmeler KM, Soliman PT, Frumovitz M. Fertility preservation in patients with early cervical cancer: radical trachelectomy. Gynecol Oncol. 2008;110:S25–8.
11. Noyes N, Abu-Rustum NR, Ramirez PT, Plante M. Options in the management of fertility-related issues after radical trachelectomy in patients with early cervical cancer. Gynecol Oncol. 2009;114:117–20.
12. Seli E, Tangir J. Fertility preservation options for female patients with malignancies. Curr Opin Obstet Gynecol. 2005;17:299–308.
13. Haie-Meder C, Morice P, Castiglione M. Cervical cancer: ESMO clinical recommendations for diagnosis, treatment and follow-up. Ann Oncol. 2009;20:iv27–8.
14. Pahisa J, Alonso I, Torné A. Vaginal approaches to fertility-sparing surgery in invasive cervical cancer. Gynecol Oncol. 2008;110:S29–32.
15. Makar AP, Tropé C. Fertility preservation in gynecologic cancer. Acta Obstet Gynecol Scand. 2001;80:794–802.
16. Ueda M, Ueki K, Kanemura M, et al. Conservative excisional laser conization for early invasive cervical cancer. Gynecol Oncol. 2004;95:231–324.
17. Tseng CJ, Horng SG, Soong YK, Hsueh S, Hsieh CH, Lin HW. Conservative conization for microinvasive carcinoma of the cervix. Am J Obstet Gynecol. 1997;176:1009–10.
18. Bisseling KC, Bekkers RL, Rome RM, Quinn MA. Treatment of microinvasive adenocarcinoma of the uterine cervix: a retrospective study and review of the literature. Gynecol Oncol. 2007;107:424–30.
19. McHale MT, Le TD, Burger RA, Gu M, Rutgers JL, Monk BJ. Fertility sparing treatment for in situ and early invasive adenocarcinoma of the cervix. Obstet Gynecol. 2001;98:726–31.
20. Kyrgiou M, Koliopoulos G, Martin-Hirsch P, Arbyn M, Prendiville W, Paraskevaidis E. Obstetric outcomes after conservative treatment for intraepithelial or early invasive cervical lesions: systemic review and meta-analysis. Lancet. 2006;367:489–98.
21. Einstein MH, Park KJ, Sonodo Y, et al. Radical vaginal versus abdominal trachelectomy for stage IB1 cervical cancer: a comparison of surgical and pathological outcomes. Gynecol Oncol. 2009;112:72–7.
22. Rodriguez-Macias Wallberg KA, Keros V, Hovatta O. Clinical aspects of fertility preservation in female patients. Pediatr Blood Cancer. 2009;53:254–60.
23. Ramirez PT, Schmeler KM, Malpica A, Soliman PT. Safety and feasibility of robotic radical trachelectomy in patients with early-stage cervical cancer. Gynecol Oncol. 2010;116:512–5.

24. Roy M, Plante M. Pregnancies after radical vaginal trachelectomy for early-stage cervical cancer. Am J Obstet Gynecol. 1998;179:1491–6.
25. Oktay K, Sönmezer M. Fertility preservation in gynecologic cancers. Curr Opin Oncol. 2007;19:506–11.
26. Abu-Rustum NR, Neubauer N, Sonoda Y, et al. Surgical and pathological outcomes of fertility-sparing radical abdominal trachelectomy for FIGO stage IB1 cervical cancer. Gynecol Oncol. 2005;111:261–4.
27. Cibula D, Sláman J, Fischerova D. Update on abdominal radical trachelectomy. Gynecol Oncol. 2008;111: S111–5.
28. Burnett AF. Radical trachelectomy with laparoscopic lymphadenectomy: review of oncologic and obstetrical outcomes. Curr Opin Obstet Gynecol. 2006;18:8–13.
29. Rodriguez M, Guimares O, Rose PG. Radical abdominal trachelectomy and pelvic lymphadenectomy with uterine conservation and subsequent pregnancy in the treatment of early cervical cancer. Am J Obstet Gynecol. 2001;185:370–4.
30. Ungár L, Pálfalvi L, Hogg R, et al. Abdominal radical trachelectomy: a fertility-preserving option for women with early cervical cancer. BJOG. 2005;112:366–9.
31. Olawaiye A, Del Carmen M, Tambouret R, Goodman A, Fuller A, Duska LR. Abdominal radical trachelectomy: success and pitfalls in a general gynecologic oncology practice. Gynecol Oncol. 2009;112:506–10.
32. Beiner ME, Hauspy J, Rosen B, Murphy J, et al. Radical vaginal trachelectomy vs. radical hysterectomy for small early stage cervical cancer: a matched case-control study. Gynecol Oncol. 2008;110: 168–71.
33. Marchiole P, Benchaib M, Buenerd A, Lazlo E, Dargent D, Mathevet P. Oncological safety of laparoscopic-assisted vaginal radical trachelectomy (LARVT or Dargent's operation): a comparative study with laparoscopic-assisted vaginal radical hysterectomy (LARVH). Gynecol Oncol. 2007;106:132–41.
34. Sonoda Chi DS, Carter J, Barakat RR, Abu-Rustum NR. Initial experience with Dargent's operation: the radical vaginal trachelectomy. Gynecol Oncol. 2008;108:214–9.
35. Diaz JP, Sonoda Y, Leitao MM, et al. Oncologic outcome of fertility-sparing radical trachelectomy versus radical hysterectomy for stage IB1 cervical carcinoma. Gynecol Oncol. 2005;111:255–60.
36. Hertel H, Köhler C, Grund D, et al. For the German Association of Gynecologic Oncologists (AGO). Gynecol Oncol. 2006;103:506–611.
37. Bernardini M, Barrett J, Seaward G, Covens A. Pregnancy outcomes in patients after radical trachelectomy. Am J Obstet Gynecol. 2003;189:1378–82.
38. Plante M, Renaud MC, Hoskins IA. Vaginal radical trachelectomy: a valuable fertility-preserving option in the management of early-stage cervical cancer. A series of 50 pregnancies and review of the literature. Gynecol Oncol. 2005;98:3–10.
39. Mathevet P, de Kaszon L, Dargent D. Fertility preservation in early cervical cancer. Gynecol Obstet Fertil. 2003;31:706–12.
40. Schlaerth JB, Spirtos NM, Schlaerth AC. Radical trachelectomy and pelvic lymphadenectomy with uterine preservation in the treatment of cervical cancer. Am J Obstet Gynecol. 2003;188:29–34.
41. Nishio H, Fujii T, Kameyama K, et al. Abdominal radical trachelectomy as a fertility-sparing procedure in women with early-stage cervical cancer in a series of 61 women. Gynecol Oncol. 2009;115:51–5.
42. Bader AA, Tamussino KF, Moinfar F, Bjelic-Radisic V, Winter R. Isolated recurrence at the residual uterine cervix after abdominal trachelectomy for early cervical cancer. Gynecol Oncol. 2005;99:785–7.
43. Palfalvi L, Ungár L, Boyle DC, Del Priore G, Smith JR. Announcement of healthy baby boy born after radical abdominal trachelectomy. Int J Gynecol Cancer. 2003;13:250.
44. Cibula D, Ungár L, Pálfalvi L, Binó B, Kuzel D. Laparoscopic abdominal radical trachelectomy. Gynecol Oncol. 2005;9:707–9.
45. Kim JH, Park JY, Kim DY, Kim YT, Nam JH. Fertility-sparing laparoscopic radical trachelectomy for young women with early stage cervical cancer. BJOG. 2010;117:340–7.
46. Lin J, Lam SK, Cheung TH. Sarcoma botryoides of the cervix treated with limited surgery and chemotherapy to preserve fertility. Gynecol Oncol. 1995;58: 270–3.
47. Bernal KL, Fahmy L, Remmenga S, Bridge S, Bridge J, Baker J. Embryonal rhabdomyosarcoma (sarcoma botryoides) of the cervix presenting as a cervical polyp treated with fertility-sparing surgery and adjuvant chemotherapy. Gynecol Oncol. 2004;95:243–6.
48. Gruessner SEM, Omwandho COA, Dreyer T, et al. Management of stage I cervical sarcoma botryoides in childhood and adolescence. Eur J Pediatr. 2004;163: 452–6.
49. Zeisler H, Mayerhofer K, Joura EA, Bancher-Todesca D, Kainz C, Breitenecker G. Embryonal rhabdomyosarcoma of the uterine cervix: case report and review of the literature. Gynecol Oncol. 1998;69:78–83.
50. Behtash N, Mousavi A, Tehranian A, Khanafshar N, Hanjani P. Embryonal rhabdomyosarcoma of the uterine cervix: case report and review of the literature. Gynecol Oncol. 2003;91:452–5.
51. Zanetta G, Rota SM, Lissoni A, Chiari S, Bratina G, Mangioni C. Conservative treatment followed by chemotherapy with doxorubicin and ifosfamide for cervical sarcoma botryoides in young females. Br J Cancer. 1990;80:403–6.
52. Scaravalla G, Simeone S, Dell'Aversana Orabona G, et al. Case report of a sarcoma botryoides of the uterine cervix in fertile age and literature review. Arch Gynecol Obstet. 2009;280:863–6.
53. Fawole AO, Babarinsa IA, Ogunbiyi JO, Familusi F, Adewole IF. Sarcoma botryoides in a seven year old: successful chemotherapeutic management. J Obstet Gynecol. 1999;19:92–3.
54. Martelli H, Oberlin O, Rey A, et al. Conservative treatment for girls with non-metastatic rhabdomyosarcoma of the genital tract: a report from the Study

Committee of the International Society of Pediatric Oncology. J Clin Oncol. 1999;17:2117–22.
55. Schmidt KT, Larsen EC, Andersen CY, Andersen AN. Risk of ovarian failure and fertility preserving methods in girls and adolescents with a malignant disease. BJOG. 2010;117:163–74.
56. Zanetta G, Rota S, Chiari S, Benazzi C, Bratina G, Mangioni C. Behavior of borderline tumors with particular interest to persistence, recurrence, and progression to invasive carcinoma: a prospective study. J Clin Oncol. 2001;19:2658–64.
57. Park JY, Kim DY, Kim JH, Kim YM, Kim YT, Nam JH. Surgical management of borderline ovarian tumors: the role of fertility-sparing surgery. Gynecol Oncol. 2009;113:75–82.
58. Poncelet C, Fauvet R, Boccara J, Daraï E. Recurrence after cystectomy for borderline ovarian tumors: results of a French multicenter study. Ann Surg Oncol. 2006;13:565–70.
59. Koskas M, Madelenat P, Yazbeck C. Tumeur borderline de l'ovarie: comment préserver la fertilité? Gynecol Obstet Fertil. 2009;37:942–50.
60. Morice P, Camatte S, El-Hassan J, Pautier P, Duvillard P, Castaigne D. Clinical outcomes and fertility after conservative treatment of ovarian borderline tumors. Fertil Steril. 2001;75:92–6.
61. Huang JYJ, Buckett WM, Gilbert L, Tan SL, Chain RC. Retrieval of immature oocytes followed by in vitro maturation and vitrification: a case report on a new strategy of fertility preservation in women with borderline ovarian malignancy. Gynecol Oncol. 2007;105:542–4.
62. Swanton A, Bankhead CR, Kehoe S. Pregnancy rates after conservative treatment for borderline ovarian tumours: a systematic review. Eur J Obstet Gynecol Reprod Biol. 2007;135:3–7.
63. Burks RT, Sherman ME. Micropapillary serous carcinoma of the ovary. A distinctive low-grade carcinoma related to serous borderline tumors. Am J Surg Pathol. 1996;20:1319–30.
64. Fauvet R, Poncelet C, Boccara J, Descamps P, Fondrinier E, Daraï E. Fertility after conservative treatment for borderline ovarian tumors: a French multicenter study. Fertil Steril. 2005;83:284–90.
65. Bell KA, Smith Sehdev AE, Kurman RJ. Refined diagnostic criteria for implants associated with ovarian atypical proliferative serous tumors (borderline) and micropapillary serous tumors. Am J Surg Pathol. 2001;25:419–32.
66. Camatte S, Rouzier R, Boccara-Dekeyser J, et al. Pronostic et fertilité après traitement conservateur d'une tumeur ovarienne à la limite de la malignité: revue d'une série continue de 68 cas. Gynecol Obstet Fertil. 2002;30:583–91.
67. Morris RT, Gershenson DM, Silva EG, Follen M, Morris M, Wahrton JT. Outcome and reproductive function after conservative surgery for borderline ovarian tumors. Obstet Gyecol. 2001;95:541–7.
68. Laurent I, Uzan C, Gouy S, Pautier P, Duvillard P, Morice P. Results after conservative treatment of serous borderline tumours of the ovary with stromal microinvasion but without micropapillary pattern. BJOG. 2009;116:860–2.
69. Romagnole C, Gadducci A, Sartori W, Zola P, Maggino T. Management of borderline ovarian tumors: results of an Italian multicenter study. Gynecol Oncol. 2006;101:255–60.
70. Boran N, Cil AP, Tulunay G, et al. Fertility and recurrence results of conservative surgery for borderline ovarian tumors. Gynecol Oncol. 2005;97:845–51.
71. Rao GG, Skinner EN, Gehrig PA, Duska LR, Miller DS, Schorge JO. Fertility-sparing surgery for ovarian low malignant potential tumors. Gynecol Oncol. 2005;98:263–6.
72. Chan JK, Lin YG, Loizzi V, Ghobriel M, DiSaia PJ. Borderline ovarian tumors in reproductive-age women. Fertility-sparing surgery and outcome. J Reprod Med. 2003;48:756–60.
73. Serracchioli R, Venturoli S, Colombo FM, Govoni F, Missiroli S, Bagnoli A. Fertility and tumor recurrence rate after conservative laparoscopic management of young women with early-stage borderline ovarian tumors. Fertil Steril. 2001;76:999–1004.
74. Uzan C, Kane A, Rey A, Gouy S, Duvillard P, Morice P. Outcomes after conservative treatment of advanced-staged serous borderline tumors of the ovary. Ann Oncol. 2010;21:55–60.
75. Fortin A, Morice P, Thoury A, Camatte S, Dhainaut C, Madelenat P. Impact of infertility drugs after treatment of borderline ovarian tumors: results of a retrospective multicenter study. Fertil Steril. 2007;87:591–6.
76. Park JY, Kim DY, Suh DS, et al. Outcomes of fertility-sparing surgery for invasive epithelial ovarian cancer: oncologic safety and reproductive outcomes. Gynecol Oncol. 2008;110:345–53.
77. Colombo N, Chiari S, Maggioni A, Bocciolone L, Torri V, Mangioni C. Controversial issues in the management of early epithelial ovarian cancer: conservative surgery and role of adjuvant therapy. Gynecol Oncol. 1994;55:S47–51.
78. Jemal A, Siegel R, Ward E, Hao Y, Xu J, Thun MJ. Cancer Statistics, 2009. CA Cancer J Clin. 2009;59:225–49.
79. Zanagnola V, Sartori W, Trussardi E, Pasinetti B, Maggino T. Preservation of ovarian function, reproductive ability and emotional attitudes in patients with malignant ovarian tumors. Eur J Obstet Gynecol Reprod Biol. 2005;123:235–43.
80. Schilder JM, Thompson AM, DePriest PD, et al. Outcome of reproductive age women with stage IA or IC invasive epithelial ovarian cancer treated with fertility-sparing therapy. Gynecol Oncol. 2002;87:1–7.
81. McHale MT, DiSaia PJ. Fertility-sparing treatment of patients with ovarian cancer. Compr Ther. 1999;25:144–50.
82. Duska LR, Chang YC, Flynn CE, et al. Epithelial ovarian carcinoma in the reproductive age group. Cancer. 1999;85:2623–9.

83. Wright JD, Shah M, Mathew L, et al. Fertility preservation in young women with epithelial ovarian cancer. Cancer. 2009;115:4118–26.
84. Schlaerth AC, Dennis SC, Poynor EA, Barakat RR, Brown CL. Long-term survival after fertility-sparing surgery for epithelial ovarian cancer. Int J Gynecol Cancer. 2009;19:1199–204.
85. Kwon YS, Hahn HS, Kim TJ, et al. Fertility preservation in patients with early epithelial ovarian cancer. J Gynecol Oncol. 2009;20:44–7.
86. Han LY, Kipps E, Kaye SB. Current treatment and clinical trials in ovarian cancer. Expert Opin Investig Drugs. 2010;19:521–34.
87. Morice P, Camatte S, Wicart-Poque F, et al. Results of conservative treatment in epithelial ovarian carcinoma. Cancer. 2001;92:2412–8.
88. Satoh T, Hatae M, Watanabe Y, et al. Outcomes of fertility-sparing surgery for stage I epithelial ovarian cancer: a proposal for patient selection. J Clin Oncol. 2010;28:1727–32.
89. Morice P, Leblanc E, Rey A, et al. Conservative treatment in epithelial ovarian cancer: results of a multicenter study of the GCCLCC (Groupe des Chirurgiens de Centre de Lutte Contre le Cancer) and SFOG (Société Française d'Oncologie Gynécologique). Hum Reprod. 2005;20:1379–85.
90. Zanetta G, Rota SM, Lissoni A, Chiari S, Bratina G, Mangioni C. Conservative treatment followed by chemotherapy with doxorubicin and ifosfamide for cervical sarcoma botryoides in young females. Br J Cancer. 1999;80:403–6.
91. Mahdavi A, Pejovic T, Nezhat F. Induction of ovulation and ovarian cancer: a critical review of the literature. Fertil Steril. 2006;85:819–26.
92. Meirow D. Ovarian injury and modern options to preserve fertility in female cancer patients treated with high dose radio-chemotherapy for hemato-oncological neoplasia and other cancers. Leuk Lymphoma. 1999;33:65–76.
93. Maltaris T, Seufert R, Fischl F, et al. The effect of cancer treatment on female fertility and strategies for preserving fertility. Eur J Obstet Gynecol Reprod Biol. 2007;130:148–55.
94. Tarumi W, Suzuki N, Takahashi N, et al. Ovarian toxicity of paclitaxel and effect of fertility in the rat. J Obstet Gynaecol Res. 2009;35:414–20.
95. Tangir J, Zelterman D, Ma W, Schwartz PE. Reproductive function after conservative surgery and chemotherapy for malignant germ cell tumors of the ovary. Obstet Gynecol. 2003;101:251–7.
96. Yoo SC, Kim WY, Yoon JH, Chang SJ, Chang KH, Ryu HS. Young girls with malignant ovarian germ cell tumors can undergo normal menarche and menstruation after fertility-preserving surgery and adjuvant chemotherapy. Acta Obstet Gynecol Scand. 2010;89:126–30.
97. Nishio S, Ushijima K, Fukui A, et al. Fertility-preserving treatment for patients with malignant germ cell tumors of the ovary. J Obstet Gynaecol Res. 2006;32:416–21.
98. Roth LM, Talerman A. Recent advances in the pathology and classification of ovarian germ cell tumors. Int J Gynecol Pathol. 2006;25:305–20.
99. Zanetta G, Bonazzi C, Cantù MG, et al. Survival and reproductive function after treatment of malignant germ cell ovarian tumors. J Clin Oncol. 2001;19:1015–20.
100. Schwartz PE. Combination chemotherapy in the management of ovarian germ cell malignancies. Obstet Gynecol. 1984;64:564–72.
101. Schwartz PE. Surgery of germ cell tumors of the ovary. Forum Trends Exper Clin Med. 2000;10:355–65.
102. Schwartz PE, Chambers SK, Chambers JT, Kohorn E, McIntosh S. Ovarian germ cell malignancies: the Yale University experience. Gynecol Oncol. 1992;45:26–31.
103. Edraki B, Schwartz PE. Fertility and reproductive potential of patients with endodermal sinus tumors of the ovary treated conservatively. CME J Gynecol Oncol. 1997;2:144–7.
104. Low JJH, Perrin LC, Crandon AJ, Hacker NF. Conservative surgery to preserve ovarian function in patients with malignant ovarian germ cell tumors. Cancer. 2000;89:391–8.
105. Tangir J, Schwartz PE. Embryonal carcinoma: diagnosis and management during pregnancy and fertility after diagnosis. CME J Gynecol Oncol. 2002;7:225–7.
106. Kumar S, Shah JP, Bryant CS, et al. The prevalence and prognostic impact of lymph node metastasis in malignant germ cell tumors of the ovary. Gynecol Oncol. 2008;110:125–32.
107. Gershenson DM, Miller AM, Champion VL, et al. Reproductive and sexual function after platinum-based chemotherapy in long-term ovarian germ cell tumor survivors: a Gynecologic Oncology Group study. J Clin Oncol. 2007;25:2792–7.
108. Fishman DA, Schwartz PE. Current approaches to diagnosis and treatment of ovarian germ cell malignancies. Curr Opin Obstet Gynecol. 1994;6:99–104.
109. Green D, Zevon M, Lowrie G, Seigelstein N, Hall B. Congenital anomalies in children of patients who received chemotherapy for cancer in childhood and adolescence. N Eng J Med. 1991;325:141–6.
110. Bats AS, Larousserie F, Le Frère Belda MA, Metzger U, Lécuru F. Tumeurs non épithéliales malignes de l'ovaire. Gynecol Obstet Fert. 2009;37:627–32.
111. Colombo N, Parma G, Zanagnolo V, Insinga A. Management of ovarian stromal cell tumors. J Clin Oncol. 2007;25:2944–51.
112. Schneider DT, Calaminus G, Harms D, Göbel U. Ovarian sex cord-stromal tumors in children and adolescents. J Reprod Med. 2005;50:439–46.
113. Gershenson DM. Fertility-sparing surgery for malignancies in women. J Natl Cancer Inst Monogr. 2005;34:43–7.
114. Brown J, Sood AK, Deavers MT, Milojevic L, Gershenson DM. Patterns of metastasis in sex cord-stromal tumors of the ovary: can routine staging

lymphadenectomy be omitted? Gynecol Oncol. 2009;113:86–90.
115. Koulouris CR, Penson RT. Ovarian stromal and germ cell tumors. Semin Oncol. 2009;36:126–36.
116. Powell JL, Johnson NA, Bajley CL, Otis CN. Management of advanced juvenile granulosa cell tumor of the ovary. Gynecol Oncol. 1993;48:119–23.
117. Young RH, Scully RE. Well-differentiated ovarian Sertoli-Leydig cell tumors: a clinicopathological analysis of 23 cases. Int J Gynecol Pathol. 1984;3: 277–90.
118. Zanagnolo V, Pasinetti B, Sartori E. Clinical review of 63 cases of sex cord-stromal tumors. Eur J Gynaecol Oncol. 2004;25:431–8.
119. Evans AT, Gaffey TA, Malkasian GD, Annegers JF. Clinicopathologic review of 118 granulosa and 82 theca cells tumors. Obstet Gynecol. 1980;55:231–8.
120. Rhattanachaiyanont M, Angsuwathana S, Techatrisak K, Tanmahasamut P, Indhavivadhana S, Leerasiri P. Clinical and pathological responses of progestin therapy for non-atypical endometrial hyperplasia: a prospective study. J Obstet Gynaecol Res. 2005;31: 98–106.
121. Ercan CM, Duru NK, Sakinici M, Alanbay I, Karasahin KE, Baser I. Successful twin pregnancy achieved by assisted reproductive technology in a patient with polycystic ovary syndrome with complex atypical endometrial hyperplasia treated with levonorgestrel-releasing intrauterine system. Gynecol Endocrinol. 2010;26:125–8.
122. Varma R, Soneja H, Bhatia K, et al. The effectiveness of a levonorgestrel-releasing intrauterine system (LNG-IUS) in the treatment of endometrial hyperplasia – a long-term follow-up study. Eur J Obstet Gynecol Reprod Biol. 2008;139:169–75.
123. Rodriguez MR, Warden M, Darney PD. Intrauterine progestins, progesterone antagonists and receptor modulators: a review of gynecologic applications. Am J Obstet Gynecol. 2010;202:420–8.
124. Wildermeersch D, Janssens D, Pylyser K, et al. Management of patients with non-atypical and atypical endometrial hyperplasia with levonorgestrel-releasing intrauterine system: long-term follow-up. Maturitas. 2007;57:210–3.
125. Kaku T, Yoshikawa H, Tsuda H, et al. Conservative therapy for adenocarcinoma and atypical endometrial hyperplasia of the endometrium in young women: central pathologic review and treatment outcome. Cancer Lett. 2001;167:39–48.
126. Ørbo A, Arnes M, Hancke C, Vereide AB, Pettersen I, Larsen K. Treatment results of endometrial hyperplasia after prospective D-score classification. A follow-up study comparing the effect of LNG-IUD and oral progestins versus observation only. Gynecol Oncol. 2008;111:68–73.
127. Le Digabel JF, Gariel C, Catala L, Dhainaut C, Madelenat P, Descamps P. Hyperplasies atypiques et carcinomes de l'endomètre de stade I chez la femme jeune désirant une grossesse: le traitement conservateur est-il envisagable? Résultats d'une etude multicentrique française. Gynecol Obstet Fertil. 2006;34:27–33.
128. Yazbeck C, Dhainaut C, Thoury A, Driguez P, Madelenat P. Traitement conservateur du cancer et des hyperplasies atypiques de l'endomètre. Gynecol Obstet Fertil. 2004;32:433–41.
129. Horn LC, Schnurrbusch U, Bilek K, Hentschel B, Einenkel J. Risk of progression in complex and atypical endometrial hyperplasia: clinicopathologic analysis in cases with and without progestogen treatment. Int J Gynecol Cancer. 2004;14:348–53.
130. Randall TC, Kurman RJ. Progestin treatment of atypical hyperplasia and well-differentiated carcinoma of the endometrium in women under age 40. Obstet Gynecol. 1997;90:434–40.
131. Gadducci A, Spirito N, Barooni E, Tana R, Genazzani AR. The fertility-sparing treatment in patients with endometrial atypical hyperplasia and early endometrial cancer: a debated therapeutic option. Gynecol Endocrinol. 2009;25:683–91.
132. Kresowik J, Ryan GL, Van Voorhis BJ. Progression of atypical endometrial hyperplasia to endometrial adenocarcinoma despite intrauterine progesterone treatment with the levonorgestrel-releasing intrauterine system. Obstet Gynecol. 2008;111:547–9.
133. Trimble CL, Kauderer J, Zaino R, et al. Concurrent endometrial carcinoma in women with a biopsy diagnosis of atypical endometrial hyperplasia: a Gynecologic Oncology Group study. Cancer. 2006;106:812–9.
134. Yu M, Yang JX, Wu M, Lang J, Huo Z, Shen K. Fertility-preserving treatment in young women with well-differentiated endometrial carcinoma and severe atypical hyperplasia of endometrium. Fertil Steril. 2009;92:2122–4.
135. Qi X, Zhao W, Duan Y, Li Y. Successful pregnancy following insertion of a levonorgestrel-releasing intrauterine system in two infertile patients with complex atypical endometrial hyperplasia. Gynecol Obstet Invest. 2008;65:266–8.
136. Elizur SE, Beiner ME, Korach J, Weiser A, Ben-Baruch G, Dor J. Outcome of in vitro fertilization treatment in infertile women conservatively treated for endometrial adenocarcinoma. Fertil Steril. 2007;88:1562–7.
137. Azim A, Oktay K. Letrozole for ovulation induction and fertility preservation by embryo cryopreservation in young women with endometrial carcinoma. Fertil Steril. 2007;88:657–64.
138. Kim YB, Holschneider CH, Ghosh K, Nieberg RK, Montz FJ. Progestin alone as primary treatment of endometrial carcinoma in premenopausal women. Cancer. 1997;79:320–7.
139. Ushijima K, Yahata H, Yoshikawa H, et al. Multicenter phase II study of fertility-sparing treatment with medroxyprogesterone acetate for endometrial carcinoma and atypical hyperplasia in young women. J Clin Oncol. 2007;25:2798–803.
140. Gotlieb WH, Beiner ME, Shalmon B, et al. Outcome of fertility-sparing treatment with progestins in

young patients with endometrial cancer. Obstet Gynecol. 2003;102:718–25.
141. Ramirez PT, Frumowitz M, Bodurka DC, Sun CC, Levenback C. Hormonal therapy for the management of grade 1 endometrial adenocarcinoma: a literature review. Gynecol Oncol. 2004;95:133–8.
142. Thigpen JT, Brady MF, Alvarez RD, et al. Oral medroxyprogesterone acetate in the treatment of advanced or recurrent endometrial carcinoma: a dose-response study by the gynecologic oncology group. J Clin Oncol. 1999;17:1736–44.
143. Creasman WT, Morrow CP, Bundy BN, Homesley HD, Graham JE, Heller PB. Surgical pathologic spread patterns of endometrial cancer (a Gynecologic Oncology Group Study). Cancer. 1987;60:2035–41.
144. Dhar KK, NeedhiRajan T, Koslowski M, Woolas RP. Is levonorgestrel intrauterine system effective for treatment of early endometrial cancer? Report of four cases and review of the literature. Gynecol Oncol. 2005;97:924–7.
145. Kothari R, Seamon L, Cohn D, Fowler J, O'Malley DM. Stage IV endometrial cancer after failed conservative management: a case report. Gynecol Oncol. 2008;111:579–82.
146. Giannopoulos T, Butler-Manuel S, Tailor A. Levonorgestrel-releasing intrauterine system (LNG-IUS) as a therapy for endometrial carcinoma. Gynecol Oncol. 2004;95:762–4.
147. Vandenput I, Van Eygen K, Moerman P, Vergote I, Amant F. Ineffective attempt to preserve fertility with a levonorgestrel-releasing intrauterine device in a young woman with endometrioid endometrial adenocarcinoma: a case report and review of the literature. Eur J Gynaecol Oncol. 2009;30:313–6.
148. Signorelli M, Caspani G, Bonazzi C, Chiappa V, Perego P, Mangioni C. Fertility-sparing treatment in young women with endometrial cancer or atypical complex hyperplasia: a prospective single-institution experience of 21 cases. BJOG. 2009;116:114–8.
149. Yamazawa K, Hirai M, Fujito A, et al. Fertility-preserving treatment with progestins, and pathologic criteria to predict responses in young women with endometrial cancer. Hum Reprod. 2007;22: 1953–8.
150. Curtis KM, Marchbanks PA, Peterson HB. Neoplasia with use of intrauterine devices. Contraception. 2007;75:S60–9.
151. Wang CB, Wang CJ, Huang HJ, et al. Fertility-preserving treatment in young patients with endometrial adenocarcinoma. Cancer. 2002;94:2192–3209.
152. Sardi J, Anchezar Henry JP, Paniceres G, Gomez Rueda N, Vighi S. Primary hormonal treatment for early endometrial carcinoma. Eur J Gynaecol. 1998;19:565–6.
153. Hahn HS, Yoon SG, Hong JS, et al. Conservative treatment with progestins and pregnancy outcomes in endometrial cancer. Int J Gynecol Cancer. 2009;19:1068–73.
154. Vinker S, Shani A, Open M, Fenig E. Conservative treatment of adenocarcinoma of the endometrium in young patients. Is it appropriate? Eur J Obstet Gynecol. 1999;83:63–5.
155. Ferrandini G, Zannoni GF, Gallotta V, Foti E, Mancusco S, Scambia G. Progression of conservatively treated endometrial carcinoma after full term pregnancy: a case report. Gynecol Oncol. 2005;99:215–7.
156. Mazzon I, Corrado G, Masciullo V, Morricone D, Ferrandina G, Scambia G. Conservative surgical management of stage IA endometrial carcinoma for fertility preservation. Fertil Steril. 2010;93:1286–9.
157. Montz FJ, Bristow RE, Bovicelli A, Tomacruz R, Kurman RJ. Intrauterine progesterone treatment of early endometrial cancer. Am J Obstet Gynecol. 2002;186:651–7.
158. Vilos G, Edris F, Al-Mubarak A, Ettler HC, Hollett-Caines J, Abu-Rafea B. Hysteroscopic surgery does not adversely affect the long-term prognosis of women with endometrial adenocarcinoma. J Minim Invasive Gynecol. 2007;14:205–10.
159. D'Angelo E, Prat J. Uterine sarcomas: a review. Gynecol Oncol. 2010;116:131–9.
160. Li AJ, Giuntoli RL, Drake R, et al. Ovarian preservation in stage I low-grade endometrial stromal sarcomas. Obstet Gynecol. 2005;106:1304–8.
161. Giuntoli RL, Metzinger DS, DiMarco CS, et al. Retrospective review of 208 patients with leiomyosarcoma of the uterus: prognostic indicators, surgical management, and adjuvant therapy. Gynecol Oncol. 2003;89:460–9.
162. Schwartz PE, Kelly MG. Malignant transformation of myomas: myth or reality? Obstet Gynecol Clin North Am. 2006;33:183–98.
163. Salman MC, Guler OT, Kucukali T, Karaman N, Ayhan A. Fertility-saving surgery for low-grade uterine leiomyosarcoma with subsequent pregnancy. Int J Gynaecol Obstet. 2007;98:160–1.
164. Yan L, Tian Y, Fu Y, Zhao X. Successful pregnancy after fertility-preserving surgery for endometrial stromal sarcoma. Fertil Steril. 2010;93:269.e1–e3.
165. Chu MC, Mor G, Lim C, Zheng W, Parkash V, Schwartz PE. Low-grade endometrial stromal sarcoma: hormonal aspects. Gynecol Oncol. 2003;90:170–6.
166. Shah JP, Bryant CS, Kumar S, Ali-Fehmi R, Malone JM, Morris RT. Lymphadenectomy and ovarian preservation in low-grade endometrial stromal sarcoma. Obstet Gynecol. 2008;112:1102–8.
167. Chan JK, Kawar NM, Shin JY, et al. Endometrial stromal sarcoma: a population-based analysis. Br J Cancer. 2008;99:1210–5.
168. Schilder JM, Hurd WW, Roth LM, Sutton GP. Hormonal treatment of an endometrial stromal nodule followed by local excision. Obstet Gynecol. 1999;93:805–7.
169. Koskas M, Morice P, Yazbeck C, Duvillard P, Walker F, Madelenat P. Conservative management of low-grade endometrial stromal sarcoma followed by pregnancy and severe recurrence. Anticancer Res. 2009;29:4147–50.

170. Kagami S, Kashimura M, Toki N, Katuhata Y. Myxoid leiomyosarcoma of the uterus with subsequent pregnancy and delivery. Gynecol Oncol. 2002;85:538–42.
171. Lissoni A, Cormio G, Bonazzi C, et al. Fertility-sparing surgery in uterine leiomyosarcoma. Gynecol Oncol. 1998;70:348–50.
172. Friedrich M, Riffel B, Schillinger H, et al. Uterine leiomyosarcoma with subsequent pregnancy. Zentralbl Gynakol. 1998;120:341–6.
173. Bonney V. The fruits of conservatism. J Obstet Gynecol Br Empire. 1937;44:1–12.
174. Van Dinh T, Woodruff JD. Leiomyosarcoma of the uterus. Am J Obstet Gynecol. 1982;144:817–23.
175. Cormio G, Loizzi V, Carriero C, Scardigno D, Putignano G, Selvaggi L. Conservative management of uterine leiomyosarcoma: report of a failure. Eur J Gynaecol Oncol. 2009;30:206–7.
176. Guntupalli SR, Ramirez PT, Anderson ML, Milam MR, Bodurka DC, Malpica A. Uterine smooth muscle tumor of uncertain malignant potential: a retrospective analysis. Gynecol Oncol. 2009;113: 324–6.
177. Behtash N, Behnamfar F, Hamedi B, Ramezanzadeh F. Term delivery following successful treatment of choriocarcinoma with brain metastases, a case report and review of literature. Arch Gynecol Obstet. 2009;279:579–81.
178. Schorge JO, Goldstein DP, Bernstein MR, Berkowitz RS. Gestational trophoblastic disease. Curr Treat Options Oncol. 2000;1:169–75.
179. Horowitz NS, Goldstein DP, Berkowitz RS. Management of gestational trophoblastic neoplasia. Semin Oncol. 2009;36:181–9.
180. Thirumagal B, Sinha D, Raghavan R, Bhatti N. Gestational trophoblastic neoplasia: are we compliant with the standards? J Obstet Gynecol. 2009;29: 434–6.
181. Lipata F, Parkash V, Talmor M, et al. Precise DNA genotyping diagnosis of hydatidiform mole. Obstet Gynecol. 2010;115:784–94.
182. Sebire NJ, Lindsay I. Current issues in the histopathology of gestational trophoblastic tumors. Fetal Pediatr Pathol. 2010;29:30–44.
183. Berkowitz RS, Tuncer ZS, Bernstein MR, Goldstein DP. Management of gestational trophoblastic diseases: subsequent pregnancy experience. Semin Oncol. 2000;27:678–85.
184. Machtinger R, Gotlieb WH, Korach J, et al. Placental site trophoblastic tumor: outcomes of five cases including fertility preserving management. Gynecol Oncol. 2005;96:56–61.
185. Leiserowitz GS, Webb MJ. Treatment of placental site trophoblastic tumor with hysterotomy and uterine reconstruction. Obstet Gynecol. 1996;88: 696–9.
186. Numnum TM, Kilgore LC, Conner MG, Straughn JM. Fertility sparing therapy in a patient with placental site trophoblastic tumor: a case report. Gynecol Oncol. 2006;103:1141–3.
187. Tsuji Y, Tsubamoto H, Hori M, Ogasawara T, Koyama K. Case of PSTT treated with chemotherapy followed by open uterine tumor resection to preserve fertility. Gynecol Oncol. 2002;87:303–7.
188. Baergen RN, Rutgers JL, Young RH, Osann K, Scully RE. Placental site trophoblastic tumor: a study of 55 cases and review of the literature emphasizing factors of prognostic significance. Gynecol Oncol. 2006;100:511–20.
189. Lo C, Low I, Tan AL, Baranyai J. Epitheliod trophoblastic tumor: a case report. Int J Gynecol Cancer. 2006;16:1473–6.
190. Case AM, Wilson S, Colgan TJ, Greenblatt EM. Fertility-sparing surgery, with subsequent pregnancy, in persistent gestational trophoblastic neoplasia. Hum Reprod. 2001;16:360–4.
191. Pfeffer PE, Sebire N, McIndoe A, Lim A, Seckl MJ. Fertility-sparing partial hysterectomy for placental-site trophoblastic tumor. Lancet Oncol. 2007;8: 744–6.
192. Goto S, Ino K, Mitsui T, et al. Survival rates of patients with choriocarcinoma treated with chemotherapy without hysterectomy: effects of anticancer agents on subsequent births. Gynecol Oncol. 2004;93:529–35.
193. Newlands ES, Mulholland PJ, Holden L, Seckl MJ, Rustin GJS. Etoposide and cisplatin/etoposide, methotrexate, and actinomycin D (EMA) chemotherapy for patients with high-risk gestational trophoblastic tumors refractory to EMA/ cyclophosphamide and vincristine chemotherapy and patients presenting with metastatic placental site trophoblastic tumors. J Clin Oncol. 2002;18:854–9.
194. Cutillo G, Cignini P, Pizzi G, et al. Conservative treatment of reproductive and sexual function in young woman with squamous carcinoma of the vagina. Gynecol Oncol. 2006;103:234–7.
195. Otton GR, Nicklin JL, Dickie GJ, et al. Early-stage vaginal carcinoma: an analysis of 70 patients. Int J Gynecol Cancer. 2004;14:304–10.
196. Stock RG, Chen ASJ, Seski J. A 30-year experience in the management of primary carcinoma of the vagina: analysis of prognostic factors and treatment modalities. Gynecol Oncol. 1995;56:45–52.
197. Fujita K, Aoki Y, Tanaka K. Stage I squamous cell carcinoma of vagina complicating pregnancy: successful conservative treatment. Gynecol Oncol. 2005;98:513–5.
198. Matthews KS, Numnum TM, Conner MG, Barnes M. Fertility-sparing radical abdominal trachelectomy for clear cell adenocarcinoma of the upper vagina: a case report. Gynecol Oncol. 2007; 105: 820–2.
199. Renaud MC, Plante M, Grégoire J, Roy M. Primitive clear cell carcinoma of the vagina treated conservatively. J Obstet Gynaecol Can. 2009;31:54–6.
200. Shepherd ES, Lowe DA, Shepherd JH. Targeted selective trachelo-coplectomy for preservation of fertility in a young woman with vaginal clear cell carcinoma. J Obstet Gynaecol. 2010;30:420–1.

201. Hudson CN, Findlay WS, Roberts H. Successful pregnancy after radical surgery for diethylstilboestrol (DES) – related vaginal adenocarcinoma. Case report. Br J Obstet Gynaecol. 1988;95:818–9.
202. Couvreux-Dif D, Lhommé C, Querleu D, Castaigne D, Verhaeghe Y. Cancer vulvaire et grossesse: à propos de deux cas et revue de la litterature. J Gynecol Obstet Biol Reprod. 2003;32:46–50.
203. Serkies K, Wysocka B, Emerich J, Hrabowska M, Jassem J. Salvage hemipelvis radiotherapy with fertility preservation in an adolescent with recurrent vulvar carcinoma. Gynecol Oncol. 2002;85:381–3.
204. Arvas M, Köse F, Gezer A, Demirkiran F, Tulunay G, Kösebay D. Radical versus conservative surgery for vulvar carcinoma. Int J Gynecol Obstet. 2005;88: 127–33.
205. Dicken CL, Lieman HJ, Dayal AK, Mutyala S, Einstein MH. A multidisciplinary approach to fertility-sparing therapy for a rare vulvar tumor. Fertil Steril. 2010;93:267.e5–e7.
206. Palmer JE, Tidy JA. Pregnancy following vulvar squamous cell carcinoma: a report of two cases. J Gynecol Oncol. 2009;20:254–6.
207. De Hullu JA, Hollema H, Lolkema S, et al. Vulvar carcinoma. The price of less radical surgery. Cancer. 2002;95:2331–8.
208. Preti M, Ronco G, Ghiringhello B, Michelletti L. Recurrent squamous cell carcinoma of the vulva: clinicopathologic determinants identifying low-risk patients. Cancer. 2000;88:1869–76.
209. Chan JK, Sugiyama V, Tajalli TR, et al. Conservative clitoral preservation surgery in the treatment of vulvar squamous cell carcinoma. Gynecol Oncol. 2004;95:152–6.
210. Hacker NF, Van der Velden J. Conservative management of early vulvar cancer. Cancer. 1993;71:1673–7.

Third-Party Reproduction and Adoption After Cancer: Practical and Emotional Considerations

15

Dorothy A. Greenfeld

Abstract

Young women are increasingly treated successfully for cancer. When the treatment or the cancer itself leaves them with limited or no reproductive capacity, but not without the desire for children, they increasingly seek motherhood through third-party reproduction, such as oocyte donation or gestational surrogacy, or through adoption. Fortunately, there are a number of resources available for guidance and support in making the transition to parenthood through third-party reproduction or adoption.

Keywords

Third-party reproduction after cancer treatment • Adoption after cancer treatment • Fertility after cancer treatment • Oocyte donation • Gestational surrogacy

Due to ongoing advances in cancer treatment, young men and women increasingly survive their diagnosis and treatment. Unfortunately, the very treatment that leads to their survival often also destroys their future fertility. Oncologists, in their efforts to treat and preserve the lives of cancer patients, have been slow to recognize the importance of future parenthood to their young patients and fertility preservation is a subject that frequently goes unaddressed. Hubner and Glazer discussed the complexities of a counseling session where, in the turmoil and sense of crisis that surround the initial diagnosis and treatment of cancer, concerns about future fertility are often overshadowed and obscured. They argue that such counseling at the onset of cancer treatment is psychologically important because cancer survivors who are uninformed about the treatment's effect on future fertility may later experience deep regrets. Also, they note that patients may welcome a discussion of future fertility since it implies a belief in their survival and recovery [1].

In the past decade, the American Society of Clinical Oncology, the Practice Committee of the American Society for Reproductive Medicine, and the Royal College of Obstetrics and Gynecologists have developed guidelines on fertility preservation that address the importance of counseling cancer patients about possible fertility

D.A. Greenfeld (✉)
Department of Obstetrics and Gynecology, and Reproductive Sciences, Yale University School of Medicine, 333 Cedar Street, New Haven, CT 06510, USA
e-mail: dorothy.greenfeld@yale.edu

E. Seli and A. Agarwal (eds.), *Fertility Preservation: Emerging Technologies and Clinical Applications*,
DOI 10.1007/978-1-4419-1783-6_15,

impairment due to treatment as well as the importance of providing them with information on options for fertility preservation [2–4]. Through efforts of this sort, the significance of fertility and fertility preservation is increasingly recognized by health professionals, yet it is still the case that too few receive this counseling. In addition, options for fertility preservation are not accessible to all patients and not all clinicians have the knowledge or the skills necessary to address these issues during the critical period taken up with diagnosis and initial treatment [5].

In some patients, fertility preservation is not an option because of "treatment-related or theoretical concerns" [6]. Furthermore, not all patients choose to participate in such treatments when they are offered and available. Of 65 patients referred to one program for fertility preservation, Klock et al. reported that 18 females declined embryo, oocyte, or tissue cryopreservation [7]. Though fertility preservation is not available or applicable to all young cancer patients and though not all patients choose to participate, counseling that the loss of fertility and reproductive capacity does not mean the loss of opportunity for parenthood *should be available to all.* Advances in cancer treatment and reproductive medicine increasingly provide the opportunity for parenthood after cancer through third-party reproduction, such as oocyte donation and gestational surrogacy. Adoption provides another family-building opportunity for cancer survivors. This chapter addresses the practical and emotional considerations of third-party reproduction and adoption after cancer.

The Importance of Parenthood to Cancer Survivors

Studies have shown that most young people who survive cancer hope to have children and that the importance they placed on family closeness was increased as a result of the cancer itself [8–10]. Patients who have not been informed prior to cancer treatment about fertility loss and fertility preservation often subsequently experience considerable distress about the matter [10]. Given that the drive to reproduce has been described as central to an individual's core gender identity, self-concept, and body image, whether or not one wants to have and raise children, this is an understandable reaction [11]. Moreover, the diagnosis of infertility itself causes patients to think about their reasons for wanting to have children [12]. Hoffman and Hoffman studied motivations for parenthood in a normal population. Motivational areas of importance specified by subjects were: the associated adult status and social identity; expansion of self; their moral values; group ties and affection; stimulation and fun; achievement and creativity; power and influence; social comparison; and economic utility [13]. To a cohort of infertile couples in another study, the primary motivators were less about social issues and values and more about their sense of personal happiness and well-being [14].

Third-Party Reproduction as a Strategy for Parenthood After Cancer

Third-party reproduction is a term used to describe the use of gametes (oocytes or sperm) or a uterus provided by a third person in order to help infertile couples achieve parenthood. Oocyte donation and gestational surrogacy, two aspects of third-party reproduction increasingly utilized by women who have lost reproductive capacity because of cancer and/or its treatment, are addressed in this chapter (Table 15.1).

Oocyte Donation

The first recipient of egg donation was a 25-year-old woman who achieved pregnancy by having a single oocyte donated to her by a 29-year-old woman undergoing in vitro fertilization (IVF) [15]. Since that initial success, oocyte donation has become increasingly acceptable and an integral part of assisted reproduction programs around the world. The technology offers new hope to women infertile as a result of premature ovarian failure, surgically removed ovaries, or

Table 15.1 Resources for couples considering third-party reproduction

Donor recipients and offspring
Donor Conception Network (England), http://www.dcnetwork.org
Donor Conception Support Group (Australia), http://www.dcsg.org.au
Surrogacy
The American Surrogacy Center, Inc. (offers a Web site with information regarding surrogacy), www.surrogacy.com
Childlessness Overcome through Surrogacy (COTS): An organization from the UK providing information regarding surrogacy and IVF, http://www.surrogacy.org.uk/
Surrogate Parenting Services, www.surrogateparenting.com
The Organization for Parents through Surrogacy (OPTS), http://www.opts.com/

advanced reproductive age. Cancer survivors who have had their ovaries surgically removed or who have diminished ovarian reserve as a result of treatment with chemotherapy or radiation are increasingly considered appropriate candidates for this treatment [6].

A growing number of children around the world are conceived through oocyte donation. Two reports, one from the International Committee for Monitoring Assisted Reproductive Technology (ICMART) and the other from the Society for Assisted Reproductive Technology (SART), reflect these growing numbers. In the year 2000, 49 countries reported that 14,877 fresh cycles of oocyte donation had been completed resulting in 6,839 babies [16]. In the USA alone that same year, 7,581 fresh cycles were completed [17]. While oocyte donation is an ever-increasing phenomenon worldwide, the practices, policies, and laws that guide it vary greatly from country to country. For example, Germany, Italy, Japan, Kenya, Latvia, the Netherlands, Norway, Portugal, Sierra Leone, Slovenia, Switzerland, Togo, and Turkey have legally banned oocyte donation [18, 19].

Other countries, while allowing egg donation, have strict rules regarding anonymity and donor compensation. In France, for example, oocyte donation is legal as long as it is anonymous and free. There, as in Belgium, the so-called practice known as personalized anonymous donation involves the exchange of oocytes from a donor recruited by one patient with those of a donor recruited by another patient in order to ensure anonymity between donors and recipients [20]. Donor compensation is also illegal in Canada and the UK, but so is *donor anonymity*. A regulatory body established in Canada, the Assisted Reproduction Agency established in 2004, eliminated donor anonymity and established a donor registry [21]. In 2005, the Human Fertilisation and Embryology Authority (HFEA) in the UK eliminated donor compensation and anonymity and established a registry, where offspring could access information about the donor when they reach 16 years of age [22].

Oocyte donation is not regulated in the USA, although most programs follow the recommended guidelines on gamete donation issued by the American Society for Reproductive Medicine [23]. Program policies on anonymity and donor compensation vary across the country, but a majority of programs accept both anonymous and known donors, and anonymous donors are generally compensated. Unique to the USA are oocyte donor agencies developed specifically for the purpose of recruiting healthy young women to donate their eggs. Interested recipients can review donor profiles through the Internet. The Ethics Committee of the ASRM issued a statement recommending that donors should not be compensated more than $10,000, but many agency donors receive greater amounts [24, 25]. However, in a study of factors relating to donors' willingness to donate again, Klock et al. reported that 40% would definitely not donate again if not compensated [26]. In another follow-up study from the USA, 50 women who had previously donated stated that their primary motivation was financial although participants with children were more likely to say that they would be willing to donate without compensation [27].

Practical Considerations for Couples Undergoing Oocyte Donation

Couples considering oocyte donation who are hindered by the laws and policies of the country they live in increasingly resort to "reproductive tourism." This relatively recent phenomenon refers to infertile couples who travel across national and international borders to receive assisted reproduction not available to them in their own country or state [28]. Many travel to programs that offer legal oocyte donation with a greater choice of donors and donor characteristics. Typical attributes of a donor program that intended parents crossing national boundaries to reach include those that offer a choice between known and anonymous donation and provide choices regarding certain donor traits, such as ethnicity, physical characteristics, and level of education and, in some cases, the ability to meet the donor. Typically, there are three types of oocyte donors.

Anonymous donors are recruited and screened by the medical clinic or by an oocyte donor agency. Recipient couples are typically provided with the donor's health history, eye color, hair color, height, weight, blood type, ancestry, educational level, occupation, interests, and outcome of previous donations. Some programs provide only the donor's baby photos while others show current photographs.

Known donors are already known to the intended parents and are usually recruited by them. Typically, known donors are sisters, other relatives, or friends of the intended mother. Known donors are screened by the medical clinic. In addition to meeting screening criteria, donors need to demonstrate that they are not being coerced in any way to participate and screening includes an assessment of the quality of the relationship between donor and intended parents.

Identified donors are young women who are willing to meet with the intended parents prior to donation. They are recruited by the intended parents themselves or by an independent agency or medical facility. They are screened by the medical facility or by the agency. The intended parents and the donor come to a mutual arrangement regarding aspects of the donation, such as whether the donor would be willing to meet with offspring in the future [29].

Demographics of anonymous donors in studies from the USA have changed somewhat over time. In 1993, Lessor et al. described the typical donor as 26 years old, married with one or two children, and had 2 years of college [30]. More recent studies found donors more likely to be single with no children, and having somewhat higher college education. Women in all studies were without psychopathology, and post donation reported a positive experience and expressed a willingness to donate again [26, 31–33].

Recipients using identified donors have not been studied yet, but research has looked at comparisons between known and anonymous choices among intended parents. The majority of recipients in a study from the USA chose anonymous donors as their first choice. Motivations for those who used a known donor include dissatisfaction with the treatment program's selection of donors; wanting to meet the donor; and the desire to provide their children with information about the donor. Recipients using anonymous donors were more concerned about the donor's health history and family background while those who used known donors were satisfied with the information they had about the donor. The donor's resemblance to the intended mother was important to recipients from both groups [34].

Emotional Considerations for Couples Undergoing Oocyte Donation

For many women who have survived cancer and its treatment, the possibility of becoming pregnant through oocyte donation presents a truly miraculous opportunity for motherhood. The typical experience for infertile women seeking oocyte donation involves a series of losses – loss of fertility, loss of feelings of normality, and loss of a genetic link to the child [21] – but cancer survivors in many cases have grieved early in their treatment and some may welcome the non-genetic link, especially if their cancer has some genetic component [9]. Nevertheless, making the decisions to choose to use a donor, whether known or anonymous, entails a significant psychological adjustment. Intended mothers who have a known donor may be at once extremely grateful and somewhat threatened. She may struggle with the idea that her friend (or relative) is, along with her partner, creating a baby for her to carry. Women using anonymous donors often fear the unknown aspects of the process, such as a sense of loss connected with the possibility of having a baby who may not look like her. She may worry that she will not be able to love the child or that the child will reject her and want to search for the donor. Disclosure is often another issue of concern: Should we tell others that we used a donor? Should we tell the child? Will the child want to search for the donor? Those choosing to become parents through oocyte donation often benefit from reading as much as they can about the process, meeting with others who have chosen this form of parenting, and meeting with a mental health professional who specializes in third-party reproduction to ease those fears and make a comfortable transition to parenthood through oocyte donation.

Gestational Surrogacy

The practice of surrogacy, where a woman carries a pregnancy for another woman who cannot, has been described most notably in the Old Testament, where Hagar gave birth to a baby for Sarah and Abraham. That method, where the surrogate is genetically related to the baby she gives birth to and ultimately relinquishes the child to the father and his wife or partner, is known as *traditional surrogacy*. In early times, this was accomplished through sexual intercourse. In modern times, the surrogate is artificially inseminated with the sperm of the intended father. The intended mother then adopts the resulting baby. This form of surrogacy entered into contemporary society in the 1970s and many contractual agencies soon formed to recruit surrogates to fulfill the role [35]. With the advent of IVF, *gestational surrogacy*, a process in which the surrogate has embryos from the intended parents transferred to her uterus and gives birth to a baby unrelated to her, came into practice. By 2005, there were approximately 6,000 surrogate births in the USA and 300 surrogate births in Great Britain, the majority of which were from gestational surrogates [35]. Gestational surrogacy also exists in several other countries, including Australia, Canada, Finland, and Brazil. Gestational surrogacy is the focus of this discussion.

The first successful pregnancy, where embryos from an infertile couple were transferred to the uterus of a surrogate through IVF, was in 1985 [36]. Reasons for using a gestational surrogate include patients with a nonfunctional uterus; those with excessive medical risk associated with pregnancy; and failed prior attempts at conception [37]. Typically, cancer survivors who require a surrogate are women who do not have a functional uterus due to cancer and its treatment or who have been advised against carrying a pregnancy due to health risks.

Practical Considerations for Couples Undergoing Gestational Surrogacy

As in oocyte donation, laws and policies concerning gestational surrogacy vary from country to county and from state to state. Some countries do not allow surrogacy in any form. Other countries do not allow financial compensation for the gestational surrogate [35]. As a result, gestational surrogacy is another possible cause for "reproductive tourism." When considering working with a gestational surrogate, it is prudent for the intended parents to determine the legality of the process in the place *where the surrogate resides*. All surrogacy cases need to have a legal contract and in many places it is possible to have a pre-birth order which makes it possible for both intended parents, and not the surrogate, to be on the birth certificate.

Gestational surrogates are sometimes relatives or friends of the intended mothers who have volunteered to carry a pregnancy for them. More often, they are women who have been recruited through an agency and matched to the intended parents. Studies of gestational surrogates from the USA and the UK reveal similar motivations and characteristics. They are typically married, in their late 20s, have an average of two children, have an average 2 years of college education, and have been raised in the Christian/Protestant faith [38–40]. They are motivated by the desire to do something important in their lives; empathy for the childless couple; desire to experience pregnancy again; financial gain; and sometimes, an opportunity to undo something in their past [38–40]. Follow-up studies of gestational surrogates post surrogacy reveal that they do not regret their involvement with surrogacy and their satisfaction was correlated with a respectful and comfortable relationship with the parents. Surrogates did not experience psychological problems with the intended parents or difficulty in relinquishing the child [41, 42]. A follow-up study of couples who have become parents through surrogacy observed 42 families with 1-year-old children and found that the parents generally regarded their experience with surrogacy as positive. They typically had good relationships with their surrogates and maintained some contact with them. They had told friends and family about their experience and were planning to tell the child [43].

Emotional Considerations for Couples Undergoing Gestational Surrogacy

Even though gestational surrogacy allows cancer survivors who cannot carry a pregnancy, a chance for motherhood, for women who have longed for pregnancy and childbirth, the decision to have someone else carry the pregnancy requires an emotional adjustment. Part of that adjustment for the intended mother includes giving up a fair amount of control, putting absolute trust in another woman, often one not known to her previously and often one who lives far away. Hanafin says that the most common problems between surrogates and intended parents fall within three categories: struggles with medical issues; struggles with the relationship; and struggles with logistical surprises [35]. At times, a psychological issue stems from differences in the medical experience between women. The carrier may be quite confident about pregnancy and childbirth but may not realize the limitations and demands of assisted reproduction – specifically the very interested involvement of several strangers in decisions about the pregnancy. Controversies may evolve around issues of travel, diet, money, illness, choice of doctor, medication, bed rest, and labor and delivery options [35]. Psychological support and counseling from a mental health professional specializing in third-party reproduction can aid intended parents and gestational surrogates in preparing for the pregnancy, preparing for closure after the baby is born, and addressing the issue of future contact between surrogates and intended parents.

Adoption as a Strategy for Parenthood After Cancer

Adoption, another option for parenthood after cancer, has been a family-building option throughout the ages, but the policies and laws that guide the process have changed dramatically (Table 15.2). Adoption was a common practice historically because often children were orphaned or abandoned as a result of poverty, illness, death,

Table 15.2 Resources for couples considering adoption

Joint Council on International Children's Services, http://www.jointcouncil.org/
National Adoption Information Clearinghouse, http://www.adoption.org/adopt/national-adoption-information-clearinghouse.php
Evan B. Donaldson Adoption Institute, www.adoptioninstitute.org
Rainbow Kids – international adoption newsletter, www.rainbowkids.com
Families with Children from China, www.FWCC.org
Families for Russian and Ukrainian Adoptions, www.FRUA.org
National Adoption Center, www.adopt.com, www.adopting.org (includes link to EurAdopt, an association of adoption organizations in 12 western European countries, as well as worldwide links)
The Center for Adoption Support and Education, www.adoptionsupport.org
Department for Education and Skills (England), http://www.education.gov.uk/popularquestions/childrenandfamilies/adoption
Australian Society of Intercountry Aid for Children, www.asaic.org.au
Australians Caring for Children, www.accau.org, www.adoption.com, www.adoption.org.uk, www.adoptionuk.org

age, or marital status of the birth parents. Couples were motivated to adopt because of involuntary childlessness, in order to fulfill labor needs or expectations of religious heritage, or because they were intent on rescuing such children [44].

In more recent history, a North American philanthropic response to the plight of children left orphaned and homeless in World War II resulted in thousands of adoptions. While such children were primarily Caucasian and those adoptions occurred mostly in North America, the picture changed considerably during the Korean and Vietnam wars. International transracial adoption increased dramatically when children orphaned in these wars were adopted by families all over the world. By the end of the twentieth century, social chaos and poverty led to a greater number of adoptions from the former Soviet Union, Eastern Europe, and Latin America. Also, China's one-child policy has led to increased adoption of Chinese girls. Recently, there has been a decline in international adoptions due to several factors: countries closing adoption programs to new cases, slowing of processing of existing cases, increased restrictions on parents, and high travel costs [45].

The rights of adopted children were also given greater world attention in 1993 as a result of The Hague Convention on the Protection of Children and Cooperation in Respect of Intercountry Adoption. The convention addressed worldwide adoption issues and its purpose was to prevent abuses, such as the sale or abduction of children; to regulate standards for legal consent in an adoption; to set standards for a child's move to a new country; and to establish the status of the adopted child in the country of placement [46] Currently, more than 70 countries have agreed to the terms of this convention (www.theadoptionguide.com).

Attitudes toward adoption have changed as well. Historically, misguided perceptions of birth mothers as irresponsible and sexually promiscuous and adoptive parents as reproductive failures gave adoption a sense of social stigma and secrecy. The social stigma associated with adoption began to dissipate in the 1950s when papers began to appear in the literature stressing the importance for adult adoptees to learn more about their biological origins. As a result, many western countries now consider the needs of all parties, adoptees, adoptive parents, and birth parents alike. Birth parents are often viewed as important and interested participants in an adoption and can be actively involved in the process. Adoptive parents are more likely to be cognizant of the fact that children need to be aware of their personal heritage and may be eager for such knowledge. Adoptees are not as likely to experience adoption as socially stigmatizing and benefit from greater social acceptance and information about their birth parents and their heritage.

In developed countries, access to birth control and abortion, a growing acceptance of single parenthood, decisions to delay marriage and conception, and increased independence and autonomy for women have contributed to fewer children being placed domestically for adoption. As a result, increasing numbers of couples adopt internationally. As adoption has increased worldwide and become a much more open process, the stigma and secrecy remain only in limited cultures.

Practical Considerations for Couples Undergoing Adoption

Choices for couples considering adoption are often confusing and overwhelming. Just as they learned a new language while negotiating the travails of cancer and its treatment, cancer survivors need a whole new landscape or language to understand adoption. Information gathering is vital to this process. Individuals are encouraged to read as much as they can on all aspects of adoption, such as "how to adopt" guides, books that are usually written from the perspective of others who have adopted and from the perspective of birth parents, books about life as an adoptive parent, and other literature geared to adopted children [44].

Laws governing adoption vary greatly from state to state and country to country. Starting the process of adoption requires learning as much as one can about the laws of one's own place of residence. In addition, information gathering requires a decision-making process on the part of the couples as to what are their own choices and what is important to them in this process. For example, will they adopt domestically or internationally? Will they adopt a child of a different race, a different ethnicity from their own? How much information do they want to have about the birth parents? Do they want an exchange of non-identifying information only or do they want to meet the birth parent(s)? What will their adopted child know about his or her heritage?

Domestic adoption involves a child and adoptive parents of the same nationality and the same country of residence. In general, domestic adoption is handled through an agency or privately through an intermediary, such as a physician or a lawyer. Agency adoptions vary according to state and country, but typically may include government agencies, social service agencies that may have a religious affiliation, and private agencies that facilitate adoption. Typically, in such adoptions, birth parents and adoptive parents remain anonymous to each other although nonidentifying information is exchanged. In recent years, this process has become more open, where many agencies facilitate meetings between birth parents and adoptive parents.

Independent adoptions (also known as private adoptions) are those which occur when the couple makes an arrangement for placement of a child without help from an agency. This process requires the intended parents to take a much more active role in locating a birth mother usually through networking and advertising. Because the birth mother is usually pregnant when contacted, this affords the couples the opportunity to participate in the pregnancy and birth. The process of networking may be through talking to friends, colleagues, family, and professionals, such as lawyers and physicians; sending out resumes, letters, or brochures; obtaining contracts through the Internet; and posting announcements on bulletin boards [44]. Many couples pursuing adoption prefer to do so independently for several reasons. They may feel a greater sense of control compared to the experience of working with an adoption agency. In doing so, they may feel more involved in the decision making process and that they can avoid some of the restrictions or rules of an agency adoption. Most importantly to some, they may take home an infant whose birth they have witnessed.

International adoption involves parents whose nationality differs from that of the adopted child. Like domestic adoptions, international adoptions may be handled by an agency in the country or state of origin, an agency or orphanage in a foreign country, or an adoption intermediary, such as an attorney. International placements generally require a great deal of red tape and paperwork, so guidance from an attorney or agency or support organization can be most helpful. Fees for international adoptions are generally not greater than those for domestic adoption, and the waiting time for such adoptions can be shorter than those of domestic adoptions [44].

Emotional Considerations for Couples Undergoing Adoption

Couples considering adoption often have many fears: fear that they will not be able to love their child or that their child will reject them in favor of the birth mother; fear that the child will not be healthy or will have psychological problems. They worry that they will not be able to talk to their child about their origin; they worry that the child will be stigmatized [47]. Prospective parents through adoption benefit meeting with other adoptive families and receiving psychological support and counsel around all of these issues.

Cancer survivors face another dilemma: How will their cancer history affect their efforts to adopt? Many countries forbid adoption by cancer survivors, and others place restrictions, such as requiring the cancer survivor to wait 5 years between remission and adoption. Many survivors have adopted, however, from such countries as Columbia, Ethiopia, Guatemala, Kazakhstan, Nepal, Russia, South Korea, Ukraine, the USA, and Vietnam [48].

Conclusions

Young women are increasingly treated successfully for cancer. When the treatment or the cancer itself leaves them with limited or no reproductive capacity but not without the desire for children, they increasingly seek motherhood through third-party reproduction, such as oocyte donation or gestational surrogacy, or through adoption. Fortunately, there are a number of resources available for guidance and support in making the transition to parenthood through third-party reproduction or adoption.

References

1. Hubner MK, Glazer ES. Now on common ground: cancer and infertility in the 1990s. Cancer survivors who are uninformed about the treatment's effect on future fertility may later experience an "intolerance of regret". Infertil Reprod Med Clin North Am. 1993;4:581–96.
2. American Society of Clinical Oncology. Recommendations on fertility preservation in cancer patients. J Clin Oncol. 2006;24:2917–31.
3. Practice Committee of the American Society for Reproductive Medicine. Ovarian tissue and oocyte cryopreservation. Fertil Steril. 2004;82:993–8.
4. Royal College of Obstetrics and Gynaecologists. Storage of ovarian and prepubertal testicular tissue – report of a working party. London: The RCOG Bookshop; 2000.
5. Tschudin S, Bitzer J. Psychological aspects of fertility preservation in men and women affected by cancer and other life-threatening diseases. Hum Reprod Update. 2009;15:587–97.
6. Jeruss JS, Woodruff TK. Preservation of fertility in patients with cancer. New Engl J Med. 2009;360: 902–11.
7. Klock SC, Zhank JX, Kazer RR. Fertility preservation for female cancer patients: early clinical experience. Fertil Steril. 2010;94:149–55.
8. Schover LR, Rybicki LA, Martin BA, et al. Having children after cancer. A pilot survey of survivors' attitudes and experiences. Cancer. 1999;86:697–709.
9. Schover LR, Brey K, Lichtin A, et al. Knowledge and experience regarding cancer, infertility, and sperm banking in younger male survivors. J Clin Oncol. 2002;20:1880–9.
10. Zebrack BJ, Casillas J, Nohr L, et al. Fertility issues for young adult survivors of childhood cancer. Psychooncology. 2004;13:689–99.
11. Rosenthal MB. Psychiatric aspects of infertility and assisted reproductive technology. Infertil Reprod Med Clin North Am. 1993;4:471–82.
12. Newton CR, Hearn MT, Yuzpe AA, et al. Motives for parenthood and responses to failed in vitro fertilization: implications for counseling. J Assist Reprod Technol Genet. 1992;9:24–31.
13. Hoffman LW, Hoffman M. The value of children to parents. In: Faxcett JT, editor. Psychological perspectives on populating. New York: Basic Books; 1973. p. 19–73.
14. Van Balen F, Trimbos-Kemper TCM. Involuntary childless couples; their desire to have children and their motives. J Psychosom Obstet Gynaecol. 1995;16: 137–44.
15. Lutjen P, Trouson A, Leeton J, et al. The establishment and maintenance of pregnancy using in vitro fertilization and embryo donation in a patient with primary ovarian failure. Nature. 1984;307:174–5.
16. Adamson GD, Mouzon JD, Lancaster P, et al. World collaborative report on in vitro fertilization, 2000. Fertil Steril. 2006;85:1586–622.
17. Society for Assisted Reproductive Technology and the American Society for Reproductive Medicine. Assisted reproductive technology in the United States: 2000 results generated from the American Society for Reproductive Medicine/Society for Assisted Reproductive Technology registry. Fertil Steril. 2004; 81:1207–20.
18. Anderson N, Gianaroli L, Felbergaum R, et al. Assisted reproductive technology in Europe, 2001. Results generated European registers by ESHRE. Hum Reprod. 2005;20:1158–76.
19. Giva-Osagie OF. ART in developing countries with reference to sub-Sahara Africa. In: Vayne E, Rowe PS, Griffin PD, editors. Current practices and controversies in assisted reproduction. World Health Organization report. Geneva: World Health Organization; 2001. p. 23–30.
20. Letur-Konirsch H. Oocyte donation in France and national balance sheet (GEDO). Different European approaches. Gynecol Obstet Fertil. 2004;32:108–15.
21. Sachs PL, Burns LH. Recipient counseling for oocyte donation. In: Covington SN, Burns LH, editors. Infertility counseling: a comprehensive handbook for clinicians. 2nd ed. New York: Cambridge University Press; 2006.
22. HFEA. Human fertilisation and embryology: the regulation of donor-assisted conception. www.hfea.gov.uk. Accessed July 2011.
23. Practice Committee of the American Society for Reproductive Medicine, Practice Committee of the Society for Assisted Reproductive Technology. Guidelines for gamete and embryo donation: a Practice Committee report. Fertil Steril. 2008;90 (5 Suppl):S30–44.
24. The Ethics Committee of the American Society for Reproductive Medicine. Financial incentives for donors. Fertil Steril. 2000;74:216–20.
25. Covington SN, Gibbons WE. What is happening to the price of eggs? Fertil Steril. 2007;87:1001–4.
26. Klock SC, Stout JE, Davidson M. Psychological characteristics and factors related to willingness to donate

again among anonymous oocyte donors. Fertil Steril. 2003;79:1312–6.
27. Patrick M, Smith AL, Meyer WR, et al. Anonymous oocyte donation: a follow-up questionnaire. Fertil Steril. 2001;75:1034–5.
28. Inhorn MC, Patrizio P. Rethinking reproductive "tourism" as reproductive "exile". Fertil Steril. 2009;92: 904–6.
29. Applegarth LA, Kingsberg SL. The donor as patient: assessment and support. In: Burns LH, Covington NH, editors. Infertility counseling: a comprehensive handbook for clinicians. 2nd ed. New York: Cambridge University Press; 2006.
30. Lessor R, Cervantes N, Balmaceda J, et al. An analysis of social and psychological characteristics of women volunteering to become oocyte donors. Fertil Steril. 1993;59:65.
31. Jordan CB, Belar CD, Williams RS. Anonymous oocyte donation: a follow-up analysis of donors' experiences. J Psychosom Obstet Gynecol. 2004;25:145–51.
32. Kalfoglou AL, Geller G. A follow-up study with oocyte donors exploring their experiences, knowledge, and attitudes about the use of their oocytes and the outcome of the donation. Fertil Steril. 2000;74:660–7.
33. Zweifel JE, Rathert MA, Klock SC, et al. Comparative assessment of pre- and post-donation attitudes toward potential oocyte and embryo disposition and management among ovum donors in an oocyte donation program. Hum Reprod. 2006;21:1325–7.
34. Greenfeld DA, Klock SC. Disclosure decisions among known and anonymous oocyte donation recipients. Fertil Steril. 2004;81:1565–71.
35. Hanafin H. Surrogacy and gestational carrier participants. In: Burns LH, editor. Infertility counseling: a comprehensive handbook for clinicians. 2nd ed. New York: Cambridge University Press; 2006.
36. Utian WH, Sheean L, Godfarb JM, Kiwi R. Successful pregnancy after in vitro fertilization embryo transfer from an infertile woman to a surrogate. N Engl J Med. 1985;313(21):1351–2.
37. Serafini P. Outcome and follow up of children born after IVF – surrogacy. Hum Reprod. 2001;7:23–7.
38. Mechanic Braverman A, Corson SL. Characteristics of participants in a gestational carrier program. J Assist Reprod Genet. 1992;9:353–7.
39. Hanafin H. The surrogate mother: an exploratory study. Dissertation. Los Angeles: California School of Professional Psychology; 1984.
40. Resnick R. Surrogate mothers: relationship between early attachment and the relinquishment of the child. Dissertation. Santa Barbara: Fielding Institute; 1989.
41. Jadva V, Murray C, Lycett E, et al. Surrogacy: the experiences of surrogate mothers. Hum Reprod. 2003; 18:2196–204.
42. Reame N, Hanafin H, Kalsoglou A. Unintended consequences and informed consent: lessons from former surrogate mothers. Hum Reprod. 1999;14:361–2.
43. MacCullum F, Lycett E, Murray C, et al. Surrogacy: the experiences of the commissioning couples. Hum Reprod. 2003;18:1334–42.
44. Salzer LP. Adoption after infertility. In: Hammer-Bums L, Covington SN, editors. Infertility counseling: a comprehensive handbook for clinicians. 2nd ed. New York: Cambridge University Press; 2006.
45. The editors. What's new in adoption? Adoptive Families Magazine; 2010. www.theadoptionguide.com. Accessed May 2010.
46. Child Welfare League of America – National data analysis system. International adoption: trends and issues; 2003. http://ndas.cwla.org. Accessed Sep 2003.
47. Rosen A. Third-party reproduction and adoption in cancer patients. J Natl Cancer Inst Monogr. 2005;34: 91–3.
48. Kantrowitz M. Adoption and cancer. Cancer points 2008–2009. www.cancerpoints.com. Accessed July 2011.

Pregnancy and Cancer Treatment

16

Christine Laky and Mert Ozan Bahtiyar

Abstract

Following treatment for cancer, parents have many concerns about the effects on the developing embryo and future cancer risks to offspring and to maternal health during pregnancy. With close monitoring of maternal status during pregnancy, female cancer survivors can achieve a successful pregnancy.

Keywords

Pregnancy after cancer treatment • Cancer and pregnancy • Maternal health during pregnancy • Fertility after cancer treatment • Cancer recurrence and pregnancy

Pregnancy after cancer is a rare, but increasing, phenomenon. Caring for a pregnant woman after treatment for a malignancy may pose a therapeutic dilemma for the obstetrician, as there can be persistent effects to organs after treatment, some persisting for many years. Expectant parents can also have underlying concerns about the risks to the developing fetus and risks of recurrence of cancer as they prepare to become parents.

Delayed child bearing may also increase the likelihood of being diagnosed with a cancer prior to and during pregnancy. According to the Centers for Disease Control and Prevention (CDC), the age at first birth has been increasing over the last 30 years to an average of 25.1 years, with a range of 23–28 years of age. Between 1990 and 2002, the U.S. fertility rate declined 9% from 70.9 to 64.8 births per 1,000 women aged 15–44 years [1]. During that time period, fertility rates declined for 42 of 50 states. Nationally, for women in their thirties and forties, the birth rates were significantly higher in 2002 than in 1990. For women aged 30–34, the birth rate was 13% higher and for women aged 35–39, the rate was 31% higher. Overall, the birth rate for women in their forties is still lower than for women in their thirties, but the rate for women aged 40–44 increased 51% from 5.5 to 8.3 births per 1,000 women. Over the same time period, birth rates for women aged 45–49 years more than doubled from 0.2 to 0.5 [1].

M.O. Bahtiyar (✉)
Department of Obstetrics and Gynecology,
Yale University School of Medicine, 333 Cedar Street,
New Haven, CT 06520-8063, USA

E. Seli and A. Agarwal (eds.), *Fertility Preservation: Emerging Technologies and Clinical Applications*,
DOI 10.1007/978-1-4419-1783-6_16,

Cancer Rates

According to the National Cancer Institute, about 72,000 reproductive-age adolescents and young adults (aged 15–39) are diagnosed with cancer each year in the USA. The most commonly diagnosed cancers are: lymphoma, leukemia, germ cell tumors (including testicular cancer), melanoma, tumors of the central nervous system, sarcoma, breast, cervical, liver, and thyroid, and colorectal cancers [2].

Cervical Cancer

In 2009, approximately 11,270 new cases of cervical cancer were diagnosed and there were 4,070 deaths from cervical cancer [2]. It is the most commonly diagnosed invasive cancer in pregnancy, affecting anywhere from 1 in 2,000 to 2,500 pregnancies [3] to 1 in 1,240 [4] pregnancies. Invasive cervical cancer during pregnancy accounts for 1% of all cervical cancers diagnosed.

Breast Cancer

In 2009, there were approximately 192,370 new cases of breast cancers diagnosed in females and 40,170 deaths [2]. Male cases accounted for 1,910 new cases and 440 deaths. Pregnancy-associated breast cancer is diagnosed during pregnancy, during lactation, or up to 1 year following delivery [5]. The average age for patients diagnosed with PABC is 32–38 years. PABC has an incidence of 1 in 3,000 to 3 in 10,000 pregnancies [6].

Ovarian Cancer

In 2009, there were an estimated 21,550 new cases of ovarian cancer and 14,600 deaths [2]. Ovarian cancer complicates approximately 0.073/1,000 pregnancies [7] or 1 in 18,000 live births [8]. In one case series, malignant germ cell tumors were the most commonly diagnosed during pregnancy, mostly dysgerminomas, which comprised 40% of ovarian tumors, followed by 27% of low-malignant potential tumors and 23% invasive epithelial tumors [7]. During pregnancy, treatment for ovarian cancer can include oophorectomy and surgical staging. However, the gestational age and tumor burden determine whether the pregnancy should continue. There is also some evidence that not every adnexal mass needs to be surgical treated during pregnancy, with surgery usually reserved for patients with ultrasonographic evidence of malignancy or torsion [9].

Outside of pregnancy, treatment for ovarian cancer usually includes total abdominal hysterectomy and bilateral oophorectomy with surgical staging and tumor debulking as necessary. Fertility-sparing surgery can be performed on certain ovarian cancer types, such as sex cord-stromal tumors and low-malignant potential tumors. This usually includes unilateral oophorectomy and surgical staging.

Lymphoma

Hodgkin's: In 2009, there were approximately 8,510 new cases and 1,290 deaths [2]. Hodgkin's lymphoma accounts for about 0.5% of all cancers, but is the second most commonly diagnosed malignancy in reproductive-age people [10]. Hodgkin's lymphoma is becoming an increasingly relevant neoplasm in obstetric practice since this disease has a particular predilection in young adults of both genders. An increasing number of young women have been treated successfully for Hodgkin's lymphoma and subsequently become pregnant. Treatment for Hodgkin's lymphoma includes total lymphoid radiation, multiagent chemotherapy, or combination chemotherapy and radiation therapy. Many women become amenorrheic or experience premature ovarian failure. However, this is not universally the case and for some, amenorrhea may be temporary. Combination therapy and age greater than 25 have been associated with premature ovarian failure. Many women can have successful pregnancies following therapy.

Non-Hodgkin's Lymphoma

In 2009, there were 65,980 new cases and 19,500 deaths in the USA [2]. Non-Hodgkin's lymphoma describes the malignancies affecting the lymphoid tissue that are not Hodgkin's lymphoma or leukemia. These lymphomas are extremely rare in pregnancy. In a review of several studies, Lishner et al. found that aggressive treatment during pregnancy is associated with best maternal outcomes [11].

Leukemia

In 2009, there were 44,790 new cases and 21,870 deaths from leukemia in the USA [2]. Treatment advances have led to the decrease of the mortality rate in children and young adults with hematologic neoplasms; therefore, many are surviving to reproductive age. Acute lymphocytic leukemia is the most common acute leukemia of childhood. In one study of 35 girls treated for childhood leukemia, 16 out of 17 prepubertal girls successfully treated for ALL subsequent experienced normal sexual development and 28 out of 35 patients (80%) had normal development [12].

Thyroid Cancer

In 2009, there were 37,200 new cases and 1,630 deaths from thyroid cancer in the USA [2]. Thyroid disease preferentially affects females, and approximately 65% of patients with thyroid cancer are female. About 28% of new cases are in females aged 20–40, and it is the fastest growing cancer diagnosis [13]. Thyroid cancer is also one of the most common malignancies diagnosed during pregnancy. Surgery for treatment can usually be postponed for nonaggressive tumors until the postpartum period, but if surgery is necessary, as with other surgeries during pregnancy, it is best performed during the second trimester. Radioactive iodine therapy should be avoided during pregnancy and pregnancy should be avoided for the first 6–12 months after treatment. There appears to be no increased risk of infertility, miscarriage, stillbirths, or congenital defects, nor have cancers been identified in offspring of patients treated [13].

For women who conceive after treatment for thyroid cancer, their pregnancies should be managed similarly to patients with hypothyroidism. Usually, these patients are maintained on thyroid replacement therapy, such as levothyroxine. Thyroid-stimulating hormone and free thyroxine levels should be measured every trimester and replacement levels adjusted accordingly. Ultrasounds for fetal growth should be performed every 4–6 weeks.

Melanoma

In 2009, there were 68,720 new cases and 8,650 deaths from melanoma [2]. Melanoma is increasingly being diagnosed in young reproductive-age women. Approximately 1/3 of women with malignant melanoma are of reproductive age and 8% of cancers in pregnancy are melanomas, in one study occurring in 2.8 out of 1,000 deliveries [14]. Surgical resection is the only effective treatment for melanoma and should not be avoided in the pregnant patient; surgical resection should be performed for the usual indications in nonpregnant patients.

Testicular Cancer

In 2009, there were 8,400 new cases and 380 deaths from testicular cancer [2]. Cancer and treatment in men are discussed below.

Bone Cancer

In 2009, there were 2,570 new cases and 1,470 deaths of bone cancer in the USA [2]. Bone cancers are rare in pregnancy, and the most common cancers seen in reproductive-age women are Ewing's sarcoma, osteogenic sarcoma, and osteocytoma. Pregnancy does not affect the growth of bone cancers and bone cancer does not affect the pregnancy. Since prognosis is so poor, delaying treatment during pregnancy is not warranted.

Cancer During Pregnancy

Most cancers can and should be treated during pregnancy depending on the type, aggressiveness of the tumor, and the gestational age of the pregnancy. Cancer and treatment during pregnancy have been addressed elsewhere, but below is a synopsis of some common cancers affecting reproductive-age women and men and treatment during pregnancy.

Cervical Cancer

Cervical cancer is the most commonly diagnosed cancer during pregnancy since screening for cervical cancer is routinely performed during pregnancy. Women with stage 1A1 cervical cancer diagnosed by biopsy can be followed in pregnancy with periodic colposcopy and cytology screening and treatment can be delayed until after delivery of the infant [15]. Cervical conization during pregnancy carries increased risks as compared to when performed outside of pregnancy, including risks of hemorrhage and fetal loss. There is some discrepancy in the loss rates reported in the first trimester. Some physicians report a loss rate of up to 25%, but this was prior to 1972, when most pregnant women were subjected to cervical conization for an abnormal pap smear [16]. More recent data show much lower to nonexistent loss rates, even when performed in the first trimester [17]. Women with stage 1A2 and 1B cervical cancer can be treated either with radiation therapy or radical hysterectomy with lymphadenectomy. Case reports suggest that a moderate delay in treatment may not have any adverse outcomes and women with delay in treatment have similar outcomes compared with women treated immediately. Current recommendations are to delay treatment in women diagnosed after 20 weeks gestation and treat prior to this time. However, women should be informed of the possible risks associated with delay in treatment.

Fertility-sparing surgery with cervical conization can be performed for 1A1 cervical cancer. Subsequent pregnancies need to be managed with transvaginal ultrasound for cervical length with a baseline at 16–18 weeks gestation and then every 2–4 weeks starting at about 24 weeks gestation until about 32 weeks. Cervical cancer treated with radiation therapy to the pelvis may spare the ovaries with the intended outcome of maintaining endogenous hormonal production and maintaining sexual function. A subsequent pregnancy may be attained in a woman who is still ovulating and may have the pregnancy complications associated with prior radiation treatment to the pelvis.

Ovarian Cancer

Approximately 1 in 1,000 pregnant women have exploratory surgery, either laparotomy or laparoscopy, for an adnexal mass and 3–6% of those are malignant [18]. The cancer rate is less than the rate for nonpregnant women as ovarian malignancies occur with less frequency in younger women; most adnexal masses found in younger women are either simple cysts or corpus luteal cysts or solid benign neoplasms, such as teratomas [19]. Germ cell cancers, mainly dysgerminomas, are the most common ovarian cancer diagnosed during pregnancy [19]. Surgical management depends on the viability of the fetus and the patient's wishes to continue or terminate a pregnancy if nonviable. In some rare cases, oophorectomy and cytoreduction can be performed during pregnancy with adjuvant chemotherapy initiated after delivery. Outcomes for pregnant patients appear to be similar to those for nonpregnant patients, although treatment may need to be modified during pregnancy. CA-125 levels may not be accurate in detecting and following ovarian cancer during pregnancy. In one study, CA-125 levels were measured at 4–6 week intervals throughout pregnancy and in the puerperium. The results showed two peaks in CA-125 levels above the normal range in the first trimester (and again in the immediate postpartum period) [20].

Breast Cancer

Breast cancer occurs in about 1.9 cases per 10,000 live births [21] and this accounts for 0.2–3.8% of breast cancers diagnosed in young women. This incidence may increase as more women delay childbearing until their later years. Treatment during pregnancy should not be any different than for nonpregnant women. Treatment needs to be individualized depending on the gestational age, stage of the disease, and the wishes of the patient and her family. Abortion is not usually recommended, but can be considered during treatment planning. Treatment needs to be coordinated with the patient's obstetrician, a perinatalogist, and the oncology team.

Early age of first full-term birth has been associated with decreased risk of developing breast cancer in the general population, especially in women less than age 20. However, this effect has not been shown for women who are carriers of BRCA 1 and BRCA 2 mutations [22].

Treatment During Pregnancy

Regardless of the type of cancer, therapy should not be withheld from a woman based solely on her pregnancy status. The following briefly explores the effects of therapy on pregnancy. The effects of treatment for cancer during pregnancy depend on the types of agents used and the stage of fetal development. Gynecological cancers frequently occur during pregnancy; about 1 in 1,000 pregnancies are complicated by cancer and, thus, may need treatment during pregnancy [21, 23].

Chemotherapy

Teratogenic effects of chemotherapeutic agents depend on their mechanism of action and the gestational age of the fetus when administered to the patient. In general, chemotherapy started in the first 2 weeks after conception leads to either a spontaneous abortion or has no effect, an "all or none effect" often seen with radiation exposure. The blastocyst itself is resistant to teratogens, but is more vulnerable after implantation. The critical time period for teratogenic effects is during organogenesis, from 3 to 8 weeks of gestation. Organogenesis begins at about the fifth gestational week and continues through the eighth week, and by the thirteenth week organogenesis is complete, except for brain and gonadal development [24]. During this time period, exposure to the most cytotoxic agents carries a high incidence of congenital malformations.

Alkylating agents and antimetabolites have the greatest potential for causing fetal abnormalities during pregnancy, but do not appear to have a lasting effect and do not appear to affect future pregnancies (see chart). Observational reports seem to support good maternal and fetal outcomes when chemotherapy is administered in the second and third trimesters; exposure to chemotherapy in the second and third trimesters usually does not result in any major birth defects [25]. One concern around the time of delivery is the risk of maternal neutropenia and thrombocytopenia. The infant after delivery also may not be able to process and excrete metabolites and may experience some adverse reactions after delivery.

Radiation Therapy

Diagnostic radiography may need to be used during pregnancy, such as X-rays, CT scans, and fluoroscopy, to assist in the diagnosis of cancer and assist in disease staging. There are no known doses of radiation that are completely safe for the fetus, but the risks must be weighed against the benefits of diagnosis. As above, early in conception, a dose of 10 cGy or above usually results in spontaneous abortion. At all stages of pregnancy, doses of less than 5 cGy are generally considered safe. A computed tomography scan of the abdomen and pelvis with contrast delivers a mean dose of approximately 3 cGy of radiation. The risk of childhood cancer if a fetus is exposed to radiation depends on the dose; for example, at a dose of up to 5 cGy, the incidence is about 0.3–1% [26]. Magnetic resonance imaging does not use ionizing radiation and may be used in place of other imaging modalities.

Radiation therapy for cancer, especially whole pelvic radiation for cervical cancer, during pregnancy is usually lethal to the fetus and leads to intrauterine fetal death followed by spontaneous abortion. Standard doses for cancer therapy are around 4,500 cGy which leads to sterility as a result of direct injury to the ovaries and uterus. Although the thresholds for permanent sterility have not been established, the dose needed to induce ovarian failure seems to be related to the patient's age. For example, a dose of 600 cGy (6 Gy) may induce menopause or cause permanent ovarian failure in women 40 years of age or a dose up to 2,000 cGy (20 Gy) for women treated in childhood [27]. Oophoropexy may mitigate these effects if the ovaries are placed out of the pelvis prior to initiation of treatment, but uterine effects cannot be avoided.

Surgery

It may still be necessary for a pregnant woman to undergo surgical procedures during pregnancy for a variety of reasons. If surgery is required during pregnancy, it is best performed during the second trimester. In one study, of the 5,405 pregnant women undergoing nonobstetric surgical procedures, most were done during the second trimester and rates of miscarriage and stillbirth were no different than the baseline rates [28]. There were also no differences in congenital anomalies, but an increased risk of preterm birth has been reported in surgeries performed in the first trimester, which was also related to longer surgical times [29].

Laparoscopic surgery is feasible during pregnancy and is also best performed in the second trimester. Laparoscopic surgeries most often performed during pregnancy are appendectomy, cholecystectomy, and to evaluate adnexal masses. Surgical procedures are similar as in nonpregnant women, with some exceptions. Pregnant women should be placed in a leftward tilt to reduce compression on the aorta and the inferior vena cava. Port placement may need to be modified to avoid the pregnant uterus. A Veress needle should be avoided as there is an increased risk of puncture to the uterus, so open technique should be used to achieve pneumoperitoneum.

Common Types of Chemotherapeutic Agents

In this section, we discuss the commonly used chemotherapeutic agents in treating cancers. This section concentrates on the effect on the female reproductive organs, to include the ovary, fallopian tube, and uterus, and addresses possible long-term consequences to future fertility and pregnancies. Many chemotherapeutic agents are used in the treatment of gynecologic cancers that are not always amenable to fertility-sparing surgeries or therapies.

Alkylating and Alkylating-Like Agents

Alkylating agents are used in the treatment of breast and ovarian cancer, as well as other common cancers, such as lymphoma and acute lymphocytic leukemia. These agents can also be used in the treatment of inflammatory and autoimmune disorders, such as systemic lupus erythematosus (SLE), that commonly affect reproductive-age women. These agents include cyclophosphamide, nitrogen mustard, ifosfamide, chlorambucil, and busulfan. Alkylating-like agents include the platinum analogs, cisplastin, and carboplatin, as well as nitrosureas and dacarbazine, as well as others. Alkylating agents work by interfering with accurate base pairing by binding to the N-7 position of guanine and other DNA sites, which results in the inhibition of RNA, DNA, and protein synthesis [30]. The main side effect for almost all of these agents is myelosuppression.

Alkylating agents are commonly used for the treatment of ovarian cancer and other gynecologic cancers; for instance, dacarbazine is used for the treatment of uterine sarcomas. Alkylating agents in particular have lasting effects on gonadal function and can cause long-term amenorrhea and azospermia, especially if used for prolonged therapy. If menses and ovulation return, there is little evidence of long-term harm to the subsequently

Table 16.1 Treatment of cancer during pregnancy

Treatment	Effect on fetus	Effect on mother
Chemotherapy[a]		
Alkylating agent	Cleft palate, absent digits, eye abnormalities, ear and facial anomalies, hypoplastic limbs, IUGR, pancytopenia	Cytopenia, gonadal toxicity, including male gonadal toxicity, premature ovarian failure, ovarian fibrosis, bladder toxicity
Antimetabolites	Cranial dysostosis, hypertelorism, wide nasal bridge, micrognathia, ear abnormalities, possible IUFD	Hepatotoxicity, myelosuppression, pulmonary damage, nephrotoxicity, stomatitis
Anthracycline antibiotics	Unclear, possible increased pregnancy complications, such as IUGR and IUFD with idarubicin and epirubicin use	Cardiac toxicity, pulmonary toxicity
Vinca alkaloids	ASD, extremity abnormalities (one reported absent radii and fifth digits). IUGR, IUFD, preterm delivery	Preeclampsia, neuropathy
Taxanes	Unclear	Motor neuropathy, alopecia
Platinum agents	Hearing loss, ventriculomegaly, IUGR, IUFD	Neurotoxicity, ototoxicity, myelosuppression
Radiation[b]	Microcephaly, possible childhood cancer	Infertility, gastrointestinal toxicity, injury to skin and mucosa, fibrosis, strictures, cystitis

[a]Data from Cardonick and Iacobucci [72]
[b]Data from Greskovich and Macklis [73]

developing follicle. Subsequent pregnancies have been reported and the risks to the developing pregnancy are minimal, unless the patient conceives within 1 year of cessation of therapy, during which time there is an increased risk of low birth weight [31]. These agents are highly toxic to the developing embryo in animal studies, although there is limited data in humans. These agents are classified as pregnancy class D medications and current recommendations are to be avoided in the first trimester (Table 16.1).

Antimetabolites

Antimetabolites are used in the treatment of ovarian, breast, and cervical cancer, but most commonly for gestational trophoblastic neoplasia (GTN). Antimetabolites are also used in the treatment of nonmalignant conditions, such as rheumatoid arthritis, SLE, and ectopic pregnancies.

Antimetabolites have structures that resemble analogs of normal purines and pyrimidines. These agents interact with vital intracellular enzymes, leading to their inactivation [30, 32]. Specifically, methotrexate is a folate antimetabolite that inhibits DNA synthesis by binding to dihydrofolate reductase. Other common antimetabolites include 5-fluorouracil, hydroxyurea, and gemcitabine. Generally, these antineoplastic agents affect actively proliferating cells and are classed as cell cycle-specific agents.

Women who have been exposed to antimetabolites only, especially methotrexate, appear to do well with a subsequent pregnancy. However, if methotrexate has been used to treat gestational trophoblastic disease, it is imperative that the disease is cleared prior to a recurrent pregnancy. Disease recurrence is monitored by measuring beta-hCG levels, which cannot be usefully monitored during a pregnancy. For severe disease, it is recommended that a woman be followed with undetectable levels for at least 1 year prior to attempting conception. For an ectopic pregnancy treated with low-dose methotrexate with 1 or 2 treatments, a woman should wait about 3–4 months before attempting conception again. There seems to be no adverse affects to pregnancies after methotrexate therapy; however, one study showed possible increase in spontaneous abortion and adverse maternal outcome in women who conceived within 6 months of completion of methotrexate therapy for GTN [33]. Men treated with methotrexate should also wait about 3 months before attempting conception. There also seems to be no long-term adverse effects on sperm production.

SLE is an autoimmune disorder that typically affects reproductive-age women. Again, it is generally recommended that chemotherapy for SLE be discontinued prior to conception. The chemotherapeutic effects on a subsequent pregnancy are usually minimal, as noted above. However, the underlying cause that needed to be treated can make a subsequent pregnancy difficult and pregnant women with SLE should be managed by a high-risk obstetrician. A pregnancy can exacerbate underlying SLE which includes a worsening of lupus nephritis. Women with SLE are at higher risk for developing severe preeclampsia, especially in the periviable period, intrauterine growth restriction, and a fetal congenital heart block. Rheumatoid arthritis has little effect on a developing pregnancy and pregnancy usually does not exacerbate rheumatoid arthritis.

Antimetabolites are considered FDA pregnancy class X agents and are contraindicated in all stages of pregnancy. There is a risk of methotrexate embryopathy in the first trimester [34], although there is evidence of minimal risk at 6–8 weeks postconception with low-dose methotrexate [35].

Antitumor Antibiotics

Antibiotics are typically derived from soil fungi. These antibiotics work by inserting between DNA base pairs, a process known as intercalation [36]. A secondary mechanism that may contribute to their antitumor properties is the formation of free radicals, which are able to damage DNA, RNA, and vital proteins. Anthracyclines include doxorubicin (Adriamycin), liposomal doxorubicin (Doxil), and daunorubicin (Danomycin). Other antitumor antibiotics are bleomycin, mitomycin C, and mithramycin. Gynecologic cancers typically treated with antitumor antibiotics include ovarian, breast, and endometrial cancer; however, many nongynecologic cancers are also treated with these agents, such as lymphoma.

One common side effect of doxorubicin is cardiotoxicity; so for this reason, the maximum tolerable cumulative dose is 450–500 mg/m^2 of ideal body surface area [37]. Cardiac effects of doxorubicin and other anthracycline antibiotics can be both early and late in onset and are dose-dependent. Early onset of heart failure peaks at 3 months after the last dose and symptomatic heart failure can occur up to 10 years after the last anthracycline use. Women who become pregnant are at risk for developing a cardiomyopathy during pregnancy. The effects of this on pregnancy can be minimal to needing admission for cardiac support and possible immediate delivery. Women with cardiomyopathy also need to be managed by a high-risk obstetrician. Prior to conception, women who have been treated with antitumor antibiotics should have an echocardiogram demonstrating baseline cardiac function and should be seen by a cardiologist. During pregnancy, serial growth ultrasounds should be performed at 4–6 week intervals to document adequate fetal growth. If a pregnant woman becomes symptomatic with shortness of breath and tachypnea, a maternal echocardiogram should be performed immediately.

Plant Alkaloids

The most common plant alkaloids are vincristine and vinblastine derived from *Vinca rosea*, the common periwinkle plant. Other plant alkaloids are used commonly in gynecologic malignancies, such as epipodophyllotoxins and paclitaxel. Vinca alkaloids work by binding to intracellular microtubular proteins, specifically tubulin [30, 38]. This disrupts microtubule assembly destroying the mitotic spindle, which arrests cells in mitosis, hence these are also cell cycle-specific.

Topoisomerase-1 Inhibitors

The mechanism of action of this class of antineoplastic agents is inhibition of the enzyme, topoisomerase-1, which is important for DNA transcription, replication, and repair [39]. Topetecan is the most commonly used medication in this class. The most common side effect is myelosuppression.

Radiation Therapy

Radiation therapy is a major therapy in the treatment of many cancers, especially gynecologic malignancies, such as cervical cancer and invasive vaginal cancer. Treatment for cervical cancer requires irradiation to the pelvis and can have profound effects on pelvic organs. However, fertility is limited after treatment with whole pelvic radiation. Radiation therapy is also used in the treatment for other cancers that can affect reproductive-age women, such as breast cancer and lymphoma. Treatment in these cases would not be directly to the pelvic organs, but there can be residual spread. The radiation dose decreases by one-fourth for every radius of spread.

One of the main effects of radiation on pelvic organs is uterine fibrosis. A subsequent pregnancy, if achieved, can be complicated by recurrent miscarriage, intrauterine growth restriction, malpresentation, and preterm labor. The muscle fibers of the uterus become fibrotic from radiation therapy and tend to be not as (stretchable) for an enlarging uterus. Placental implantation is also hampered by damaged blood vessels and decreased blood flow and leads to decreased oxygen and nutrient exchange through the placenta.

The effects of radiation therapy also depend on the area of the body that was irradiated and the distance of the pelvic organs from the main radiation source. According to the inverse square law, the intensity of a radiation beam in a nonabsorbing medium decreases as the inverse of the square of the distance, i.e., the intensity at twice the distance is one-fourth the original intensity [40].

$$\mathrm{I} = \Delta \mathrm{p} / 4\pi \mathrm{d}^2,$$

where I is the intensity, Δp is the number of photons emitted by a source, $4pd^2$ is the area of a sphere with radius, and d is the distance or

$$\mathrm{I}_2 = \mathrm{I}_1(\mathrm{d}_1 / \mathrm{d}_2)^2,$$

where I_2 is the new intensity with the change in distance from d_1 to d_2.

Pregnancy After Therapy

After Chemotherapy

The effects of chemotherapy on the ovaries are age- and dose-dependent, as well as dependent on the type of chemotherapeutic agent used and the duration of therapy. For example, alkylating agents are associated with ovarian dysfunction; daily cyclophosphamide treatment has been associated with amenorrhea in women under age 40 if treated for greater than a year [41] and there is a 50% incidence of amenorrhea within 1 month of starting cyclophosphamide [42]. Combination chemotherapy has been shown to more extensively cause premature ovarian failure. Strategies to preserve fertility while undergoing chemotherapy have been addressed elsewhere in this book.

A common concern for cancer survivors of reproductive age is the effect of previous chemotherapy exposure to the developing fetus once a pregnancy occurs. A common side effect of chemotherapeutic agents is amenorrhea and loss of ovarian function. One large study from Norway compared hospital registry from cancer patients to birth registries to calculate probability of first-time parenthood by age 35. The results found that the rate was similar among males (63–64% in both cancer and noncancer survivors), but lower among female cancer survivors (66 vs. 79%) [43].

Once pregnant, the effects of prior chemotherapeutic agents on the developing fetus are minimal as the follicle that became the mature egg was likely not affected by the prior chemotherapy. Until ovulation, primordial follicles are quiescent, and as seen above many chemotherapeutic agents are cell cycle-specific and act on active and dividing cells. After treatment has ended, assuming menstruation has resumed, the patient will be experiencing normal ovulation of eggs that are unaffected. There is no evidence that chemotherapeutic agents directly damage uterine tissue. For male cancer survivors, infertility can also be a significant issue (see below).

There appears to be no increased risk of congenital or chromosomal anomalies in the offspring of patients treated for childhood cancers. These risks are not increased whether treated with chemotherapy, radiation, or both. One early study compared three groups of women; pregnancy before therapy, pregnancy during therapy, and pregnancy after therapy. There was no significant difference in birth defects among the groups, but there appeared to be an increase in low-birth weight infants in the group who conceived and delivered after therapy [31].

One study by Byrne et al. compared 2,198 children born to 1,062 and matched to 4,544 children born to 2,032 controls who were siblings of the cancer survivors. There was no significant difference in anomalies and the incidence of genetic disease was the same for both groups – 3%, which is similar to the background rate of anomalies in the general population which is about 3.6% [44]. Another study compared the incidence of congenital anomalies in children born to either a mother or father treated for cancer vs. matched controls. There was no association with congenital anomalies and any type of cancer treatment in either parent [45].

The current recommendations are to wait at least 1 year after cessation of therapy prior to conception, especially after exposure to alkylating agents, to ensure that risks are decreased to the background risks. There are generally no recommendations for increased fetal monitoring for pregnant women who have been exposed to chemotherapeutic agents.

Cardiac toxicity is a known complication of chemotherapeutic agents, even in patients with normal hearts. Anthracyclines are most commonly associated with long-term cardiovascular complications, to include arrhythmias, cardiomyopathy, and acute coronary artery vasopasm. Women with a history of exposure to these and other cardiotoxic drugs should have a cardiac evaluation prior to pregnancy. This may include electrocardiogram, echocardiogram, and possibly a 24-h Holter monitor. A preconception consultation with a cardiologist should also be obtained. Any cardiac or respiratory symptoms in a pregnant woman with prior exposure should be evaluated immediately. One study of 37 women with 72 pregnancies showed favorable outcomes in 29 women with pre-pregnancy ejection fraction of ≥30% and less favorable outcomes in patients who had a pre-pregnancy ejection fraction of less than 30% [46].

Another concern for parents who are cancer survivors is the risk of cancer in their offspring. Their offspring are not at increased risk for cancer unless the initial cancer was part of an inherited syndrome. In one study, the risk of cancer was evaluated among 5,847 children of 14,652 survivors and found the incidence to be 1.3% which was not significantly higher than the risk to the general population [47]. Another case series found a 0.30% incidence of cancer in the offspring of cancer survivors vs. 0.23% in matched controls which is not significantly different [48].

After Radiation

Although the studies for pregnancy after chemotherapy are reassuring, the same do not appear to be as reassuring after radiation treatment. One major study of 427 pregnancies of 20 weeks or greater duration after the treatment of Wilms' tumor found an increased risk of congenital anomalies in women after radiation treatment. This study found that the risk of congenital anomalies in nonirradiated women was 3.2% and that the risk to women of irradiated partners was 3.3% [49]. However, there was an increased risk of 10–10.4% of congenital anomalies in women who received flank radiation, regardless of the dose. Most of the anomalies were isolated, including ventricular septal defect, which is the most common congenital malformation. Larger studies do not seem to confirm these findings.

After radiation therapy, a pregnancy may be difficult to achieve if the woman has become infertile or subfertile due to damage to the ovaries. As noted above, larger doses of radiation are needed to produce permanent ovarian failure in women under the age of 40. The ovaries of younger females, prepubertal and adolescent, are more resistant to radiation-induced damage, especially if the irradiation does not involve the abdomen or pelvis [50]. The effect of radiation on the ovaries is also dose-dependent. The LD_{50}, which is the dose needed to permanently destroy 50% of oocytes, is less than 2 Gy [51].

Other long-term effects of pelvic radiation therapy to the female genital tract include vaginal atrophy, narrowing, shortening, or agglutination, making intercourse painful and also decreasing sexual desire [52]. Chronic cystitis, proctosigmoiditis, enteritis, and bowel obstruction can also make intercourse painful and interfere with sexual function.

Another cause of reduced fertility and sexual function in childhood cancer survivors is cranial irradiation, which may work by interrupting the hypothalamic-pituitary-gonadal axis by direct effect on the hypothalamus [53]. Pituitary cells seem more resistant to irradiation. One study of 593 long-term survivors of childhood acute lymphocytic anemia found an increased rate of infertility in those treated with whole brain radiation suggesting infertility by disrupting gonadotropin release [54].

Pregnancies that are achieved after pelvic, flank, or abdominal radiation should be monitored closely with the input of a perinatologist. Prior pelvic irradiation appears to be associated with higher risk of miscarriage, malpresentation, preterm labor and birth, intrauterine growth restriction, and abnormal placentation, such as accreta. The dose of radiation to the uterus can range from 35 to 70 Gy depending on the cancer being treated, such as lymphomas or sarcomas [55]. Women exposed to radiation therapy during childhood have a greater chance of uterine damage, especially if exposed prior to puberty and at early ages [56].

Delanian and Lefaix have described the radiation-induced fibroatrophic process with three stages: (1) vascular phase characterized by chronic inflammation; (2) fibroblastic phase, where tissue undergoes fibrosis; and (3) matrix phase, which involves tissue fibrotic atrophy [57]. They also described the histologic effects on the uterus at the endometrial and myometrial levels. Arteriolar sclerosis, mild fibrosis, and muscular atrophy were the histologic features seen in the myometrium [57]. This can cause decreased elasticity of the uterus from injury of the smooth muscle which decreases the ability of the uterus to expand and stretch to accommodate a developing pregnancy and decreases the volume of the uterus. This can lead to such problems as preterm labor and delivery of a premature infant, and can also cause such problems as persistent malpresentation, such as breech presentation. However, a prophylactic cerclage is not indicated in these cases.

Radiation therapy may also injure the endometrial lining of the uterus. The histologic characteristics seen at the endometrial level include necrosis and atrophy, with fibrosis of the superficial layer and teleangectasic vessels in the inner layer [57]. This can prevent normal decidualization and result in abnormal placentation, such as placenta accreta and percreta. There is also some evidence for direct injury to the uterine vessels [58]. Radiation can damage these vessels and cause fibrin deposition, necrosis, or sclerosis which may decrease blood flow and cause intrauterine growth restriction.

Women who have received pelvic or abdominal radiation should have increased fetal surveillance during pregnancy because of the above. Serial ultrasounds for growth should be performed every 4–6 weeks starting in the mid-second through the third trimesters. Evaluation of fetal positioning should also be performed. Placentation can be evaluated by ultrasound at this time as well. Venous placental lakes and loss of boundary between the placenta and uterine wall are two signs on sonography that are suspicious for placenta accreta. Magnetic resonance imaging is a modality that is proving useful to evaluate placentation that does not increase radiation exposure to either the mother or the fetus. A prophylactic cervical cerclage is not recommended for the prevention of preterm birth. In contrast, exposure to radiation causes cervical stenosis and damaged tissue should not be operated on unnecessarily. If there is a suspicion of preterm labor, evaluation can be performed by evaluating contractions using a tocometer and measuring cervical length to check for cervical shortening. A fetal fibronectin test can be performed between 22 and 34 weeks estimated gestational weeks. A negative test is reassuring, but a positive test is not necessarily indicative of an impending preterm birth.

Cardiac toxicity is also possible after radiation treatment, especially treatment to the thorax. Cardiac injury includes acute pericarditis during treatment and delayed pericarditis that can be either acute or chronic and present as either an effusion or constriction. Damage to the pericardium or myocardium can result in fibrosis with decrease in elasticity of the muscle. Radiation-induced cardiomyopathy can also be present even in the absence of significant pericardial disease and can also present for the first time during pregnancy. Other injuries include coronary artery disease, conduction defects, and valvular injuries.

Breast Cancer and Pregnancy

Breast cancer in women under the age of 35 is uncommon; however, first-birth rate for women aged 30–34 and 35–39 is increasing as women are delaying childbearing. Breast cancer is now found to complicate 1 in 3,000 pregnancies [2]. Breast cancer during pregnancy has been addressed elsewhere. The prognosis for a woman diagnosed with breast cancer during pregnancy is poor as it is usually diagnosed in later stages and with a higher tumor burden.

Pregnancy after breast cancer is not contraindicated, but patients need to be aware of risks of a future pregnancy. Current recommendations in managing breast cancer include prophylactic oophorectomy, especially in women with estrogen-receptor-positive tumors to eliminate the ovarian source of estrogen with the thought of preventing or delaying cancer recurrence. The theory is that a subsequent pregnancy would also stimulate a recurrence secondary to high levels of circulating estrogen. However, a retrospective review by Gelber in 2001 reviewed the survival outcomes for women who did and did not become pregnant after treatment for early-stage breast cancer and found a higher survival rate in women who had achieved a subsequent pregnancy [59]. Although this does not conclude that pregnancy protects against recurrence, it is reasonable to conclude that a subsequent pregnancy is safe and does not increase risk of recurrence.

Prior to conception, breast cancer survivors should undergo an extensive evaluation to look for recurrence or metastases. This evaluation should include bone and liver scans, chest X-rays, and mammography of opposite breast if conserved. Because most recurrences happen within 2 years after the initial treatment is completed, it is also reasonable to recommend waiting at least 2–3 years before attempting a subsequent pregnancy [60]. The patient should also be observed closely during pregnancy to include frequent breast exams. Breastfeeding after breast-conserving therapy is also not contraindicated and breastfeeding has a known protective effect, which may be protective in the contralateral breast. There is also some evidence that there is a transient increase in the risk of breast cancer in the first 3–4 years after delivery [60].

Gestational Trophoblastic Neoplasia

GTN is a pregnancy-related malignancy that arises from either a molar or partial hydatidiform molar pregnancy. These pregnancies can manifest as either a missed abortion, a spontaneous abortion, or can be seen concurrently with a normal pregnancy. GTN can also follow from a nonmolar missed abortion or spontaneous abortion or can follow a normal pregnancy. The incidence of hydatidiform mole varies widely from geographic regions; in North American and European countries, the incidence is about 1 in 1,000–1,500 pregnancies, whereas in Latin American and Asian countries the incidence is about 1 in 12–500 pregnancies [33]. About 20% of complete moles and 2–4% of partial moles develop into malignant GTN; however, GTN following complete moles is more likely to become metastatic. The incidence of choriocarcinoma also follows the same geographic distribution with an incidence of 2–5 in 100,000 pregnancies in North America and Europe and a higher incidence in Asia and Latin America of 100 in 100,000 pregnancies [33]. Malignant diseases include GTN, choriocarcinoma, and placental-site trophoblastic tumors.

Because of the risk of GTN following molar pregnancies, human chronic gonadotropin (β-hCG) levels should be followed to undetectable levels and women should not have a subsequent pregnancy until persistent GTN has been ruled out. The current recommendation is to follow β-hCG levels weekly until undetectable, then monthly for six months following a partial mole, and for a year following a complete mole. GTN can be diagnosed by persistent elevations in β-hCG levels after the end of a pregnancy. GTN can be treated with chemotherapy and is very sensitive to methotrexate therapy. Single-agent therapy can be accomplished using methotrexate with or without leucovorin, dactinomycin, etopside, or 5-fluorouracil. Multiagent chemotherapy can be used for disease refractory to single-agent therapy. For patients after chemotherapy for GTN, one study of a total of 2,657 subsequent pregnancies resulted in 77% full-term births, 5% premature deliveries, 1% stillbirths, and 14% spontaneous abortions. Although stillbirth rate is slightly increased over the baseline rate, the congenital anomaly rate remained 1.8% which is consistent with the background rate [61].

Prior to a subsequent pregnancy, the patient should be evaluated preconceptually to ensure that she does not have persistent or recurrent disease. This can be accomplished by checking β-hCG levels. The patient should also have an ultrasound evaluation and diagnostic radiography of the chest. The current recommendation is to delay a subsequent pregnancy for 6 months to a year after completion of pregnancy. After conception, the patient should be followed closely to look for symptoms of recurrence disease or for side effects of chemotherapy.

Risks of Cancer Recurrence

Pregnancy generally does not increase the risk of cancer recurrence, with one notable exception of GTN. One of the main risk factors for GTN, besides extremes of age, includes previous GTN. Breastfeeding has a protective effect against developing a future breast cancer and should also have a protective effect in a woman who has had a previous breast cancer. Women who have been treated with fertility-sparing surgery during pregnancy for ovarian or cervical cancer should have a completion procedure and treatment for the primary cancer and should be managed by a gynecologic oncologist. Baseline risks of recurrence are the same for the primary cancer.

Risks of Radiation and Chemotherapy on Male Fertility

Testicular cancer is the most common solid tumor affecting reproductive-age males, especially between the ages of 15 and 35, and accounts for 1% of all cancers in men. Although the incidence of testicular cancer has been increasing, the 5-year survival rate is over 95%. According to the National Cancer Institute [62], the overall incidence of testicular germ cell tumors among American men rose 44% (from 3.35 to 4.84 per 100,000 men) between 1973 and 1998; the incidence of seminoma rose 62% while the incidence of nonseminomatous germ cell tumors rose only 24%. Approximately 85% of seminomas are present with stage 1 disease and can be treated with different modalities.

Stage 1 disease can be managed with radical orchiectomy and careful surveillance with radiation or chemotherapy reserved for recurrent disease. Options also include adjuvant radiation therapy or adjuvant chemotherapy using platinum-based agents. One large, multicenter trial in Europe randomized patients to either radiation therapy or chemotherapy and found equivalent relapse rates [63]. Patients treated with chemotherapy seem to have an 80–90% long-term relapse-free rate for if they were initially complete responders [64].

The two major reproductive toxicities associated with chemotherapy treatment of male germ cell tumors are oligo- and azospermia and Leydig cell dysfunction, although many men may actually have oligospermia at the time of diagnosis or after orchiectomy, unrelated to chemotherapy treatment [65]. Leydig cells produce androgens, such as testosterone. In one study, 27% of treated men had azospermia and 86% had Leydig cell dysfunction, but also showed that sperm counts increase 2–3 years after treatment and return to normal in about 50% of patients [66]. However, it also seems that about 25% of men remain permanently azospermic [67].

Treatment of androgen-sensitive tumors with gonadotropin-releasing hormone agonists, such as leuprolide or goserelin, is associated with testicular atrophy which may be permanent if treatment exceeds 2 years [68]. Testicular tissue is one of the most radiosensitive tissue and low doses of radiation can cause dysfunction. For instance, a dose of 15 cGy can transiently disrupt spermatogenesis, a dose of 600 cGy can permanently destroy germinal tissue, and doses of 2,000–3,000 cGy can cause Leydig cell dysfunction [69]. Erectile dysfunction (ED) can also be a major obstacle in achieving a subsequent pregnancy; ED is also seen in patients treated with radiation therapy for prostate cancer and may be higher than the rate of ED seen in patients treated with prostatectomy [70].

Risks to future pregnancies are generally minimal as sperm can be cryopreserved easily prior to treatment. Most seminomas tend to be unilateral, but there appears to be pretreatment impairment of spermatogenesis in about 50% of men with seminomas. The contralateral testicle can be shielded during therapy, and although there is a transient decrease in sperm counts following treatment, levels usually rise to pretreatment levels. Initial recommendations were to wait at least 6 months following therapy before attempting conception to allow exposed sperm to be replaced by healthy nonexposed sperm. In one Norwegian study, about 80% of males exposed to either radiation or chemotherapy trying to conceive without cryopreserved sperm achieved paternity [71]. There appeared to be no adverse outcomes on the subsequent pregnancies.

Conclusions

Following treatment for cancer, parents have many concerns about the effects on the developing embryo and future cancer risks to offspring and to maternal health during pregnancy. With close monitoring of maternal status during pregnancy, female cancer survivors can achieve a successful pregnancy.

References

1. Sutton P, Mathews T. Trends in characteristics of births by State: United States, 1990, 1995, and 2000–2002. Natl Vital Stat Rep. 2004;52(19):1–152.
2. National Cancer Institute. http://www.cancer.gov/cancertopics/aya/types
3. Shivvers S, Miller D. Preinvasive and invasive breast and cervical cancer prior to or during pregnancy. Clin Perinatol. 1997;24(2):369–89.
4. Hacker N, Berek J, Lagasse L, Charles E, Savage E, Moore J. Carcinoma of the cervix associated with pregnancy. Obstet Gynecol. 1982;59(6):735–46.
5. Creasy RK, Resnick R, Iams JD. Creasy and Resnick's maternal-fetal medicine: principles and practice. 6th ed. Philadelphia: Elsevier; 2009. p. 891.
6. Hahn K, Johnson P, Gordon N, Kuerer H, Middleton L, Ramirez M, et al. Treatment of pregnant breast cancer patients and outcomes of children exposed to chemotherapy in utero. Cancer. 2006;107(6):1219–26.
7. Zhao X, Huang H, Lian L, Lang J. Ovarian cancer in pregnancy: a clinicopathologic analysis of 22 cases and review of the literature. Int J Gynecol Cancer. 2006;16(1):8–15.
8. Creasman W, Rutledge F, Smith J. Carcinoma of the ovary associated with pregnancy. Obstet Gynecol. 1971;38(1):111–6.
9. Schmeler K, Mayo-Smith W, Peipert J, Weitzen S, Manuel M, Gordinier M. Adnexal masses in pregnancy: surgery compared with observation. Obstet Gynecol. 2005;105(5 Pt 1):1098–103.
10. Azim HJ, Pavlidis N, Peccatori F. Treatment of the pregnant mother with cancer: a systematic review on the use of cytotoxic, endocrine, targeted agents and immunotherapy during pregnancy. Part II: hematological tumors. Cancer Treat Rev. 2010;36(2):110–21.
11. Lishner M, Zemlickis D, Sutcliffe S, Koren G. Non-Hodgkin's lymphoma and pregnancy. Leuk Lymphoma. 1994;14(5–6):411–3.
12. Siris E, Leventhal B, Vaitukaitis J. Effects of childhood leukemia and chemotherapy on puberty and reproductive function in girls. N Engl J Med. 1976;294(21):1143–6.
13. Ying A, Huh W, Bottomley S, Evans D, Waguespack S. Thyroid cancer in young adults. Semin Oncol. 2009;36(3):258–74.
14. Smith R, Randall P. Melanoma during pregnancy. Obstet Gynecol. 1969;34(6):825–9.
15. Creasy RK, Resnick R, Iams JD. Creasy and Resnick's maternal-fetal medicine: principles and practice. 6th ed. Philadelphia: Elsevier; 2009. p. 888.
16. Averette H, Nasser N, Yankow S, Little W. Cervical conization in pregnancy. Analysis of 180 operations. Am J Obstet Gynecol. 1970;106(4):543–9.
17. Hannigan E, Whitehouse Hr, Atkinson W, Becker S. Cone biopsy during pregnancy. Obstet Gynecol. 1982;60(4):450–5.

18. Whitecar M, Turner S, Higby M. Adnexal masses in pregnancy: a review of 130 cases undergoing surgical management. Am J Obstet Gynecol. 1999;181(1): 19–24.
19. Leiserowitz G, Xing G, Cress R, Brahmbhatt B, Dalrymple J, Smith L. Adnexal masses in pregnancy: how often are they malignant? Gynecol Oncol. 2006; 101(2):315–21.
20. Spitzer M, Kaushal N, Benjamin F. Maternal CA-125 levels in pregnancy and the puerperium. J Reprod Med. 1998;43(4):387–92.
21. Smith L, Danielsen B, Allen M, Cress R. Cancer associated with obstetric delivery: results of linkage with the California cancer registry. Am J Obstet Gynecol. 2003;189(4):1128–35.
22. Kotsopoulos J, Lubinski J, Lynch H, Klijn J, Ghadirian P, Neuhausen S, et al. Age at first birth and the risk of breast cancer in BRCA1 and BRCA2 mutation carriers. Breast Cancer Res Treat. 2007; 105(2):221–8.
23. Oehler M, Wain G, Brand A. Gynaecological malignancies in pregnancy: a review. Aust N Z J Obstet Gynaecol. 2003;43(6):414–20.
24. Sorosky J, Sood A, Buekers T. The use of chemotherapeutic agents during pregnancy. Obstet Gynecol Clin North Am. 1997;24(3):591–9.
25. Avilés A, Díaz-Maqueo J, Talavera A, Guzmán R, García E. Growth and development of children of mothers treated with chemotherapy during pregnancy: current status of 43 children. Am J Hematol. 1991; 36(4):243–8.
26. Giles D, Hewitt D, Stewart A, Webb J. Malignant disease in childhood and diagnostic irradiation in utero. Lancet. 1956;271(6940):447.
27. Sklar C. Maintenance of ovarian function and risk of premature menopause related to cancer treatment. J Natl Cancer Inst Monogr. 2005;34:25–7.
28. Mazze R, Källén B. Reproductive outcome after anesthesia and operation during pregnancy: a registry study of 5405 cases. Am J Obstet Gynecol. 1989; 161(5):1178–85.
29. Kort B, Katz V, Watson W. The effect of nonobstetric operation during pregnancy. Surg Gynecol Obstet. 1993;177(4):371–6.
30. Berek JS, Hacker NF. Berek and Hacker's gynecologic oncology. 5th ed. Philadelphia: Lippincott, Williams, and Wilkins; 2010.
31. Mulvihill J, McKeen E, Rosner F, Zarrabi M. Pregnancy outcome in cancer patients. Experience in a large cooperative group. Cancer. 1987;60(5): 1143–50.
32. Berek JS, Hacker NF. Berek and Hacker's gynecologic oncology. 5th ed. Philadelphia: Lippincott, Williams, and Wilkins; 2010. p. 79.
33. Altieri A, Franceschi S, Ferlay J, Smith J, La Vecchia C. Epidemiology and aetiology of gestational trophoblastic diseases. Lancet Oncol. 2003;4(11):670–8.
34. Bawle E, Conard J, Weiss L. Adult and two children with fetal methotrexate syndrome. Teratology. 1998; 57(2):51–5.
35. Kozlowski R, Steinbrunner J, MacKenzie A, Clough J, Wilke W, Segal A. Outcome of first-trimester exposure to low-dose methotrexate in eight patients with rheumatic disease. Am J Med. 1990;88(6):589–92.
36. Berek JS, Hacker NF. Berek and Hacker's gynecologic oncology. 5th ed. Philadelphia: Lippincott, Williams, and Wilkins; 2010. p. 77.
37. Von Hoff D, Layard M, Basa P, Davis HJ, Von Hoff A, Rozencweig M, et al. Risk factors for doxorubicin-induced congestive heart failure. Ann Intern Med. 1979;91(5):710–7.
38. Hacker JSBNF. Berek and Hacker's gynecologic oncology. 5th ed. Philadelphia: Lippincott, Williams, and Wilkins; 2010. p. 80.
39. Pizzolato J, Saltz L. The camptothecins. Lancet. 2003;361(9376):2235–42.
40. Washington CM, Leaver D. Principles and practice of radiation therapy. 3rd ed. St. Louis: Mosby Elsevier; 2010. p. 289–90.
41. Uldall P, Kerr D, Tacchi D. Sterility and cyclophosphamide. Lancet. 1972;1(7752):693–4.
42. Warne G, Fairley K, Hobbs J, Martin F. Cyclophosphamide-induced ovarian failure. N Engl J Med. 1973;289(22):1159–62.
43. Magelssen H, Melve K, Skjaerven R, Fosså S. Parenthood probability and pregnancy outcome in patients with a cancer diagnosis during adolescence and young adulthood. Hum Reprod. 2008;23(1): 178–86.
44. Byrne J, Rasmussen S, Steinhorn S, Connelly R, Myers M, Lynch C, et al. Genetic disease in offspring of long-term survivors of childhood and adolescent cancer. Am J Hum Genet. 1998;62(1):45–52.
45. Agha M, Williams J, Marrett L, To T, Zipursky A, Dodds L. Congenital abnormalities and childhood cancer. Cancer. 2005;103(9):1939–48.
46. Bar J, Davidi O, Goshen Y, Hod M, Yaniv I, Hirsch R. Pregnancy outcome in women treated with doxorubicin for childhood cancer. Am J Obstet Gynecol. 2003;189(3):853–7.
47. Sankila R, Olsen J, Anderson H, Garwicz S, Glattre E, Hertz H, et al. Risk of cancer among offspring of childhood-cancer survivors. Association of the Nordic Cancer Registries and the Nordic Society of Paediatric Haematology and Oncology. N Engl J Med. 1998; 338(19):1339–44.
48. Mulvihill J, Myers M, Connelly R, Byrne J, Austin D, Bragg K, et al. Cancer in offspring of long-term survivors of childhood and adolescent cancer. Lancet. 1987;2(8563):813–7.
49. Green D, Peabody E, Nan B, Peterson S, Kalapurakal J, Breslow N. Pregnancy outcome after treatment for Wilms tumor: a report from the National Wilms Tumor Study Group. J Clin Oncol. 2002;20(10):2506–13.
50. Abeloff MD, Armitage JO, Niederhuber JE, Kastan MB, McKenna WG. Abeloff's clinical oncology. 4th ed. Philadelphia: Elsevier; 2008. p. 1002.
51. Wallace W, Thomson A, Kelsey T. The radiosensitivity of the human oocyte. Hum Reprod. 2003;18(1): 117–21.

52. Washington CM, Leaver D. Principles and practice of radiation therapy. 3rd ed. St. Louis: Mosby Elsevier; 2010. p. 819–20.
53. Abeloff MD, Armitage JO, Niederhuber JE, Kastan MB, McKenna WG. Abeloff's clinical oncology. 4th ed. Philadelphia: Elsevier; 2008. p. 1001.
54. Boughton B. Childhood cranial radiotherapy reduces fertility in adulthood. Lancet Oncol. 2002;3(6):330.
55. Revelli A, Rovei V, Racca C, Gianetti A, Massobrio M. Impact of oncostatic treatments for childhood malignancies (radiotherapy and chemotherapy) on uterine competence to pregnancy. Obstet Gynecol Surv. 2007;62(12):803–11.
56. Bath L, Critchley H, Chambers S, Anderson R, Kelnar C, Wallace W. Ovarian and uterine characteristics after total body irradiation in childhood and adolescence: response to sex steroid replacement. Br J Obstet Gynaecol. 1999;106(12):1265–72.
57. Delanian S, Lefaix J. The radiation-induced fibroatrophic process: therapeutic perspective via the antioxidant pathway. Radiother Oncol. 2004;73(2):119–31.
58. Critchley H, Buckley C, Anderson D. Experience with a "physiological" steroid replacement regimen for the establishment of a receptive endometrium in women with premature ovarian failure. Br J Obstet Gynaecol. 1990;97(9):804–10.
59. Gelber S, Coates A, Goldhirsch A, Castiglione-Gertsch M, Marini G, Lindtner J, et al. Effect of pregnancy on overall survival after the diagnosis of early-stage breast cancer. J Clin Oncol. 2001;19(6): 1671–5.
60. Helewa M, Lévesque P, Provencher D, Lea R, Rosolowich V, Shapiro H, et al. Breast cancer, pregnancy, and breastfeeding. J Obstet Gynaecol Can. 2002;24(2):164–80. quiz 81–4.
61. Garner E, Lipson E, Bernstein M, Goldstein D, Berkowitz R. Subsequent pregnancy experience in patients with molar pregnancy and gestational trophoblastic tumor. J Reprod Med. 2002;47(5):380–6.
62. McGlynn K, Devesa S, Sigurdson A, Brown L, Tsao L, Tarone R. Trends in the incidence of testicular germ cell tumors in the United States. Cancer. 2003; 97(1):63–70.
63. Oliver R, Mason M, Mead G, von der Maase H, Rustin G, Joffe J, et al. Radiotherapy versus single-dose carboplatin in adjuvant treatment of stage I seminoma: a randomised trial. Lancet. 2005;366(9482):293–300.
64. Dearnley D, Horwich A, Ahern R. Combination chemotherapy with bleomycin, etopside and cisplatin (BEP) for metastatic testicular teratoma: long-term follow-up. Eur J Cancer. 1991;27:684.
65. Abeloff MD, Armitage JO, Niederhuber JE, Kastan MB, McKenna WG. Abeloff's clinical oncology. 4th ed. Philadelphia: Elsevier; 2008. p. 1739.
66. Hansen S, Berthelsen J, von der Maase H. Long-term fertility and Leydig cell function in patients treated for germ cell cancer with cisplatin, vinblastine, and bleomycin versus surveillance. J Clin Oncol. 1990; 8(10):1695–8.
67. Stephenson W, Poirier S, Rubin L, Einhorn L. Evaluation of reproductive capacity in germ cell tumor patients following treatment with cisplatin, etoposide, and bleomycin. J Clin Oncol. 1995;13(9): 2278–80.
68. Smith JJ, Urry R. Testicular histology after prolonged treatment with a gonadotropin-releasing hormone analogue. J Urol. 1985;133(4):612–4.
69. Ash P. The influence of radiation on fertility in man. Br J Radiol. 1980;53(628):271–8.
70. Zelefsky M, Eid J. Elucidating the etiology of erectile dysfunction after definitive therapy for prostatic cancer. Int J Radiat Oncol Biol Phys. 1998;40(1): 129–33.
71. Brydøy M, Fosså S, Klepp O, Bremnes R, Wist E, Wentzel-Larsen T, et al. Paternity following treatment for testicular cancer. J Natl Cancer Inst. 2005; 97(21):1580–8.
72. Cardonick E, Iacobucci A. Use of chemotherapy during human pregnancy. Lancet Oncol. 2004;5:283–91.
73. Greskovich JF, Macklis RM. Radiation therapy in pregnancy: risk calculation and risk minimization. Semin Oncol. 2000;27(6):633–45.

Fertility Preservation Strategies in Healthy Women

17

Enrique Soto and Alan B. Copperman

Abstract

Reproductive endocrinologists are faced with the challenge of using various technological and laboratory advances in the field to help healthy women preserve fertility for a number of years until conception is desired. Cryopreservation of oocytes has the advantage that it surmounts the possible ethical, religious, and legal limitations of the cryopreservation of embryos.

Further improvements in laboratory technologies, like vitrification and in vitro maturation of oocytes or ovarian tissue, are likely to replace embryo cryopreservation as the main strategy for fertility preservation in the near future.

Keywords

Fertility preservation in women • Reproductive endocrinology • Cryopreservation of oocytes • Reproductive technology • Vitrification • In vitro maturation of oocytes

Women and couples over the past decade have been increasingly deferring conception. Population studies demonstrate that the mean age of childbearing has increased as much as 3 years over the period of 1980–2001 [1]. The motives for which women and couples defer fertility are often multifactorial and complex. Some of the possible causes are the economical constraints of reproducing at a young age, a competitive work environment, the economic slowdown, females' perceived career threats [2], and for some patients, not having a partner or declining to use donor sperm when conception is desired. Hence, reproductive endocrinologists are confronted with the challenge of using various technological and laboratory advances to help healthy women preserve fertility until conception is desired. The current chapter describes the current techniques available for fertility preservation in healthy women.

A.B. Copperman (✉)
Department of Obstetrics, Gynecology and Reproductive Science, Division of Reproductive Endocrinology and Infertility, Mount Sinai Hospital/Mount Sinai School of Medicine, 281 Avenue C, Apt 6-E, New York 10009, NY, USA

E. Seli and A. Agarwal (eds.), *Fertility Preservation: Emerging Technologies and Clinical Applications*,
DOI 10.1007/978-1-4419-1783-6_17,

Fertility and Age

It is well-documented that fertility and "ovarian age" are closely correlated with maternal chronological age. Female fecundity, which is defined as the ability to produce offspring, decreases as maternal age increases. It is well-accepted that women are born with approximately 2 million oogonia, but that number decreases throughout their life span until menopause is reached [3]. In fact, the vast majority of oogonia are lost through atresia or cell death rather than through ovulation. Recently, it has been hypothesized that oogonial stem cells might continue after birth [4], as evidenced by the finding of active germ stem cells in adult mouse ovaries [5]. In a recent report, data from ovarian histological studies was taken to create a mathematical model to calculate the ovarian reserve at any given age. Importantly, the authors found that in the majority of patients, the rate of recruitment of nongrowing follicles increases from birth until approximately 14 years of age and then decreases until menopause [6].

The intimate relationship between female chronological age and fecundity has been demonstrated in prospective demographic studies [7], in studies that measure markers of ovarian reserve like basal antral follicle count (BAFC) [4], and studies of patients undergoing in vitro fertilization (IVF). In a prospective study, the percentage of infertility was 8% for women aged 19–26 years, 13–14% for women aged 27–34 years, and 18% for women aged 35–39 years [7]. A study showed that follicle numbers decline with age in a biexponential fashion, with an accelerated decline seen after a critical number of 25,000 at age of 37.5 year. This accelerated rate of atresia lowers the follicle population to 1,000 by 51 years, corresponding to the median age of menopause in the general population [4]. Additionally, there is a similar decline in the success of IVF as the age of patient increases. In 2007, the percentage of fresh embryo transfers from nondonor oocytes resulting in live births in the USA was 46.1 for patients <35 years of age and showed a progressive decline to 16.4 in patients 41–42 years of age [8].

Other factors that play an important role in the relationship between female chronological age and fecundity include the known increased risk in spontaneous abortion as maternal age increases [9]. Aneuploidy, the primary cause of spontaneous abortion, increases with female chronological age. The etiology of aneuploidy in patients of advanced age is due in part to the age-related decline in the cohesive ties that hold the chromosomes together coupled with the intrinsic inability of oocytes to detect and repair the resulting error [10]. This process can also be exacerbated by exposure to agents that further impair detection and repair of these errors [10].

Options for Fertility Preservation

Recent advances in assisted reproductive technologies [2] have allowed expansion of the options available for fertility preservation in healthy women. In this chapter, we describe the benefits, caveats, and other factors associated with each of these options.

Embryo Cryopreservation

Cryopreservation revolutionized assisted reproductive technology following its introduction in the early 1980s [11]. This laboratory technique allowed multiple embryo transfers resulting from a single oocyte retrieval, and thus an improved cumulative effectiveness of IVF. According to the Society for Assisted Reproductive Technology (SART) 2008 data, the percentage of thawed embryo transfers that resulted in live births was 35.6% for patients <35 years and 29.5% for patients 35–37 years and the percentage decreased as age increased, being 19.3% for patients 41–42 years. The number of embryos transferred per cycle was 2.1–2.3 in each age group [8]. Embryo cryopreservation was additionally the first technique clinically available for fertility preservation.

In addition to its application for fertility preservation, embryo cryopreservation has been clinically useful as an adjunct both in patients requiring preimplantation genetic diagnosis

(PGD) and in those at risk of developing ovarian hyperstimulation syndrome (OHSS) while undergoing ovulation induction or IVF.

Complete genome hybridization (CGH) has been used in patients requiring PGD in order to select the embryos best suited for transfer [12, 13]. Other technologies, including single nucleotide polymorphism microarrays and real-time quantitative polymerase chain reaction, are able to yield results within the window of implantation which can potentially make this technology available for clinical use in fresh embryo cycles in the near future [14].

OHSS can be a serious complication of ovulation induction and IVF, and has a reported prevalence ranging from 0.5 to 5% in its severe form [15]. The incidence of OHSS appears to be directly related to human chorionic gonadotropin (hCG) levels [16]. In patients at risk of OHSS, cryopreservation of embryos instead of fresh embryo transfer may be preferential, and can prevent severe OHSS in nearly all cases [17]. In these cases, the patient would be allowed to recover from OHSS and a frozen cycle of IVF is then undertaken following resolution of symptoms. The pregnancy rate in subsequent frozen cycles of these patients has been reported in some series to be up to 50% or comparable to fresh cycles reported from the same center [18].

The process to obtain embryos for cryopreservation for fertility preservation is similar to that in a cycle of IVF for planned fresh embryo transfer. Controlled ovarian hyperstimulation is achieved with gonadotropins and the oocytes are retrieved vaginally under ultrasound guidance. IVF or intracytoplasmic sperm injection is performed as indicated. Depending on the quantity and quality of embryos available and upon confirmation of written consent of the patient or couple, viable embryos can be cryopreserved. Two common methods of cryopreservation are "slow freezing" and "vitrification." The goal of cryopreservation is to obtain a high post-thaw survival and viability of the embryos in an attempt to optimize implantation, pregnancy, and live birth rates. We briefly describe both techniques. The review by Varghese et al. [19] provides a comprehensive description of these techniques.

Cryopreservation Techniques: Slow Freezing and Vitrification

Slow Freezing

Successful post-thaw survival of embryos (mouse) was described in 1972 by Whittingham et al. [20]. Survival of embryos required slow cooling (0.3–2°C/min) until a temperature of −196°C was achieved and slow warming (4–25°C/min). After cryopreservation of embryos for 8 days, successful pregnancies that led to full-term living mice were achieved. The application of this technology to human assisted reproductive technology led to the first human pregnancy after cryopreservation and thaw of an 8-cell embryo in 1983 [11].

Cryoprotectants, like propanediol and dimethyl sulfoxide (DMSO), are used during slow freezing protocols in an attempt to avoid the formation of intracellular crystals. Slow freezing attempts to lower the temperature slow enough to allow the embryo to lose the intracellular water osmotically to avoid the formation of crystals. Despite these measures, the embryo can be damaged by intracellular ice formation [21] or by osmotic shock that may take place during cellular shrinkage [22]. The ice front may additionally cause physical intracellular damage caused by contact or by intracellular and extracellular gas bubble formation [23]. The zona pellucida [24] and the intracellular organelles can, thus, be damaged in the cryopreservation process.

From a practical perspective, slow freezing poses a substantial cost to the IVF laboratory given the expensive programmable freezing equipment and time required by the embryologists to perform this type of cryopreservation (at least 90 min). Despite these limitations, slow freezing is currently the most widely used method for embryo cryopreservation worldwide.

Vitrification

Vitrification is a cryopreservation method that entails a rapid cooling rate of embryos while using high concentrations of cryoprotectants. Development of mouse embryos cryopreserved by vitrification was reported in the 1980s [25], with the first studies in humans reported by

Fahy [26]. The process of vitrification has the enormous advantage, in that it practically eliminates ice crystal formation and thus minimizes the risk of physical damage to the embryo.

The intracellular water in embryos is largely replaced by cryoprotectants upon direct contact and the embryos are plunged directly into the liquid nitrogen. The cooling rate achieved by this method is between 15,000 and 30,000°C/min and the liquid intracellular water is transformed directly into a glassy vitrified state [27].

In contrast with slow freezing, vitrification takes only a few seconds to achieve the desired freezing temperature and does not require expensive equipment. The costs, however, can also be considerable, as the freezing and thawing media and learning curve must be considered.

Slow Freezing vs. Vitrification

In a recent meta-analysis of available randomized controlled trials (RCTs) by Kolibianakis et al., vitrification was shown to have better post-thawing survival rates both for cleavage-stage embryos (odds ratio [OR]: 6.35, 95% confidence interval [CI] [28]: 1.14–35.26) and for blastocysts (OR: 4.09, 95% CI: 2.45–6.84). Post-thawing blastocyst development of embryos cryopreserved in the cleavage stage was also higher in vitrification in comparison to slow freezing (OR: 1.56, 95% CI: 1.07–2.27). However, there were no significant differences in clinical pregnancy rates between the two methods [29]. Importantly, this conclusion was drawn from six RCTs and only one had data of live birth rates.

Furthermore, a RCT from Balaban et al. showed that day 3 embryos had better post-thaw survival, embryo metabolism, and development to blastocyst after vitrification in comparison to slow freezing. More embryos survived the vitrification procedure (222/234, 94.8%) than slow freezing (206/232, 88.7%; $P<0.05$), pyruvate uptake was significantly greater in the vitrification group (12.05 ± 0.97 vs. 7.50 ± 0.52 pmol/embryo/h; $P<0.01$), and development to the blastocyst was also higher following vitrification (134/222, 60.3%) than following freezing (106/206, 49.5%; $P<0.05$) [30].

Even though there is a definitive need for more studies to evaluate embryo survival and live birth rate after vitrification, this technique is proving to have many benefits over slow freezing and has the potential to alter cryopreservation methodologies worldwide.

Mature Oocyte Cryopreservation

Historically, human oocyte cryopreservation began in the 1980s with isolated and sporadic reports of successful pregnancies using cryopreserved oocytes [31, 32].

This method has many potential advantages over cryopreservation of embryos in the area of elective fertility preservation. Probably, the most attractive feature of this technique is that it allows a patient to extend her fertility without the need for a male partner or a sperm donor at the time of decision to cryopreserve. Another important advantage of this technique is that it surmounts the ethical, religious, and legal limitations surrounding embryo cryopreservation in some countries [28].

Similar to embryo cryopreservation, the process of obtaining mature oocytes for cryopreservation requires ovarian stimulation with gonadotropins and vaginal oocyte retrieval. The oocytes are then cryopreserved by either slow freezing or vitrification as outlined above.

Pregnancy rates after transfer of embryos derived from vitrified oocytes are encouraging (one recent report quotes up to 38%) [33]. Vitrification seems to achieve better cleavage rates (84%) compared to slow freezing/thawing (71%).

In 2008, the American Society for Reproductive Medicine (ASRM) stated that "oocyte cryopreservation presently should be considered an experimental technique only to be performed under investigational protocol under the auspices of an institutional review board (IRB)" and that "neither ovarian tissue nor oocyte cryopreservation should be marketed or offered as a means to defer reproductive aging" [34]. In a more recent statement, however, the ASRM stated that oocyte cryopreservation had "great promise for applications in oocyte donation and fertility preservation" [35]. Some authors have questioned the experimental label given to oocyte cryopreservation by ASRM given that the recent progress in the field has allowed many clinics to achieve outcomes comparable to IVF with embryo transfer [36].

The statement might need to be soon readdressed, as more clinical data becomes available. An example of a study that is of important clinical value is the HOPE Registry, which is a phase IV observation 5-year study that started enrolling reproductive-age women in 2008 who have thawed frozen oocytes for subsequent use for IVF and embryo transfer [37].

Immature Oocytes and Ovarian Tissue Cryopreservation

Another experimental procedure for fertility preservation is cryopreservation of immature oocytes within ovarian tissue. This technique has the advantage that it does not require ovarian stimulation with gonadotropins and that a large pool of oocytes can be cryopreserved in comparison to cryopreservation of mature oocytes. Currently, there are many factors that complicate the development of this technique. Some of the limitations are that: (a) surgical procedure is required to obtain the ovarian tissue and (b) requires either transplantation of the tissue into the patient or in vitro maturation of oocytes. The transplantation of ovarian tissue can either be orthotopic (in the pelvis) or heterotopic (to another site, typically the forearm).

The clinical data for this technique in humans is limited, but at least 16 live births have been reported in the literature [38–40]. The latest case reported is of a twin pregnancy achieved after ovarian tissue transplantation, controlled ovarian stimulation, and oocyte vitrification with embryo culture and replacement [39].

The experience in transplantation of ovarian tissue in humans and in vitro maturation of oocytes is currently limited and further studies are required to evaluate the role of ovarian tissue cryopreservation in the future of the field of fertility preservation.

In 2008, the ASRM stated, "ovarian tissue cryopreservation and transplantation is experimental. Future research in larger numbers of patients will determine whether acceptable longevity can be achieved with both pelvic and forearm ovarian cortical transplant procedures and whether fertility reliably can be restored" [34]. Aside from its technical difficulties and experimental nature, this fertility preservation option has theoretic value to patients with malignancies. Transvaginal oocyte retrieval and cryopreservation of oocytes, however, appears to continue to be the method of choice for fertility preservation in healthy patients.

Limitations and Ethical Considerations

As outlined above, some of the most important limitations for fertility preservation surround embryo cryopreservation. In addition to the ethical considerations, embryo cryopreservation has a variety of legal implications worldwide. In 2004, the Italian Parliament enacted a law that prohibits the cryopreservation of embryos stating that "no fertilization procedure can produce embryos in excess of three, and that all fertilized pre-embryos must be implanted simultaneously" [41].

With nearly 400,000 embryos stored in the USA alone (in 2003) [42], embryo abandonment is a major problem for IVF clinics worldwide. The ASRM states that "it is ethically acceptable to consider embryos to have been abandoned if more than 5 years have passed since contact with a couple, diligent efforts have been made by telephone and registered mail to contact the couple at their last known address, and no written instruction from the couple exists concerning disposition" [43].

When confronted with embryo abandonment, the IVF clinic faces the burden of either discarding the embryos vs. continuing to cryopreserve the embryos despite the economical and ethical implications on doing so. In the guidelines of its Ethics Committee, the ASRM also states that "a program's willingness to store embryos does not imply an ethical obligation to store them indefinitely" and that "If a program reasonably determines under this standard that embryos have been abandoned (as above), the Ethics Committee concludes that the program may dispose of the embryos by removal from storage and thawing without transfer. In no case without prior consent, should embryos deemed abandoned be donated to other couples or be used in research" [43]. In a recent study, an embryo abandonment (defined as failure of contact between the facility and the couple of >2 years) rate of 1% was described despite a multimodal outreach effort to contact the patient, spouse, and emergency family contacts by telephone, correspondence, and search in a public database, including death records and Internet directory portals [44].

An important limitation for the use of different fertility preservation options is the cost associated with the procedure to retrieve oocytes and IVF in case of embryo cryopreservation and the costs associated with cryopreservation in the USA. The storage cost alone for oocytes is currently approximately $300 per year [45].

Furthermore, the experimental nature and ASRM label of the newer methods for fertility preservation pose a limitation to their widespread application and development. The assisted reproductive team should be familiar and comply with the assisted reproductive laws that apply to the jurisdiction and act according to the general ethical principles.

Approach to the Patient and Counseling

A consultation for fertility preservation with a reproductive endocrinologist requires a thorough history, physical exam, and ancillary tests. The physician should focus the history on the factors that are likely to affect the reproductive potential of the patient or couple at the time of evaluation and planned conception. Factors that are critical to consider are age of the patient (and partner if applicable), parity, social habits, and general health of the patient determined after taking a detailed medical and surgical history. Medical conditions that merit special considerations include carcinoma (especially endometrial, ovarian, or of the breast), history of chemotherapy or radiation, and any condition in which pregnancy would pose a significant risk on the patient.

In addition to a complete physical exam, an ultrasound to evaluate the ovaries, BAFC, uterus, and the pelvic cavity is helpful to assess the anatomy and ovarian reserve. According to the protocol of each IVF center, ancillary blood tests (including day 3 follicular stimulation hormone, antimullerian hormone, etc.) can also be measured as adjuncts to predict ovarian responsiveness to stimulation.

After taken into account the patient's age, medical conditions, current partner status, and plans for future conception, the options for fertility preservation applicable are then described in detail. Risks, benefits, and outcomes of each of these options need to be described in detail at this time. In adherence to the ethical and professional principles, data regarding post-thaw embryo or oocyte survival and pregnancy must include data from the actual IVF clinic to offer an honest and realistic expectation to the patient.

Informed consents should adhere to ASRM guidelines and include a detailed description of the fate of the cryopreserved oocytes or embryos in case of abandonment, death of patient or partner, or other unforeseen situations that may pose a conflict for all parties involved.

Given the decline in fertility and oocyte quality as the patient ages, embryos or oocytes should ideally be cryopreserved when the patient is <35 years. However, many requests for oocyte freezing are from women aged 36–40 years and older [45].

Additionally, there is no reliable mathematical algorithm to assist in determining the number of oocytes that are to be cryopreserved to achieve the best chance of pregnancy [45].

Conclusions

Reproductive endocrinologists are faced with the challenge of using various technological and laboratory advances in the field to help healthy women preserve fertility for a number of years until conception is desired. Cryopreservation of oocytes has the advantage that it surmounts the possible ethical, religious, and legal limitations of the cryopreservation of embryos.

Further improvements in laboratory technologies, like vitrification and in vitro maturation of oocytes or ovarian tissue, are likely to replace embryo cryopreservation as the main strategy for fertility preservation in the near future.

References

1. Hvidtfeldt UA, Gerster M, Knudsen L, Keiding N. Are low Danish fertility rates explained by changes in timing of births? Scand J Public Health. 2010;38(4): 426–33.
2. Willett LL, Wellons MF, Hartig JR, et al. Do women residents delay childbearing due to perceived career threats? Acad Med. 2010;85(4):640–6.
3. Baker TG. A quantitative and cytological study of germ cells in human ovaries. Proc R Soc Lond B Biol Sci. 1963;158:417–33.
4. Faddy MJ, Gosden RG, Gougeon A, Richardson SJ, Nelson JF. Accelerated disappearance of ovarian follicles in mid-life: implications for forecasting menopause. Hum Reprod. 1992;7(10):1342–6.
5. Johnson J, Canning J, Kaneko T, Pru JK, Tilly JL. Germline stem cells and follicular renewal in the postnatal mammalian ovary. Nature. 2004;428(6979): 145–50.
6. Wallace WH, Kelsey TW. Human ovarian reserve from conception to the menopause. PLoS One. 2010;5(1):e8772.
7. Dunson DB, Baird DD, Colombo B. Increased infertility with age in men and women. Obstet Gynecol. 2004;103(1):51–6.
8. Society for Assisted Reproductive Technology. 2008. https://www.sartcorsonline.com/rptCSR_PublicMultYear.aspx?ClinicPKID=0. Accessed April 2010.
9. Barron SL. The epidemiology of human pregnancy. Proc R Soc Med. 1968;61(11 Part 2):1200–6.
10. Jones KT. Meiosis in oocytes: predisposition to aneuploidy and its increased incidence with age. Hum Reprod Update. 2008;14(2):143–58.
11. Trounson A, Mohr L. Human pregnancy following cryopreservation, thawing and transfer of an eight-cell embryo. Nature. 1983;305(5936):707–9.
12. Wells D, Delhanty JD. Comprehensive chromosomal analysis of human preimplantation embryos using whole genome amplification and single cell comparative genomic hybridization. Mol Hum Reprod. 2000; 6(11):1055–62.
13. Wells D, Alfarawati S, Fragouli E. Use of comprehensive chromosomal screening for embryo assessment: microarrays and CGH. Mol Hum Reprod. 2008;14(12): 703–10.
14. Treff NR, Su J, Tao X, Levy B, Scott Jr RT. Accurate single cell 24 chromosome aneuploidy screening using whole genome amplification and single nucleotide polymorphism microarrays. Fertil Steril. 2010;94(6): 2017–21.
15. Sills ES, McLoughlin LJ, Genton MG, Walsh DJ, Coull GD, Walsh AP. Ovarian hyperstimulation syndrome and prophylactic human embryo cryopreservation: analysis of reproductive outcome following thawed embryo transfer. J Ovarian Res. 2008;1(1):7.
16. MacDougall MJ, Tan SL, Jacobs HS. In-vitro fertilization and the ovarian hyperstimulation syndrome. Hum Reprod. 1992;7(5):597–600.
17. Tiitinen A, Husa LM, Tulppala M, Simberg N, Seppala M. The effect of cryopreservation in prevention of ovarian hyperstimulation syndrome. Br J Obstet Gynaecol. 1995;102(4):326–9.
18. Fitzmaurice GJ, Boylan C, McClure N. Are pregnancy rates compromised following embryo freezing to prevent OHSS? Ulster Med J. 2008;77(3):164–7.
19. Varghese AC, Nagy ZP, Agarwal A. Current trends, biological foundations and future prospects of oocyte and embryo cryopreservation. Reprod Biomed Online. 2009;19(1):126–40.
20. Whittingham DG, Leibo SP, Mazur P. Survival of mouse embryos frozen to −196 degrees and −269 degrees C. Science. 1972;178(59):411–4.
21. Leibo SP, McGrath JJ, Cravalho EG. Microscopic observation of intracellular ice formation in unfertilized mouse ova as a function of cooling rate. Cryobiology. 1978;15(3):257–71.
22. Oda K, Gibbons WE, Leibo SP. Osmotic shock of fertilized mouse ova. J Reprod Fertil. 1992;95(3): 737–47.
23. Ashwood-Smith MJ, Morris GW, Fowler R, Appleton TC, Ashorn R. Physical factors are involved in the destruction of embryos and oocytes during freezing and thawing procedures. Hum Reprod. 1988;3(6):795–802.
24. Van Den Abbeel E, Van Steirteghem A. Zona pellucida damage to human embryos after cryopreservation and the consequences for their blastomere survival and in-vitro viability. Hum Reprod. 2000; 15(2):373–8.
25. Rall WF, Wood MJ, Kirby C, Whittingham DG. Development of mouse embryos cryopreserved by vitrification. J Reprod Fertil. 1987;80(2):499–504.
26. Fahy GM. Vitrification: a new approach to organ cryopreservation. Prog Clin Biol Res. 1986;224:305–35.
27. Son WY, Tan SL. Comparison between slow freezing and vitrification for human embryos. Expert Rev Med Devices. 2009;6(1):1–7.
28. Oktay K, Cil AP, Zhang J. Who is the best candidate for oocyte cryopreservation research? Fertil Steril. 2010;93(1):13–5.
29. Kolibianakis EM, Venetis CA, Tarlatzis BC. Cryopreservation of human embryos by vitrification or slow freezing: which one is better? Curr Opin Obstet Gynecol. 2009;21(3):270–4.
30. Balaban B, Urman B, Ata B, et al. A randomized controlled study of human day 3 embryo cryopreservation by slow freezing or vitrification: vitrification is

associated with higher survival, metabolism and blastocyst formation. Hum Reprod. 2008;23(9):1976–82.
31. Chen C. Pregnancy after human oocyte cryopreservation. Lancet. 1986;1(8486):884–6.
32. Chen C. Pregnancies after human oocyte cryopreservation. Ann N Y Acad Sci. 1988;541:541–9.
33. Smith GD, Serafini PC, Fioravanti J, et al. Prospective randomized comparison of human oocyte cryopreservation with slow-rate freezing or vitrification. Fertil Steril. 2010;94(6):2088–95.
34. Practice Committee of American Society for Reproductive Medicine; Practice Committee of Society for Assisted Reproductive Technology. Ovarian tissue and oocyte cryopreservation. Fertil Steril. 2008;90(5 Suppl):S241–6.
35. ASRM Practice Committee response to Rybak and Lieman. elective self-donation of oocytes. Fertil Steril. 2009;92(5):1513–4.
36. Noyes N, Boldt J, Nagy ZP. Oocyte cryopreservation: is it time to remove its experimental label? J Assist Reprod Genet. 2010;27:69–74.
37. Ezcurra D, Rangnow J, Craig M, Schertz J. The HOPE registry: first US registry for oocyte cryopreservation. Reprod Biomed Online. 2008;17(6):743–4.
38. Donnez J, Dolmans MM. Cryopreservation of ovarian tissue: an overview. Minerva Med. 2009;100(5):401–13.
39. Sanchez-Serrano M, Crespo J, Mirabet V, et al. Twins born after transplantation of ovarian cortical tissue and oocyte vitrification. Fertil Steril. 2010;93(1):268e11–3.
40. Silber SJ. Fresh ovarian tissue and whole ovary transplantation. Semin Reprod Med. 2009;27(6):479–85.
41. Boggio A. Italy enacts new law on medically assisted reproduction. Hum Reprod. 2005;20(5):1153–7.
42. Hoffman DI, Zellman GL, Fair CC, et al. Cryopreserved embryos in the United States and their availability for research. Fertil Steril. 2003;79(5):1063–9.
43. Ethics Committee of the American Society for Reproductive Medicine. Disposition of abandoned embryos. Fertil Steril. 2004;82 Suppl 1:S253.
44. Walsh AP, Tsar OM, Walsh DJ, Baldwin PM, Shkrobot LV, Sills ES. Who abandons embryos after IVF? Ir Med J. 2010;103(4):107–10.
45. Molloy D, Hall BA, Ilbery M, Irving J, Harrison KL. Oocyte freezing: timely reproductive insurance? Med J Aust. 2009;190(5):247–9.

Approach to Fertility Preservation in Adult and Prepubertal Females

18

Kenny A. Rodriguez-Wallberg
and Kutluk Oktay

Abstract

Fertility preservation is recognized as an issue of great importance to patients diagnosed with cancer. Infertility is emotionally painful for many patients and female patients may see infertility as a loss of their femininity. The available evidence suggests that infertility resulting from cancer treatment may be still more devastating and is often associated with psychosocial distress. In Western countries, women are delaying initiation of childbearing to later in life. Since the incidence of most cancers increases with age, delayed childbearing results in more female cancer survivors interested in fertility preservation. Fertility preservation is often possible in adult female patients before starting their gonadotoxic treatments. Embryos, oocytes, or ovarian tissue can be cryopreserved and stored until the time when the patients are disease-free and wish to start a family. In prepubertal girls, the only options are still experimental and most challenging. Because of differences in clinical presentation of the main disease, expected treatment, age of the patient, existence of a partner, and the time available before the onset of cancer treatment, each case is unique and therefore requires individual consideration with regard to the options for fertility preservation.

Keywords

Fertility preservation • Prepubertal female fertility preservation • Infertility • Cancer treatment and infertility • Infertility and cancer treatment • Reproductive counseling

K. Oktay (✉)
Professor of Obstetrics & Gynecology, Medicine, and Cell Biology & Anatomy, Director, Division of Reproductive Medicine & Institute for Fertility Preservation
e-mail: kutluk.oktay@kutlukoktay.com

E. Seli and A. Agarwal (eds.), *Fertility Preservation: Emerging Technologies and Clinical Applications*,
DOI 10.1007/978-1-4419-1783-6_18,

Fertility preservation is recognized as an issue of great importance to patients diagnosed with cancer [1]. Infertility is emotionally painful for many patients and female patients may see infertility as a loss of their femininity. The available evidence suggests that infertility resulting from cancer treatment may be still more devastating and is often associated with psychosocial distress [2, 3]. In Western countries, women are delaying initiation of childbearing to later in life. Since the incidence of most cancers increases with age, delayed childbearing results in more female cancer survivors interested in fertility preservation.

Fertility preservation is often possible in adult female patients before starting their gonadotoxic treatments. Embryos, oocytes, or ovarian tissue can be cryopreserved and stored until the time when the patients are disease-free and wish to start a family. In prepubertal girls, the only options are still experimental and most challenging. Because of differences in clinical presentation of the main disease, expected treatment, age of the patient, existence of a partner, and the time available before the onset of cancer treatment, each case is unique and therefore requires individual consideration with regard to the options for fertility preservation.

Need for Timely Reproductive Counseling

To safeguard fertility potential, fertility preservation options should be offered and performed before patients commence cancer treatments requiring early referral to a reproductive medicine expert and close collaboration between the oncology team and the reproductive medicine unit [1]. When the patient is a child, parents of children undergoing gonadotoxic cancer treatments should receive counseling by reproductive health care providers at an appropriate specialized center about short- and long-term side effects of the planned cancer treatment on the reproductive system and about their options [4]. Counseling should also include the option of deciding not to take fertility-preserving measures and should include information on alternatives to become a parent through adoption and gamete donation [5].

Saving Fertility Potential

Shielding to Reduce Gonadal Damage by Radiation Therapy

Radiation results in dose-related damage of the gonads by the reduction in the nonrenewable primordial follicle pool in women. The extent of the damage in females depends on the dose, fractionation schedule, and irradiation field [6, 7]. The degree and persistence of the damage in female patients are also influenced by age at the time of exposure to radiotherapy as younger women have a greater reserve of primordial follicles than older women and they may, thus, have a higher remaining primordial pool after a cancer treatment [8]. Table 18.1 presents a compilation of current knowledge on the impact of radiation in ovarian function of women [9]. In pediatric patients, failure in pubertal development may be the first sign of gonadal failure.

Shielding to reduce radiation to the reproductive organs when possible is the standard medical procedure currently offered to female patients. When shielding of the gonadal area is not possible, the surgical fixation of the ovaries far from the radiation field, oophoropexy (ovarian transposition), may be considered. It is estimated that this procedure significantly reduces the risk of ovarian failure by about 50% and those patients

Table 18.1 Radiotherapy protocols with high or intermediate impact on ovarian function

High risk of amenorrhea in women
Total body irradiation (TBI) for bone marrow transplant/stem cell transplant
Pelvic or whole abdominal radiation dose ≥6 Gy in adult women
Pelvic or whole abdominal radiation dose ≥10 Gy in postpubertal girls
Pelvic radiation or whole abdominal dose ≥15 Gy in prepubertal girls
Intermediate risk
Pelvic or whole abdominal radiation dose 5–10 Gy in postpubertal girls
Pelvic or whole abdominal radiation dose 10–15 Gy in prepubertal girls
Craniospinal radiotherapy dose ≥25 Gy

Modified from Rodriguez-Wallberg and Oktay [9]

may retain some menstrual function and fertility [1]. Failure of this procedure seems to be related to scattered radiation and damage of the blood vessels that supply the ovaries [1].

Impact of Radiation of the Uterus

Because of the age-related reduction in ovarian reserve, younger women are less sensitive to radiation-induced ovarian damage than older women. However, even if ovarian function is maintained and pregnancy is achieved, younger patients at the time of radiotherapy may present with restricted blood flow and impaired uterine growth as consequences of the irradiation of the uterus and, therefore, have a high risk of spontaneous abortion, premature labor, and low-birth-weight offspring [10, 11].

Conservative Oncologic Surgery

Fertility-sparing surgery may be an option for selected patients. Indications for fertility-sparing surgery often include a well-differentiated low-grade tumor in its early stages or with low malignant potential. Table 18.2 summarizes current indications for fertility-sparing interventions among women of fertile age [12]. A detailed discussion of fertility preservation options in women with gynecologic cancers is presented in this book by Richter and Schwartz (see Chap. 14).

Although conservative surgery offers the opportunity to preserve reproductive organs, it offers no guarantee of pregnancy or live birth. Causes of subfertility may be present in the patients and a number of those patients may further require assisted reproduction treatments [13].

Saving Fertility Potential Before Gonadotoxic Chemotherapy Treatments

Knowledge of the risk of gonadal damage caused by the different chemotherapeutic agents is essential. Table 18.3 summarizes the gonadotoxic impact of chemotherapy agents on the ovary. Women's undeveloped oocytes and pregranulosa cells of primordial follicles are particularly sensitive to alkylating agents, and ovarian failure is common after such treatment.

Because of a reduction of the primordial follicle pool with aging, older women have a higher risk of developing ovarian failure and permanent infertility after a cancer treatment compared with younger women. Some female patients recover a certain degree of ovarian function following chemotherapy or radiotherapy treatments and they should be recommended not to delay childbearing for too many years [9]. However, owing to the toxicity of cancer treatments on growing oocytes, they should be advised to avoid conception in the 6–12 month period immediately following completion of treatment [14]. This is due to the higher risk of teratogenesis during or immediately following chemotherapy; nevertheless, DNA integrity has shown to return over time after a cancer treatment [15].

Cryopreservation of Embryos

For adult women, embryo cryopreservation after controlled ovulation stimulation, oocyte retrieval, and in vitro fertilization (IVF) treatment is a well-established procedure. However, this procedure can only be offered if the woman has a partner or if it is possible to use a sperm donor, which may not be allowed in some countries. IVF may require 2–6 weeks depending on the women's menstrual cycle phase at the time of planning the treatment. Transfer of frozen/thawed embryos today is a clinical routine in conventional IVF programs worldwide and it has been used for nearly 25 years. Intact embryos after thawing have the same implantation potential as fresh embryos and this can lead to a 59% pregnancy rate and a 26% live birth rate [16].

Stimulation Protocols for Women with Breast Cancer

Breast cancer is the most frequent cancer occurring at reproductive age. With current treatments, more than 80% of women below the age of 40 suffering from breast cancer are successfully treated.

Table 18.2 Fertility-sparing interventions in female patients

Diagnosis	Type of surgery	Description	Obstetric outcome	Oncologic outcome
Cervical cancer stage IA1, 1A2, 1B1	Radical vaginal trachelectomy	Laparoscopic pelvic lymphadenectomy. Vaginal resection of the cervix and surrounding parametria keeping the corpus of the uterus and the ovaries intact	Spontaneous pregnancies described in up to 70%. Risk of second-trimester pregnancy loss and preterm delivery	Rates of recurrence and mortality are comparable to those described for similar cases treated by means of radical hysterectomy or radiation therapy
Borderline ovarian tumors FIGO stage I	Unilateral oophorectomy	Removal of the affected ovary only, keeping in place the unaffected one and the uterus	Pregnancies have been reported and favorable obstetric outcome	Oncologic outcome is comparable with the more radical approach of removing both ovaries and the uterus. Recurrence 0–20 vs. 12–58% when only cystectomy was performed
Ovarian epithelial cancer stage I, grade 1	Unilateral oophorectomy	Removal of the affected ovary only, keeping in place the unaffected one and the uterus	Pregnancies have been reported and favorable obstetric outcome	7% recurrence of the ovarian malignancy and 5% deaths
Malignant ovarian germ cell tumors/sex cord stromal tumors	Unilateral oophorectomy	Removal of the affected ovary only	Pregnancies have been reported and favorable obstetric outcome	Risk of recurrence similar to historical controls
Endometrial adenocarcinoma Grade 1, stage 1A (without myometrial or cervical invasion)	Hormonal treatment with progestational agents for 6 months	Follow-up with endometrial biopsies every 3 months	Pregnancies have been reported	Recurrence rate 30–40%. Five percent recurrence during progesterone treatment

From Rodriguez-Macias Wallberg et al. [12], with permission

Table 18.3 Chemotherapy agents and their risk to induce ovarian failure in women

High risk
Cyclophosphamide
Ifosfamide
Melphalan
Busulfan
Nitrogen mustard
Procarbazine
Chlorambucil
Intermediate risk
Cisplatin
Adriamycin
Low risk
Bleomycin
Actinomycin D
Vincristine
Methotrexate
5-Fluorouracil

Although the majority of breast cancers present in young women are diagnosed at early stages, they present with a high prevalence of ductal infiltration and most of these patients are likely to undergo adjuvant systemic chemotherapy and in some cases adjuvant endocrine therapy as well. The adjuvant endocrine therapy with tamoxifen, usually recommended for at least 5 years for patients with endocrine-sensitive tumors, may further delay the possibility of pregnancy and reduce the likelihood that breast cancer survivors will conceive during their most fertile years, even if they do not receive a gonadotoxic treatment.

Because of the elevation of circulating estradiol levels during conventional ovulation induction for IVF, alternative and potentially safer protocols, including natural cycle IVF or the use of tamoxifen and aromatase inhibitors to reduce estrogen exposure, have been introduced [17]. Natural cycle IVF does not give more than one oocyte or embryo per cycle and has a high rate of cycle cancelation. The aromatase inhibitor letrozole's stimulation protocols seem to offer a better fertility outcome with regard to the number of oocytes and embryos obtained, when compared to tamoxifen stimulation protocols [17]. Short-term follow-up after stimulation protocols with letrozole has not shown any detrimental effect on survival in women with breast cancer undergoing ovarian stimulation [18]. Further improvements in letrozole stimulation protocols combined with gonadotropins for breast cancer patients have been recently reported. In vitro maturation (IVM) of immature yielded oocytes has increased the number of oocytes and embryos cryopreserved for these patients [19]. Triggering of oocyte maturation with gonadotropin-releasing hormone (GnRH) agonists decreases the post-trigger estradiol exposure as well as ovarian hyperstimulation syndrome risks [20]. Aromatase inhibitors are contraindicated during pregnancy. However, data indicates that fertility treatment with letrozole is safe and its use before conception does not seem to have increased risks for the fetus [21].

Cryopreservation of Oocytes

Freezing either mature or immature oocytes, although still experimental, is the only option for female fertility preservation that can be offered to single cancer patients with no male partner and to those who do not wish to use a sperm donor. This procedure may also be recommended to older teenagers in selected cases. However, as the procedure usually involves ovarian stimulation in order to obtain mature oocytes, some cancer patients may not have enough time to complete a stimulation cycle before starting cancer treatment. The recent development of vitrification techniques for cryopreservation of oocytes has shown improved success rates in survival of oocytes, fertilization rates, and pregnancy rates, approaching that of fresh oocytes [22, 23].

When there is not enough time for ovarian stimulation, another strategy is to retrieve oocytes without ovarian stimulation and to freeze them at an immature stage or to mature them in vitro (IVM). Immature oocytes survive cryopreservation better than mature metaphase II oocytes [24]. After thawing, they can be matured in vitro and fertilized. IVM of oocytes is at an experimental stage and needs further development [25]. Only a few fertility centers worldwide offer treatments by using this technique.

Ovarian Tissue Freezing and Transplantation

Cryopreservation of ovarian cortex for future ovarian tissue transplantation is a new promising method for fertility preservation [26, 27]. A vast majority of eggs exist in the ovarian cortex within primordial follicles. Ovarian cortical tissue can be harvested by an outpatient laparoscopy without any significant delay of cancer therapy. This procedure does not require ovarian stimulation.

Cryopreservation of human ovarian tissue has been a feasible method since 1996, with functioning tissue after thawing [28]. Follicles in slices of cryopreserved-thawed tissue successfully survive long-term in vitro organ culture and transplantation to immunodeficient mice, and mature oocytes have been found in the xenografts [29, 30]. Slow-programmed freezing has been the method of choice for cryopreservation of ovarian tissue, but poor survival of ovarian stroma is the main limitation of this method. The recently described technique of vitrification for freezing of ovarian tissue seems to improve the viability of all compartments of the tissue with a survival rate of follicles similar to that after slow freezing, much better integrity of ovarian stroma, and undamaged morphology of blood vessels [31].

Ovarian tissue can later be transplanted orthotopically [32] or heterotopically [33]. Ovarian tissue freezing is the only viable option in children as sexual maturity is not required. There have been hundreds of patients undergoing ovarian tissue freezing, but only a small percentage of these have returned for ovarian transplantation.

Autotransplantation is only possible if absence of malignant cells in the graft is confirmed. In hematological malignancies, detection of cancer cells in ovarian tissue should be performed using immunohistochemistry or the polymerase chain reaction, as well as by other methods that allow identification of malignant residual cells as xenotransplantation [34].

In Vitro Culture of Ovarian Follicles

In vitro culture and maturation of follicles within a piece of thawed tissue can be an option to obtain mature oocytes for fertilization. It would be especially useful in patients with hematological and ovarian malignancies, where the risk of retransmission of malignancy is high. In vitro culture and maturation of human ovarian follicles started experimentally over a decade ago [29] and although many improvements have been reported [35–37], it is still at the developmental stage.

Saving Fertility Potential in Prepubertal Girls

Fertility preservation options for prepubertal patients are the most challenging. As attempts to retain the reproductive capacity of children who are to undergo cancer treatment are relatively new, most realistic options are still in an experimental phase, such as gonadal tissue cryopreservation for transplantation or in vitro culture and maturation of ovarian follicles.

Cryopreservation of ovarian tissue is the only option for fertility preservation in prepubertal girls. It can be carried out immediately after the diagnosis of malignant disease by an outpatient laparoscopy or, if an operation is imminent, via laparotomy. Although the timing of ovarian biopsy is crucial and it is important to carry out cryopreservation before high-risk gonadotoxic therapy, young women, adolescents, and girls have normally an abundant number of primordial follicles and attempts to cryopreserve ovarian tissue may still be worthwhile after the first courses of chemotherapy if the procedure was not possible before [38, 39]. As described for adult women, the tissue should be cryopreserved and stored for a future transplantation or in vitro culture until the time when the patient is disease-free and is hoping to start a family.

Cryopreservation of gonadal tissue offers hope to childhood cancer patients and their families; however, it also raises several medical questions. The normality of imprinted genes of cryobanked oocytes matured in vitro has yet to be verified experimentally. Ovarian tissue cryopreservation and transplantation have shown not to interfere with proper genomic imprinting in mice pups [40], but additional studies in other animal models are needed.

Gonadal Protection by Drugs

Although data collected from observational studies suggest that ovarian protection may be achieved through ovarian suppression via GnRH agonists or antagonists during chemotherapy [41], the bulk of the data from experimental studies is still inconclusive. The fact that prepubertal children with cancer still develop ovarian failure after chemotherapy suggests that this treatment is of limited benefit [42]. Primordial follicles, which make up the ovarian reserve, do not express gonadotropin receptors and hence hormonal manipulation is not likely to affect them [43]. Unfortunately, at this time, the only long-term prospective randomized study recently reported no benefit from gonadal suppression in preserving fertility [44]. Those results are supported by two small randomized studies after short-term follow-up [45, 46]. There are, however, agents in development, such as sphingosine-1-phosphate (S1P) agonist FTY720, which may in the future be used to pharmacologically prevent chemo- and radiotherapy-induced germ cell death [47].

Case Presentations

Case 1

A single 37-year-old woman with a diagnosis of breast cancer successfully treated by lumpectomy. The patient is not going to receive chemotherapy, but she is planned for a 5-year treatment with tamoxifen.

- This patient should be referred to a fertility specialist for counseling. At patient's age, fertility has already started to diminish, and after completion of a 5-year course of adjuvant endocrine therapy with tamoxifen, the patient might be too old to have a chance to conceive spontaneously.
- The patient is single and for single women, there is the option of egg freezing. If the women are willing to use a sperm donor, they may get their eggs fertilized and frozen as embryos.
- The patient should be stimulated for IVF with protocols using drugs that reduce estrogen exposure, i.e., letrozole in combination with gonadotropins.
- The IVF success rates with frozen-thawed eggs have improved. They are now closer to the success rates of standard IVF techniques utilizing fresh eggs.

Case 2

A 28-year-old woman with Hodgkin's lymphoma planned to start chemotherapy (BEACOPP regimen) within a few days and thereafter mediastinal radiotherapy.

- This patient should be referred rapidly to a fertility specialist for counseling.
- The only option in this case is ovarian tissue freezing for future transplantation. This procedure does not require ovarian stimulation and can be performed within few days, before start of chemotherapy is planned. The ovarian tissue harvesting is performed by an outpatient laparoscopy, and the frozen tissue can later be transplanted.
- Hodgkin's lymphoma does not involve the ovaries, and retransplantation is considered safe.
- Ovarian tissue freezing is the only viable option in young girls, women for whom hormone stimulation is contraindicated, and women who need to start cancer treatment with no delay.

Case 3

A 10-year-old prepubertal girl with acute lymphoblastic leukemia planned for bone marrow transplantation conditioned by total body irradiation and chemotherapy.

- This patient and the parents should be referred rapidly to a fertility specialist for counseling. The indicated treatment has a very high possibility to induce ovarian failure.
- The only option for this patient is the cryopreservation of ovarian tissue.

- Due to the systemic malignancy, retransplantation is currently contraindicated.
- Methods for in vitro culture of oocytes from the frozen-thawed tissue are under development.

References

1. Lee SJ, Schover LR, Partridge AH, Patrizio P, Wallace WH, Hagerty K, et al. American Society of Clinical Oncology recommendations on fertility preservation in cancer patients. American Society of Clinical Oncology. J Clin Oncol. 2006;24(18):2917–31.
2. Schover LR, Rybicki LA, Martin BA, Bringelsen KA. Having children after cancer. A pilot survey of survivors' attitudes and experiences. Cancer. 1999;86(4): 697–709.
3. Rosen A, Rodriguez-Wallberg KA, Rosenzweig L. Psychosocial distress in young cancer survivors. Semin Oncol Nurs. 2009;25(4):268–77.
4. The Ethics Committee of the American Society of Reproductive Medicine. Fertility preservation and reproduction in cancer patients. Fertil Steril. 2005;83(6):1622–8.
5. Hollen PJ, Hobbie WL. Establishing comprehensive specialty follow-up clinics for long-term survivors of cancer: providing systematic physiologic and psychosocial support. Support Care Cancer. 1995;3(1):40–4.
6. Gosden RG, Wade JC, Fraser HM, Sandow J, Faddy MJ. Impact of congenital or experimental hypogonadotrophism on the radiation sensitivity of the mouse ovary. Hum Reprod. 1997;12(11):2483–8.
7. Speiser B, Rubin P, Casarett G. Aspermia following lower truncal irradiation in Hodgkin's disease. Cancer. 1973;32(3):692–8.
8. Wallace WH, Anderson RA, Irvine DS. Fertility preservation for young patients with cancer: who is at risk and what can be offered? Lancet Oncol. 2005;6(4):209–18.
9. Rodriguez-Wallberg KA, Oktay K. Fertility preservation medicine: options for young adults and children with cancer. J Pediatr Hematol Oncol. 2010;32(5): 390–6.
10. Wo JY, Viswanathan AN. Impact of radiotherapy on fertility, pregnancy, and neonatal outcomes in female cancer patients. Int J Radiat Oncol Biol Phys. 2009;73(5):1304–12.
11. Green DM, Sklar CA, Boice Jr JD, Mulvihill JJ, Whitton JA, Stovall M, et al. Ovarian failure and reproductive outcomes after childhood cancer treatment: results from the Childhood Cancer Survivor Study. J Clin Oncol. 2009;27(14):2374–81.
12. Rodriguez-Macias Wallberg KA, Keros V, Hovatta O. Clinical aspects of fertility preservation in female patients. Pediatr Blood Cancer. 2009;53(2):254–60.
13. Wong I, Justin W, Gangooly S, Sabatini L, Al-Shawaf T, Davis C, et al. Assisted conception following radical trachelectomy. Hum Reprod. 2009;24(4):876–9.
14. Meirow D, Epstein M, Lewis H, Nugent D, Gosden RG. Administration of cyclophosphamide at different stages of follicular maturation in mice: effects on reproductive performance and fetal malformations. Hum Reprod. 2001;16(4):632–7.
15. Simon B, Lee SJ, Partridge AH, Runowicz CD. Preserving fertility after cancer. CA Cancer J Clin. 2005;55(4):211–28.
16. Marrs RP, Greene J, Stone BA. Potential factors affecting embryo survival and clinical outcome with cryopreserved pronuclear human embryos. Am J Obstet Gynecol. 2004;190(6):1766–71.
17. Oktay K, Buyuk E, Libertella N, Akar M, Rosenwaks Z. Fertility preservation in breast cancer patients: a prospective controlled comparison of ovarian stimulation with tamoxifen and letrozole for embryo cryopreservation. J Clin Oncol. 2005;23(19):4347–53.
18. Azim AA, Costantini-Ferrando M, Oktay K. Safety of fertility preservation by ovarian stimulation with letrozole and gonadotropins in patients with breast cancer: a prospective controlled study. J Clin Oncol. 2008;26(16):2630–5.
19. Oktay K, Buyuk E, Rodriguez-Wallberg KA, Sahin G. In vitro maturation improves oocyte or embryo cryopreservation outcome in breast cancer patients undergoing ovarian stimulation for fertility preservation. Reprod Biomed Online. 2010;20(5):634–8.
20. Oktay K, Türkçüoğlu I, Rodriguez-Wallberg KA. GnRH agonist trigger for women with breast cancer undergoing fertility preservation by aromatase inhibitor/FSH stimulation. Reprod Biomed Online. 2010;20(6):783–8.
21. Tulandi T, Martin J, Al-Fadhli R, Kabli N, Forman R, Hitkari J, et al. Congenital malformations among 911 newborns conceived after infertility treatment with letrozole or clomiphene citrate. Fertil Steril. 2006; 85(6):1761–5.
22. Oktay K, Cil AP, Bang H. Efficiency of oocyte cryopreservation: a meta-analysis. Fertil Steril. 2006; 86(1):70–80.
23. Grifo JA, Noyes N. Delivery rate using cryopreserved oocytes is comparable to conventional in vitro fertilization using fresh oocytes: potential fertility preservation for female cancer patients. Fertil Steril. 2010;93(2):391–6.
24. Boiso I, Martí M, Santaló J, Ponsá M, Barri PN, Veiga A. A confocal microscopy analysis of the spindle and chromosome configurations of human oocytes cryopreserved at the germinal vesicle and metaphase II stage. Hum Reprod. 2002;17(7):1885–91.
25. Oktay K, Demirtas E, Son WY, Lostritto K, Chian RC, Tan SL. In vitro maturation of germinal vesicle oocytes recovered after premature luteinizing hormone surge: description of a novel approach to fertility preservation. Fertil Steril. 2008;89(1):228. e19–22.
26. Oktay K, Oktem O. Ovarian cryopreservation and transplantation for fertility preservation for medical indications: report of an ongoing experience. Fertil Steril. 2010;93(3):762–8.

27. von Wolff M, Donnez J, Hovatta O, Keros V, Maltaris T, Montag M, et al. Cryopreservation and autotransplantation of human ovarian tissue prior to cytotoxic therapy – a technique in its infancy but already successful in fertility preservation. Eur J Cancer. 2009;45(9):1547–53.
28. Hovatta O, Silye R, Krausz T, Abir R, Margara R, Trew G, et al. Cryopreservation of human ovarian tissue using dimethylsulphoxide and propanediol-sucrose as cryoprotectants. Hum Reprod. 1996;11(6): 1268–72.
29. Hovatta O, Silye R, Abir R, Krausz T, Winston RM. Extracellular matrix improves survival of both stored and fresh human primordial and primary ovarian follicles in long-term culture. Hum Reprod. 1997;12(5): 1032–6.
30. Van den Broecke R, Liu J, Handyside A, Van der Elst JC, Krausz T, Dhont M, et al. Follicular growth in fresh and cryopreserved human ovarian cortical grafts transplanted to immunodeficient mice. Eur J Obstet Gynecol Reprod Biol. 2001;97(2):193–201.
31. Keros V, Xella S, Hultenby K, Pettersson K, Sheikhi M, Volpe A, et al. Vitrification versus controlled-rate freezing in cryopreservation of human ovarian tissue. Hum Reprod. 2009;24(7):1670–83.
32. Oktay K, Karlikaya G. Ovarian function after transplantation of frozen, banked autologous ovarian tissue. N Engl J Med. 2000;342(25):1919.
33. Oktay K, Buyuk E, Veeck L, Zaninovic N, Xu K, Takeuchi T, et al. Embryo development after heterotopic transplantation of cryopreserved ovarian tissue. Lancet. 2004;363(9412):837–40.
34. Meirow D, Hardan I, Dor J, Fridman E, Elizur S, Ra'anani H, et al. Searching for evidence of disease and malignant cell contamination in ovarian tissue stored from hematologic cancer patients. Hum Reprod. 2008;23(5):1007–13.
35. Telfer EE, McLaughlin M, Ding C, Thong KJ. A two-step serum-free culture system supports development of human oocytes from primordial follicles in the presence of activin. Hum Reprod. 2008;23(5):1151–8.
36. Xu M, Kreeger PK, Shea LD, Woodruff TK. Tissue-engineered follicles produce live, fertile offspring. Tissue Eng. 2006;12(10):2739–46.
37. Xu M, Barrett SL, West-Farrell E, Kondapalli LA, Kiesewetter SE, Shea LD, et al. In vitro grown human ovarian follicles from cancer patients support oocyte growth. Hum Reprod. 2009;24(10):2531–40.
38. Poirot CJ, Martelli H, Genestie C, Golmard JL, Valteau-Couanet D, Helardot P, et al. Feasibility of ovarian tissue cryopreservation for prepubertal females with cancer. Pediatr Blood Cancer. 2007;49(1): 74–8.
39. Abir R, Ben-Haroush A, Felz C, Okon E, Raanani H, Orvieto R, et al. Selection of patients before and after anticancer treatment for ovarian cryopreservation. Hum Reprod. 2008;23(4):869–77.
40. Sauvat F, Capito C, Sarnacki S, Poirot C, Bachelot A, Meduri G, et al. Immature cryopreserved ovary restores puberty and fertility in mice without alteration of epigenetic marks. PLoS One. 2008;3(4): e1972.
41. Blumenfeld Z, Avivi I, Eckman A, Epelbaum R, Rowe JM, Dann EJ. Gonadotropin-releasing hormone agonist decreases chemotherapy-induced gonadotoxicity and premature ovarian failure in young female patients with Hodgkin lymphoma. Fertil Steril. 2008;89: 166–73.
42. Yap JK, Davies M. Fertility preservation in female cancer survivors. J Obstet Gynaecol. 2007;27(4): 390–400.
43. Oktay K, Briggs D, Gosden RG. Ontogeny of follicle-stimulating hormone receptor gene expression in isolated human ovarian follicles. J Clin Endocrinol Metab. 1997;82(11):3748–51.
44. Behringer K, Wildt L, Mueller H, Mattle V, Ganitis P, van den Hoonaard B, et al. No protection of the ovarian follicle pool with the use of GnRH-analogues or oral contraceptives in young women treated with escalated BEACOPP for advanced-stage Hodgkin lymphoma. Final results of a phase II trial from the German Hodgkin Study Group. Ann Oncol. 2010;21(10):2052–60.
45. Waxman JH, Ahmed R, Smith D, Wrigley PF, Gregory W, Shalet S, et al. Failure to preserve fertility in patients with Hodgkin's disease. Cancer Chemother Pharmacol. 1987;19(2):159–62.
46. Ismail-Khan R, Minton S, Cox C, Sims I, Lacevic M, Gross-King M, et al. Preservation of ovarian function in young women treated with neoadjuvant chemotherapy for breast cancer: a randomized trial using the GnRH agonist (triptorelin) during chemotherapy. J Clin Oncol. 2008;26:524.
47. Zelinski MB, Murphy M, Lawson M, Toscano N, Fanton J, Tilly JL. Intraovarian delivery of a sphingosine-1-phosphate (S1P) agonist, FTY720, prior to ovarian X-irradiation yields fertilization, embryonic development and pregnancy in rhesus monkeys. In: Abstracts of the scientific oral & poster sessions program supplement, American Society for reproductive medicine 64th annual meeting. Fertil Steril. 2008;90 Suppl 1:S1.

Part III

Fertility Preservation in Men

Spermatogenesis and Testicular Function

19

Ciler Celik-Ozenci

Abstract

Spermatogenesis is characterized by three specific phases: proliferation and differentiation, meiosis, and morphogenesis of haploid germ cells which takes approximately 1 month in mice and 2 months in humans. Sex determination occurs in the fetal gonads during development. Germ cells enter meiosis, and thus commit to the oogenic pathway, or stay away from meiosis, enter a state of quiescence and commit to the spermatogenic pathway. The main clinical consequence of defects in testicular differentiation during fetal periods is likely to be infertility. Another outcome of this condition in humans is increased susceptibility to testicular germ cell tumors, which have been proposed to originate from impaired or delayed germ cell differentiation during fetal testis development. Chemotherapy and radiotherapy, even in low doses, may have a detrimental effect on the seminiferous epithelium and disrupt spermatogenesis in both children and adults. The extent of damage to germ cells depends on the class of chemotherapeutic agent, dosage, spermatogenetic stage targeted as well as the original pretreatment fertility potential of the patient. Cytotoxic therapy influences spermatogenesis at least temporarily and in some cases permanently. Germ cells are also very sensitive to irradiation while the Leydig cells are more resistant owing to their slower rate of turnover. Radiotherapy affects sperm concentration; moreover, irradiation increases sperm DNA damage.

Keywords

Spermatogenesis • Testicular function • Testicular germ cell development • Gonad development in the womb • Chemotherapy and radiation effects on spermatogenesis

C. Celik-Ozenci (✉)
Department of Histology and Embryology, School of Medicine, Akdeniz University, Campus, 07070 Antalya, Turkey
e-mail: cilerozenci@akdeniz.edu.tr

E. Seli and A. Agarwal (eds.), *Fertility Preservation: Emerging Technologies and Clinical Applications*, DOI 10.1007/978-1-4419-1783-6_19,

Testicular germ cell development consists of three fundamental periods: (1) the primordial germ cells (PGCs), the first cells of the germ line lineage in the embryo, (2) spermatogonial stem cells (SSCs), (3) and the spermatozoon, respectively. The first two periods cover the fetal and neonatal periods leading to the formation of the SSCs and include two main types of germ cells, PGCs and gonocytes. PGC differentiates into gonocytes which form type A spermatogonia. The third period is the spermatogenic cycle, a highly regulated chain of events, including mitosis and meiosis, and a differentiation process, such as spermiogenesis, starting with the formation of differentiating spermatogonia and ending with the production of spermatozoa [1]. Accordingly, the course of this chapter, testicular differentiation and spermatogenesis, will be clarified by the following three phases: fetal, neonatal–prepubertal, and postpubertal stages of testis development.

Bipotential Gonads

The primordium of the gonads, during the early stages of fetal life, is undifferentiated; later on sex-determining genes switch on or switch off the expression of genes that contribute to gonadal differentiations which induce the development of the undifferentiated gonads into a testis or an ovary [2]. Induction and patterning of the testis occurs over a brief window of time.

The origins of the gonads form as genital ridges that comprise thickened coelomic epithelium and are invaded by PGCs [3]. In mice, PGCs are first observed at days 7–8 of pregnancy, in the extra-embryonic mesoderm (Fig. 19.1). The specification of PGCs from pluripotent cells of the embryonic epiblast is dependent on bone morphogenetic protein (BMP) and WNT3 signaling molecules [4–6]. While undergoing repeated divisions, PGCs migrate through the hindgut and dorsal mesentery and arrive at the gonadal primordium on days 10–11 of pregnancy [3]. In humans, by the fourth week of gestation, the PGCs proliferate and migrate from the endoderm of the yolk sac into the undifferentiated gonads, which become morphologically distinct during the seventh week of gestation [7]. Now, the migrated PGCs in the gonads are called gonocytes.

Proliferating coelomic epithelial cells give rise to somatic cells of the differentiating testis which start to proliferate and migrate into the gonads (Fig. 19.1). They fill the gonads and then surround the germ cells, which leads to the formation of sex cords. When the somatic cell proliferation is inhibited, testis cord formation, which is important for the Sertoli cell maintenance, is altered [8]. In the gonad primordium, Sertoli cell precursors form the seminiferous cords at days 12–13 of pregnancy in mice. Within the newly developed cords, Sertoli cells start dividing again. At this stage, it becomes clear whether the gonad becomes a testis or ovary. In the testis, gonocytes proliferate until days 13–14 of pregnancy and then arrest their cell cycles. Their proliferation does not start again until a few days after birth.

Vertebrate gonads are composed of cortex and medulla. The distribution of germ cells in the sex cords is gender-specific [2]. During gonad differentiation into the ovary, germ cells stay in the peripheral region. Thus, the cortex develops better than the medulla and the sex cords disappear. During gonad differentiation into the testis, however, germ cells migrate with the developing sex cords into the gonads, and thus the medulla prevails over the cortex, the rudiment of which transforms into the tunica albuginea.

The organization of the testis structure is also dependent on the migration of mesonephric cells into the gonads (Fig. 19.1). Recent findings indicate that the great majority of cells immigrating into the gonads are endothelial cells [9], but not peritubular myoid cells (PTMCs) as suggested earlier [10–13]. Notably, the testis cord formation occurs at the same time as the development of the testis-specific vascular structure, the coelomic vessel and its branches [9, 14, 15]. Additionally, the onset of steroidogenesis is also critical for the developing testis.

In the ovary, cell proliferation, mesonephric cell migration, testis cord formation, and testis-specific vascular development, as well as steroidogenesis,

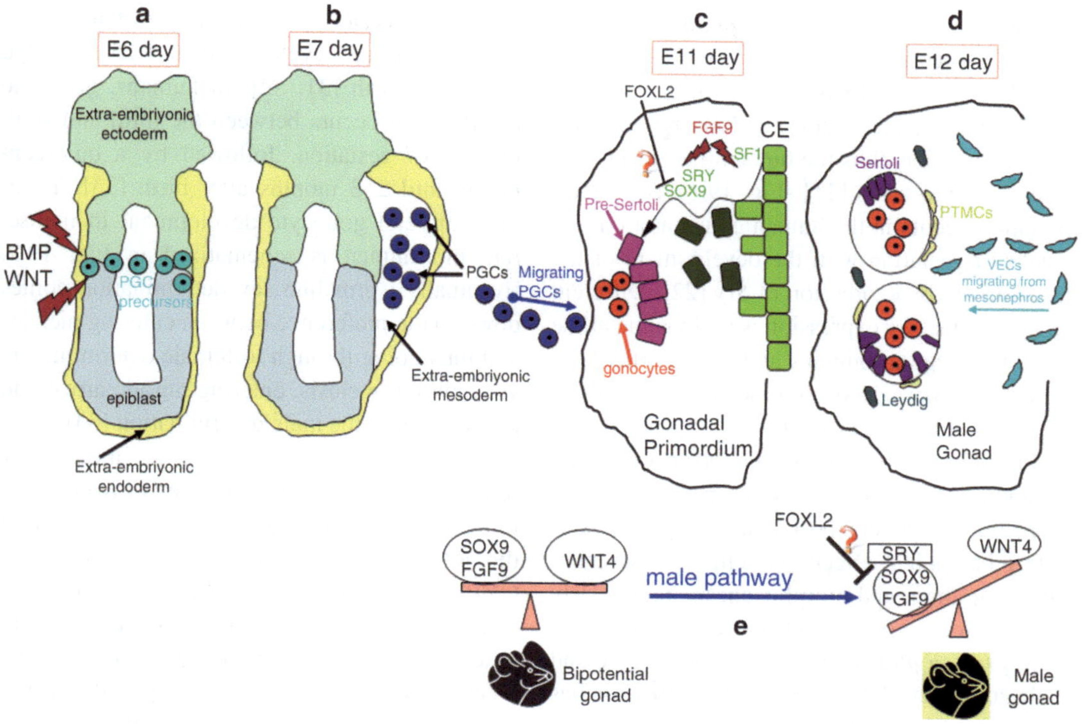

Fig. 19.1 Testicular differentiation during fetal period. (**a**) Bipotential stages cover PGC precursors derived from proximal epiblast cells under the influence of factors, such as BMP and WNT. (**b**) These precursor cells migrate to the extra-embryonic mesoderm, where they are called as PGCs. They continue their migration through the hindgut and dorsal mesentery and reach at the gonadal primordium. (**c**) The migrated PGCs in the gonads now are called gonocytes. Under some factors, such as SF1 and FGF9, coelemic epithelial cells that are indicated as *light green squares*, proliferate and migrate near the gonocytes. They start Sry and Sox9 expression, differentiate into pre-Sertoli cells and start surrounding the gonocytes. A recent question, whether inhibition Foxl2 would contribute to male-specific gene expression (Sry and Sox9) thus regulates Sertoli cell differentiation, needs further investigation. (**d**) Pre-Sertoli cells in the seminiferous cords start their proliferation again and then are called as Sertoli cells. These cells express AMH thus are immature and stop their proliferation and AMH expression while entering puberty. PTMCs surround the seminiferous cords and fetal Leydig cells are seen in the interstitial area between the cords. Vascular endothelial cells migrating from mesonephros contribute to seminiferous cord formation. (**e**) Tight control of the balance between male- and female-specific factors regulates bipotential gonad differentiation into testis. *PGC* primordial germ cell; *CE* coelemic epithelium; *PTMCs* peritubular myoid cells; *VECs* vascular endothelial cells

are inhibited. Primitive sex cords in the ovary disappear and transform into cysts, in which germ cells enter meiosis [2].

Proliferation of Somatic and Germ Cells

Testis determination in most mammals is regulated by a genetic hierarchy initiated by the sex-determining region Y (*Sry*) gene. Sry expression is well described in mice sex determination. About 10 day post coitus (dpc), gonads are bipotential but, at around 10.5 dpc and under the effect of the *Sry* gene, the XY gonad primordia start developing as testes and, in the absence of *Sry*, the XX gonad primordia begin to develop as ovaries [16] (Fig. 19.1). Proliferation, which expands the population of Sry-positive cells in the XY gonad, starts at 11.5 dpc and is evident by 12 dpc [8, 17]. *Sry* expression reaches a peak at 11.5 dpc and terminates before 12.5 dpc [18–20]. In the absence of proper expression of *Sry*, pre-supporting cells develop as granulosa cells [16]. Owing to the proliferation of somatic cells, the testis size becomes twice the size of the ovary after the sex-determining period [2].

Along with the upregulation of Sry-box 9 (*Sox9*) expression, Sry-positive cells turn into Sertoli cells [20]. In mice, *Sry* works together with steroidogenic factor 1 (*SF1*) and activates *Sox9* expression, thus the bipotential gonads differentiate into testis [21] (Fig. 19.1). Sex determination occurs in the human gonad at around 6 weeks of gestation with the development of the testis driven by expression of *Sry* [22]. Different from rodents, *Sry* expression is still visible at the postnatal stage in humans and pigs, particularly in humans; *Sry* is also expressed in germ cells as well as Sertoli cells [23, 24].

Fibroblast growth factors (FGFs) are expressed throughout embryonic development and in a more limited pattern postnatally [25]. FGF signaling has potent effects on cellular differentiation, migration, and morphogenesis, particularly *Fgf9* is necessary for Sertoli cell commitment (Fig. 19.1). *Fgf9* is expressed in both female and male gonads at 11.5 dpc, but *Fgf9* expression has disappeared in the female and enhanced in the male gonad shortly after *Sry* is expressed [26]. Fgf9 promotes the earliest phase of proliferation in coelomic epithelial SF1-positive cells via surface receptor Fgf receptor 2 (Fgfr2) [17, 27]. Fgfr2 is initially found in the cell membrane of coelomic epithelial cells. Fgfr2 seems to mediate proliferation at the surface of the male gonad. On the other hand, it is localized in the nuclei of cells dispersed in the interior of the male gonad but not in the female gonad, thus this receptor in nuclei can induce Sertoli cell differentiation [15, 28].

Recent studies challenge the view that early ovarian development is a default pathway switched on passively by the absence of *Sry* [29]. Interestingly, it has been shown that the mouse ovary continuously represses male-specific genes, from embryonic stages to adulthood [30]. Remarkably, the absence of the key ovarian transcription factor Forkhead box protein L2 (*Foxl2*) induces the activation of repressed male-specific genes during development, postnatally and even during adulthood (Fig. 19.1). Ablation of *Foxl2* in the adult transdifferentiates the supporting cells of the ovary to characteristics of the supporting cells of the testis.

As soon as gonocytes reach the developing male gonad on 10–11 dpc, they continue to proliferate for a short period of time and arrest in G0/G1 phase of the cell cycle at about 12.5–13.5 dpc until after birth [31, 32]. In humans, gonocyte proliferation occurs between the third and sixth months of gestation, followed by a quiescent phase until 2–3 months after birth [33]. Timelines of fetal gonocyte development in mouse, rat, and human is schematized in Fig. 19.2. Eventually, germ line sex determination comes down to the preference between entering meiosis and thus going through to female determination, vs. avoiding meiosis, entering mitotic arrest and progressing to the male determination [34].

It is apparent that meiosis is initiated by retinoic acid (RA), the active derivative of vitamin A, and is avoided at least largely because of the RA-degrading action of a P450 enzyme, CYP26B1 [35–39]. CYP26B1 is initially expressed in developing mice gonads of both sexes, but is then greatly upregulated in the testis and downregulated in the ovary, and thus has been determined as a male-specific meiosis-preventing factor [36, 37, 40]. In addition to the degradation of RA in the male fetal gonad, non-retinoid-secreted factors inhibiting both meiosis and mitosis and produced by the testicular somatic cells during fetal and neonatal life have been found [41]. Using a coculture model in which an undifferentiated female gonad is cultured with a fetal or neonatal testis, it has been demonstrated that the testis prevented the initiation of meiosis and induced male germ cell differentiation in the female gonad.

The source of RA is the mesonephroi to which the gonads are attached [36]. In mice gonads, RA seems to act by inducing the expression of stimulated by RA gene 8 (*Stra8*) [36, 37], a gene considered to be a premeiotic marker [40, 42, 43]. Although a role for RA in initiating meiosis in humans has not been demonstrated yet, meiosis initiates from gonad–mesonephros connection in many mammalian species, including humans [44, 45].

Formation of Sex Cords

The formation of testis cords, which is essential for the maintenance of male testis development,

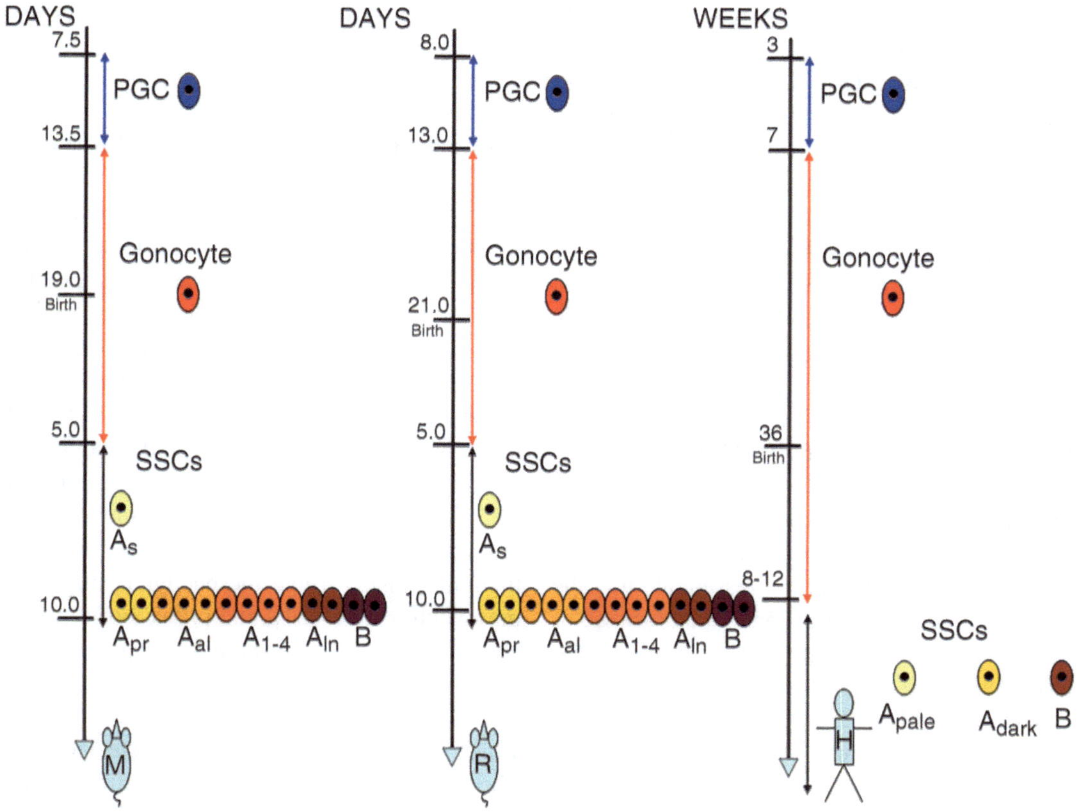

Fig. 19.2 Comparison of the time-lines of fetal and neonatal germ cell development in mouse, rat, and human. *PGCs* primordial germ cells; *SSCs* spermatogonial stem cells; A_s A_{single} spermatogonia; A_{pr} A_{paired} spermatogonia; A_{al} $A_{aligned}$ spermatogonia; $A_{1\text{–}4}$ $A_{1\text{–}4}$ spermatogonia; A_{In} $A_{intermediate}$ spermatogonia; *B* type B spermatogonia

takes place between 11.5 and 12.5 dpc in mice [2]. The creation of Sertoli cell aggregations by their proliferation is related to the expression of extracellular matrix (ECM) and membrane proteins. At this time, the synthesis of cytokeratins begins in Sertoli cells, while desmin is downregulated and becomes synthesized in PTMCs that surround the testis cords [46, 47]. An interaction between the peritubular and Sertoli cells is necessary for basal lamina formation and thus for maintaining the testis cords. It has recently been shown that PTMCs are induced within the gonads and differentiate from interstitial cells derived from the coelomic epithelium [48].

Proteases and their inhibitors, such as testatin, vanin-1, protease nexin-1 (Pn-1) and matricellular protein secreted protein, acidic and rich in cysteine (SPARC), may play key roles in gonadal differentiation. Testatin is specifically expressed in differentiating Sertoli cells in mice. The expression of vanin-1 and Pn-1 is upregulated during sex determination in male gonads [49]. Vanin-1 can regulate the cell-adhesion process, which is essential for proper migration and testis cord formation. Pn-1 may be responsible for the maintenance of the integrity of the basal lamina. SPARC is male-specifically expressed during testis development [50] and is localized intracellularly in Sertoli cells, germ cells, and in Leydig cells. SPARC binds to growth factors, platelet-derived growth factors (PDGFs) and FGFs, thereby blocks their action, inhibits cell division and differentiation and alters testis cord and coelomic vessel formation [51, 52].

Testis-specific vascular formation occurs at the same time with testis cord formation [14, 15, 53]. Individual endothelial cells, releasing from the mesonephric vascular plexus located near the gonad–mesonephros border, migrate into the developing testis. They form the wall of the coelomic vessel under the coelomic epithelium, along the cranio-caudal axis of the male gonad. The inhibition of the development of coelomic vessel branches results in the absence of testis cords, which shows that vascular structure is required for the initiation of cord division and also for maintaining the spaces between testis cords [9].

Migration of Mesonephric Cells

Inward migration of mesonephric cells is required for proper cord formation and for maintaining the Sertoli cell differentiation, essential for the growth of the male gonad [54]. The blockage of migration before 11.5 dpc in mice results in the disorganization of testis cords, as well as the coelomic vessel, whereas after 12.5 dpc mesonephric cell migration is not essential for further testis development [13].

PDGFs are diffusible extracellular ligands and they support mesonephric cell migration via their receptors (PDGFRα and PDGFRβ) both of which are expressed in the male gonad. PDGFRα is expressed in gonadal cells throughout the interstitium, and PDGFR is associated with the vasculature [55]. Mutation of *PDGFRa* disrupts signaling between cords and interstitium, thus causes disrupted morphogenesis and lack of fetal Leydig cells (FLCs) [56]. It seems that one of the phenotypes is a secondary result of the other.

Vascular endothelial growth factor (VEGF) may be another key factor in testis development, since blocking its action prevents endothelial cell migration in the testis, as well as coelomic vessel formation and testis cord development [57]. In addition to VEGF, neurotropin (NT) signaling appears to be another required growth factor for proper migration mechanism [58]. Although all of the NT family members and most of their receptors were detected in the embryonic male gonad, neurotropin 3 (NT3) and its receptor neurotrophic tyrosine kinase receptor type (TrkC) were detected in the majority [59].

Steroidogenesis in Embryonic Testis

Sex steroids are crucial regulators of sexual differentiation in males dissimilar to females [60]. Since Leydig cells are the primary major producers of these steroid hormones, maintenance of the normal functions of these cells determines normal development of the testis. FLCs appear in the embryonic gonad shortly after testis determination, and they are present in the mouse testis from 12.5 dpc until they disappear after birth [2]. These cells probably arise from multiple embryonic tissues, including the coelomic epithelium, gonadal ridge mesenchyme, migrating mesonephric cells [61], and probably from fetal adrenal cells [62]. In the human embryonic testis, the precursors of FLCs become functionally active as early as after 6–7 weeks of gestation, at which time testosterone can be detected [63]. The Leydig cells proliferate and differentiate progressively and constantly until they are mature before 19 week of gestation followed by regression [64].

Leydig cell differentiation requires two signaling molecules: PDGF secreted by Sertoli cells and acting via PDGFRα, and Sertoli cells secreted Desert Hedgehog (Dhh) morphogen acting via the Patched 1 (Ptch1) receptor [56, 65]. Although most data concerning the role of Dhh in the regulation of FLCs has been obtained in rodents, there are also reports on the significant role of Dhh signaling in the development of normal testicular phenotype in humans [66, 67]. Besides, PDGF pathway may be involved in the control of Leydig cell development in humans, similar to what has been observed in rodents [68]. Moreover, two additional factors, the transcription factor GATA-4 and the ligands of the insulin-like growth factor (IGF) system, may regulate FLC differentiation both in humans and rodents [69–72]. All of these factors seem to work together to control the differentiation, maturation, and regression of human FLCs.

From Birth to Puberty

From Neonatal Gonocyte to Spermatogonia: Gonocyte Proliferation, Differentiation, and Migration

Testicular gonocytes are the precursors of the SSCs and they correspond to the stage between the PGC and the SSC. However, it is worth mentioning that recent studies have shown that not all but only a fraction of the gonocytes directly differentiates into differentiating spermatogonia [73]. Quiescent fetal gonocytes resume proliferation after birth, and some migrate to the peripheral basement membrane of the seminiferous tubules, where they are termed as spermatogonia.

Neonatal gonocyte proliferation occurs between day post partum (dpp)1 and 4 in the mouse and dpp3 to 4 in the rat [33]. Proliferation in fetal and neonatal gonocytes seems to be differently regulated [74]. From the few studies that have questioned gonocyte proliferation, several factors have emerged as important, including PDGF-BB, 17β-estradiol (E2), leukemia inhibitory factor (LIF), and RA [33]. PDGF and E2 have been shown to be produced by Sertoli cells and are both required to activate gonocyte proliferation at dpp3 [75]. After 3 days in coculture with Sertoli cells, LIF increases the proportion of proliferating rat gonocytes of dpp1, whereas this effect is not seen when gonocytes of dpp3 are cultured [76]. RA is capable of stimulating neonatal rat gonocyte proliferation at dpp3 [77]. Moreover, RA can induce the number of proliferating human fetal gonocytes from 6- to 10-week-old fetuses after 4 days in organ culture [78].

Differences between a neonatal gonocyte and a spermatogonium are their morphological appearances and their different locations within the seminiferous cord, gonocytes being large spherical cells at the center of the cords while spermatogonia are smaller half-moon shaped cells positioned along the basement membrane [33]. In fact, location by itself is not a sufficient criterion and many of the proteins expressed in gonocytes are also present in spermatogonia. Recent advances in the characterization of SSCs have revealed new gene markers that might help discrimination between the two stages.

One of these genes Stra8, an indicator of gonocyte to spermatogonia transition, is expressed in spermatogonia and premeiotic cells, but not in postmeiotic or Sertoli cells [79]. A progressive increase of Stra8 in the germ cells of the first spermatogonial wave, starting with a few positive spermatogonia at dpp5, followed by an increasing number of positive preleptotenes and early leptotene spermatocytes over time has been shown in postnatal testis. Recently, RA has been identified as a key regulator of gonocyte differentiation in rats [80] and in mice [81]. RA makes this action by reducing the proportion of SSCs via decreasing growth factor receptor alpha 1 (GFRα1) transcript and by increasing the proportion of type A spermatogonia via increasing c-kit mRNA in these cells [80]. It may also be presumed that the same mechanisms might have a role in human gonocyte differentiation.

In order to differentiate, gonocytes should migrate from the center of the seminiferous cords to the basement membrane [82]. c-kit is transiently expressed in subsets of gonocytes presenting pseudopods and seems to regulate gonocyte migration toward the base of the cords [83]. After SSCs locate at the basement membrane of the cords c-kit is downregulated in these nonmigrating cells. ADAM-integrin-tetraspanin complexes, which are expressed in gonocytes, have been proposed as potential mediators of gonocyte migration [82]. A disintegrin and a metalloprotease domain (ADAM) 1 and ADAM 2 appear at dpp1 (ADAM1) and dpp2 (ADAM2) mice gonocytes [84]. Recently, Basciani et al. observed that a subset of the gonocytes migrated to the basement membrane regardless of the PDGFR and c-kit inhibitor, imatinib [85]. Migrated gonocytes could proliferate and differentiate, thus they could correspond to a pool of gonocytes that will differentiate into SSCs. Thus, at least two populations of neonatal gonocytes could coexist in the newborn testis [86–88]. First, gonocytes which will become SSCs and will populate the SSC niche. Second, a group of gonocytes that will differentiate to A pair stage of spermatogonia.

SSCs: Self Renewal, Differentiation, and SSC Niche

Self-renewal and differentiation of SSCs, a subpopulation of type A spermatogonia laying on the basement membrane of seminiferous tubules, are key events for normal spermatogenesis. A better understanding of molecular mechanisms in this population may help to distinguish factors which contribute to male infertility and testicular cancer.

There are several subtypes of A spermatogonia in rodents, including the A_{single} (A_s), the A_{paired} (A_{pr}), the $A_{aligned}$ (A_{al}), and the A_{1-4} spermatogonia based on differences in their morphology and phenotype [89, 90]. A_{single} spermatogonia act as stem cells, whereas interconnected spermatogonia are committed to differentiation and lose their stem cell potential. In mice, after mitotic divisions, cell cysts are formed by intercellular bridges between the cells (A_{al} spermatogonia) and they most probably enter meiosis synchronously. The A_{al} spermatogonia, in turn, give rise to several generations of spermatogonia, including type A_{1-4}, intermediate, and type B spermatogonia [89] (Fig. 19.2). This model, established on the basis of the morphological observation of fixed specimens, has provided the theoretical basis for germ line stem cell biology, regardless of the animal species. However, a logical shortcoming of this model is that it has not been derived from the behavior of cells. Type B cells generate spermatocytes, spermatids, and mature sperm.

In humans, A_{dark} and A_{pale} spermatogonia are distinguished by Clermont [91–94] (Fig. 19.2). It has been considered that A_{dark} cells are reserve stem cells, while the A_{pale} cells are renewing stem cells, a concept which is also adopted for monkey spermatogonia. An adult rhesus testis contains roughly equal numbers of A_{dark} and A_{pale} spermatogonia [95], where A_{dark} appear to fulfill the role of a "reserve stem cell." However, the functional identity of the human SSCs is still unknown. For extensive reading about key signaling molecules and their physiological roles in regulating cell fate decisions of spermatogonia, you may read the review by He et al. [96].

Intrinsic and extrinsic factors tightly control self-renewal and differentiation of SSCs (Fig. 19.3). Among the intrinsic regulators is promyelocytic leukemia zinc finger (Plzf), a transcriptional repressor expressed in the $A_s/A_{pr}/A_{al}$ population. In mice lacking Plzf, spermatogenesis is initially established but is then gradually lost within months, indicating that Plzf is crucial for the maintenance of stem cell activity rather than stem cell founding [97, 98]. Recently, it has been discovered that another intrinsic factor, Nanos2, plays a central role in the maintenance of self-renewal in mice SSCs [99, 100]. The researchers showed that Nanos2 is expressed in a majority of A_s and A_{pr} spermatogonia and its constitutive expression inhibits spermatogenic differentiation. On the other hand, depletion of Nanos2 causes a loss of stem cell maintenance.

In the animal body, self-renewal and differentiation of stem cells are under the control of a specialized microenvironment called the stem cell niche [101–103]. Existing knowledge suggests that the $A_s/A_{pr}/A_{al}$ population is localized at the vicinity of the blood vessels and interstitium around the seminiferous tubules and that they leave this niche upon differentiation into A_1 spermatogonia and migrate to the lumen of the tubule [87, 104, 105]. Although it needs further investigation; this region could be specialized to provide a microenvironmental niche for SSCs. On the other hand, factors, which have been secreted extracellularly, play essential roles in the stem cell-niche interactions.

Glial cell line-derived neurotrophic factor (GDNF), secreted by Sertoli cells, regulates the self-renewal and differentiation of mice SSCs in a dose-dependent manner. In vivo, GDNF overexpression causes the accumulation of undifferentiated spermatogonia, and on the contrary, GDNF ablation results in the reduction of spermatogonia numbers [106], suggesting that GDNF is vital for the self-renewal and maintenance of SSCs. GDNF acts via a receptor complex containing GDNF family receptor alpha-1 (GFRA1) and rearranged during transfection (RET), both of which are expressed in SSCs, Apr cells [107–109], and A_{al} spermatogonia [110]. Loss of GDNF/GFRA1 signaling triggers differentiation of SSCs into A_{1-4} spermatogonia [109].

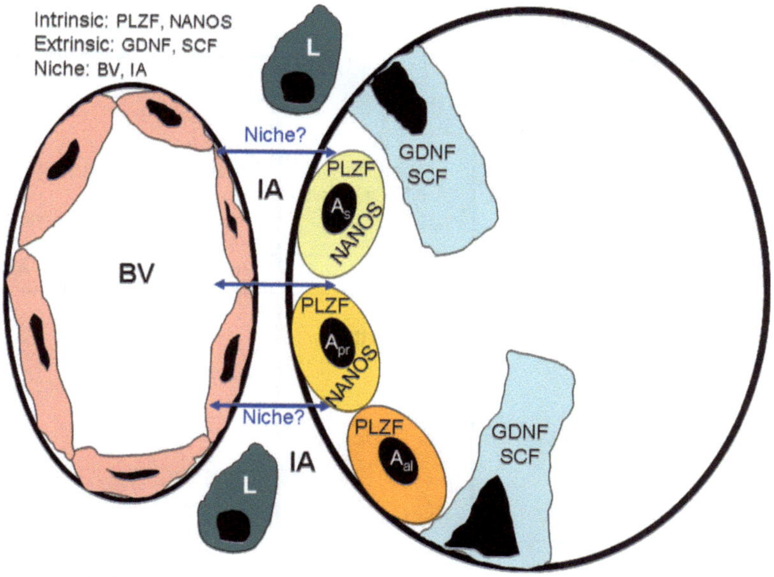

Fig. 19.3 Factors that control self-renewal and differentiation of SSCs. Some important intrinsic factors derived from spermatogonia, and extrinsic factors secreted from Sertoli cells are known to regulate the stem cell pool and differentiation. Since the testicular stem cell population at the base of the seminiferous tubules is localized close to blood vessels and interstitial area, it is possible to argue that this niche might contribute to their proliferation and differentiation processes as well. *BV* blood vessel; *L* Leydig cell; *IA* interstitial area

Stem cell factor (SCF, KIT ligand, steel factor) may play a diverse role in regulating the fate of mice type A_{1-4} spermatogonia [111]. After birth, there was a striking elevation in SCF transcript in whole mice testes at day 5, corresponding to the time when proliferation of Type A spermatogonia is initiated. SCF protein was present within Sertoli cells at 1–7 dpp and in the adult [112] suggesting that a significant change in Sertoli cell function occurs at this time, which is probably linked to the transition of gonocytes to spermatogonia and expansion of the latter population by mitosis.

SSC-specific marker has not been identified for any species so far [113]. However, the combined expression profiles of multiple markers provide information about stem, progenitor, and differentiating spermatogonia in rodents that may be useful for identifying primate SSCs. Expression of rodent spermatogonial markers (GFRA1, PLZF, NGN3, and KIT) were recently identified in the adult rhesus testis, and it has been found that most A_{dark} and almost 50% of A_{pale} exhibit the phenotype GFRA1$^+$, PLZF$^+$, NGN3$^-$, and KIT$^-$ those could be either stem or A_{al} progenitor type spermatogonia. Based on the conservation of molecular markers from rodents to primates, it has been proposed that the stem cell pool is considerably larger in rhesus than mice testes. In contrast to the large SSC pool, the relative small size of the progenitor pool (GFRA1$^+$, PLZF$^+$, NGN3$^+$, KIT$^-$) has been identified in adult macaques compared to rodents. Thus, it appears that rodents may have few SSCs and more progenitors while rhesus testes may have more SSCs and fewer progenitors.

When comparative analysis of Plzf, SSC maintenance marker, was carried out in mice, rhesus monkeys, and human testes, it was found that Plzf staining in human testes was restricted to some cells on the basement membrane of the seminiferous epithelium similar to mice and monkeys and the number of Plzf cells was more comparable to monkeys than mice. By analyzing expressions of POU5F1, TSPY, and KIT, other differences in marker expression have been reported between rodent and human spermatogonia, suggesting that the dynamics of the SSC pools in humans are similar to monkeys [114].

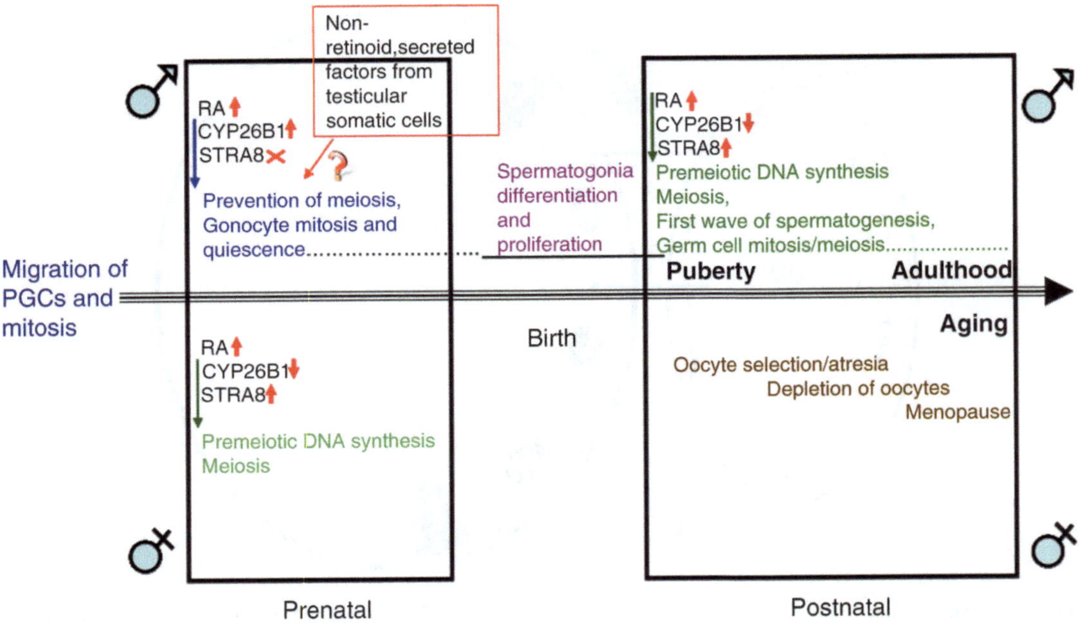

Fig. 19.4 Schematic representation of meiosis regulation during prenatal and postnatal stages in male and female

Initiation of Meiosis, Building the Barrier, and Entering Puberty

Many components of the meiosis are shared between males and females during meiotic divisions; however, as mentioned in earlier sections of this chapter, the timing and regulation of meiosis differ between the sexes [115, 116]. In females, meiotic divisions initiate during a brief window in embryonic development, whereas meiotic initiation in males is first observed at puberty and then reappears repeatedly and continuously throughout adulthood. In male mice, in addition to the mitotically active spermatogonia Stra8 is expressed in preleptotene spermatocytes, the most advanced cell type before meiotic prophase [43, 117]. It has been discovered that similar to the initiation of ovarian germ cell meiosis, Stra8 regulates meiotic initiation through RA induction of Stra8 in juvenile mice testes, phenomenon so prominently shared between two sexes, suggesting that the molecular control of premeiosis S-phase entry is conserved and gender-neutral [39] (Fig. 19.4).

The proliferation of type B spermatogonia results in the production of primary spermatocytes that enter and undergo the first meiotic division [118]. Meiosis is composed of prophase, metaphase, anaphase, and telophase. Prophase consists of the following defined cells: preleptotene spermatocytes, the first cells to be formed; leptotene spermatocytes, where chromosomes become more apparent as long threads; zygotene spermatocytes, where chromosomal pairing and synapsis is initiated; and pachytene spermatocytes, where pairing of chromosomes is completed. Desynapsis occurs in diplotene spermatocytes and chromosomes partially separate. Diakinesis involves shortening of chromosomes with each one composed of two chromatids. The synaptonemal complex is a proteinaceous scaffold that connects homologous chromosomes along their entire length to form synapsed bivalents during meiotic prophase I of spermatocytes. This complex mediates the synapsis of homologous chromosomes. Prophase is followed by metaphase where attachment of paired chromatids to the equator of the spindle takes place. Anaphase involves movement of paired chromatids to opposite poles of the spindle, and telophase gives rise to daughter cells with a haploid number of chromosomes with diploid DNA content called secondary spermatocytes. These cells have a short half life and they go through the second meiotic division to form spermatids with a haploid DNA content and a haploid chromosomal number. The duration of spermatocyte maturation takes around 21 days in rats and 25 days in humans [119].

Spermatogenesis requires coordination of germ cell movement and differentiation which essentially depends on interactions between Sertoli and germ cells. The leptotene spermatocytes must cross the blood-testis barrier (BTB) before the G2–MI transition of their cell cycle and enter the adluminal compartment of the seminiferous epithelium [120]. The movement of germ cells across the seminiferous epithelium requires distinct kinases and phosphatases. Cyclins and cyclin-dependent kinases, polo-like and aurora kinases, and the mitogen-activated protein kinase (MAPK) signaling pathway regulate the transit of primary spermatocytes across the BTB and contribute to BTB remodeling during germ cell divisions. Additionally, transforming growth factor β (TGF-β) can promote cell cycle arrest thus misregulation of its activity can affect cell division, disturb cell adhesion and increase cell movement. Cell cycle events are likely to participate in the movement of leptotene spermatocytes across the BTB. Cytokines and testosterone also have important roles during this process. Nevertheless, additional proteins and signaling cascades that regulate the entry of differentiating leptotene spermatocytes into the adluminal compartment are yet to be determined.

Puberty and Sperm Production

Spermatogenesis can be divided into premeiotic, meiotic, and postmeiotic phases. The premeiotic phase is characterized by an increase in cell numbers as a result of mitotic divisions of diploid spermatogonia. The meiotic prophase then leads to the formation of haploid round spermatids. The postmeiotic phase (spermiogenesis) involves the morphogenetic events that are necessary for sperm production.

Spermiogenesis is a metamorphosis process involving the maturation and differentiation of the early haploid male gamete to a mature spermatozoon. The spermiogenetic period takes about 2–3 weeks in mice [121] and 5–6 weeks in humans [91, 122]. During spermiogenesis of round spermatids, the nucleus takes on a more compact shape, the mitochondria are rearranged, the flagellum develops and the acrosome is generated [1].

Striking chromatin remodeling and genome reorganization takes place during the postmeiotic maturation of round spermatids. The regulation of gene expression in postmeiotic male germ cells is governed by specific mechanisms unique to these cells. In cAMP-dependent signaling pathway, gene expression is mediated primarily by two molecules: the cAMP-response element binding protein (CREB) and the cAMP-responsive element modulator (CREM) [123]. The transcriptional activator CREM is highly expressed in postmeiotic cells [124–126], and CREM deficiency results in the lack of postmeiotic cell-specific gene expression. CREM mutant male mice show postmeiotic arrest at the first step of spermiogenesis, where late spermatids were completely absent [127]. CREM activity is regulated through interactions with a germ cell-specific, the CREM phosphorylation-independent transcriptional coactivator, the activator of CREM in testis (ACT) [128]. The ability of ACT to regulate CREM activity is controlled by a germ cell-specific kinesin, Kif17b, which regulates the subcellular distribution of ACT. The CREM gene is expressed also in human germ cells and a switch from the expression of repressors to activators is present in normospermic men [129]. Conversely, in patients showing a testicular pattern of round spermatid maturation arrest only, CREM repressors are expressed. Thus, it has been proposed that CREM plays a role also in human spermatogenesis and that the absence of "the CREM switch" can be associated to spermatogenic arrest.

During spermiogenesis, spermatids repackage their DNA with highly basic arginine and cysteine rich protamines, but in almost all species a small residue of histone-bound DNA is kept and gains access to the ooplasm [130]. Despite the fundamental nature of these events, the molecular basis of the driving elements controlling them is not known. The entry of core histones alongside modified histones and histone variants to the egg leaves room for DNA and histone-based epigenetic signals that may be important for subsequent embryonic development. Recent evidence, arising from a closer examination of the composition of histone-bound and protamine-bound domains in human and murine spermatozoa, indicates that the former chromatin is indeed the more significant contributor to a novel epigenetic signal in these cells [131–135].

Although the sperm nucleus is transcriptionally quiet, it contains diverse RNA populations, mRNAs, antisense, and miRNAs that have been transcribed throughout spermatogenesis [136]. There is also an endogenous reverse transcriptase that may be activated under certain circumstances. It is now commonly accepted that sperm can deliver some RNAs to the oocyte at fertilization. Some recent data are provided, supporting the view that analyzing the profile of spermatozoal RNAs could be useful for the assessment of male fertility. The importance of paternal RNAs for spermatogenesis and subsequently for male fertility is yet to be determined.

Conclusions

Spermatogenesis is characterized by three specific phases: proliferation and differentiation, meiosis, and morphogenesis of haploid germ cells which takes approximately 1 month in mice and 2 months in humans. Sex determination occurs in the fetal gonads during development. Germ cells enter meiosis, and thus commit to the oogenic pathway, or stay away from meiosis, enter a state of quiescence and commit to the spermatogenic pathway. The main clinical consequence of defects in testicular differentiation during fetal periods is likely to be infertility. Another outcome of this condition in humans is increased susceptibility to testicular germ cell tumors, which have been proposed to originate from impaired or delayed germ cell differentiation during fetal testis development.

Chemotherapy and radiotherapy, even in low doses, may have a detrimental effect on the seminiferous epithelium and disrupt spermatogenesis in both children and adults [137]. Many chemotherapeutic drugs cross the BTB and act by damaging rapidly proliferating cells, therefore exerting their gonadotoxic effect. Exposure to chemotherapy may also result in the induction of sperm DNA fragmentation [138]. The extent of damage to germ cells depends on the class of chemotherapeutic agent, dosage, spermatogenetic stage targeted as well as the original pretreatment fertility potential of the patient. Cytotoxic therapy influences spermatogenesis at least temporarily and in some cases permanently. Recovery of spermatogenesis depends on the drugs used and on the cumulative dose given. Germ cells are also very sensitive to irradiation while the Leydig cells are more resistant owing to their slower rate of turnover [139]. Radiotherapy affects sperm concentration; moreover, irradiation increases sperm DNA damage [140].

In conclusion, the seeds of functional testes are built at prenatal life, progress during prepubertal ages, are upregulated after puberty and continue lifelong in males. Fortunately, mechanisms regulating these events are being discovered by researchers perpetually and much more still remain to be uncovered. The gonadotoxic side effect of cancer therapy is one of the challenges yet to be conquered. Since prepubertal age is not a silent period in testicular development, the effects of chemotherapy and radiotherapy on the prepubertal gonads seem just as detrimental as later in life.

References

1. Russell LDER, Sinha HAP, Clegg ED. Mammalian spermatogenesis. In: Russell LD, Ettlin RA, Sinha HAP, Clegg ED, editors. Histological and histopathological evaluation of the testis. Clearwater: Cache River Press; 1990. p. 1–38.
2. Piprek RP. Molecular and cellular machinery of gonadal differentiation in mammals. Int J Dev Biol. 2010;54(5):779–86.

3. McLaren A. Primordial germ cells in the mouse. Dev Biol. 2003;262(1):1–15.
4. Surani MA, Hayashi K, Hajkova P. Genetic and epigenetic regulators of pluripotency. Cell. 2007;128(4):747–62.
5. Saitou M. Specification of the germ cell lineage in mice. Front Biosci. 2009;14:1068–87.
6. Ohinata Y, Ohta H, Shigeta M, et al. A signaling principle for the specification of the germ cell lineage in mice. Cell. 2009;137(3):571–84.
7. Shalet SM. Normal testicular function and spermatogenesis. Pediatr Blood Cancer. 2009;53(2):285–8.
8. Schmahl J, Capel B. Cell proliferation is necessary for the determination of male fate in the gonad. Dev Biol. 2003;258(2):264–76.
9. Combes AN, Wilhelm D, Davidson T, et al. Endothelial cell migration directs testis cord formation. Dev Biol. 2009;326(1):112–20.
10. Buehr M, Gu S, McLaren A. Mesonephric contribution to testis differentiation in the fetal mouse. Development. 1993;117(1):273–81.
11. Martineau J, Nordqvist K, Tilmann C, et al. Male-specific cell migration into the developing gonad. Curr Biol. 1997;7(12):958–68.
12. Merchant-Larios H, Moreno-Mendoza N, Buehr M. The role of the mesonephros in cell differentiation and morphogenesis of the mouse fetal testis. Int J Dev Biol. 1993;37(3):407–15.
13. Tilmann C, Capel B. Mesonephric cell migration induces testis cord formation and Sertoli cell differentiation in the mammalian gonad. Development. 1999;126(13):2883–90.
14. Brennan J, Karl J, Capel B. Divergent vascular mechanisms downstream of Sry establish the arterial system in the XY gonad. Dev Biol. 2002;244(2):418–28.
15. Brennan J, Capel B. One tissue, two fates: molecular genetic events that underlie testis versus ovary development. Nat Rev Genet. 2004;5(7):509–21.
16. Koopman P, Gubbay J, Vivian N, et al. Male development of chromosomally female mice transgenic for Sry. Nature. 1991;351(6322):117–21.
17. Schmahl J, Eicher EM, Washburn LL, et al. Sry induces cell proliferation in the mouse gonad. Development. 2000;127(1):65–73.
18. Hacker A, Capel B, Goodfellow P, et al. Expression of Sry, the mouse sex determining gene. Development. 1995;121(6):1603–14.
19. Bullejos M, Koopman P. Spatially dynamic expression of Sry in mouse genital ridges. Dev Dyn. 2001;221(2):201–5.
20. Sekido R, Bar I, Narvaez V, et al. SOX9 is up-regulated by the transient expression of SRY specifically in Sertoli cell precursors. Dev Biol. 2004;274(2):271–9.
21. Sekido R, Lovell-Badge R. Sex determination involves synergistic action of SRY and SF1 on a specific Sox9 enhancer. Nature. 2008;453(7197):930–4.
22. Houmard B, Small C, Yang L, et al. Global gene expression in the human fetal testis and ovary. Biol Reprod. 2009;81(2):438–43.
23. Parma P, Pailhoux E, Cotinot C. Reverse transcription-polymerase chain reaction analysis of genes involved in gonadal differentiation in pigs. Biol Reprod. 1999;61(3):741–8.
24. Salas-Cortes L, Jaubert F, Barbaux S, et al. The human SRY protein is present in fetal and adult Sertoli cells and germ cells. Int J Dev Biol. 1999;43(2):135–40.
25. Itoh N. The Fgf families in humans, mice, and zebrafish: their evolutional processes and roles in development, metabolism, and disease. Biol Pharm Bull. 2007;30(10):1819–25.
26. Kim Y, Kobayashi A, Sekido R, et al. Fgf9 and Wnt4 act as antagonistic signals to regulate mammalian sex determination. PLoS Biol. 2006;4(6):e187.
27. Karl J, Capel B. Sertoli cells of the mouse testis originate from the coelomic epithelium. Dev Biol. 1998;203(2):323–33.
28. Kim Y, Bingham N, Sekido R, et al. Fibroblast growth factor receptor 2 regulates proliferation and Sertoli differentiation during male sex determination. Proc Natl Acad Sci USA. 2007;104(42):16558–63.
29. Veitia RA. FOXL2 versus SOX9: a lifelong “battle of the sexes”. Bioessays. 2010;32(5):375–80.
30. Uhlenhaut NH, Jakob S, Anlag K, et al. Somatic sex reprogramming of adult ovaries to testes by FOXL2 ablation. Cell. 2009;139(6):1130–42.
31. Hilscher B, Hilscher W, Bulthoff-Ohnolz B, et al. Kinetics of gametogenesis. I. Comparative histological and autoradiographic studies of oocytes and transitional prospermatogonia during oogenesis and prespermatogenesis. Cell Tissue Res. 1974;154(4):443–70.
32. McLaren A. Meiosis and differentiation of mouse germ cells. Symp Soc Exp Biol. 1984;38:7–23.
33. Culty M. Gonocytes, the forgotten cells of the germ cell lineage. Birth Defects Res C Embryo Today. 2009;87(1):1–26.
34. Bowles J, Koopman P. Sex determination in mammalian germ cells: extrinsic versus intrinsic factors. Reproduction. 2010;139(6):943–58.
35. Baltus AE, Menke DB, Hu YC, et al. In germ cells of mouse embryonic ovaries, the decision to enter meiosis precedes premeiotic DNA replication. Nat Genet. 2006;38(12):1430–4.
36. Bowles J, Knight D, Smith C, et al. Retinoid signaling determines germ cell fate in mice. Science. 2006;312(5773):596–600.
37. Koubova J, Menke DB, Zhou Q, et al. Retinoic acid regulates sex-specific timing of meiotic initiation in mice. Proc Natl Acad Sci USA. 2006;103(8):2474–9.
38. MacLean G, Li H, Metzger D, et al. Apoptotic extinction of germ cells in testes of Cyp26b1 knockout mice. Endocrinology. 2007;148(10):4560–7.
39. Anderson EL, Baltus AE, Roepers-Gajadien HL, et al. Stra8 and its inducer, retinoic acid, regulate meiotic initiation in both spermatogenesis and oogenesis in mice. Proc Natl Acad Sci USA. 2008;105(39):14976–80.
40. Menke DB, Page DC. Sexually dimorphic gene expression in the developing mouse gonad. Gene Expr Patterns. 2002;2(3–4):359–67.

41. Guerquin MJ, Duquenne C, Lahaye JB, et al. New testicular mechanisms involved in the prevention of fetal meiotic initiation in mice. Dev Biol. 2010;346(2):320–30.
42. Bouillet P, Oulad-Abdelghani M, Vicaire S, et al. Efficient cloning of cDNAs of retinoic acid-responsive genes in P19 embryonal carcinoma cells and characterization of a novel mouse gene, Stra1 (mouse LERK-2/Eplg2). Dev Biol. 1995;170(2):420–33.
43. Oulad-Abdelghani M, Bouillet P, Decimo D, et al. Characterization of a premeiotic germ cell-specific cytoplasmic protein encoded by Stra8, a novel retinoic acid-responsive gene. J Cell Biol. 1996;135(2):469–77.
44. Byskov AG. The role of the rete ovarii in meiosis and follicle formation in the cat, mink and ferret. J Reprod Fertil. 1975;45(2):201–9.
45. Bendsen E, Byskov AG, Andersen CY, et al. Number of germ cells and somatic cells in human fetal ovaries during the first weeks after sex differentiation. Hum Reprod. 2006;21(1):30–5.
46. Virtanen I, Kallajoki M, Narvanen O, et al. Peritubular myoid cells of human and rat testis are smooth muscle cells that contain desmin-type intermediate filaments. Anat Rec. 1986;215(1):10–20.
47. Frojdman K, Paranko J, Virtanen I, et al. Intermediate filaments and epithelial differentiation of male rat embryonic gonad. Differentiation. 1992;50(2):113–23.
48. Cool J, Carmona FD, Szucsik JC, et al. Peritubular myoid cells are not the migrating population required for testis cord formation in the XY gonad. Sex Dev. 2008;2(3):128–33.
49. Grimmond S, Van Hateren N, Siggers P, et al. Sexually dimorphic expression of protease nexin-1 and vanin-1 in the developing mouse gonad prior to overt differentiation suggests a role in mammalian sexual development. Hum Mol Genet. 2000;9(10):1553–60.
50. Wilson MJ, Bowles J, Koopman P. The matricellular protein SPARC is internalized in Sertoli, Leydig, and germ cells during testis differentiation. Mol Reprod Dev. 2006;73(5):531–9.
51. Yan Q, Sage EH. SPARC, a matricellular glycoprotein with important biological functions. J Histochem Cytochem. 1999;47(12):1495–150653.
52. Brekken RA, Sage EH. SPARC, a matricellular protein: at the crossroads of cell-matrix communication. Matrix Biol. 2001;19(8):816–27.
53. Coveney D, Cool J, Oliver T, et al. Four-dimensional analysis of vascularization during primary development of an organ, the gonad. Proc Natl Acad Sci USA. 2008;105(20):7212–7.
54. Capel B, Albrecht KH, Washburn LL, et al. Migration of mesonephric cells into the mammalian gonad depends on Sry. Mech Dev. 1999;84(1–2):127–31.
55. Cool J, Capel B. Mixed signals: development of the testis. Semin Reprod Med. 2009;27(1):5–13.
56. Brennan J, Tilmann C, Capel B. Pdgfr-alpha mediates testis cord organization and fetal Leydig cell development in the XY gonad. Genes Dev. 2003;17(6):800–10.
57. Cool J. Testis morphogenesis requires VEGF mediated endothelial migration via a novel mechanism of vascular remodelling. In: Fifth international symposium on vertebrate sex determination. Kona, Hawai; 2009.
58. Cupp AS, Kim GH, Skinner MK. Expression and action of neurotropin-3 and nerve growth factor in embryonic and early postnatal rat testis development. Biol Reprod. 2000;63(6):1617–28.
59. Levine E, Cupp AS, Skinner MK. Role of neurotropins in rat embryonic testis morphogenesis (cord formation). Biol Reprod. 2000;62(1):132–42.
60. Svechnikov K, Landreh L, Weisser J, et al. Origin, development and regulation of human Leydig cells. Horm Res Paediatr. 2010;73(2):93–101.
61. Habert R, Lejeune H, Saez JM. Origin, differentiation and regulation of fetal and adult Leydig cells. Mol Cell Endocrinol. 2001;179(1–2):47–74.
62. O'Shaughnessy PJ, Baker PJ, Johnston H. The foetal Leydig cell – differentiation, function and regulation. Int J Androl. 2006;29(1):90–5. discussion 105–8.
63. Tapanainen J, Kellokumpu-Lehtinen P, Pelliniemi L, et al. Age-related changes in endogenous steroids of human fetal testis during early and midpregnancy. J Clin Endocrinol Metab. 1981;52(1):98–102.
64. Haider SG. Cell biology of Leydig cells in the testis. Int Rev Cytol. 2004;233:181–241.
65. Pierucci-Alves F, Clark AM, Russell LD. A developmental study of the Desert hedgehog-null mouse testis. Biol Reprod. 2001;65(5):1392–402.
66. Canto P, Vilchis F, Soderlund D, et al. A heterozygous mutation in the desert hedgehog gene in patients with mixed gonadal dysgenesis. Mol Hum Reprod. 2005;11(11):833–6.
67. Fowler PA, Cassie S, Rhind SM, et al. Maternal smoking during pregnancy specifically reduces human fetal desert hedgehog gene expression during testis development. J Clin Endocrinol Metab. 2008;93(2):619–26.
68. Gnessi L, Basciani S, Mariani S, et al. Leydig cell loss and spermatogenic arrest in platelet-derived growth factor (PDGF)-A-deficient mice. J Cell Biol. 2000;149(5):1019–26.
69. Ketola I, Pentikainen V, Vaskivuo T, et al. Expression of transcription factor GATA-4 during human testicular development and disease. J Clin Endocrinol Metab. 2000;85(10):3925–31.
70. Bielinska M, Seehra A, Toppari J, et al. GATA-4 is required for sex steroidogenic cell development in the fetal mouse. Dev Dyn. 2007;236(1):203–13.
71. Rouiller-Fabre V, Lecref L, Gautier C, et al. Expression and effect of insulin-like growth factor I on rat fetal Leydig cell function and differentiation. Endocrinology. 1998;139(6):2926–34.
72. Berensztein EB, Baquedano MS, Pepe CM, et al. Role of IGFs and insulin in the human testis during post natal activation: differentiation of steroidogenic cells. Pediatr Res. 2008;63(6):662–5.
73. Yoshida S, Sukeno M, Nakagawa T, et al. The first round of mouse spermatogenesis is a distinctive

program that lacks the self-renewing spermatogonia stage. Development. 2006;133(8):1495–505.
74. Ohmura M, Naka K, Hoshii T, et al. Identification of stem cells during prepubertal spermatogenesis via monitoring of nucleostemin promoter activity. Stem Cells. 2008;26(12):3237–46.
75. Thuillier R, Wang Y, Culty M. Prenatal exposure to estrogenic compounds alters the expression pattern of platelet-derived growth factor receptors alpha and beta in neonatal rat testis: identification of gonocytes as targets of estrogen exposure. Biol Reprod. 2003;68(3):867–80.
76. De Miguel MP, De Boer-Brouwer M, Paniagua R, et al. Leukemia inhibitory factor and ciliary neurotropic factor promote the survival of Sertoli cells and gonocytes in coculture system. Endocrinology. 1996; 137(5):1885–93.
77. Livera G, Rouiller-Fabre V, Durand P, et al. Multiple effects of retinoids on the development of Sertoli, germ, and Leydig cells of fetal and neonatal rat testis in culture. Biol Reprod. 2000;62(5):1303–14.
78. Lambrot R, Coffigny H, Pairault C, et al. Use of organ culture to study the human fetal testis development: effect of retinoic acid. J Clin Endocrinol Metab. 2006;91(7):2696–703.
79. Giuili G, Tomljenovic A, Labrecque N, et al. Murine spermatogonial stem cells: targeted transgene expression and purification in an active state. EMBO Rep. 2002;3(8):753–9.
80. Wang Y, Culty M. Identification and distribution of a novel platelet-derived growth factor receptor beta variant: effect of retinoic acid and involvement in cell differentiation. Endocrinology. 2007;148(5): 2233–50.
81. Zhou Q, Li Y, Nie R, et al. Expression of stimulated by retinoic acid gene 8 (Stra8) and maturation of murine gonocytes and spermatogonia induced by retinoic acid in vitro. Biol Reprod. 2008;78(3):537–45.
82. Tres LL, Kierszenbaum AL. The ADAM-integrin-tetraspanin complex in fetal and postnatal testicular cords. Birth Defects Res C Embryo Today. 2005;75(2):130–41.
83. Orth JM, Qiu J, Jester WF, et al. Expression of the c-kit gene is critical for migration of neonatal rat gonocytes in vitro. Biol Reprod. 1997;57(3):676–83.
84. Rosselot C, Kierszenbaum AL, Rivkin E, et al. Chronological gene expression of ADAMs during testicular development: prespermatogonia (gonocytes) express fertilin beta (ADAM2). Dev Dyn. 2003; 227(3):458–67.
85. Basciani S, De Luca G, Dolci S, et al. Platelet-derived growth factor receptor beta-subtype regulates proliferation and migration of gonocytes. Endocrinology. 2008;149(12):6226–35.
86. Shinohara T, Orwig KE, Avarbock MR, et al. Germ line stem cell competition in postnatal mouse testes. Biol Reprod. 2002;66(5):1491–7.
87. Yoshida S, Sukeno M, Nabeshima Y. A vasculature-associated niche for undifferentiated spermatogonia in the mouse testis. Science. 2007;317(5845):1722–6.
88. Seandel M, Falciatori I, Shmelkov SV, et al. Niche players: spermatogonial progenitors marked by GPR125. Cell Cycle. 2008;7(2):135–40.
89. de Rooij DG, Grootegoed JA. Spermatogonial stem cells. Curr Opin Cell Biol. 1998;10(6):694–701.
90. Yoshida S. Stem cells in mammalian spermatogenesis. Dev Growth Differ. 2010;52(3):311–7.
91. Clermont Y. The cycle of the seminiferous epithelium in man. Am J Anat. 1963;112:35–51.
92. Clermont Y. Renewal of spermatogonia in man. Am J Anat. 1966;118(2):509–24.
93. Clermont Y. Spermatogenesis in man. A study of the spermatogonial population. Fertil Steril. 1966;17(6): 705–21.
94. Clermont Y. Kinetics of spermatogenesis in mammals: seminiferous epithelium cycle and spermatogonial renewal. Physiol Rev. 1972;52(1): 198–236.
95. Marshall GR, Plant TM. Puberty occurring either spontaneously or induced precociously in rhesus monkey (Macaca mulatta) is associated with a marked proliferation of Sertoli cells. Biol Reprod. 1996;54(6):1192–9.
96. He Z, Kokkinaki M, Dym M. Signaling molecules and pathways regulating the fate of spermatogonial stem cells. Microsc Res Tech. 2009;72(8):586–95.
97. Buaas FW, Kirsh AL, Sharma M, et al. Plzf is required in adult male germ cells for stem cell self-renewal. Nat Genet. 2004;36(6):647–52.
98. Costoya JA, Hobbs RM, Barna M, et al. Essential role of Plzf in maintenance of spermatogonial stem cells. Nat Genet. 2004;36(6):653–9.
99. Sada A, Suzuki A, Suzuki H, et al. The RNA-binding protein NANOS2 is required to maintain murine spermatogonial stem cells. Science. 2009; 325(5946):1394–8.
100. Suzuki H, Sada A, Yoshida S, et al. The heterogeneity of spermatogonia is revealed by their topology and expression of marker proteins including the germ cell-specific proteins Nanos2 and Nanos3. Dev Biol. 2009;336(2):222–31.
101. Spradling A, Drummond-Barbosa D, Kai T. Stem cells find their niche. Nature. 2001;414(6859): 98–104.
102. Ohlstein B, Kai T, Decotto E, et al. The stem cell niche: theme and variations. Curr Opin Cell Biol. 2004;16(6):693–9.
103. Morrison SJ, Spradling AC. Stem cells and niches: mechanisms that promote stem cell maintenance throughout life. Cell. 2008;132(4):598–611.
104. Chiarini-Garcia H, Hornick JR, Griswold MD, et al. Distribution of type A spermatogonia in the mouse is not random. Biol Reprod. 2001;65(4):1179–85.
105. Chiarini-Garcia H, Raymer AM, Russell LD. Non-random distribution of spermatogonia in rats: evidence of niches in the seminiferous tubules. Reproduction. 2003;126(5):669–80.
106. Meng X, Lindahl M, Hyvonen ME, et al. Regulation of cell fate decision of undifferentiated spermatogonia by GDNF. Science. 2000;287(5457):1489–93.

107. Hofmann MC, Braydich-Stolle L, Dym M. Isolation of male germ-line stem cells; influence of GDNF. Dev Biol. 2005;279(1):114–24.
108. Naughton CK, Jain S, Strickland AM, et al. Glial cell-line derived neurotrophic factor-mediated RET signaling regulates spermatogonial stem cell fate. Biol Reprod. 2006;74(2):314–21.
109. He Z, Jiang J, Hofmann MC, et al. Gfra1 silencing in mouse spermatogonial stem cells results in their differentiation via the inactivation of RET tyrosine kinase. Biol Reprod. 2007;77(4):723–33.
110. Tokuda M, Kadokawa Y, Kurahashi H, et al. CDH1 is a specific marker for undifferentiated spermatogonia in mouse testes. Biol Reprod. 2007;76(1):130–41.
111. Yoshinaga K, Nishikawa S, Ogawa M, et al. Role of c-kit in mouse spermatogenesis: identification of spermatogonia as a specific site of c-kit expression and function. Development. 1991;113(2):689–99.
112. Loveland KL, Schlatt S. Stem cell factor and c-kit in the mammalian testis: lessons originating from Mother Nature's gene knockouts. J Endocrinol. 1997;153(3):337–44.
113. Hermann BP, Sukhwani M, Hansel MC, et al. Spermatogonial stem cells in higher primates: are there differences from those in rodents? Reproduction. 2010;139(3):479–93.
114. Dym M, Kokkinaki M, He Z. Spermatogonial stem cells: mouse and human comparisons. Birth Defects Res C Embryo Today. 2009;87(1):27–34.
115. Hunt PA, Hassold TJ. Sex matters in meiosis. Science. 2002;296(5576):2181–3.
116. Morelli MA, Cohen PE. Not all germ cells are created equal: aspects of sexual dimorphism in mammalian meiosis. Reproduction. 2005;130(6):761–81.
117. Zhou Q, Li Y, Nie R, et al. Expression of stimulated by retinoic acid gene 8 (Stra8) and maturation of murine gonocytes and spermatogonia induced by retinoic acid in vitro. Biol Reprod. 2008;79:35–42.
118. Hermo L, Pelletier RM, Cyr DG, et al. Surfing the wave, cycle, life history, and genes/proteins expressed by testicular germ cells. Part 1: background to spermatogenesis, spermatogonia, and spermatocytes. Microsc Res Tech. 2010;73(4):241–78.
119. Cobb J, Handel MA. Dynamics of meiotic prophase I during spermatogenesis: from pairing to division. Semin Cell Dev Biol. 1998;9(4):445–50.
120. Lie PP, Cheng CY, Mruk DD. Coordinating cellular events during spermatogenesis: a biochemical model. Trends Biochem Sci. 2009;34(7):366–73.
121. Oakberg EF. Duration of spermatogenesis in the mouse and timing of stages of the cycle of the seminiferous epithelium. Am J Anat. 1956;99(3):507–16.
122. Heller CG, Clermont Y. Spermatogenesis in man: an estimate of its duration. Science. 1963;140:184–6.
123. Don J, Stelzer G. The expanding family of CREB/CREM transcription factors that are involved with spermatogenesis. Mol Cell Endocrinol. 2002;187(1–2):115–24.
124. Foulkes NS, Mellstrom B, Benusiglio E, et al. Developmental switch of CREM function during spermatogenesis: from antagonist to activator. Nature. 1992;355(6355):80–4.
125. Foulkes NS, Schlotter F, Pevet P, et al. Pituitary hormone FSH directs the CREM functional switch during spermatogenesis. Nature. 1993;362(6417):264–7.
126. Delmas V, van der Hoorn F, Mellstrom B, et al. Induction of CREM activator proteins in spermatids: down-stream targets and implications for haploid germ cell differentiation. Mol Endocrinol. 1993;7(11):1502–14.
127. Nantel F, Monaco L, Foulkes NS, et al. Spermiogenesis deficiency and germ-cell apoptosis in CREM-mutant mice. Nature. 1996;380(6570):159–62.
128. Hogeveen KN, Sassone-Corsi P. Regulation of gene expression in post-meiotic male germ cells: CREM-signalling pathways and male fertility. Hum Fertil (Camb). 2006;9(2):73–9.
129. Peri A, Serio M. The CREM system in human spermatogenesis. J Endocrinol Invest. 2000;23(9):578–83.
130. Miller D, Brinkworth M, Iles D. Paternal DNA packaging in spermatozoa: more than the sum of its parts? DNA, histones, protamines and epigenetics. Reproduction. 2010;139(2):287–301.
131. Gardiner-Garden M, Ballesteros M, Gordon M, et al. Histone- and protamine-DNA association: conservation of different patterns within the beta-globin domain in human sperm. Mol Cell Biol. 1998;18(6):3350–6.
132. Wykes SM, Krawetz SA. The structural organization of sperm chromatin. J Biol Chem. 2003;278(32):29471–7.
133. Moulson CL, Fong LG, Gardner JM, et al. Increased progerin expression associated with unusual LMNA mutations causes severe progeroid syndromes. Hum Mutat. 2007;28(9):882–9.
134. Arpanahi A, Brinkworth M, Iles D, et al. Endonuclease-sensitive regions of human spermatozoal chromatin are highly enriched in promoter and CTCF binding sequences. Genome Res. 2009;19(8):1338–49.
135. Hammoud SS, Nix DA, Zhang H, et al. Distinctive chromatin in human sperm packages genes for embryo development. Nature. 2009;460(7254):473–8.
136. Dadoune JP. Spermatozoal RNAs: what about their functions? Microsc Res Tech. 2009;72(8):536–51.
137. Mitchell RT, Saunders PT, Sharpe RM, et al. Male fertility and strategies for fertility preservation following childhood cancer treatment. Endocr Dev. 2009;15:101–34.
138. Morris ID. Sperm DNA damage and cancer treatment. Int J Androl. 2002;25(5):255–61.
139. Shalet SM, Tsatsoulis A, Whitehead E, et al. Vulnerability of the human Leydig cell to radiation damage is dependent upon age. J Endocrinol. 1989;120(1):161–5.
140. Stahl O, Eberhard J, Jepson K, et al. Sperm DNA integrity in testicular cancer patients. Hum Reprod. 2006;21(12):3199–205.

Impact of Radiotherapy and Chemotherapy on the Testis

20

Carolina Ortega and Herman Tournaye

Abstract

Nowadays, anticancer treatments involving either chemotherapy or radiotherapy have become increasingly successful and the survival rates have dramatically improved in the last 30 years from less than 20% to nearly 80%. However, according to the doses used and the duration of the treatment, 10–100% of surviving cancer patients will show reduced semen parameters and overall 15–30% will remain sterile on the long term. Interindividual variability exists in the spermatogenetic recovery after any gonadotoxic treatment making any individual fertility prognosis virtually impossible. Furthermore, while the initial treatment is known when starting cancer therapy, eventually the treatment regimen may be changed making any assessment of the risk for sterility even more difficult. Therefore, sperm cryopreservation should be offered routinely to all male patients exposed to gonadotoxic treatments. Age must not be a discriminative parameter since different methods to obtained semen sample are available to obtain sperm from postpubertal patients. When patients have been cured, assisted reproduction, e.g., intracytoplasmic sperm injection can be used to offer patients the best chances to father their genetically own children. In prepubertal boys, testicular stem cell banking may be offered.

Keywords

Radiotherapy and the testis • Chemotherapy and the testis • Infertility in the male • Testis and chemotherapy • Fertility in men • Pregnancy after cancer treatment

H. Tournaye (✉)
Centre for Reproductive Medicine, University Hospital,
Dutch-speaking Brussels Free University,
Laarbeeklaan 101, 1090 Brussels, Belgium

E. Seli and A. Agarwal (eds.), *Fertility Preservation: Emerging Technologies and Clinical Applications*,
DOI 10.1007/978-1-4419-1783-6_20, © Springer Science+Business Media, LLC 2012

Thanks to improvements in therapeutic regimens, around 70–75% of children treated for cancer will survive and grow to adulthood [1]. It has been estimated that approximately 200,000 people living in the USA [2] and one in 715 people in the UK are childhood cancer survivors [3]. Approximately, half of childhood cancers are hematologic malignancies (leukemia and lymphoma) and the anticipated long-term survival for children with these disorders is nowadays exceeding 75% [4]. Testicular cancer represents about 1% of all cancers in men, but it is the most common solid organ tumor between 25 and 35 years of age with an incidence of 3–6 new cases occurring per 100,000 males per year in Western society and up to 9–10 new cases per year in Germany and Scandinavian countries [5]. The use with chemotherapy and radiotherapy combined with surgical techniques has enabled the cure of 90% of the cases [6] to the extent that testicular cancer therapy is one of the outstanding successes in medicine today [7–9].

Unfortunately, during cancer treatment different organ systems are affected, including cardiac, pulmonary, renal, hepatic, and endocrine system. Since the quality of life has become an important factor, one of the major concerns of patients treated for cancer is infertility related to either the malignancy itself or more frequently, the treatment.

Adverse Effects on the Testis as an Endocrine Organ

The testis is highly susceptible to the toxic effects of radiation and chemotherapy at all ages of life. It is one of the most radiosensitive tissues, and damage can be caused by direct irradiation or from scattered irradiation at adjacent tissues [10]. After chemotherapy, testicular damage is drug specific and dose related [11, 12]. The age, at which chemotherapy and radiotherapy are carried out, is equally important. It has been suggested that the germinal epithelium of the prepuberal testis is less susceptible to damage than the adult testis [13]. However, if the doses of chemotherapy are calculated on a per square meter and calculating radiation doses to the gonad given during childhood and adolescence, some chemotherapy agents and radiotherapy doses that produce nonreversible azoospermia in those patients, seems to be the same as those for adults [14]. Two important endocrinologically active cells are found in the testis, i.e., the Sertoli cells (SCs) and the Leydig cells (LCs).

Effects on Sertoli Cell Function

The SC has a specific role during fetal life in the testicular development and sexual differentiation by secreting anti-Müllerian hormone which ensures the regression of the Müllerian ducts and by secreting testosterone to develop a male phenotype. The SC proliferates during two specific periods of life, i.e., during the fetal and neonatal period (up to 12–18 months postpartum) and during the pubertal period. After the neonatal period, the SC enters into a quiescent period which changes into an active replicating period at puberty in order to support further spermatogenesis during adult life [15]. During puberty, SC switches from an immature proliferative stage to a mature nonproliferative stage. Adult testis size and daily spermatozoa production are related to the number of SCs determining as such the number of germ cells that can be produced [16, 17].

Cytotoxic treatments can only cause damage to the SC population when these cells are replicating, and as a result in adults SCs are radio- and chemotherapy resistant [16, 18].

Yet in a recent case report, possible Sertoli-cell damage because of chemotherapy has been reported [19]. The authors demonstrated that a patient who underwent chemotherapy during puberty shows SC immunopositive for CK-18 during adulthood in approximately 13% of SC in tubules with impaired spermatogenesis. While SCs are relatively resistant to chemotherapy, the presence of CK-18 may reflect a dedifferentiation of SC as a consequence of testicular damage produced by chemotherapy treatment.

Effects on Leydig Cell Function

The LC produces both testosterone and insulin-like factor 3, playing a role in downstream events of masculinization and in the descent of the testes into the scrotum in fetal life [20, 21]. At puberty, the production of testosterone increases thereby inducing the secondary sex differentiation. Testosterone output then gradually decreases from age 30 [22].

Chemotherapy may have a direct toxic effect on the LC. It has been reported that after chemotherapy 59–75% of men have increased luteinizing hormone (LH) levels in the presence of low or normal testosterone [23–26]. Some of these men with subnormal testosterone levels may also show reduced bone density, body composition, and disturbance of quality of life [27, 28]. However, there is an evidence of recovery of LC function during the first 10 years after chemotherapy [29].

Regarding the effect of radiotherapy, LC function may be more prone to damage from irradiation in prepubertal life compared with adulthood [30]. Patients undergoing single dose of above 0.75 Gy [13] produce significant rises in LH which is also seen with fractionated doses of radiation of above 2 Gy [31], although values gradually return to normal levels over 30 months and testosterone values are not altered with these doses [28]. However, doses above 30 Gy produce significant decrease in testosterone serum values [30].

Adverse Effects on the Testis as an Exocrine Organ

Spermatogenesis, the exocrine function of the testis, starts at puberty and continues throughout adult life. Pluripotent spermatogonial stem cell undergoes self-renewal, apoptosis, and differentiation into mature spermatozoa [17]. On average, the total number of germ cells during childhood varies from 13 to 83×10^6 [32]. Two different populations of spermatogonial stem cells exist, i.e., the A_{dark} which are reserve stem cells with a low mitotic activity, and the A_{pale} which are the true male germline progenitor cells with a high mitotic activity. A_{pale} can differentiate into B spermatogonia around 4–5 years of age after which their number slowly increases until the age of 8–9 years. While type B spermatogonia are being produced, the true stem cell population gets refueled since half of the sister cells remain A_{pale} spermatogonia [33]. Then, a marked spermatogonial proliferation period starts. The development of the testes can lead to postpubertal sperm counts of up to several hundred million (or more) sperm per ml of semen depending on the adult testicular volume reached [34].

Effects of Radiotherapy on the Exocrine Function

Radiotherapy can be applied either by electromagnetic radiation (X-rays) and corpuscular radiation (electrons) produced by a linear accelerator or by rays generated by the decay of the cobalt 60 radioisotope. It is well documented, that impairment of spermatogenesis following singledose radiation depends on the dose and the age in which the patient underwent radiation therapy but impairment on the number of spermatozoa amount is always observed: oligozoospermia can be observed with doses as low as 0.1 Gy due to damage at the spermatogonial stage while azoospermia is observed when the patient is irradiated with doses above 2 Gy. With doses of 2–3 Gy, the spermatocytes are being damaged and azoospermia is observed 60 days after the onset of the treatment and at doses of 4–6 Gy, damage occurs up to the spermatid stage, and thus azoospermia is observed earlier [13]. Recovery of spermatogenesis is observed depending on the total dose of irradiation, on the survival of the spermatogonial stem cells of both "true" and "reserve" compartments and their ability to further differentiate. Hence, a complete recovery of sperm concentration is seen after 9–18 months with doses below 1 Gy [35], after 30 months with doses between 2 and 3 Gy and after 5 years with doses above 4 Gy [13]. However, doses above 20 Gy causing also LC damage are irreversible [35].

Many patients, however, will undergo fractionated radiotherapy which induces more gonadal toxicity compared to single-dose radiotherapy. Several studies showed that patients who received fractionated irradiation with doses between 1 and 3 Gy became azoospermic, and the recovery of spermatogenesis following the treatment was observed only in few patients [36–38]. Radiation doses of 1.2 Gy are considered as a threshold for permanent testicular damage without recovery in most of the patients. Yet in patients with spermatogenesis recovering after radiotherapy, the frequencies of numerical and structural chromosomal abnormalities were found to be increased in relation to the radiation dose used [39].

Total body irradiation is associated with even higher gonadal toxicity. Eighty percent of patients who underwent TBI after stem cell transplantation will have permanent gonadal failure [40] with only 0.5–1% of patients who received, respectively, 12 and 10 Gy or more during TBI, achieving fatherhood [41].

Effects of Chemotherapy on the Exocrine Function

Nowadays, different types of chemotherapeutical agents are used for cancer treatment. They interfere with DNA and RNA synthesis, inhibit protein synthesis or prevent microtubule functions, all vital for successful cell division. Effects of chemotherapy depend on the dose and the agent used with alkylating agents being most aggressive, e.g., procarbazine and cyclophosphamide.

Testicular damage caused by cytotoxic drugs was first described in 1948, when azoospermia was reported in 27 of 30 men following the treatment with nitrogen mustard [42]. The use of chemotherapy and radiotherapy produce a combination of destruction of the proliferating germ cell pool and inhibition of further differentiation of the survival germs cells. Different radiotherapy and chemotherapy regimens vary widely in their effect on fertility and can lead to effects from temporary oligozoospermia to permanent azoospermia. Damage to the other aspects of sperm function, such as loss of motility or morphological abnormalities, are less pronounced. If spermatozoa are produced after therapy, their motility and percentage that exhibit normal morphology are restored to pretreatments levels [43, 44]. However, 15–30% remains sterile in the long term [45]. The American Society for Clinical Oncology has published a document including an overview of the anticipated sperm quality after various chemotherapeutic drugs and guidelines on fertility preservation promoting cryobanking whenever possible [46]. A working party of the UK Royal Colleges in 2007 recommended that all patients who required anticancer treatment should be fully informed about the potential gonadotoxic side effects at the time of the diagnosis and before potentially gonadotoxic treatments [47]. Unfortunately, up to 40% of adult patients are not informed about the sterilizing side effects of their treatment [48] and the same figure is reported in children cancer [49]. The latter survey also shows that in the subpopulation of children at high risk for sterility only in 14% of cases the problem of sterility is discussed with the parents [49].

When cytotoxic drugs affect replicating spermatogonial stem cells, DNA damage can probably be repaired and the risk for transmissible mutations may be low. The cytotoxic agents also affect the differentiating germinal cells. If the damage is produced at the spermatocyte stage, DNA damage can be repaired. When chemotherapy affects at spermatids or testicular spermatozoa transmissible gene mutations and chromosomal aberrations can result because at these stages, DNA damage cannot be repaired [14].

Pharmacological Prevention

Methods to prevent the sterilizing side effects of radio- and chemotherapy on sperm production and improve gonadal function after cancer treatment are extremely important and now more emphasis is put on the quality of life after the cancer cure.

Several biochemical and biological approaches have been studied in animal models to protect the testis against cytotoxic agents [50]. These methods focus on hormonal modulation to prevent or reverse damage to the germ line during cancer treatment. This strategy is based on the observation that in rodents, germinal epithelium cells are more resistant to cytotoxic agents in their nonreplicating state rather than in their proliferative state. Suppression therapy with gonadotropin and testosterone in rats before or after cytotoxic agents considerably improve spermatogenesis and fertility [51]. Some clinical trials have been performed in men using suppression hormonal therapy. However, this treatment in men did not improve sperm count and fertility as in rats [52–55]. Hence, the application of these procedures to humans remains uncertain and more studies must be done.

Prevention by Gamete Preservation

A large number of cancer patients survive after radio and chemotherapy treatments. These patients expect the quality of life, including fatherhood. The American Society of Clinical Oncology (ASCO) recommends that fertility treatments must be discussed with patients and advices depend on the patient's age, disease, prognosis, and time interval before treatment. Hence, all patients who undergo anticancer treatment may be considered for cryobanking. Semen cryopreservation should be performed preferentially before starting any treatment. However, semen sample can be taken even when treatment has already started [56]. One single semen sample, even of limited quality, is sufficient to perform several intracytoplasmic sperm injections (ICSIs) cycles and urgency for starting chemotherapy has become an unlikely excuse.

Semen cryostorage has been shown to be an effective strategy and is recommended by the National Institute for Health and Clinical Excellence (NICE) and ASCO [46, 57]. In adolescent boys, sperm can be obtained by masturbation [58].

In the situation of ejaculation failure, the search for spermatozoa in a urine sample could be proposed [59, 60] and eventually semen can be obtained using penile vibratory stimulation or rectal electrostimulation technique under general anesthesia. When spermatozoa cannot be obtained using the techniques described above, a fine needle aspiration (FNA) or testicular sperm extraction (TESE) can be performed.

In prepubertal boys having immature testis, there are no options currently available for fertility preservation at diagnosis and options remain experimental. Testicular biopsy (TESE) and spermatogonial stem cell cryopreservation might be an option for those prepubertal patients [61, 62].

In a mouse model, it has been shown that testicular germ cells transplanted into a sterile mice, can restore fertility and reproduction in vivo [63, 64].

One of the mayor concerns, however, is the possibility of contamination of testicular tissue or spermatogonial stem cells that could be transplanted back and reexposing him to cancer later on [65]. Nowadays, there is no safe technology available preventing transmission of malignant cells by this novel strategy [66].

Attitudes of Physicians, Parents, and Patients Toward Preservation

Fertility problems have a major impact on the future quality of life of patients and physicians and parents are aware of it. However, discussing the storage of sperm of an adolescent with cancer is a challenging aspect of pediatric oncology care. A study carried out in The Netherlands showed that physicians always wanted a separated discussion with the adolescent because of the sensitive nature of the topic [67]. Ginsberg et al. demonstrated that parents of prepubertal boys are willing to pursue testicular cryopreservation at diagnosis even if the diagnosis is stressful. Factors, such as religion, finance, ethics and experimental nature of cryopreservation, did not play a major role in decision-making [62]. Concerning child participation during the decision-making process, a study carried out by de Vries et al. parents and physicians do not have the same opinion. All physicians agree that children should participate in the decision-making; however, some parents doubted whether the issue should be discussed with their son [67]. And although children can have some ideas about the effects of cancer treatment on them fertility, normally their parents are more worried about preserving reproductive function. Regarding the in vitro techniques that can be used in patients who have been treated for cancer, a questionnaire sent to oncologists in Minnesota showed that 74% of the responding oncologists were unaware of recent advances in assisted reproduction and did not know about the existence of ICSI [68] and an Irish survey corroborates these findings [69]. Unfortunately, oncologists offer semen cryopreservation to <25% of their adolescent patients [48]. Risk of fertility is only discussed with 76% of postpubertal and 61% of prepubertal boys and in most cases discussion occurred at diagnosis, but in 7% it is delayed until during treatment [49].

Ethical, Legal, and Economic Considerations for Gamete Preservation

Patients must be informed about fertility preservation, and the need for an informed consent is mandatory. Fertility preservation in children causes a difficult ethical problem. The informed consent concerning children needs the involvement of their parents or legal guardians. When the patient is able to understand the issue (postpubertal boys), he can give an assent (permission less than full consent) together with parents consent. Regarding the TESE in adolescents, an ethical board approval may be needed before considering surgery. Patients must be informed about the financial cost of cryopreservation and know that in some cases insurances will not cover all the costs [70].

Pregnancy and Assisted Reproduction Techniques After Cancer Treatment

The introduction of ICSI has totally changed the reproductive prospects for boys and men who underwent radiotherapy and chemotherapy. Hence, cured cancer patients can now father children who are genetically their own, even with the poorest semen samples [71–77]. A study carried out by Naysmith, showed that 45% of the patients after anticancer treatment could potentially benefit from ICSI [74]. In sexually mature men, frozen-thawed spermatozoa obtained by TESE before cytotoxic treatments can be successfully used performing ICSI [78, 79] even if the sperm has extremely poor characteristics of count, motility, and morphology [61]. In prepubertal boys, cryopreservation of mature and immature sperm before starting anticancer treatment is feasible. Once the child is cured from the cancer, thawed testicular tissue can be transplanted back into his own testis or can be matured in vitro or in vivo until the stage at which he will be competent to procure normal fertilization with ICSI.

Regarding the possible congenital malformations in children of patients who underwent anticancer treatments, some authors showed that the risk is negligible [1, 80–83] though other authors demonstrated a measurable increase in the frequency of sperm chromosomal abnormalities related to cytotoxic treatments [84, 85]. The risk of producing chromosomally abnormal sperm is highest in the few weeks after completion of chemotherapy, so men should wait for 6 months after stopping their chemotherapy before conception is attempted [86]. It has been demonstrated that men treated with chemotherapy for Hodgkin's disease or testicular cancer had an increased aneuploidy in their spermatozoa [87–89]. Patients with unilateral testicular cancer show a high prevalence of chromosomal abnormalities of spermatogenesis, which might be observed in contralateral testis [90] because of the presence of the cancer itself. However, children fathered by patients treated for cancer, did not show more congenital disorders [82]. Hence, given the absence of any clinical evidence of chromosomal abnormalities in offspring born from men undergoing chemotherapy or men who had chemotherapy, even patients who already started their cancer treatment can be offered cryobanking [56]. Additionally, in order to control for aneuploidy, preimplantation genetic diagnosis (PGD) may be suggested whenever assisted reproduction is performed using sperm banked during chemotherapy.

Conclusions

Currently, the effectiveness of anticancer treatments has improved the survival rates tremendously and as a result most of the patients are cured. The quality of life has now become an important issue, including the posttreatment fertility status. However, depending on the dose and type of drugs used, radiotherapy and chemotherapy can produce gonadal toxicity ranging from temporary oligozoospermia to permanent azoospermia.

The testis has both an endocrine and an exocrine function. The endocrine output depends on an adequate LC function. For the exocrine function, i.e., spermatogenesis, both a functional SC and a spermatogonial stem cell pool are necessary. Spermatogonial stem cells are undifferentiated cells that give rise to the spermatogenic cells and, finally, the spermatozoa. Even though there is a continual loss of differentiated cells, the spermatogenic cell lineage maintains its cell number; thanks to the adult spermatogonial stem cells, which produce both new stem cells (self-renewal) and differentiating cells.

LCs are rather resistant to chemo- and radiotherapy. While elevated LH indicates LC dysfunction, most men will show normal testosterone levels and will retain a normal bone density. While routine testosterone supplementation after gonadotoxic treatment is not indicated, men should have endocrine monitoring after cancer treatment and in selected cases testosterone substitution may be required. LCs seem more vulnerable to chemo- and radiotherapy in prepubertal life.

Also SCs show a good resistance to chemo- and radiotherapy, except in the postnatal and pubertal period when their numbers are increasing through mitosis. Increases in FSH are due to indirect effects through loss of the germ cell pool.

The extent of the gonadotoxicity on this germ cell is strongly related to the nature of the specific agents that are used and their dose or to the intensity of the radiation and the place of the body where it was administered. Transient reductions of sperm count can occur even after mild forms of chemotherapy or low doses of gonadal radiation, due to the destruction of the sensitive differentiating spermatogonia. Stronger chemotherapeutic regimes or higher doses of gonadal irradiation, however, lead to prolonged reduction in sperm count or complete azoospermia. Whether sperm production will eventually recover depends on the survival of the spermatogonial stem cells and the integrity of their ability to differentiate. High doses of radiation to the testis (>2.5 Gy) cause DNA damage and cell death. The most damaging chemotherapeutic agents in the adult man are the alkylating agents.

Even though the gonadotoxic effect of most anticancer treatments is known, it is difficult to predict the final effect on the fertility potential of the patient. There remain important interindividual differences in response to the treatment and even if a regimen with low gonadotoxicity is started, it is possible that eventually a more gonadotoxic treatment has to be administered because of earlier treatment failure.

Patients undergoing anticancer therapy must be informed about their future fertility status and oncologist must discuss at diagnosis the different options available to preserve fertility. If the patient is a child, this information should be given to the parents or legal guardians.

Cryobanking should be offered and performed preferentially before starting any treatment. Otherwise, semen cryopreservation during the first months of treatment can be considered. When the patient is prepubertal, spermatogonial stem cell banking, although still being in a preclinical phase of development, can be considered.

References

1. Grundy R, Gosden RG, Hewitt M, et al. Fertility preservation for children treated for cancer (1): scientific advances and research dilemmas. Arch Dis Child. 2001;84(4):355–9. Review.
2. Hollen PJ, Hobbie WL. Establishing comprehensive specialty follow-up clinics for long-term survivors of cancer. Providing systematic physiological and psychosocial support. Support Care Cancer. 1995;3(1):40–4.
3. SIGN 76. The long-term follow up of children treated for cancer. www.sign.com. Accessed Nov 2004.
4. Fallat ME, Hutter J, American Academy of Pediatrics Committee on Bioethics, et al. Preservation of fertility

in pediatric and adolescent patients with cancer. Pediatrics. 2008;121(5):e1461–9. Review.
5. Huyghe E, Matsuda T, Thonneau P. Increasing incidence of testicular cancer worldwide: a review. J Urol. 2003;170(1):5–11. Review.
6. Fosså SD, Abyholm T, Normann N, et al. Post-treatment fertility in patients with testicular cancer. III. Influence of radiotherapy in seminoma patients. Br J Urol. 1986;58(3):315–9.
7. Presti JC, Herr HW, Carroll PR. Fertility and testis cancer. Urol Clin North Am. 1993;20(1):173–9. Review.
8. Dearnaley D, Huddart R, Horwich A. Regular review: managing testicular cancer. BMJ. 2001;322(7302): 1583–8. Review.
9. Laguna MP, Pizzocaro G, Klepp O, et al. EAU guidelines on testicular cancer. Eur Urol. 2001;40(2): 102–10.
10. Howell SJ, Shalet SM. Spermatogenesis after cancer treatment: damage and recovery. J Natl Cancer Inst Monogr. 2005;34:12–7. Review.
11. Pryzant RM, Meistrich ML, Wilson G, et al. Long-term reduction in sperm count after chemotherapy with and without radiation therapy for non-Hodgkin's lymphomas. J Clin Oncol. 1993;11(2):239–47.
12. da Cunha MF, Meistrich ML, Fuller LM, et al. Recovery of spermatogenesis after treatment for Hodgkin's disease: limiting dose of MOPP chemotherapy. J Clin Oncol. 1984;2(6):571–7.
13. Rowley MJ, Leach DR, Warner GA, et al. Effect of graded doses of ionizing radiation on the human testis. Radiat Res. 1974;59(3):665–78.
14. Meistrich ML. Male gonadal toxicity. Pediatr Blood Cancer. 2009;53(2):261–6. Review.
15. Sharpe RM, McKinnell C, Kivlin C, et al. Proliferation and functional maturation of Sertoli cells, and their relevance to disorders of testis function in adulthood. Reproduction. 2003;125(6):769–84. Review.
16. Orth JM, Gunsalus GL, Lamperti AA. Evidence from Sertoli cell-depleted rats indicates that spermatid number in adults depends on numbers of Sertoli cells produced during perinatal development. Endocrinology. 1988;122(3):787–94.
17. Sharpe RM. Regulation of spermatogenesis. In: Knobil E, Neill JD, editors. The physiology of reproduction. 2nd ed. New York: Raven Press; 1994. p. 1363–434.
18. Jégou B, Sharpe RM. Paracrine mechanisms in testicular control. In: de Kretser DM, editor. Molecular biology of the male reproductive system. New York: Academic; 1993. p. 271–310.
19. Bar-Shira Maymon B, Yogev L, Marks A, et al. Sertoli cell inactivation by cytotoxic damage to the human testis after cancer chemotherapy. Fertil Steril. 2004;81(5):1391–4.
20. Hutson JM, Hasthorpe S, Heyns CF. Anatomical and functional aspects of testicular descent and cryptorchidism. Endocr Rev. 1997;18(2):259–80. Review.
21. Sharpe RM. Hormones and testis development and the possible adverse effects of environmental chemicals. Toxicol Lett. 2001;120(1–3):221–32. Review.
22. Vermeulen A. The male climaterium. Ann Med. 1993;25(6):531–4.
23. Howell SJ, Radford JA, Ryder WD, et al. Testicular function after cytotoxic chemotherapy: evidence of Leydig cell insufficiency. J Clin Oncol. 1999;17(5): 1493–8.
24. Hansen SW, Berthelsen JG, von der Maase H. Long-term fertility and Leydig cell function in patients treated for germ cell cancer with cisplatin, vinblastine, and bleomycin versus surveillance. J Clin Oncol. 1990;8(10):1695–8.
25. Palmieri G, Lotrecchiano G, Ricci G, et al. Gonadal function after multimodality treatment in men with testicular germ cell cancer. Eur Endocrinol. 1996;134(4):431–6.
26. Stuart NS, Woodroffe CM, Grundy R, et al. Long-term toxicity of chemotherapy for testicular cancer – the cost of cure. Br J Cancer. 1990;61(3):479–84.
27. Holmes SJ, Whitehouse RW, Clark ST, et al. Reduced bone mineral density in men following chemotherapy for Hodgkin's disease. Br J Cancer. 1994;70(2):371–5.
28. Howell SJ, Shalet SM. Effect of cancer therapy on pituitary-testicular axis. Int J Androl. 2002;25(5): 269–76. Review.
29. Howell SJ, Radford JA, Ryder WD, Shalet SM. Testicular function after cytotoxic chemotherapy: evidence of Leydig cell insufficiency. J Clin Oncol. 1999;17(5):1493–8.
30. Shalet SM, Tsatsoulis A, Whitehead E, et al. Vulnerability of the human Leydig cell to radiation damage is dependent upon age. J Endocrinol. 1989;120(1):161–5.
31. Shapiro E, Kinsella TJ, Makuch RW, et al. Effects of fractionated irradiation of endocrine aspects of testicular function. J Clin Oncol. 1985;3(9):1232–9.
32. Müller J, Skakkebaek NE. Quantification of germ cells and seminiferous tubules by stereological examination of testicles from 50 boys who suffered from sudden death. Int J Androl. 1983;6(2):143–56.
33. Ehmcke J, Wistuba J, Schlatt S. Spermatogonial stem cells: questions, models and perspectives. Hum Reprod Update. 2006;12(3):275–82.
34. World Health Organisation. WHO manual for the standardized investigation, diagnosis and management of the infertile male. Cambridge: Cambridge University Press; 2000.
35. Colpi GM, Contalbi GF, Nerva F, et al. Testicular function following chemo-radiotherapy. Eur J Obstet Gynecol Reprod Biol. 2004;113 Suppl 1:S2–6. Review.
36. Speiser B, Rubin P, Casarett G. Aspermia following lower truncal irradiation in Hodgkin's disease. Cancer. 1973;32(3):692–8.
37. Centola GM, Keller JW, Henzler M, et al. Effect of low-dose testicular irradiation on sperm count and fertility in patients with testicular seminoma. J Androl. 1994;15(6):608–13.
38. Hahn EW, Feingold SM, Nisce L. Aspermia and recovery of spermatogenesis in cancer patients following incidental gonadal irradiation during treatment: a progress report. Radiology. 1976;119(1):223–5.

39. Martin RH, Hildebrand K, Yamamoto J, et al. An increased frequency of human sperm chromosomal abnormalities after radiotherapy. Mutat Res. 1986;174: 219–25.
40. Socié G, Salooja N, Cohen A, et al. Late effects working party of the European study group for blood and marrow transplantation. Nonmalignant late effects after allogeneic stem cell transplantation. Blood. 2003;101:3373–85.
41. Sanders JE, Hawley J, Levy W, et al. Pregnancies following high-dose cyclophosphamide with or without high-dose busulfan or total body irradiation and bone marrow transplantation. Blood. 1996;87:3045–52.
42. Spitz S. The histological effects of nitrogen mustards on human tumors and tissues. Cancer. 1948;1(3):383–98.
43. Meistrich ML, Wilson G, Brown BW, et al. Impact of cyclophosphamide on long-term reduction in sperm count in men treated with combination chemotherapy for Ewing and soft tissue sarcomas. Cancer. 1992;70(11):2703–12.
44. Gandini L, Sgrò P, Lombardo F, et al. Effect of chemo- or radiotherapy on sperm parameters of testicular cancer patients. Hum Reprod. 2006;21(11):2882–9.
45. Schrader M, Müller M, Straub B, et al. The impact of chemotherapy on male fertility: a survey of the biologic basis and clinical aspects. Reprod Toxicol. 2001;15(6):611–7. Review.
46. Lee SJ, Schover LR, Partridge AH, et al. American Society of Clinical Oncology recommendations on fertility preservation in cancer patients. J Clin Oncol. 2006;24(18):2917–31. Erratum in: J Clin Oncol. 2006;24(36):5790.
47. Royal College of Physicians, Royal College of Radiologist, Royal College of Obstetricians and Gynaecologist. The effects of cancer treatment on reproductive functions. Guidance on management. Report of a working party. 2007. http://www.rcog.org.uk/resources/public/pdf/EffectCancerReprod.pdf
48. Schover LR, Brey K, Lichtin A, et al. Oncologists' attitudes and practices regarding banking sperm before cancer treatment. J Clin Oncol. 2002;20(7): 1890–7.
49. Anderson RA, Weddell A, Spoudeas HA, et al. Do doctors discuss fertility issues before they treat young patients with cancer? Hum Reprod. 2008;23(10):2246–51.
50. Meistrich ML, Zhang Z, Porter KL, Bolden-Tiller OU, Shetty G. Prevention of adverse effects of cancer treatment on the germline. In: Anderson D, Brinkworth MH, editors. Male-mediated developmental toxicity. Cambridge: Royal Society of Chemistry; 2007. p. 114–23.
51. Shetty G, Meistrich M. Hormonal approaches to preservation and restoration of male fertility after cancer treatment. J Natl Cancer Inst Monogr. 2005;34:36–9. Review.
52. Johnson DH, Linde R, Hainsworth JD, et al. Effect of a luteinizing hormone releasing hormone agonist given during combination chemotherapy on post-therapy fertility in male patients with lymphoma: preliminary observations. Blood. 1985;65:832–6.
53. Waxman JH, Ahmed R, Smith D, et al. Failure to preserve fertility in patients with Hodgkin's disease. Cancer Chemother Pharmacol. 1987;19:159–62.
54. Redman JR, Bajorunas DR. Suppression of germ cell proliferation to prevent gonadal toxicity associated with cancer treatment. In: Workshop on psychosexual and reproductive issues affecting patients with cancer. New York: American Cancer Society; 1987. p. 90–4.
55. Masala A, Faedda R, Alagna S, et al. Use of testosterone to prevent cyclophosphamide-induced azoospermia. Ann Intern Med. 1997;126:292–5.
56. Carson SA, Gentry WL, Smith AL, et al. Feasibility of semen collection and cryopreservation during chemotherapy. Hum Reprod. 1991;6(7):992–4.
57. National Institute for Health and Clinical Excellence. Fertility assessment and treatment for people with fertility problems. 2004. www.rcog.org.uk/resources/Public/pdf/Fertility_summary.pdf
58. Kliesch S, Behre HM, Jürgens H, et al. Cryopreservation of semen from adolescent patients with malignancies. Med Pediatr Oncol. 1996;26(1):20–7.
59. Bahadur G, Ling KL, Hart R, et al. Semen production in adolescent cancer patients. Hum Reprod. 2002; 17(10):2654–6.
60. Menon S, Rives N, Mousset-Siméon N, et al. Fertility preservation in adolescent males: experience over 22 years at Rouen University Hospital. Hum Reprod. 2009;24(1):37–44.
61. Tournaye H, Goossens E, Verheyen G, et al. Preserving the reproductive potential of men and boys with cancer: current concepts and future prospects. Hum Reprod Update. 2004;10(6):525–32. Review.
62. Ginsberg JP, Carlson CA, Lin K, Hobbie WL, et al. An experimental protocol for fertility preservation in prepubertal boys recently diagnosed with cancer: a report of acceptability and safety. Hum Reprod. 2010;25(1):37–41.
63. Brinster RL, Zimmermann JW. Spermatogenesis following male germ-cell transplantation. Proc Natl Acad Sci U S A. 1994;91(24):11298–302.
64. Brinster RL, Avarbock MR. Germline transmission of donor haplotype following spermatogonial transplantation. Proc Natl Acad Sci U S A. 1994;91(24): 11287–9.
65. Jahnukainen K, Hou M, Petersen C, et al. Intratesticular transplantation of testicular cells from leukemic rats causes transmission of leukemia. Cancer Res. 2001;61(2):706–10.
66. Geens M, Van de Velde H, De Block G, et al. The efficiency of magnetic-activated cell sorting and fluorescence-activated cell sorting in the decontamination of testicular cell suspensions in cancer patients. Hum Reprod. 2007;22(3):733–42.
67. de Vries MC, Bresters D, Engberts DP, et al. Attitudes of physicians and parents towards discussing infertility risks and semen cryopreservation with male adolescents diagnosed with cancer. Pediatr Blood Cancer. 2009;53(3):386–91.
68. Zapzalka DM, Redmon JB, Pryor JL. A surve of oncologists regarding sperm cryopreservation and

assisted reproductive techniques for male cancer patients. Cancer. 1999;86:1812–7.
69. Allen C, Keane D, Harrison RF. A survey of Irish consultants regarding awareness of sperm freezing and assisted reproduction. Ir Med J. 2003;96(1):23–5.
70. Deepinder F, Agarwal A. Technical and ethical challenges of fertility preservation in young cancer patients. Reprod Biomed Online. 2008;16(6): 784–91.
71. Chen SU, Ho HN, Chen HF, et al. Pregnancy achieved by intracytoplasmic sperm injection using cryopreserved semen from a man with testicular cancer. Hum Reprod. 1996;11(12):2645–7.
72. Hallak J, Sharma RK, Thomas Jr AJ, et al. Why cancer patients request disposal of cryopreserved semen specimens posttherapy: a retrospective study. Fertil Steril. 1998;69(5):889–93.
73. Lass A, Akagbosu F, Abusheikha N, et al. A programme of semen cryopreservation for patients with malignant disease in a tertiary infertility centre: lessons from 8 years' experience. Hum Reprod. 1998;13(11):3256–61.
74. Naysmith TE, Blake DE, Harvey VJ, et al. Do men undergoing sterilizing cancer treatments have a fertile future? Hum Reprod. 1998;13(11):3250–5.
75. Tournaye H. Storing reproduction for oncological patients. Mol Cell Endocrinol. 2000;169(1–2):133–6. Review.
76. Ginsburg ES, Yanushpolsky EH, Jackson KV. In vitro fertilization for cancer patients and survivors. Fertil Steril. 2001;75(4):705–10.
77. Horne G, Atkinson A, Brison DR, et al. Achieving pregnancy against the odds: successful implantation of frozen-thawed embryos generated by ICSI using spermatozoa banked prior to chemo/radiotherapy for Hodgkin's disease and acute leukaemia. Hum Reprod. 2001;16(1):107–9.
78. Hovatta O, Foudila T, Siegberg R, et al. Pregnancy resulting from intracytoplasmic injection of spermatozoa from a frozen-thawed testicular biopsy specimen. Hum Reprod. 1996;11(11):2472–3.
79. Oates RD, Mulhall J, Burgess C, et al. Fertilization and pregnancy using intentionally cryopreserved testicular tissue as the sperm source for intracytoplasmic sperm injection in 10 men with non-obstructive azoospermia. Hum Reprod. 1997;12(4):734–9.
80. Byrne J, Mulvihill JJ, Connelly RR, et al. Reproductive problems and birth defects in survivors of Wilms' tumor and their relatives. Med Pediatr Oncol. 1988;16(4):233–40.
81. Senturia YD, Peckham CS. Children fathered by men treated with chemotherapy for testicular cancer. Eur J Cancer. 1990;26(4):429–32.
82. Hawkins MM, Draper GJ, Smith RA. Cancer among 1,348 offspring of survivors of childhood cancer. Int J Cancer. 1989;43(6):975–8.
83. Li FP, Fine W, Jaffe N, et al. Offspring of patients treated for cancer in childhood. J Natl Cancer Inst. 1979;62(5):1193–7.
84. Brandriff BF, Meistrich ML, Gordon LA, et al. Chromosomal damage in sperm of patients surviving Hodgkin's disease following MOPP (nitrogen mustard, vincristine, procarbazine, and prednisone) therapy with and without radiotherapy. Hum Genet. 1994;93(3):295–9.
85. Jenderny J, Röhrborn G. Chromosome analysis of human sperm. I. First results with a modified method. Hum Genet. 1987;76(4):385–8.
86. Revel A, Revel-Vilk S. Pediatric fertility preservation: is it time to offer testicular tissue cryopreservation? Mol Cell Endocrinol. 2008;282(1–2):143–9. Review.
87. Genescà A, Benet J, Caballín MR, et al. Significance of structural chromosome aberrations in human sperm: analysis of induced aberrations. Hum Genet. 1990;85(5):495–9.
88. Monteil M, Rousseaux S, Chevret E, et al. Increased aneuploid frequency in spermatozoa from a Hodgkin's disease patient after chemotherapy and radiotherapy. Cytogenet Cell Genet. 1997;76(3–4):134–8.
89. Robbins WA, Meistrich ML, Moore D, et al. Chemotherapy induces transient sex chromosomal and autosomal aneuploidy in human sperm. Nat Genet. 1997;16(1):74–8.
90. Gurgan T, Salman C, Demirol A. Pregnancy and assisted reproduction techniques in men and women after cancer treatment. Placenta. 2008;29(Suppl B): 152–9. Review.

Impact of Paternal Exposure to Gonadotoxins on Embryo and Offspring and the Male Evaluation

21

Kathleen Hwang, Paul Gittens, Desiderio Avila Jr., and Larry I. Lipshultz

Abstract

There are several routes of exposure to gonadotoxins for the male. Exposures originate from lifestyle choices, medications, and treatment for malignancies, as well as occupational and chemical hazards. Regardless of the route, the impact of paternal exposure to gonadotoxins on subsequent fertility is significant as it not only affects the individual, but may also affect future offspring. Given the increasing popularity of in vitro fertilization (IVF)/intracytoplasmic sperm injection (ICSI) and the vast numbers of procedures performed each year throughout the world, continued research on the safety of IVF/ICSI and IVF itself is crucial. Patients can only truly give informed consent when they are properly educated as to all the associated risks.

Keywords

Gonadotoxin effect on embryo and offspring • Chemotherapy and radiation effects on male fertility • Cancer treatment effects on gonads

The failure to conceive within 1 year occurs in approximately 15% of couples [1]. Approximately 50% of problems related to conception is either caused entirely by the male or is a combined problem with the male and his partner.

L.I. Lipshultz (✉)
Scott Department of Urology, Baylor College of Medicine, O'Quinn Medical Tower, 6624 Fannin Street, Suite 1700, Houston, TX 77030, USA

Gonadotoxic Exposure

There are several routes of exposure to gonadotoxins for the male. Exposures originate from lifestyle choices, medications, and treatment for malignancies, as well as occupational and chemical hazards. Regardless of the route, the impact of paternal exposure to gonadotoxins on subsequent fertility is significant as it not only affects the individual, but may also affect future offspring.

E. Seli and A. Agarwal (eds.), *Fertility Preservation: Emerging Technologies and Clinical Applications*, DOI 10.1007/978-1-4419-1783-6_21, © Springer Science+Business Media, LLC 2012

Gonadotoxic Effects of Cancer Treatment

With drastic improvements in the treatment of childhood cancers, more attention has been given to the quality of life of cancer survivors. Furthermore, cancer and its effect on fertility have become a major focus of extensive research over the past decade. This amplified interest is due to the increased incidence of cancer among the reproductive age group as well as improvements in long-term survival rates [2]. Cancer treatment encompasses three main treatment modalities: surgery, radiation therapy, and chemotherapy.

Testicular cancer unlike other malignancies exert a direct effect on spermatogenesis through the destruction of surrounding tissue, altering local blood flow which increases intrascrotal temperature not only in the affected testicle but the contralateral one as well [3]. These tumors can also produce hormones (human chorionic gonadotropin and alpha feto-protein) and effect spermatogenesis. Radical orchiectomy remains the mainstay of treatment for testicular tumors with partial orchiectomies recently becoming more popular as a method of preserving hormonal and sperm function in select patients [4]. Petersen et al. reported a 50% decrease in the sperm concentration during the first few months after radical orchiectomy, and found that 10% of patients with sperm prior to surgery will become azoospermic [5].

As an adjunct to radical orchiectomy, a subset of patients may proceed to a retroperitoneal lymph node dissection, which potentially causes further impairment in fertility. With this node dissection, despite template modifications and intentional nerve sparing techniques, damage to the sympathetic ganglia responsible for emission and ejaculation can still occur. For these patients who are anejaculatory after surgery, surgical sperm harvest and percutaneous extraction for assisted reproduction is an option; however, less invasive options include electroejaculation to stimulate antegrade ejaculation, post-ejaculate urinalysis as well as α-adrenergic therapy can be used [6, 7].

The effect of radiation therapy is entirely dependent on the dosage, treatment field and fractionation schedule [8]. Radiation doses start to adversely affect spermatogenesis at 0.1–1.2 Gy, with irreversible damage at doses >4 Gy [9]. However, higher doses (>20 Gy) may also affect the Leydig cells resulting in reduced testosterone and raised serum gonadotropins [10]. Although the recovery of spermatogenesis can take up to 9 years after treatment, updated radiation technologies with more efficient dosages and mandated gonadal shielding has allowed for a faster recovery of sperm production. Howell and Shalet reported complete recovery within 12 months after treatment with doses of ≤1 Gy, after 30 months with doses of 2–3 Gy, and >5 years with doses >4 Gy [11]. In addition, irradiation may increase sperm DNA damage that persists for up to 2 years after treatment, affecting fertility rates even after recovery of spermatogenesis [9].

Chemotherapeutic medications have a negative effect on fertility, although some agents are more gonadotoxic than others. Similarly to radiation therapy, the toxic effect on spermatogenesis is dependent on dosage, frequency, and pretreatment semen quality; however, recovery is also contingent upon the mechanism of action of each agent as well as the synergistic effect of multiple agent regimens [12]. Table 21.1 reviews the effect of standard regimens on gonadal function [12, 13]. The most gonadotoxic agents are the nitrogen mustard derivatives, such as busulphan, and alkylating agents, such as cyclophosphamide and isophosphamide which result in permanent azoospermia in 80–90% of cases [14]. Treatment of Hodgkin's disease with the nonalkylating combination of adriamycin, bleomycin, vinblastine, and dacarbazine revolutionized chemotherapy regimens, with a sperm recovery rate of ~90% within 1–5 years posttreatment while still maintaining therapeutic efficacy [15]. Germ cell tumors are typically treated with the regimen of bleomycin, etoposide and cisplatin (BEP), with patients demonstrating recovery rates back to baseline of 50% within 2 years and 80% within 8 years of treatment [11].

Numerous groups have examined the incidence of sperm aneuploidy after administration of chemotherapy by sperm fluorescent in situ hybridization (FISH). Multiple studies examining the

Table 21.1 Cytotoxic agents

Group	Agent	Mechanism of action	Gonadotoxicity
Alkylating agents	Cyclophosphamide Ifosfamide Carmustine Busulfan	Adds alkyl groups to DNA, altering structure and function	Prolonged azoospermia
Antibiotics	Actinomycin-D	Binds to DNA inhibiting RNA synthesis	Possible azoospermia
	Bleomycin	Single and double strand breaks in DNA	Temporary reduction in sperm count
	Doxorubicin	Inhibits progression of topoisomerase II dependent DNA	Temporary reduction in sperm count; may cause prolonged azoospermia when used with other agents
Antimetabolites	Fluorouracil 6-Mercaptopurine Thioguanine Methotrexate	Pyrimidine analog Purine analog Antifolate	Temporary reduction in sperm count with standard regimens, although additive effects are possible
Plant derivatives	Vincristine Vinblastine Etoposide	Inhibits formation of microtubules Inhibits topoisomerase II	Temporary reduction in sperm count
Platinum analogs	Carboplatin Cisplatin	Forms DNA complex and interferes with repair mechanisms	Prolonged azoospermia
Miscellaneous	Interferon-α Prednisone	Stimulates macrophages and NK cells Suppresses immune response	No known affects on spermatogenesis

Data from Giwercman et al. [11] and Lee et. al. [12]

effect of BEP on patients treated for testicular cancer demonstrated increased sperm aneuploidy that typically returned to baseline within 2 years after treatment [16, 17]. Patients treated for Hodgkin's lymphoma receiving either adriamycin, bleomycin, vincristine, and dacarbazine or novantrone, oncovin, velban, and prednisone had an increased incidence of sperm disomy and diploidy after therapy that typically returned to baseline months after treatment [18]. Advances in chemotherapy regimens continue to strive for treatment efficacy while minimizing side effects, particularly gonadal toxicity.

Effects of Chemical Exposures

Male reproductive health in the general population has garnered increasing attention due to reports of a time-related decline of semen quality, increased urogenital anomalies, and increased rates of testicular cancer [19–21]. The possibility that environmental chemicals may be partially responsible for these adverse outcomes has been a real concern since the discovery in 1977 of the severe spermatotoxic effect of dibromochloropropane (DBCP), a nematocide widely used in the USA and Central America, among workers at a chemical plant [22]. Furthermore, Anway et al. demonstrated that there are transgenerational effects of chemicals, where exposure of the parental gametes to chemicals may confer an increased risk of altered semen quality in the offspring [23]. Although the list of potential environmental gonadotoxicants is exhaustive, we focus on phthalates, pesticides, bisphenol A (BPA), and glycol ethers.

Phthalates

Phthalates are among the most widely used man-made chemicals released into the environment over the last several decades. They are primarily used as plasticizers in the manufacture of flexible vinyl, which is found in medical devices, toys, floor and wall coverings, personal care products, and food packaging. The ubiquitous use of phthalates results in exposure to these compounds mainly through diet, medical devices, and consumer products. Humans quickly metabolize phthalates to their respective di- and monoesters; however, their metabolites have been found in the urine of more than 95% of men and women that have been investigated [24]. There have been several epidemiological studies addressing the male reproductive toxicity of phthalates. Duty et al. [25] observed an inverse dose-response relationship between phthalate metabolites and sperm concentration and motility. Another group also demonstrated an association between phthalate metabolite exposure and increased levels of DNA damage in sperm [26].

Pesticides

Pesticides form a large group of heterogeneous chemicals which are used to control insect, weed, rodent, and fungi populations. There are several epidemiological studies from across the globe on men exposed to contemporary-use pesticides during agricultural work. In a recent Japanese study, pesticide sprayers exposed primarily to organophosphates showed spraying season dependent reductions in motile sperm compared with unexposed controls [27]. One of the first studies demonstrating a link between specific biomarkers of environmental exposure to pesticides and biomarkers of male reproduction was performed in the USA [28]. Men demonstrated an inverse dose-response relationship between herbicide exposure (DBCP) and sperm concentration, motility, and morphology.

Men exposed to pesticides have also been examined for sperm aneuploidy with FISH. Recio et al. [29] demonstrated that male agricultural workers exposed to organophosphate pesticides with increased serum concentrations had an increased frequency of sperm aneuploidy. However, a study conducted by Smith et al. [30] failed to demonstrate an association between exposure to miscellaneous pesticides and sperm aneuploidy in men with low levels of exposure. There are numerous indications that certain pesticides may impair semen quality; however, the direct affect noted with pesticides, such as DBCP, fortunately are no longer as evident with contemporary-use pesticides.

Bisphenol A

BPA is a chemical with weak estrogenic activity and widely used in polycarbonate plastic products, epoxy resins, sealants, and the lining of food cans. BPA has been reported to have estrogenic effects in animal models, such as decreased sperm production, stimulation of prolactin release, and promotion of cell proliferation in a breast cancer cell line [31, 32]. BPA has been shown to leech from products due to incomplete polymerization and degradation by exposure to high temperatures [33]. BPA levels have been measured in the serum and urine of adults, breast milk, fetal plasma, as well as amniotic fluid [34–36]. Levels of BPA in the serum have also been correlated with circulating androgen levels, and found to be elevated in women with polycystic ovarian syndrome [37].

Women carrying fetuses with karyotype abnormalities, furthermore, were found to have elevated maternal serum BPA levels [38] while BPA levels were found to be 3 times higher in 45 women with three or more consecutive first-trimester miscarriages [39]. The harmful potential of BPA has made such an indelible impact that BPA has been electively removed from the manufacturing of baby bottles sold in the USA by the six largest manufacturers.

Glycol Ethers

Glycol ethers are a family of alkyl derivatives of ethylene and propylene glycol. They are most commonly found in industrial uses, such as semiconductors, photographic film, paints, textile dyes, varnishes, printing inks, and hydraulic fluids. The oxidative metabolites of glycol ethers, particularly methoxyacetic acid (MAA), have been shown to be teratogenic and embryotoxic. In animal studies, MAA has altered gene expression and protein kinase activity, increased oxidative stress, and enhanced transcription of nuclear receptors for progesterone, estrogen, and androgens [40, 41]. The deleterious effects of glycol ether exposure have been reported for years. Over two decades ago, shipyard workers exposed to glycol ethers in paint fumes were found to have increased incidences of oligospermia [42]. More recently, men attending fertility clinics were found to have low motile sperm counts directly related to glycol ether exposure [43].

Lifestyle Factors

Lifestyle exposures, including ethanol, tobacco, recreational drug use, and caffeine have all been studied in relationship to male reproductive health and the focus has more recently been directed at semen quality and subsequent fertility. These substances have a substantial impact because of their frequent use and global exposure.

Ethanol is one of the most abused substances worldwide. It has been shown to have deleterious effects at all levels of the male reproductive system. Ethanol interferes with feedback mechanisms of the hypothalamus-pituitary-gonadal (HPG) axis resulting in impairment of production and secretion of adequate levels of luteinizing hormone and follicle-stimulating hormone [44]. Additionally, ethanol has been associated with significant decreases in serum testosterone levels, increased rates of plasma testosterone clearance, increased levels of plasma steroid hormone-binding globulin (SHBG) and increased levels of estrogen [45]. These effects are dose dependent and are potentially irreversible in chronic abusers, particularly increased levels of SHBG [46]. Ethanol has demonstrated direct effects on spermatogenesis which have manifested as a reduction in seminiferous tubular diameter with subsequent damage to the germinal epithelium, reduction in sperm count, reduction in progressive motility, reduction in sperm with normal morphology, as well as an increase in reactive oxygen species (ROS) [47, 48]. While few studies have addressed the effects of paternal ethanol exposure on offspring, animal studies have demonstrated significantly reduced fecundity in peripubertal ethanol-fed fathers [49].

Similarly, one-fifth of Americans *smoke* and nearly 50% of nonsmokers are exposed to secondhand smoke. Shockingly, 25% of Americans between the ages of 18 and 24 smoke and the number continues to rise [50]. Smoking is a lifestyle hazard for both the active and passive smoker. While much is known of the carcinogenic properties of tobacco and its resultant effects on organs, such as lung and bladder, the impact on fertility still remains less defined. Numerous studies, however, have well established the fact that toxins in cigarette smoke reach the male reproductive system and their effects are mainly due to their direct interaction with the components of seminal fluid [51]. This interaction has led to a greater presence of ROS, increased leukocytes and round cells, as well as a higher frequency of DNA fragmentation in comparison to nonsmokers [52–54]. In smokers, either male or female, there was noted to be a significant delay of over 6 months in natural conception in comparison to nonsmokers [55]. Nicotine and its metabolites are present in the spermatozoa of smokers, and furthermore, these toxicants are found in embryos resulting from in vitro fertilization (IVF) cycles with male smokers [56].

Marijuana is the most common illicit drug in the USA, and its abuse and dependence by users are steadily increasing [57]. It has been well established that the active components of marijuana have a negative impact on sperm function and fertilizing ability since it leads to oligospermia, an association with pyospermia as well as the development of gynecomastia [58, 59]. This is incredibly important, given the prevalence of marijuana use particularly in the young, as well as the more recent development of legalization of cannabinoid use for the treatment of specific medical conditions in certain countries. The effect of marijuana use on male fertility has to be taken into careful account.

Multiple researchers have examined the effect of *smoking* on sperm aneuploidy. Despite inconsistent results, combined data from each study demonstrated a significant increase in sex chromosome and autosome aneuploidy [60]. While controlling for smoking and alcohol consumption, Robbins et al. [61] found an increase in sex chromosome disomy and diploidy with high levels of *caffeine* intake while Rubes et al. [62] did not find an increase in either when compared to controls. Although numerous articles demonstrate deleterious effects of *alcohol* on semen parameters, only one article examined alcohol consumption by sperm FISH and found an association between increased alcohol consumption and sex chromosome disomy [61]. The potential confounding effects of smoking, caffeine, and alcohol intake in each of these articles makes drawing any broad conclusions difficult without rigorously controlled prospective studies.

Medications

The usage of certain drugs, and specific drug classes, may contribute to infertility. Antihypertensives are a class of drugs that has well-documented side effects, including reduced sperm quality. Spironolactone, a potassium-sparing diuretic, has proven antiandrogen activity and shown to reduce semen quality in men on chronic therapy [63]. Calcium channel blockers have also demonstrated a reversible functional deficit in sperm quality and a significant interference with fertilizing capability, although there was no change in sperm production noted [64]. Antibiotics are another class of drugs that have notoriously had significant side effects. Historically, nitrofurantoin given chronically at high doses (10 mg/kg/day) to young males has been reported to result in early maturation arrest; however, this has not been seen in clinical practice [65]. In vitro studies have also highlighted the potential to decrease fertility with the use of erythromycin, tetracycline, and gentamicin [66]. Again, there does not appear to be clinically supported relevance.

Other medications that have negatively affected fertility include sulfasalazine and colchicine, leading to reversible defects in sperm concentration and motility [67, 68]. Baumgartner et al. [69] described two men with a significant increase in sperm aneuploidy who had chronically ingested diazepam for greater than 6 months. Another group reported that finasteride was associated with increased sperm diploidy and sex chromosome disomy, which did not improve 1 year after cessation of the drug [70].

Male Evaluation

When performing the male evaluation for infertility, it is important to simulate the work up of any medical patient. A thorough history should be performed, addressing the amount of time in which the couple has attempted conception, previous pregnancies, intercourse timing, use of lubrication and erectile function. A query of medical conditions and surgical conditions should also be noted. Special attention should be paid to a developmental history, recent febrile illness, a history of viral orchitis, broncho-pulmonary illnesses, undescended testes, childhood cancers and treatments, genitourinary tract infections, and congenital or genetic abnormalities. Scrotal, prostatic, spinal, inguinal, and retroperitoneal surgery should be highlighted along with a family history of both infertility and genetic abnormalities. A medication history including the use of anabolic steroids, either past or present, should be elicited. Finally, occupational and environmental exposures, such as contact with toxins, chemicals, radiation, ethanol intake, and smoking should be noted.

The physical exam should include a general description of the patient, absence or presence of facial and pubic hair, gynecomastia and skeletal structure. The examination should include a complete penile exam, which includes the length of the phallus and position of the meatus. The scrotum should be examined for the size and consistency of the testes, presence and consistency of the epididymis, vas deferens, and the presence of a varicocele(s). The digital rectal exam should also begin with an examination of the position of the anus, sphincter tone, and a thorough prostate exam evaluating its size, consistency, and the presence of midline cysts or dilation of the seminal vesicles.

Spermatogenesis is a complex step-wise process, which relies on an adequate and functional signaling pathway. A testosterone and FSH should be obtained to evaluate the HPG axis. If abnormal values are discovered, then more extensive blood tests should be implemented which may include estradiol (if the patient is obese), LH, prolactin, TSH, SHBG, and cortisol. Although endocrinopathies account for less than 3% of infertility cases, a thorough hormonal evaluation is essential when appropriate.

Male infertility can be divided into two main components: inadequate sperm production or insufficient sperm delivery. The patient should have at least two semen samples analyzed and referenced to well-established World Health Organization (WHO) criteria. In the absence of sperm, the specimen should be centrifuged, the pellet evaluated, and a quantitative fructose performed. The physician may also adjunctively order one of many additional studies, i.e., strict morphology, DNA integrity testing, and ROS quantitative, if suggested by abnormalities in the semen analysis (SA).

Integrating the history, physical examination, SA, and endocrine profile are steps toward establishing the diagnosis of the infertile male. By utilizing this systematic testing, the patient can be further characterized as primary, secondary, or tertiary testicular failure; obstructed vs. nonobstructed; or having abnormalities of sperm quantity and quality. During this process, an effort is made to identify not only the underlying conditions and harmful exposures leading to male infertility, but also to determine the potential effectiveness of specific vs. empiric therapies, as well as likely candidacy for assisted reproductive techniques (ART).

Semen Analysis

The most important aspect of the initial laboratory evaluation is the SA. It is recommended that at least two semen analyses be obtained with similar abstinence periods. Wide variations in semen parameters between specimens from the same patient are possible. Reports indicate that sperm density increases 25% per day for the first 4 days of abstinence prior to an SA, but other parameters such as motility and morphology remain stable [71].

A word of caution when classifying a man as being fertile, subfertile, or sterile based solely on semen parameters. While it may be true that a man with azoospermia is considered sterile, there is a wide range of overlap between semen parameters of fertile compared to infertile men [72]. The WHO has published reference values for semen, including volume, pH, sperm concentration, total sperm count, motility, morphology, and others (Table 21.2) [73, 74]. Medical professionals and reproductive specialists throughout the world have increasingly utilizing these reference values in the evaluation of the infertile couple.

Table 21.2 WHO semen parameter reference values

Semen parameter	Range
Volume	≥1.5 mL
Sperm concentration	≥15 M/mL
Total sperm count	≥39 M
Motility	≥40% total motility ≥28% progressive (a+b) motility
Morphology	≥4% by strict criteria
Vitality	≥58% sperm viable
WBC	<1 M/mL

From Cooper et al. [74], with permission of Oxford University Press

Semen samples should be obtained correctly by either masturbation or ejaculation into a specialized nonlatex spermatocidal-free collection condom. Collection of the sample can be performed in the office, or it may be collected at home and brought in for processing. In the case of the latter, the sample must be stored at room temperature and delivered within an hour of collection.

Once received, the ejaculated specimen is allowed to liquefy for 20–30 min. Gentle pipetting of the sample can aid in situations when there is incomplete liquefaction. This same technique is used to homogenize the sample prior to performing sperm counts. Alternately, one may use mechanical agitation with a vortex for several seconds. Viscosity is considered normal if semen exits the pipette drop by drop, and abnormal if it strands.

Microscopic Examination

Microscopic survey of semen must be performed prior to actual sperm counting. This is done to identify round cells, debris, and bacteria. Contamination of the specimen during collection as well as urethral or prostatic bacteria can result in the presence of bacteria in the semen. Significant infection is seen if concentration exceeds 1,000 bacteria per millimeter [75]. The presence of round cells in the semen can represent either immature germ cells or leukocytes [76]. Of note, up to 20% of infertile men have leukocytes present in the semen and of these only 20% are associated with significant bacterial colonization [77].

Sperm Counting

The preferred method for determining sperm density ("counts") is by using commercially available Neubauer hemocytometer counting chamber slides. Older methods required dilution of the specimen, which confounded a technique with an already inherently high technical error. Despite using counting chambers there still exists a significant variation in results between each type of chamber, whether computer based or manually determined [78, 79]. In general, one should count at least 200 sperm for an adequate determination of sperm concentration. When using microscopes that integrate a grid for counting, all sperm is counted within the grid and the concentration is calculated as a function of a coefficient specific to each chamber [80].

Sperm Motility

Percent motility has been shown to be the most important parameter when correlating semen samples to pregnancy outcomes [81, 82]. It is also the most difficult part of the manual evaluation of the SA and introduces a significant amount of subjectivity across technicians. One method involves a subjective estimation of motility from surveying several fields by the technician and averaging those estimates to produce a motility percentage.

A more objective and accepted method is to count motile and nonmotile (NM) sperm in each grid and average the percentages to produce a motility value. This can be done with a manual hematology cell counter. Drawbacks of this method are that it is inherently time consuming and that overestimation of motility can occur. This is due to the fact that as the technician moves along the grid, very motile sperm may progress to other areas of the grid and, theoretically, be double counted.

Methods to neutralize the inherent difficulty of counting motile sperm have been proposed [80]. Once the semen sample is homogenized, an aliquot is prepared in parallel for counting. In one aliquot only the NM sperm are counted. In the other, an immobilizing agent is used, usually water, after which all sperms (T) are counted. Motility can be calculated by using the formula T-NM.

Strict Sperm Morphology

The traditional evaluation of sperm morphology classifies sperm as normal if they do not fit into one of several defined categories, although the majority of sperm in an ejaculate are neither uniform nor symmetrical, displaying large variation in shape and size. A common classification scheme designates sperm as normal (oval), amorphous, tapered, duplicated, and immature. However, more strict criteria have been developed to identify "normal" spermatozoa [83, 84]. Kruger et al. reported that using his strict criteria, patients with less than 4% normal forms had a fertilization rate of 7.6% of oocytes in comparison to the usual >50% in patients with 4–14% normal forms [83].

For determining strict criteria, a minimum of 100–200 spermatozoa are counted on a stained slide using 100× oil-immersion magnification. An eyepiece reticule is initially recommended for measuring the sperm head and tail size (length and width), although this may not be necessary with an experienced andrology technician. Using the strict morphological criteria, a normal spermatozoon is characterized by a smooth oval head, 4.0–6.0 μm in length and 2.4–3.5 μm in width. The acrosome must be well defined, covering 40–70% of the sperm head. There cannot be any mid piece or tail defects. Finally, there should be no cytoplasmic droplets greater than half the size of the sperm head. Utilizing the strict criteria method, a spermatozoon that might be considered "borderline" would be classified as abnormal [84].

Although strict morphology criteria have been widely accepted, its clinical usefulness remains an area of controversy. Sperm morphology was not a reliable predictor in selecting sperm without chromosomal aberrations [85] while normal morphology was not a good indicator of genetically normal sperm [86]. Guzick et al. reported the existence of significant overlap between fertile and infertile men; and sperm morphology, concentration, or motility was not a powerful discriminator between the two groups [72].

While the establishment of high quality control standards in evaluating morphology may improve the clinical utility of morphology determination, couples should be counseled with caution to pursue ART based solely on an abnormal strict morphology.

Reactive Oxygen Species

Oxidative stress is a recognized etiology of male infertility [87]. ROS in the form of superoxide anions, hydrogen peroxide, and the hydroxyl-free radical are formed as a by-product of oxygen metabolism. The presence of excess ROS can cause oxidative damage to lipids, proteins, and DNA [88–90]. Abnormal ROS formation is found in up to 40% of infertile patients [91], with some reports suggesting an inverse relationship between seminal ROS levels and spontaneous pregnancy outcomes of infertile couples [92]. Many studies have attempted to define the relationship between seminal ROS and IVF [93, 94] but have met with conflicting results. Nevertheless, a growing body of knowledge on ROS and fertility makes testing for oxidants in the semen an important part of the infertile male evaluation.

There are various methods to detect seminal ROS, including chemiluminescence, nitroblue tetrazolium test, cytochrome C reduction test, flow cytometry, electron spin resonance, and xylenol orange-based assay. We discuss the chemiluminescence method, as this is the most commonly used.

The chemiluminescence assay utilizes a luminometer to measure chemical reactions between ROS found in human semen and a chemiluminescent probe, such as luminol or lucigenin. Luminol is an uncharged particle that is cell membrane permeable and therefore can react extracellularly and intracellularly with hydrogen peroxide, hydroxyl anions, and superoxide anions. In contrast, lucigenin is a positively charged particle that is membrane impermeable and reacts with superoxide anions in the extracellular space [95].

It is important to remember that the majority of the assays for chemiluminescence measure total oxidative stress and do not generally distinguish between ROS produced by leukocytes and that produced by spermatozoa. Moreover, there are other factors that can alter these assay results and must be considered.

Leukocyte contamination in the semen has been shown to impact negatively on fertility [77]. Leukocytes have been shown to be responsible for a significant proportion of ROS activity in the semen [96, 97]; therefore, these assays should be coupled with selective leukocyte removal strategies if pyospermia is present [98]. Not doing so would lead to a falsely elevated ROS value. Other factors that may spuriously increase ROS results include repeated centrifugation of the sample [99] and the use of certain oxidase-containing buffers for sample preparation [100].

In contrast, prolonged time from preparation to analysis of the sample can artificially decrease the ROS identifiable in the semen [101]. For this reason, it is recommended that testing be performed within an hour of sperm preparation [102]. Finally, poor liquefaction of the sample can interfere with the normal oxidative process resulting in a falsely lower ROS value [100].

DNA Damage

High levels of sperm DNA damage can negatively impact reproduction. Sperm samples from infertile men have been shown to have significantly more DNA damage than their fertile counterparts [103–105]. In addition, DNA damage has been correlated with poor outcomes from intrauterine inseminations [106]. Multiple reports have also implicated DNA damage with poor IVF results

[107, 108], although newer studies have not corroborated these findings [109–111]. Finally, some evidence suggests that DNA fragmentation can be a cause of early embryo death, poor embryo progression and poor implantation [112, 113]. In general, damage to sperm DNA occurs during intratesticular development as well as during the maturation and transport process that takes place outside the testis.

It is known that spermatogenesis is controlled by selective apoptosis. Abnormal sperm is tagged for apoptosis in the same manner that all other cells are marked for programmed cell death. Evidence suggests that a malfunction in this process allows sperm with DNA damage to be transported in the ejaculate, a process referred as abortive apoptosis [114, 115]. Protamine deficiency has been identified as another primary testicular cause of sperm DNA damage, and this deficiency is frequently seen in infertile men compared to fertile counterparts [116]. In addition, certain polymorphisms in the protamine gene have been implicated in male infertility and sperm DNA damage [117].

Excessive ROS in the ejaculate correlates with increasing sperm DNA damage [88, 97, 118]. Fortunately, there is evidence suggesting that reduction in ROS levels with antioxidant therapies can decrease sperm DNA damage [103, 119]. Studies have also implicated clinically significant varicoceles as a cause of sperm DNA damage [120]. Recent reports suggest improvement if DNA damage following microsurgical varicocelectomy [121–123]. Other external factors implicated in sperm DNA damage include smoking and genital tract inflammation [124, 125].

Since assisted reproductive technologies are now commonly used to circumvent virtually all types of male infertility, it is important to understand the differences between ejaculated, epididymal, and testicular sperm and their respective levels of DNA damage. O'Connell and colleagues found that testicular sperm had fewer DNA mutations and fragmentations when compared to epididymal sperm in preparation for IVF/intracytoplasmic sperm injection (ICSI) [126]. When comparing testicular sperm to ejaculated sperm, Greco and colleagues found that there was significantly lower DNA fragmentation in the testicular sperm. In addition, they reported improved pregnancy rates using testicular sperm compared to ejaculated sperm [127]. If IVF/ICSI is to be performed using sperm with high DNA damage consideration should be given to testicular sperm extraction (TESE) only after less invasive treatments for known causes of DNA damage have failed.

Common Genetic Abnormalities in the Male

Over the last several decades, advances in the field of infertility have enabled physicians to better evaluate the genetic causes of male factor infertility. It is important that patients, who are found to have genetic abnormalities or those who are suspected, be counseled on the risk to the fetus by either a genetic counselor or an experienced and knowledgeable physician. Proper risk assessment should be discussed with the patient as well as proposals for pre- and postconception genetic testing. Limitations in currently available genetic testing are highlighted in cases, where there is a high suspicion of a genetic cause, but current testing yields negative results, underscoring the importance of education and counseling for these couples.

Males who are found to have impaired semen quality should be referred to a male infertility specialist prior to the couple undergoing ART. The objective of the evaluation is to identify the etiology the infertility, realizing, however, that a significant percentage of these cases will remain idiopathic. Causes of abnormal semen can be "genetic," hormonal, mechanical or the result of concurrent disease or cancer [128, 129]. The physician should next attempt to either enhance the patient's sperm concentration for natural conception, intrauterine insemination or if necessary, proceed with extraction of sperm for IVF/ICSI. The infertility specialist should also be aware of common genetic abnormalities and proper testing protocols for suspected patients.

In patients with genetic abnormalities undergoing ICSI, preimplantation genetic testing can be used in conjunction with IVF/ICSI. Preimplantation genetic diagnosis (PGD), which involves the removal and analysis of genetic material from the embryo, is indicated in patients that have a risk of transmitting genetic disorders or mutations to their offspring. PGD can also be used in combination with other prenatal diagnostic testing, such as chorionic villous sampling, amniocentesis, blood testing, and ultrasound. Risks should be thoroughly discussed, such as increased risk of chromosomal abnormalities, prenatal mortality, multiple gestations, low birth weight, and preterm delivery with IVF/ICSI as well as the possibility of misdiagnosis and subsequent fetal loss with current diagnostic techniques [130–133].

CFTR: Absence of the Vas Deferens

The vas deferens can be palpated easily in the majority of males during the physical exam. In cases in which there is unilateral or bilateral absence of the vas deferens further evaluation is warranted. These patients also have a higher incidence of an abnormal or missing renal unit. Furthermore, 25% with unilateral vasal agenesis (CUAVD) and 10% with bilateral (CBAVD) will have abnormal seminal vesicles found on ultrasound [134]. Mutations in the cystic fibrosis transmembrane conductance regulator (CFTR) gene are responsible for many cases of CBAVD and CUAVD. Approximately 2/3 of men with CBAVD will have a mutation in the CFTR gene, and this abnormality effects 1–2% of men presenting with infertility [135]. However, 10–40% of men that are found to have CBAVD do not have subsequent mutations with routine CFTR gene testing [136]. In cases where there are positive physical findings and a negative test, the female partner should be tested. The majority of males with either CBAVD or CUAVD have normal semen parameters. Sperm can be retrieved from these patients via microscopic epididymal sperm aspiration (MESA), TESE or percutaneous testicular sperm aspiration (PESA), and ART can be used.

Klinefelter's Syndrome

Identifiable chromosomal abnormalities account for approximately 5% of male factor infertility [137]. In patients that are found to be azoospermic, the rate increases threefold [138]. Aneuploidy occurs in instances in which there are superfluous or insufficient numbers of chromosomes. The most common syndrome associated with aneuploidy is Klinefelter's disease, which occurs in 1:500 births and is found in 15% of males with infertility. The most common chromosomal arrangement in Klinefelter's disease is nonmosaic 47,XXY or mosaic 47,XXY/46X. These patients commonly have some degree of spermatogenic dysfunction that ranges from severe oligospermia to azoospermia. Testicular sperm retrieval techniques can be utilized for ICSI. Chromosomal abnormalities can be identified by a karyotype of peripheral leukocytes. Klinefelter's patients should be counseled on the increased risk of autosomal and sex chromosomal abnormalities in children conceived through IVC/ICSI [139].

Robertsonian Translocations

Robertsonian translocations are the most common form of balanced chromosomal rearrangement found in humans. The disorganization of the chromosomes can cause a loss of genetic material and/or abnormal signaling. The most common chromosomes involved are chromosomes 13, 14, 15, 21, 22, and occur in approximately 1 in 1,000 births [140]. Carriers can appear phenotypically normal; however, these patients are frequently present with oligospermia or azoospermia. These patients are also at higher risk for repeat miscarriages, failed IVF cycles and have a higher risk for offspring to be affected with unbalanced Robertsonian translocations, such as trisomy 21 and 13 [141]. Individuals with significant risk factors are recommended to undergo sperm FISH and PGD.

Y Chromosome Deletion

Y chromosome deletions occur in 5–10% of infertile males presenting with non-obstructive sperm deficiencies. The Y chromosome contains vital components needed for male differentiation and sperm function. The azoospermia factor region (AZF) on the long arm of the Y chromosome (Yq) is responsible for sperm development. The AZF region is subdivided by location into AZFa, AZFb, and AZFc, which correspond to proximal, middle, and distal portions of the chromosome [142]. Deletions in these locations are responsible for varying degrees of spermatogenic dysfunction. Entire microdeletions of AZFa or AZFb regions of the Y chromosome portend an exceptionally poor prognosis in sperm retrieval, such that microscopic sperm extraction is predictably negative [143].

Whereas microdeletions in AZFc alone can result in a variety of phenotypes which include oligospermia and azoospermia, depending on the severity of the deletion [143]. Deletions in the Y(q) are too small to be detected with a karyotype and thus are termed microdeletions. Identification using polymerase chain reaction techniques to analyze sequence tagged sites is currently used. Indications for testing AZF microdeletions are sperm concentrations less than 5 million/mL. Importantly, male offspring of patients with Y microdeletions will inherit the abnormal gene, rendering them likely to be infertile. Thus, ICSI with PGD should be discussed.

Significance of Assisted Reproduction

Over the past three decades, the treatment of infertile couples has been revolutionized by ART, including the development of ICSI and PGD. However, unlike most therapeutic procedures used in medicine today, ARTs have not undergone the rigorous safety testing prior to clinical use. With these new technologies, we are able to overcome many types of sterility and bypass nature's protective mechanisms that were developed in evolution to prevent fertilization by defective or deficient sperm. Subsequently, large numbers of couples undergo fertility treatments without a complete understanding of their infertility or the potential long-term risks for their children.

Historically, the treatment of male factor infertility with IVF was largely unsuccessful; yet the development of ICSI revolutionized the treatment of severe male factor infertility, affording many men the opportunity to father their own biologic children. However, evidence exists with regard to risks to the mother and offspring that are significantly increased with ART, including multiple gestation, preterm delivery (even in singleton pregnancies), and congenital anomalies in offspring [144–146]. Distinctive issues arise with regard to understanding the clinical implications of these data. Although most IVF pregnancies proceed uneventfully, studies consistently identify an increased risk of problems in IVF/ICSI pregnancies and deliveries [147].

Several large studies (Table 21.3) have shown a higher risk of genitourinary, cardiovascular, musculoskeletal, and gastrointestinal defects in children conceived by IVF and IVF/ICSI [148–157]. While current data suggest that IVF/ICSI is safe, it is still important to understand the nature of the risks and establish whether they are directly related to the technology itself or as a result of genetic defects from the parents. While controversial, there are further studies suggesting that developmental delay and impaired neurologic status occur in children conceived with IVF and IVF/ICSI [158, 159]. Furthermore, reporting on congenital abnormalities is inconsistent and at times inaccurate. Simpson and Lamb [160] outlined the limitations of IVF outcomes research and focused on the significant incidence of underreporting.

Given the increasing popularity of IVF/ICSI and the vast numbers of procedures performed each year throughout the world, continued research on the safety of IVF/ICSI and IVF itself is crucial. Patients can only truly give informed consent when they are properly educated as to all the associated risks. This information should be discussed with infertile couples prior to beginning the exhaustive journey, emotionally and financially, of the process of ART.

Table 21.3 Association between congenital malformations and conception by assisted reproductive technologies

Study	Years studied	Study type and sample size	Findings
Hansen et al. [84], Australia	1993–1997	Registry, controlled: 1,138 ART children (301 IVF/ICSI, 837 IVF, 4,000 SC)	IVF and ICSI: ↑ likelihood of birth defects even after correction
Zhu et al. [85], Denmark	1997–2003	Registry/questionnaire, controlled: 64,405 children (SC: 50,897 singletons and 1,366 twins from fertile couples, 5,764 singletons and 100 twins from subfertile couples; ART: 4,588 singletons and 1,690 twins)	ART and subfertile spontaneous conceptions: ↑ congenital malformation with increasing delay to conception
Koivurova et al. [86], Finland	1990–1995	Registry, controlled: 304 IVF, 569 SC	IVF: ↑ adverse pregnancy outcomes before correction for multiplicity and cardiac malformations regardless of multiplicity
Ludwig and Kalainic [87], Germany	1998–2002	Registry, controlled: 3,372 IVF/ICSI, 30,940 SC (Mainz Birth Registry)	IVF/ICSI: ↑ risk of all adverse outcomes; RR 1.25 (95% CI 1.11–1.40)
Anthony et al. [88], The Netherlands	1995–1996	Registry, controlled: 4,224 ART, 314,605 SC	ART: ↑ cardiovascular malformation; differences minimal for any other congenital malformation
Ericson and Kallen [89], Sweden	1982–1997	Registry, controlled: 9,111 IVF, 1,690,577 SC	ICSI: ↑ risk of alimentary atresia, neural tube defects and hypospadias
Sutcliffe et al. [90], UK	1989–1994	Prospective, controlled: 91 ART (cryopreserved embryos), 83 SC	ART: differences not significant for any congenital malformation
Olson et al. [91], US	1989–2002	Registry, controlled: 1,462 ART, 8,422 SC	IVF: ↑ birth defects after correction for multiplicity
Lie et al. [92]	2005	Meta-analysis, controlled: 5,395 IVF/ICSI, 13,086 IVF	IVF/ICSI: no significant risk of any single congenital malformation
Hansen et al. [93]	2005	Meta-analysis, controlled: 28,638 ART	ART: ↑ any birth defect in all 25 studies or on analysis of seven well-designed studies

Adapted from Alukal and Lipshultz [161], reprinted with permission of Macmillan Publishers, Ltd.
ART-assisted reproductive technologies; *ICSI* intracytoplasmic sperm injection; *IVF* in vitro fertilization; *OR* odds ratio; *RR* relative risk; *SC* spontaneous conceptions

References

1. Thonneau P, Marchand S, Tallec A, et al. Incidence and main causes of infertility in a resident population (1,850,000) of three French regions (1988–1989). Hum Reprod. 1991;6:811.
2. Padron OF, Sharma RK, Thomas Jr AJ, et al. Effects of cancer on spermatozoa quality after cryopreservation: a 12-year experience. Fertil Steril. 1997;67:326.
3. Costabile RA, Spevak M. Cancer and male factor infertility. Oncology (Williston Park). 1998;12:557.
4. Shukla AR, Woodard C, Carr MC, et al. Experience with testis sparing surgery for testicular teratoma. J Urol. 2004;171:161.
5. Petersen PM, Skakkebaek NE, Rorth M, et al. Semen quality and reproductive hormones before and after orchiectomy in men with testicular cancer. J Urol. 1999;161:822.
6. Halstead LS, VerVoort S, Seager SW. Rectal probe electrostimulation in the treatment of anejaculatory spinal cord injured men. Paraplegia. 1987;25:120.
7. Kamischke A, Nieschlag E. Treatment of retrograde ejaculation and anejaculation. Hum Reprod Update. 1999;5:448.
8. Thomson AB, Critchley HO, Kelnar CJ, et al. Late reproductive sequelae following treatment of childhood cancer and options for fertility preservation. Best Pract Res Clin Endocrinol Metab. 2002;16:311.
9. Stahl O, Eberhard J, Jepson K, et al. Sperm DNA integrity in testicular cancer patients. Hum Reprod. 2006;21:3199.
10. Shalet SM, Tsatsoulis A, Whitehead E, et al. Vulnerability of the human Leydig cell to radiation damage is dependent upon age. J Endocrinol. 1989;120:161.
11. Howell SJ, Shalet SM. Spermatogenesis after cancer treatment: damage and recovery. J Natl Cancer Inst Monogr. 2005;34:12–7.
12. Giwercman A, Petersen PM. Cancer and male infertility. Baillieres Best Pract Res Clin Endocrinol Metab. 2000;14:453.
13. Lee SJ, Schover LR, Partridge AH, et al. American Society of Clinical Oncology recommendations on fertility preservation in cancer patients. J Clin Oncol. 2006;24:2917.
14. Pryzant RM, Meistrich ML, Wilson G, et al. Long-term reduction in sperm count after chemotherapy with and without radiation therapy for non-Hodgkin's lymphomas. J Clin Oncol. 1993;11:239.
15. Fossa SD, Magelssen H. Fertility and reproduction after chemotherapy of adult cancer patients: malignant lymphoma and testicular cancer. Ann Oncol. 2004;15 Suppl 4:iv259.
16. Martin RH, Ernst S, Rademaker A, et al. Analysis of sperm chromosome complements before, during, and after chemotherapy. Cancer Genet Cytogenet. 1999;108:133.
17. De Mas P, Daudin M, Vincent MC, et al. Increased aneuploidy in spermatozoa from testicular tumour patients after chemotherapy with cisplatin, etoposide and bleomycin. Hum Reprod. 2001;16:1204.
18. Tempest HG, Ko E, Chan P, et al. Sperm aneuploidy frequencies analysed before and after chemotherapy in testicular cancer and Hodgkin's lymphoma patients. Hum Reprod. 2008;23:251.
19. Carlsen E, Giwercman A, Keiding N, et al. Evidence for decreasing quality of semen during past 50 years. BMJ. 1992;305:609.
20. Paulozzi LJ. International trends in rates of hypospadias and cryptorchidism. Environ Health Perspect. 1999;107:297.
21. Huyghe E, Matsuda T, Thonneau P. Increasing incidence of testicular cancer worldwide: a review. J Urol. 2003;170:5.
22. Whorton D, Krauss RM, Marshall S, et al. Infertility in male pesticide workers. Lancet. 1977;2:1259.
23. Anway MD, Cupp AS, Uzumcu M, et al. Epigenetic transgenerational actions of endocrine disruptors and male fertility. Science. 2005;308:1466.
24. Wittassek M, Wiesmuller GA, Koch HM, et al. Internal phthalate exposure over the last two decades – a retrospective human biomonitoring study. Int J Hyg Environ Health. 2007;210:319.
25. Duty SM, Silva MJ, Barr DB, et al. Phthalate exposure and human semen parameters. Epidemiology. 2003;14:269.
26. Hauser R, Meeker JD, Singh NP, et al. DNA damage in human sperm is related to urinary levels of phthalate monoester and oxidative metabolites. Hum Reprod. 2007;22:688.
27. Kamijima M, Hibi H, Gotoh M, et al. A survey of semen indices in insecticide sprayers. J Occup Health. 2004;46:109.
28. Swan SH, Kruse RL, Liu F, et al. Semen quality in relation to biomarkers of pesticide exposure. Environ Health Perspect. 2003;111:1478.
29. Recio R, Robbins WA, Borja-Aburto V, et al. Organophosphorous pesticide exposure increases the frequency of sperm sex null aneuploidy. Environ Health Perspect. 2001;109:1237.
30. Smith JL, Garry VF, Rademaker AW, et al. Human sperm aneuploidy after exposure to pesticides. Mol Reprod Dev. 2004;67:353.
31. vom Saal FS, Cooke PS, Buchanan DL, et al. A physiologically based approach to the study of bisphenol A and other estrogenic chemicals on the size of reproductive organs, daily sperm production, and behavior. Toxicol Ind Health. 1998;14:239.
32. Steinmetz R, Brown NG, Allen DL, et al. The environmental estrogen bisphenol A stimulates prolactin release in vitro and in vivo. Endocrinology. 1997;138:1780.
33. Krishnan AV, Stathis P, Permuth SF, et al. Bisphenol-A: an estrogenic substance is released from polycarbonate flasks during autoclaving. Endocrinology. 1993;132:2279.

34. Calafat AM, Kuklenyik Z, Reidy JA, et al. Urinary concentrations of bisphenol A and 4-nonylphenol in a human reference population. Environ Health Perspect. 2005;113:391.
35. Ikezuki Y, Tsutsumi O, Takai Y, et al. Determination of bisphenol A concentrations in human biological fluids reveals significant early prenatal exposure. Hum Reprod. 2002;17:2839.
36. Sun Y, Irie M, Kishikawa N, et al. Determination of bisphenol A in human breast milk by HPLC with column-switching and fluorescence detection. Biomed Chromatogr. 2004;18:501.
37. Takeuchi T, Tsutsumi O, Ikezuki Y, et al. Positive relationship between androgen and the endocrine disruptor, bisphenol A, in normal women and women with ovarian dysfunction. Endocr J. 2004;51:165.
38. Yamada H, Furuta I, Kato EH, et al. Maternal serum and amniotic fluid bisphenol A concentrations in the early second trimester. Reprod Toxicol. 2002;16:735.
39. Sugiura-Ogasawara M, Ozaki Y, Sonta S, et al. Exposure to bisphenol A is associated with recurrent miscarriage. Hum Reprod. 2005;20:2325.
40. Bagchi G, Hurst CH, Waxman DJ. Interactions of methoxyacetic acid with androgen receptor. Toxicol Appl Pharmacol. 2009;238:101.
41. Miller RR, Carreon RE, Young JT, et al. Toxicity of methoxyacetic acid in rats. Fundam Appl Toxicol. 1982;2:158.
42. Welch LS, Schrader SM, Turner TW, et al. Effects of exposure to ethylene glycol ethers on shipyard painters: II. Male reproduction. Am J Ind Med. 1988;14:509.
43. Cherry N, Moore H, McNamee R, et al. Occupation and male infertility: glycol ethers and other exposures. Occup Environ Med. 2008;65:708.
44. Emanuele MA, Emanuele NV. Alcohol's effects on male reproduction. Alcohol Health Res World. 1998;22:195.
45. Muthusami KR, Chinnaswamy P. Effect of chronic alcoholism on male fertility hormones and semen quality. Fertil Steril. 2005;84:919.
46. Vicari E, Arancio A, Giuffrida V, et al. A case of reversible azoospermia following withdrawal from alcohol consumption. J Endocrinol Invest. 2002; 25:473.
47. Villalta J, Ballesca JL, Nicolas JM, et al. Testicular function in asymptomatic chronic alcoholics: relation to ethanol intake. Alcohol Clin Exp Res. 1997; 21:128.
48. Donnelly GP, McClure N, Kennedy MS, et al. Direct effect of alcohol on the motility and morphology of human spermatozoa. Andrologia. 1999;31:43.
49. Emanuele NV, LaPagli N, Steiner J, et al. Peripubertal paternal EtOH exposure. Endocrine. 2001;14:213.
50. Orleans CT. Increasing the demand for and use of effective smoking-cessation treatments reaping the full health benefits of tobacco-control science and policy gains – in our lifetime. Am J Prev Med. 2007;33:S340.
51. Kulikauskas V, Blaustein D, Ablin RJ. Cigarette smoking and its possible effects on sperm. Fertil Steril. 1985;44:526.
52. Sepaniak S, Forges T, Fontaine B, et al. Negative impact of cigarette smoking on male fertility: from spermatozoa to the offspring. J Gynecol Obstet Biol Reprod (Paris). 2004;33:384.
53. Trummer H, Habermann H, Haas J, et al. The impact of cigarette smoking on human semen parameters and hormones. Hum Reprod. 2002;17:1554.
54. Saleh RA, Agarwal A, Sharma RK, et al. Effect of cigarette smoking on levels of seminal oxidative stress in infertile men: a prospective study. Fertil Steril. 2002;78:491.
55. Hull MG, North K, Taylor H, et al. Delayed conception and active and passive smoking. The Avon Longitudinal Study of Pregnancy and Childhood Study Team. Fertil Steril. 2000;74:725.
56. Zenzes MT, Puy LA, Bielecki R, et al. Detection of benzo[a]pyrene diol epoxide-DNA adducts in embryos from smoking couples: evidence for transmission by spermatozoa. Mol Hum Reprod. 1999;5:125.
57. Compton WM, Grant BF, Colliver JD, et al. Prevalence of marijuana use disorders in the United States: 1991–1992 and 2001–2002. JAMA. 2004;291:2114.
58. Close CE, Roberts PL, Berger RE. Cigarettes, alcohol and marijuana are related to pyospermia in infertile men. J Urol. 1990;144:900.
59. Rossato M, Pagano C, Vettor R. The cannabinoid system and male reproductive functions. J Neuroendocrinol. 2008;20 Suppl 1:90.
60. Robbins WA, Elashoff DA, Xun L, et al. Effect of lifestyle exposures on sperm aneuploidy. Cytogenet Genome Res. 2005;111:371.
61. Robbins WA, Vine MF, Truong KY, et al. Use of fluorescence in situ hybridization (FISH) to assess effects of smoking, caffeine, and alcohol on aneuploidy load in sperm of healthy men. Environ Mol Mutagen. 1997;30:175.
62. Rubes J, Lowe X, Moore II D, et al. Smoking cigarettes is associated with increased sperm disomy in teenage men. Fertil Steril. 1998;70:715.
63. Tidd MJ, Horth CE, Ramsay LE, et al. Endocrine effects of spironolactone in man. Clin Endocrinol Oxf. 1978;9:389.
64. Benoff S, Cooper GW, Hurley I, et al. The effect of calcium ion channel blockers on sperm fertilization potential. Fertil Steril. 1994;62:606.
65. Nelson WO, Bunge RG. The effect of therapeutic dosages of nitrofurantoin (furadantin) upon spermatogenesis in man. J Urol. 1957;77:275.
66. Hargreaves CA, Rogers S, Hills F, et al. Effects of co-trimoxazole, erythromycin, amoxycillin, tetracycline and chloroquine on sperm function in vitro. Hum Reprod. 1998;13:1878.
67. Birnie GG, McLeod TI, Watkinson G. Incidence of sulphasalazine-induced male infertility. Gut. 1981; 22:452.

68. Sarica K, Suzer O, Gurler A, et al. Urological evaluation of Behcet patients and the effect of colchicine on fertility. Eur Urol. 1995;27:39.
69. Baumgartner A, Schmid TE, Schuetz CG, et al. Detection of aneuploidy in rodent and human sperm by multicolor FISH after chronic exposure to diazepam. Mutat Res. 2001;490:11.
70. Collodel G, Scapigliati G, Moretti E. Spermatozoa and chronic treatment with finasteride: a TEM and FISH study. Arch Androl. 2007;53:229.
71. Carlsen E, Petersen JH, Andersson AM, et al. Effects of ejaculatory frequency and season on variations in semen quality. Fertil Steril. 2004;82:358.
72. Guzick DS, Overstreet JW, Factor-Litvak P, et al. Sperm morphology, motility, and concentration in fertile and infertile men. N Engl J Med. 2001;345:1388.
73. World Health Organization. WHO laboratory manual for the examination of human semen and sperm-cervical mucus interaction, 4th ed. Cambridge: World Health Organization, Published on behalf of the World Health Organization by Cambridge University Press; 1999, p. 128.
74. Cooper TG, Noonan E, von Eckardstein S, et al. World Health Organization reference values for human semen characteristics. Hum Reprod Update. 2010;16:231.
75. Keck C, Gerber-Schafer C, Clad A, et al. Seminal tract infections: impact on male fertility and treatment options. Hum Reprod Update. 1998;4:891.
76. Colpi GM, Lange A. Diagnostic usefulness of study of the round cells in seminal fluid. Acta Eur Fertil. 1984;15:265.
77. Wolff H. The biologic significance of white blood cells in semen. Fertil Steril. 1995;63:1143.
78. Prathalingam NS, Holt WW, Revell SG, et al. The precision and accuracy of six different methods to determine sperm concentration. J Androl. 2006; 27:257.
79. Tomlinson M, Turner J, Powell G, et al. One-step disposable chambers for sperm concentration and motility assessment: how do they compare with the World Health Organization's recommended methods? Hum Reprod. 2001;16:121.
80. World-Health-Organization. WHO laboratory manual for the examination and processing of human semen. 5th ed. Geneva: World-Health-Organization; 2010.
81. Bostofte E, Bagger P, Michael A, et al. Fertility prognosis for infertile men: results of follow-up study of semen analysis in infertile men from two different populations evaluated by the Cox regression model. Fertil Steril. 1990;54:1100.
82. Mayaux MJ, Schwartz D, Czyglik F, et al. Conception rate according to semen characteristics in a series of 15 364 insemination cycles: results of a multivariate analysis. Andrologia. 1985;17:9.
83. Kruger TF, Menkveld R, Stander FS, et al. Sperm morphologic features as a prognostic factor in in vitro fertilization. Fertil Steril. 1986;46:1118.
84. Mortimer D, Menkveld R. Sperm morphology assessment – historical perspectives and current opinions. J Androl. 2001;22:192.
85. Celik-Ozenci C, Jakab A, Kovacs T, et al. Sperm selection for ICSI: shape properties do not predict the absence or presence of numerical chromosomal aberrations. Hum Reprod. 2004;19:2052.
86. Ryu HM, Lin WW, Lamb DJ, et al. Increased chromosome X, Y, and 18 nondisjunction in sperm from infertile patients that were identified as normal by strict morphology: implication for intracytoplasmic sperm injection. Fertil Steril. 2001;76:879.
87. Aitken RJ. A free radical theory of male infertility. Reprod Fertil Dev. 1994;6:19.
88. Aitken RJ, Baker MA, Sawyer D. Oxidative stress in the male germ line and its role in the aetiology of male infertility and genetic disease. Reprod Biomed Online. 2003;7:65.
89. Griveau JF, Le Lannou D. Reactive oxygen species and human spermatozoa: physiology and pathology. Int J Androl. 1997;20:61.
90. Henkel R, Hajimohammad M, Stalf T, et al. Influence of deoxyribonucleic acid damage on fertilization and pregnancy. Fertil Steril. 2004;81:965.
91. Iwasaki A, Gagnon C. Formation of reactive oxygen species in spermatozoa of infertile patients. Fertil Steril. 1992;57:409.
92. Aitken RJ, Irvine DS, Wu FC. Prospective analysis of sperm-oocyte fusion and reactive oxygen species generation as criteria for the diagnosis of infertility. Am J Obstet Gynecol. 1991;164:542.
93. Sukcharoen N, Keith J, Irvine DS, et al. Prediction of the in-vitro fertilization (IVF) potential of human spermatozoa using sperm function tests: the effect of the delay between testing and IVF. Hum Reprod. 1996;11:1030.
94. Hammadeh ME, Radwan M, Al-Hasani S, et al. Comparison of reactive oxygen species concentration in seminal plasma and semen parameters in partners of pregnant and non-pregnant patients after IVF/ICSI. Reprod Biomed Online. 2006; 13:696.
95. McKinney KA, Lewis SE, Thompson W. Reactive oxygen species generation in human sperm: luminol and lucigenin chemiluminescence probes. Arch Androl. 1996;36:119.
96. Aitken RJ, West KM. Analysis of the relationship between reactive oxygen species production and leucocyte infiltration in fractions of human semen separated on Percoll gradients. Int J Androl. 1990; 13:433.
97. Saleh RA, Agarwal A, Kandirali E, et al. Leukocytospermia is associated with increased reactive oxygen species production by human spermatozoa. Fertil Steril. 2002;78:1215.
98. Aitken RJ, Buckingham DW, West K, et al. On the use of paramagnetic beads and ferrofluids to assess and eliminate the leukocytic contribution to oxygen radical generation by human sperm suspensions. Am J Reprod Immunol. 1996;35:541.

99. Shekarriz M, DeWire DM, Thomas Jr AJ, et al. A method of human semen centrifugation to minimize the iatrogenic sperm injuries caused by reactive oxygen species. Eur Urol. 1995;28:31.
100. Aitken RJ, Baker MA, O'Bryan M. Shedding light on chemiluminescence: the application of chemiluminescence in diagnostic andrology. J Androl. 2004;25:455.
101. Aitken RJ, Clarkson JS. Cellular basis of defective sperm function and its association with the genesis of reactive oxygen species by human spermatozoa. J Reprod Fertil. 1987;81:459.
102. Kobayashi H, Gil-Guzman E, Mahran AM, et al. Quality control of reactive oxygen species measurement by luminol-dependent chemiluminescence assay. J Androl. 2001;22:568.
103. Kodama H, Yamaguchi R, Fukuda J, et al. Increased oxidative deoxyribonucleic acid damage in the spermatozoa of infertile male patients. Fertil Steril. 1997;68:519.
104. Spano M, Bonde JP, Hjollund HI, et al. Sperm chromatin damage impairs human fertility. The Danish First Pregnancy Planner Study Team. Fertil Steril. 2000;73:43.
105. Evenson DP, Jost LK, Marshall D, et al. Utility of the sperm chromatin structure assay as a diagnostic and prognostic tool in the human fertility clinic. Hum Reprod. 1999;14:1039.
106. Duran EH, Morshedi M, Taylor S, et al. Sperm DNA quality predicts intrauterine insemination outcome: a prospective cohort study. Hum Reprod. 2002;17:3122.
107. Sun JG, Jurisicova A, Casper RF. Detection of deoxyribonucleic acid fragmentation in human sperm: correlation with fertilization in vitro. Biol Reprod. 1997;56:602.
108. Lopes S, Sun JG, Jurisicova A, et al. Sperm deoxyribonucleic acid fragmentation is increased in poor-quality semen samples and correlates with failed fertilization in intracytoplasmic sperm injection. Fertil Steril. 1998;69:528.
109. Bungum M, Humaidan P, Axmon A, et al. Sperm DNA integrity assessment in prediction of assisted reproduction technology outcome. Hum Reprod. 2007;22:174.
110. Huang CC, Lin DP, Tsao HM, et al. Sperm DNA fragmentation negatively correlates with velocity and fertilization rates but might not affect pregnancy rates. Fertil Steril. 2005;84:130.
111. Gandini L, Lombardo F, Paoli D, et al. Full-term pregnancies achieved with ICSI despite high levels of sperm chromatin damage. Hum Reprod. 2004;19:1409.
112. Jurisicova A, Varmuza S, Casper RF. Programmed cell death and human embryo fragmentation. Mol Hum Reprod. 1996;2:93.
113. Tesarik J, Mendoza C, Greco E. Paternal effects acting during the first cell cycle of human preimplantation development after ICSI. Hum Reprod. 2002;17:184.
114. Sakkas D, Mariethoz E, St John JC. Abnormal sperm parameters in humans are indicative of an abortive apoptotic mechanism linked to the Fas-mediated pathway. Exp Cell Res. 1999;251:350.
115. Sakkas D, Moffatt O, Manicardi GC, et al. Nature of DNA damage in ejaculated human spermatozoa and the possible involvement of apoptosis. Biol Reprod. 2002;66:1061.
116. Aoki VW, Moskovtsev SI, Willis J, et al. DNA integrity is compromised in protamine-deficient human sperm. J Androl. 2005;26:741.
117. Iguchi N, Yang S, Lamb DJ, et al. An SNP in protamine 1: a possible genetic cause of male infertility? J Med Genet. 2006;43:382.
118. Tremellen K. Oxidative stress and male infertility – a clinical perspective. Hum Reprod Update. 2008;14:243.
119. Greco E, Iacobelli M, Rienzi L, et al. Reduction of the incidence of sperm DNA fragmentation by oral antioxidant treatment. J Androl. 2005;26:349.
120. Smith R, Kaune H, Parodi D, et al. Increased sperm DNA damage in patients with varicocele: relationship with seminal oxidative stress. Hum Reprod. 2006;21:986.
121. Werthman P, Wixon R, Kasperson K, et al. Significant decrease in sperm deoxyribonucleic acid fragmentation after varicocelectomy. Fertil Steril. 2008;90:1800.
122. Zini A, Blumenfeld A, Libman J, et al. Beneficial effect of microsurgical varicocelectomy on human sperm DNA integrity. Hum Reprod. 2005;20:1018.
123. Smit M, Romijn JC, Wildhagen MF, et al. Decreased sperm DNA fragmentation after surgical varicocelectomy is associated with increased pregnancy rate. J Urol. 2010;183:270.
124. Potts RJ, Newbury CJ, Smith G, et al. Sperm chromatin damage associated with male smoking. Mutat Res. 1999;423:103.
125. Pasqualotto FF, Sharma RK, Potts JM, et al. Seminal oxidative stress in patients with chronic prostatitis. Urology. 2000;55:881.
126. O'Connell M, McClure N, Lewis SE. Mitochondrial DNA deletions and nuclear DNA fragmentation in testicular and epididymal human sperm. Hum Reprod. 2002;17:1565.
127. Greco E, Scarselli F, Iacobelli M, et al. Efficient treatment of infertility due to sperm DNA damage by ICSI with testicular spermatozoa. Hum Reprod. 2005;20:226.
128. Dohle GR. Male infertility in cancer patients: review of the literature. Int J Urol. 2010;17:327.
129. van der Horst-Schrivers AN, van Ieperen E, Wymenga AN, et al. Sexual function in patients with metastatic midgut carcinoid tumours. Neuroendocrinology. 2009;89:231.
130. The practice Committee of the Society for Assisted Reproductive Technology and the Practice committee of the American Society for Reproductive Medicine. Preimplantation genetic testing: a Practice Committee opinion. Fertil Steril. 2007;88:1497.

131. Helmerhorst FM, Perquin DA, Donker D, et al. Perinatal outcome of singletons and twins after assisted conception: a systematic review of controlled studies. BMJ. 2004;328:261.
132. Jackson RA, Gibson KA, Wu YW, et al. Perinatal outcomes in singletons following in vitro fertilization: a meta-analysis. Obstet Gynecol. 2004; 103:551.
133. The ESHRE Capri Workshop Group. Intracytoplasmic sperm injection (ICSI) in 2006: evidence and evolution. Hum Reprod Update. 2007;13:515.
134. Schlegel PN, Shin D, Goldstein M. Urogenital anomalies in men with congenital absence of the vas deferens. J Urol. 1996;155:1644.
135. Anguiano A, Oates RD, Amos JA, et al. Congenital bilateral absence of the vas deferens. A primarily genital form of cystic fibrosis. JAMA. 1992;267: 1794.
136. McCallum T, Milunsky J, Munarriz R, et al. Unilateral renal agenesis associated with congenital bilateral absence of the vas deferens: phenotypic findings and genetic considerations. Hum Reprod. 2001;16:282.
137. O'Flynn O'Brien KL, Varghese AC, Agarwal A. The genetic causes of male factor infertility: a review. Fertil Steril. 2010;93:1.
138. Ferlin A, Raicu F, Gatta V, et al. Male infertility: role of genetic background. Reprod Biomed Online. 2007;14:734.
139. Denschlag D, Tempfer C, Kunze M, et al. Assisted reproductive techniques in patients with Klinefelter syndrome: a critical review. Fertil Steril. 2004;82:775.
140. Nielsen J, Wohlert M. Chromosome abnormalities found among 34,910 newborn children: results from a 13-year incidence study in Arhus, Denmark. Hum Genet. 1991;87:81.
141. Engels H, Eggermann T, Caliebe A, et al. Genetic counseling in Robertsonian translocations der(13;14): frequencies of reproductive outcomes and infertility in 101 pedigrees. Am J Med Genet A. 2008; 146A:2611.
142. Kuroda-Kawaguchi T, Skaletsky H, Brown LG, et al. The AZFc region of the Y chromosome features massive palindromes and uniform recurrent deletions in infertile men. Nat Genet. 2001;29:279.
143. Hopps CV, Mielnik A, Goldstein M, et al. Detection of sperm in men with Y chromosome microdeletions of the AZFa, AZFb and AZFc regions. Hum Reprod. 2003;18:1660.
144. Kaufman GE, Malone FD, Harvey-Wilkes KB, et al. Neonatal morbidity and mortality associated with triplet pregnancy. Obstet Gynecol. 1998;91:342.
145. Pinborg A. IVF/ICSI twin pregnancies: risks and prevention. Hum Reprod Update. 2005;11:575.
146. Klemetti R, Gissler M, Sevon T, et al. Children born after assisted fertilization have an increased rate of major congenital anomalies. Fertil Steril. 2005; 84:1300.
147. Reddy UM, Wapner RJ, Rebar RW, et al. Infertility, assisted reproductive technology, and adverse pregnancy outcomes: executive summary of a National Institute of Child Health and Human Development workshop. Obstet Gynecol. 2007; 109:967.
148. Hansen M, Kurinczuk JJ, Bower C, et al. The risk of major birth defects after intracytoplasmic sperm injection and in vitro fertilization. N Engl J Med. 2002;346:725.
149. Zhu JL, Basso O, Obel C, et al. Infertility, infertility treatment, and congenital malformations: Danish national birth cohort. BMJ. 2006;333:679.
150. Koivurova S, Hartikainen AL, Gissler M, et al. Neonatal outcome and congenital malformations in children born after in-vitro fertilization. Hum Reprod. 2002;17:1391.
151. Ludwig M, Katalinic A. Malformation rate in fetuses and children conceived after ICSI: results of a prospective cohort study. Reprod Biomed Online. 2002;5:171.
152. Anthony S, Buitendijk SE, Dorrepaal CA, et al. Congenital malformations in 4224 children conceived after IVF. Hum Reprod. 2002;17:2089.
153. Ericson A, Kallen B. Congenital malformations in infants born after IVF: a population-based study. Hum Reprod. 2001;16:504.
154. Sutcliffe AG, D'Souza SW, Cadman J, et al. Minor congenital anomalies, major congenital malformations and development in children conceived from cryopreserved embryos. Hum Reprod. 1995; 10:3332.
155. Olson CK, Keppler-Noreuil KM, Romitti PA, et al. In vitro fertilization is associated with an increase in major birth defects. Fertil Steril. 2005; 84:1308.
156. Lie RT, Lyngstadaas A, Orstavik KH, et al. Birth defects in children conceived by ICSI compared with children conceived by other IVF-methods; a meta-analysis. Int J Epidemiol. 2005;34:696.
157. Hansen M, Bower C, Milne E, et al. Assisted reproductive technologies and the risk of birth defects – a systematic review. Hum Reprod. 2005;20:328.
158. Bonduelle M, Ponjaert I, Steirteghem AV, et al. Developmental outcome at 2 years of age for children born after ICSI compared with children born after IVF. Hum Reprod. 2003;18:342.
159. Zadori J, Kozinszky Z, Orvos H, et al. Dilemma of increased obstetric risk in pregnancies following IVF-ET. J Assist Reprod Genet. 2003;20:216.
160. Simpson JL, Lamb DJ. Genetic effects of intracytoplasmic sperm injection. Semin Reprod Med. 2001;19:239.
161. Alukal JP, Lipshultz LI. Safety of assisted reproduction, assessed by risk of abnormalities in children born after use of in vitro fertilization techniques. Nat Clin Pract Urol. 2008;5(3):140–50.

Update on Sperm Banking

22

Pankaj Talwar

Abstract

Sperm banking has become a widely accepted compendium of infertility treatment and in recent times mooted as having a new role as fertility insurance. Once looked upon with improbability, this practice has established time and time again to be a successful technique of keeping the anticipation of a family alive for countless families. The motives for storage are as diverse as humans themselves. So far, no limit has been established for how long human semen can be frozen when maintained and stored in appropriate liquid nitrogen storage. Scientific literature shows, conclusively, that the sperm motility, viability, and morphology are not affected by proper long-term cryopreservation. Cryo thaw semen pregnancies have been reported after 2–3 decades of semen banking. Appropriate screening should be carried out before semen banking, till standard guidelines are available. However, the current recommendations being followed for sperm banking are guidelines by the British Andrology Society and the Practice Committee of the American Society for Reproductive Medicine.

Keywords

Sperm banking • Cryopreservation of semen • Male gamete banking • Donor screening in sperm banking • Legislation related to sperm banking

Long life and desire to leave genetic footprints in this world has been a long cherished dream of human mankind. The concept of sperm banking has fulfilled these dreams partly by being able to preserve body cells and genital tissues. The banking of the male gametes involves initial exposure of the spermatozoa to the cryoprotectant and gradually cooling them to subzero temperatures as per desired cooling curve [1]. Semen sample in suitable container is then cryopreserved in liquid nitrogen at −196° centigrade for later use. Cryo-bioreposited semen sample is then thawed, gradually warmed to the room temperature and diluted with suitable buffered media. Cryoprotectant is washed away and the thawed sample evaluated and used for insemination, intracytoplasmic sperm injection (ICSI) or for research purposes.

P. Talwar (✉)
ART Centre, Army Hospital Research and Referral, Dhaula Kuan, New Delhi 110010, India
e-mail: pankaj_1310@yahoo.co.in

E. Seli and A. Agarwal (eds.), *Fertility Preservation: Emerging Technologies and Clinical Applications*, DOI 10.1007/978-1-4419-1783-6_22,

The male gamete must maintain its macro and microarchitectural structure along with genomic integrity during the whole procedure and recover its physiological functions completely after the procedure [2].

Factors known to effect outcome of this delicate procedure depends upon the quality of the semen specimen, developmental stage at which sperms are being frozen, type of cryoprotectant being used, and the freezing protocols [3].

Background of Sperm Banking

Freezing and thawing techniques have been used to cryopreserve semen since 1776 [4]. The newer techniques have been developed with scientific advancements in measurement of temperature and the chemistry of cryopreserving solutions. Equally important were innovations made in the nineteenth century involving understandings of liquefaction of gases and potential use of such refrigerants to cool and store specimens at extremely low temperatures.

Advancements in the equipments, disposables, and awareness about sterility have also helped the cryobiologist and embryologist to further enhance the freeze thaw cycle results. This approach of cryopreserving human semen has become an acceptable procedure in assisted reproductive technology (ART) laboratories across the world.

Few of the developmental mile stones are worth mentioning:

1776 – Spallanzani – observed the effects of freezing on human sperm and subsequent recovery of motility on warming.

1866 – Montegazza – proposed semen banking for veterinary practice and for soldiers going to battle field [5].

1930 – Shettels and Jahnel S – observed that sperm survives at temperatures as low as −269°C [6, 7].

1949 – Polge, Smith, and Parkes – accidental discovery of cryoprotective properties of glycerol [8].

1963 – Sherman – reported birth of first child born after insemination by sperms which had been frozen using liquid nitrogen [9].

Principles of Cryobiology

Cryopreservation is a process which maintains cellular life for an extended period of time at low subzero temperatures. The aim of any cryopreservation protocol is to minimize cell membrane damage associated with exposure to low temperatures, regulate cell volume during the procedure, and prevent lethal intracellular ice crystal formation. This can be achieved by controlling intracellular and extracellular movement of solutes and water [10]. The outcome of freeze thaw cycle depends on various factors:

1. The cryoprotectant (extender) in which the cells are suspended during cooling and warming.
2. The rate at which the cells are cooled.
3. The temperature at which the sample is plunged into liquid nitrogen.
4. The warming rate.
5. Cryoprotectant removal after thawing.

The major physical and chemical events of cryopreservation that lead to cell membrane damage, include the loss of water from the cell due to osmolarity of the cryopreservation medium, formation of intracellular and extracellular ice crystals, the resultant increase in osmolality of the remaining liquid and reversal of the process during thawing. At least half of the motile sperm sustains cryoinjury when subjected to cryopreservation and thawing [11].

The typical effects observed in frozen-thawed spermatozoa derive from the unique nature of the cell. Human spermatozoa are relatively simple cells with a large surface area/volume ratio and have high permeability to water. This ensures rapid osmotic equilibrium in the presence of cryoprotectants. The genetic material is highly condensed and is less prone to injury to cryoprotectants unlike embryos.

Biological Behavior of the Phospholipids Membrane of the Sperm

Cell membranes of all mammalian species are composed of a phospholipid bilayer and associated membrane proteins (receptors, enzymes, etc.). Phospholipid bilayers are composed of a polar head group with hydrocarbon tails. Specific phospholipids polar heads face outward and others face inward, with hydrocarbon tails directed inward from each polar head group. Some nonbilayer lipids are associated with specific integral proteins (receptors, enzymes, etc.). This association influences the protein's functional properties [12].

Cold shock etiology involves damage to the cellular membranes and alteration in metabolic function, probably caused by changes in the arrangement of membrane constituents. When a cell membrane is cooled, a reordering of the membrane components occur which leads to increased membrane viscosity and decreased fluidity. A decrease in temperature causes a thermotropic phase transition in the membrane phospholipids from a liquid-crystalline to a gel phase, resulting in more rigid membrane structure. The occurrence of cold shock can be prevented by controlling the rate of cooling and by adding protective compounds to semen diluents [13, 14, 15].

When nonbilayer lipids associated with the membrane proteins reach their transition temperature, a phase separation occurs within the plane of the membrane, which contains the gel-forming lipids. Lipids-associated proteins may be excluded laterally into areas of the membrane, which are still in a liquid state. The movement of the proteins within the membrane leads them to associate with different membrane lipids, possibly changing the activity of the protein. Rewarming of the membrane also could result in altered lipid/protein patterns.

Changes in Volume of Sperm When Exposed to Cryoprotectants

Spermatozoa undergo several volume changes during cryopreservation and thawing. The changes in cell volume throughout freezing and thawing are reliant upon water transport across the cell's plasma membrane. The first volume change occurs when the cryoprotective agent (CPA) is added to the sperm suspension.

The sperm undergoes rapid shrinkage as intracellular water leaves the cell in response to the hyperosmotic extracellular solution. Gradually, the sperms returns to its original volume as the CPA permeate the cell. The presence of CPA in the intracellular and extracellular water lowers the freezing point of the cell and the solutes, allowing them to remain in super cooled state.

Ice formation is initiated in the extracellular medium. As more water leaves the cryopreservation medium to contribute to crystal formation, the concentration of particles in the remaining liquid increases, resulting in a continuous drop in the freezing point of the remaining liquid and an increase in extracellular osmotic pressure. With rising hyperosmolarity of the extracellular fluid, more intracellular water is drawn out of the spermatozoa leaving the cell dehydrated and reducing the risk of lethal intracellular ice formation [16].

Indications of Semen Cryopreservation

Sperm banking is the process of semen cryopreservation using well-documented protocols for future use (Table 22.1). The semen is either preserved by an individual for his own future use, which is termed as autologous sperm banking or client depositor semen cryofreezing, or by fertile donors, after screening, for the purpose of third party reproduction [17, 18]. Adequate care is taken to do phenotypic/blood group matching in these cases. Matching physical characteristics and race of the partner, hair color, texture, and eye color are mandatory (Table 22.2).

Common Indications for Autologous Semen Banking

- Malignancy patients prior to surgery, chemotherapy, or radiation therapy.
- During the performance of cold surgical sperm retrieval (SSR) techniques like testicular or epididymal aspiration of sperms in cases with

Table 22.1 Current approach to male fertility preservation

Clinical presentation	Modality of sperm harvesting	Modality of banking
Infertile normal reproductive males requiring semen banking	Masturbation/electro-ejaculation	Semen freezing
Infertile azoospermic males		
Obstructive	Sample extracted from the epididymis – percutaneous epididymal sperm aspiration (PESA)	Freezing of the epididymal aspirate either unprepared or after gentle wash and swim up
Nonobstructive	Testicular sperm extraction/ aspiration (TESE/TESA)	Freezing of the testicular tissue is done after gentle teasing in a sterile dish using fine needles Tissue is frozen either unprepared or after density gradient wash
In malignancy patients before chemotherapy or radiotherapy		
Prepubertal boys	Masturbation	Semen banking
Pubertal	Testicular tissue (multiple samples)	Freezing of the testicular tissue

Table 22.2 Indications of semen cryopreservation

Donor semen banking	Autologous semen banking
Azoospermia (hypergonadotropic – hypogonadism)	Adult patients prior to surgery, chemotherapy, or radiation therapy
Very low sperm count and couple willing for donor insemination	Soldiers/frequent travelers before they go away for overseas assignments, or due to anticipated absence on day of insemination due to commitments of work
Male partner carries a genetic defect (Huntington's Chorea, etc.)	During performance of cold surgical sperm retrieval (SSR) techniques like testicular or epididymal aspiration of sperms in cases with azoospermia
Single women donor insemination	Anticipated performance anxiety on the day of assisted reproduction technology procedure
Recurrent IVF failures/abortions	Patients before undergoing vasectomy may preserve semen as insurance to further fertility Spinal cord injured patient: Electroejaculation specimen can be cryopreserved and successfully used in assisted reproduction for spinal cord injured patient Oligozoospermic/asthenozoospermic sample: Anticipation of enriching the number of motile sperm by pooling several cryopreserved samples

azoospermia or having the inability to achieve ejaculation.

- In case of anticipated absence of an individual on the day of insemination, such as soldiers, frequent travelers and individuals with professional and other commitments.
- Anticipated performance anxiety on the day of IUI/IVF.
- Semen retrieval by electro-ejaculation in men suffering from spinal cord injury.
- Occupational hazard (risk of radiation).
- Patients before undergoing vasectomy may preserve semen as insurance to further fertility.
- Pooling of semen sample when the native semen samples have low count.

Indications for Donor Semen Banking

Male factor infertility

- Azoospermia (hypergonadotropic hypogonadism).

- Morphological abnormalities of the sperms.
- Very low sperm count and couple willing for IUI-AID.
- Partner is either impotent or has retrograde ejaculations (if sperm retrieval has failed).

When exposed to known toxins, such as lead and agents with mutagenic potential.

Male partner carries a genetic defect (Huntington's Chorea, etc.).

Rh incompatibility with isoimmunization.

Single women donor insemination.

Recurrent IVF failures/abortions.

Recent indication desirous of designer baby.

Outline of Cryoprotectants

Biochemical and Physical Aspects of Sperm Cryopreservation

Sperms subjected to cryopreservation and thawing have reduced viability, motility and fecundity. Some of the morphological damages to the membranes and acrosome associated with cryopreservation and thawing are increased plasma membrane permeability, loss of acrosomal integrity and loss of superoxide dismutase enzyme from the plasma membranes [19, 20].

All the cryopreservation protocol are based on the theory that cell membrane damage can be minimized through the addition of suitable CPAs, buffer agents, controlling the osmolality and pH of cryopreservation medium and controlling the freezing rate during the procedure [21].

A variety of extenders (cryoprotective media) exist for the cooling and cryopreservation of semen. The purpose of the extender is manifold. The media contains nutrient, a buffer, a cryoprotectant agent, and an antibiotic. A typical nutrient added is a sugar, such as glucose or sucrose, which serves to provide energy source for the sperm. Buffers are added to balance pH and osmolarity of the solution. An ideal biological buffer should have a pH value between 6 and 8 [22]. The role of the buffer in cryopreservation is to pick up hydrogen ions in the surrounding media, thereby assisting in dehydration of the cell and maintaining a neutral pH.

Basics of Cryoprotectants as Applied to Semen Banking

Cryoprotectants have been divided into two classes, with first being permeating cryoprotectants, such as dimethylsulfoxide (DMSO), propylene glycol, and glycerol, and the second being non-permeating cryoprotectants, such as sucrose, raffinose and glycine.

Glycerol has remained the cryoprotectant of choice for the preservation of spermatozoa for most species [23]. Glycerol is superior to DMSO or ethylene glycol as a cryoprotectant [24]. The protective effects of glycerol are mediated by its colligative properties, depression of freezing point, alteration of cell membrane properties by inducing changes in lipid packing structure, and the consequent lowering of electrolyte concentrations in the unfrozen fraction at any given temperature, which will help to counter the harmful "solution effects" imposed during the freezing process [25]. In order to improve cryosurvival rates, more complex diluents containing other mainly nonpermeable CPAs, such as glycine, zwitterions, citrate, and egg yolk were developed [26]. Among the earliest and best known extenders for human semen is glycerol egg yolk citrate (GEYC). Lipoproteins and phospholipid fractions of egg yolk were originally added to cryopreservation media in effort to protect the sperm from the deleterious effects of cold shock by protecting the integrity of the plasma membrane. Studies have demonstrated the binding of various lipid constituents of egg yolk to the spermatozoa membrane [27]. Commonly used human sperm preserving medium (HPSM), is a modified Tyrode's medium containing glycerol (5–7.5% v/v final volume), sucrose, glucose, and glycine as CPAs, human serum albumin as stabilizing agent and (*N*-(2-hydroxyethyl)-piperazine-*N*′-(2-ethanesulfonic acid)) (HEPES) as the buffering agent. Other commonly used cryoprotective buffer is a zwitterion buffer system containing

N-tris-(hydroxymethyl)-methyl-2-aminoethane sulfonic acid (TES) and tris-(hydroxymethyl)-aminomethane (TRIS). This TES–TRIS combination (usually abbreviated to "TEST") is most often used in conjunction with egg yolk and citrate with glycerol as the permeating cryoprotectant. In the initial report, it proved superior to glycerol alone as a cryoprotectant. This remarkable result was due largely to the very rapid freezing protocol employed, and was impossible to duplicate using standard cooling methods [28].

The addition of the cryoprotectant glycerol has been shown to increase the motile sperm cryosurvival to an average of 50%, well above the <20% cryosurvival reported without addition of glycerol. The cryoprotectant DMSO was shown to be unsuitable for sperm cryopreservation as it had lower post-thaw percent motility when compared to glycerol. Numerous studies have reported irreversible reduction in sperm motility after exposure to glycerol and other intracellular CPAs [29].

Essentials of Freeze Thaw Cycle

Decision of Semen Packaging Protocol Before the Procedure and Its Importance to the Reproductive Biologist

Semen parameters should be thoroughly assessed using WHO criterion. We should decide the outcome of the procedure which will further guide the freezing protocol. If the sample is satisfactory with good count and motility, we may freeze raw sample with the aim to carry out IUI at latter date. On the other hand, an oligospermic sample with round cells and debris should be prepared and packaged for ICSI. Finally, the methodology depends upon the experience of the reproductive biologist and the laboratory protocols.

Specimen Glycerolization

Glycerol is added directly or indirectly, as a component of a cryopreservation medium, to neat semen/prepared semen (swim up) in drop by drop fashion slowly over a period of 2–3 min, with continuous mixing of both. This step is essential to reduce toxicity of the cryoprotectant as it can cause sudden osmotic shock to the sperms. The glycerol is metabolized during the procedure with the formation of neutral lipid. It is suggested that the metabolized glycerol may contribute to the plasma membrane of the sperm increasing its stability which may lead to improved post-thaw motility [30].

Packaging of Semen After Addition of Cryopreservative

Glycerated or extended semen can be cryopreserved in various containers. Factors, which influence the decisions, include the volume of sample to be cryopreserved, ease of container labeling, handling, storage, and recovery as well as biocompatibility of the packaging material. The packaging of semen into traditional 0.25 or 0.5 mL straws requires the use of vacuum pump and filling nozzle to aspirate the semen-cryoprotectant mixture. Straws made from an ionomeric resin (CBS High Security Straws, commonly referred to as "CBS straws": CryoBioSystem, Paris, France) have shown encouraging results. Straws are available in a variety of colors suitable for the easy identification of samples, and many hundreds can be stored in plastic goblets in canisters within liquid nitrogen vessels. Disadvantages of straws include: (1) Maximum capacity of approximately 0.5 mL only. (2) Overfilled straws are prone to cracking and expelling the powder sealing plugs into the liquid nitrogen. (3) Labeling difficulties. (4) Filling difficulties. (5) A high surface/volume ratio which makes the sample very susceptible to warming shock damage resulting from exposure to ambient temperatures during handling

Cryovials made from polypropylene with either polypropylene or polyethylene screw caps have the advantages of ease of labeling and handling. These are easy to fill and store nearly 1.5 mL of the semen plus cryoprotectant mixture. Storage on aluminum canes is not good as after sometime aluminum canes lose their memory and cryovials may jump off the holder. These occupy lots of cryocan space and are not good for busy semen bank. Some amount of leakage does occur which can be prevented by the use of NUNC™ CryoFlex™ tubes. Due to more capacity, have the potential of exploding upon thawing because liquid nitrogen trapped in the vials expand to many times its volume when it converts to gaseous nitrogen. The low surface: volume ratio and thick wall of the cryovial increase the time required for samples to reach critical temperatures and thus increase the risk of damage from brief exposure to ambient temperatures. Placing cryovials in drawers in racking system is more efficient, but prone to induce fluctuation in storage temperature during retrieval. Screw-top vials do not maintain their seals, and leakage of liquid in nitrogen into a container is common, with consequent risk of rupture during thawing. *Glass vials* transmit heat more uniformly and were preferred until safety considerations for embryologist and andrologist made it preferable to use plastic straws or vials. Glass vials are very fragile thus there use is discouraged these days. *Syringes* are difficult to be loaded and sealed safely and aseptically. It has the advantage for direct insemination, but this is outweighed by the excessive amount of space needed to store multiple ejaculates.

Cooling and Warming Rates

The outcome of sperm cryosurvival is related to the rate at which the cells are cooled and warmed. A high rate of cooling does not allow time for the sufficient amount of intracellular water to diffuse out of the cells. These can form ice crystals during further cooling and cause intracellular damage. During warming, these small crystals aggregate, forming larger crystals, which is called recrystallization. Formation of these large crystals is believed to cause cell damage or death by physical disruption of the plasma membrane and/or cell organelle membranes. A relatively slow cooling rate allows excess intracellular water to diffuse out of the cells, resulting in extreme cell shrinkage, cell dehydration, and a high intracellular solute concentration. The risk of intracellular ice formation is not completely eliminated by slow cooling. While slow cooling reduces the risk of lethal intracellular ice formation, extreme changes in cell volume and high intracellular solute concentrations may damage the cell membrane [31].

Freezing

Sperm cryopreservation is accomplished using liquid nitrogen vapors for noncontrolled rate or a programmable freezer for controlled rate cooling. Comparisons of the controlled rate and noncontrolled rate freezing have revealed no differences in sperm cryosurvival or post-thaw motility. Noncontrolled rate freezing in liquid nitrogen vapor is simpler and less expensive. Regardless of the cooling process, the ultimate quality control appraisal of sperm cryopreservation is the cryosurvival of the spermatozoa determined during thawing [32].

Storage

Once specimens attain temperature of −80 to −120°C, they are immediately plunged into liquid nitrogen. After plunging, the vials are quickly transferred to a precooled, labeled aluminum cane/goblet for storage in liquid phase of liquid nitrogen tank. When freezing semen by noncontrolled rate protocol, vials may be loaded on to the aluminum cane prior to cryopreservation to eliminate the need for transfer after cryopreservation. Every effort should be made to limit the time; specimens spend out of the liquid phase of liquid nitrogen.

Straws are quickly transferred to precooled, labeled plastic goblets, snapped onto a labeled aluminum cane. Straws should be oriented in the goblet so that identifying information can be read without completely withdrawing the straw from the goblet. Aluminum canes are placed in predetermined locations within the cryostorage vessel. All storage containers should be stored in a secured room in locked/chained containers. Liquid nitrogen dewars and storage tanks are available in a variety of sizes. Dewars require manual filling while most storage tanks have an automatic filling feature. Liquid nitrogen levels in storage units should be monitored regularly at all times. It is important to appreciate the length of time cryopreserved sperm may be stored for. At −196°C, storage of sperms, even for a lengthy period of time, does not affect the survival rates. Liquid nitrogen holds specimens at a temperature (−196°C) at which there is virtually no movement of atoms or molecules. At temperatures above −130°C, atoms, and molecules are able to move. Temperatures of −90°C and above allow ice crystal growth and even short periods of exposure to such temperatures can cause lethal damage to cells. As long as the cells are maintained at −196°C, the only known potential for cell damage is degradation of deoxyribonucleic acid (DNA) caused by background radiation

Based on normal background radiation of 0.1 rad/year, it has been predicted that the male gametes should maintain its genetic integrity for over 200 years when stored at −196°C.

Thawing

Practical approach to semen thawing would be to wash the vials/straws in running water. They should be cleaned externally till the sweating disappears over a period of few minutes. Once the sample is thawed, mix the sample well with a pipette before sampling. Perform the sperm count and assess the motility as per WHO guidelines. Specimen should be processed immediately after post-thaw analysis.

Effects of Cryofreeze/Thaw Cycle

The success of cryopreservation is measured by the number of motile spermatozoa recovered post thaw. There is nearly 30–40% loss in motility in the thawed sample. Sperm membranes are destabilized by the sudden transitions of the temperature (thermal stress), the volume alterations with water and cryoprotectant shifts and subsequent exposure to changing osmolarity (osmotic stress). At a molecular level, changes in membrane organization, such as modifications of specific lipid–protein interaction, phospholipid asymmetry, and lipid composition are implicated in the loss of permeability and rigidifying effect on the membrane fluidity. Cryopreservation has deleterious effects on spermatozoa, especially on plasmalemma, acrosomes and chromatin integrity [33]. Studies have compared the recovery of post-thaw motility obtained with the various cryoprotective extenders and some of their derivatives. Results are conflicting with no obviously superior candidate emerging, but this is probably a reflection of the various cooling and thawing rates employed by the different research groups, making comparison difficult. It is recommended that additional tests of functional/fertilization ability of the sperms, such as cervical mucus penetration or zona-free hamster oocyte fusion testing, are applied to assess the potential fertility of cryopreserved spermatozoa [34].

Legislation Pertaining to the Semen Banking

Indian Council of Medical Research has issued comprehensive guidelines for assisted reproductive technology centers. These are guidelines which may be applied to any functioning semen bank. An ART clinic or a law firm or any other suitable independent organization may set up a semen bank. All donors should produce their semen samples within the collection area of the center so that the sample cannot be substituted by the semen sample of others. It is essential that there is suitable privacy, time, and environment for patients to do this. Donor records and coding of the specimens stored must be kept securely. The centers should audit their cryobanks annually. As per the ICMR guidelines, the semen bank shall not supply semen of one donor for more than ten successful pregnancies. It will be the responsibility of the ART clinic or the patient, to inform the bank about a successful pregnancy. The bank shall keep a record of all semen received, stored, and supplied, and details of the use of the semen of each donor. This record will be liable to be reviewed by the accreditation authority. A semen bank may store a semen preparation for exclusive use on the donor's wife or on any other woman designated by the donor. An appropriate charge may be levied by the bank for the storage. In the case of nonpayment of the charges when the donor is alive, the bank would have the right to destroy the semen sample or give it to a bonafide organization to be used only for research purposes. In the case of the death of the donor, the semen would become the property of the legal heir or the nominee of the donor at the time the donor gives the sample for storage to the bank. In the UK, the semen is not normally stored for longer than 10 years or beyond the age of 55 years for donors; generally, it is not stored in France for more than 5 years. Donors may express a wish to further limit the period of storage or the number of pregnancies that can be obtained from one donor. This is restricted to ten children by the same donor. Confidentiality remains a primary consideration in most countries.

Donor Screening Prior to Semen Banking

Fresh donor insemination is not recommended for the fear of transmission of common infective diseases. Donors should be tested for HIV 1 and 2, HTLV I and II antibodies, hepatitis B surface antigen, hepatitis B core antibody, hepatitis C, RPR, TP-PA, cytomegalovirus antibodies, chlamydia and gonorrhea. Some agents and diseases that can be transmitted by the seminal fluid include HIV, hepatitis B, hepatitis C, and syphilis.

Donors are screened for the infections at the time of presentation. As a donor may be in window period of an infection, it is necessary to repeat the examination for hepatitis B and HIV after an appropriate quarantine period of 180 days. If the history or physical examination indicates infection, the donor should be rejected and advised to seek appropriate medical advice. Donor should be thoroughly screened for common genetic and communicable diseases and those specific to their geographic location before their enrolment in the program.

Cross-Infection in the Semen Banks

There is a potential danger of cross-infection within the bank thus the samples must be handled and stored with paramount care. Cryopreserved semen may be spilled in the cryocan, and the infectious organism (hepatitis B virus) may survive in the liquid nitrogen with the possibility of cross-infection of other stored samples. It is recommended that samples be stored in isolation cryocans till the quarantine period of HIV, HbsAg, and HCV. Use of CBS ionomeric straws which have been hermetically sealed offers protection against cross-infection. Men with malignancy often need to bank their semen at short notice as to preclude complete prefreeze infective screening. The semen of these men could be reposited in a quarantine tank until the requisite screening had been completed. Client depositor/autologous who has the comfort of time, e.g., men considering a vasectomy, a cryobank must insist upon screening for infective pathogens as a safety precaution for the security of the semen of other men stored in the same canister. When cryopreserving semen from patients who are known to carry an infective infection, e.g. HIV and HBsAG, the samples must be stored in separate "contaminated" tanks for each recognized pathogenic organisms. In the USA and in the UK, guidelines have been published. The HEFA is moving toward a position whereby laboratories are compelled to screen donors in this way [35].

Security of the Semen Bank

The straws or vials must be clearly labeled. Inventory control is of utmost importance. Every precaution must be made to ensure that each straw or vial can be linked to the sperm source, date of cryopreservation and specimen number, canister/cane or rack/cryocan number. The cryocans have a locking facility which must be utilized and limited staff allowed access to the keys. A lot of softwares is available in the industry for easy identification of the specimen. Some examples of vial or straw identification mechanisms currently employed by sperm banks include: computerized or manual bar coding system, color coding with cryomarkers, vial caps or straw plugs, or use of adhesive labels.

Secure cryopreservation of semen requires regular maintenance of the equipment and refilling of liquid nitrogen in the cryocans. Liquid nitrogen evaporates very quickly or the cans can leak thus causing loss of precious samples. The loss of stored semen from cancer or other patients like those having spinal cord injuries and in whom semen has been retrieved with electro-ejaculation is not only difficult to quantify but is also of immense emotional value to the owner/couple. All the details and records must be stored confidentially and country specific guidelines adhered to when carrying out cryofreezing at the center. The following measures are considered customary by various accreditation authorities:

1. The levels of LN2 in cryocans that are filled manually should be monitored on a regular basis, using a cryoscale.
2. Large cryobiorepository cans should be attached to an automated cryomanifold with "auto-fill" controller.
3. Low-level temperature sensors should be installed in all cryogenic storage tanks and also, an alarm should be connected to the tanks to alert biologist about unwarranted problems.

The Future of Semen Cryopreservation

Cryopreservation of human gametes expose them to numerous stresses, including mechanical, thermal, and chemico-physiological, which can lead to compromised function of the gametes. We have come a long way since accidental discovery of glycerol as cryoprotectant for human sperms. The enigma of finding an ideal cryoprotectant would be the holy grail of the modern cryobiology. The applications of semen cryopreservation which had humble origin in veterinary practice now entails cryopreservation of stem cell-derived male sperms, though in experimental stages using vitrification. The introduction of clinical ICSI in 1992 at Brussels opened a new era in the field of Assisted Reproductive Techniques, allowing couples with severe male factor infertility to hope for a child of their own genetic origin. Nearly 2 years later, it was established that pregnancy was possible by carrying out ICSI on sperms extracted by SSR techniques. Nagy et al. showed that comparable result in terms of pregnancy rates can be obtained by performing ICSI with ejaculated, epididymal, or testicular spermatozoa, although fertilization rates were significantly higher with the ejaculated sperm. Epididymal spermatozoa can be retrieved either by microsurgery or by percutaneous needle puncture. These can be subsequently cryofreezed [36, 37]. The frequent indication for epididymal aspiration is obstructive azoospermia, and thus it is not uncommon for relatively large quantities of sperm to be obtained and subsequently used for IVF, or even intrauterine insemination (IUI). Surplus sperm may be frozen for future use. If large numbers of epididymal spermatozoa are obtained, then density gradient centrifugation is an effective method for preparing those spermatozoa for subsequent use. Testicular specimens are contaminated invariably with large amounts of red blood cells and testicular tissues; additional steps are needed to isolate a clean preparation of spermatozoa. In order to free the seminiferous tubule-bound spermatozoa, it is necessary to use either enzymatic digestants (collagenase) or mechanical methods. For the latter, testicular tissues in supportive culture medium is macerated using glass coverslips until a fine slurry of dissociated tissues is produced, and the resulting suspension can then be processed for therapeutic use.

Excess testicular spermatozoa obtained in this manner can be frozen for future use in order to avoid further surgeries [38, 39]. Testicular spermatozoa can also be obtained from a needle biopsy, although only a small amount of tissue is usually retrieved and the resulting sperm yield is proportionately low. Another method is cryopreservation of single human spermatozoa in empty human Zona shell, which is established by Jacques Cohen. A hollow sphere remains when cellular material is removed from the zona. Because it can be seen and handled microscopically both before and after cyropreservation, it is an ideal capsule for freezing individual and small groups of sperm cells [40]. The conventional semen freezing techniques have proven inefficient in recovering sperms, from men with severe oligospermia, for IVF. Hence, the concept of cryoprotectant free vitrification of semen sample was mooted. An adequate number of sperms was recovered by ultrafast freezing the prepared sample and then warming them to 37°C using copper loops as carrier device [41, 42]. Semen banking is going to encompass testicular tissue banking in greater number of patients in the years to come.

Emerging Role of Semen Banking in Onco-ART

Malignancy is one of the common diseases with approximately 50% of men facing this diagnosis during the course of their lifetime. Till date, the focus of clinicians has been timely diagnosis and appropriate treatment of the patients. With increasing favorable outcome the focus is shifting to fertility preservation [43, 44]. This, in turn, has provided many patients with the opportunity to live complete lives, allowing them to reflect on life subsequent to treatment fertility issues, such as posttreatment marriage and parenthood are considered as important by many parents and young patients. At the time of cancer diagnosis, patients and clinicians alike are often weighed down by the large number of urgent tests and procedures that must be carried out in a timely manner. In the present scenario, proactivity is desired from the oncology and the Assisted Reproductive Team alike. There is no specific consideration for semen banking in such cases except the time factor and stress they are facing. Sample may be collected as in normal healthy males. While collecting semen for banking from young boys/teenagers, the parents, child, and clinicians face socio-psychological inconvenience due to the nature of the subject and such situations may require psychological counseling [45]. Another problem we face is in young males who are not permitted to masturbate as per the religious guidelines. Such males are advised testicular tissue extraction and banking. Similar protocol is being followed for young cancer patients who do no masturbate. It is recommended to bank multiple vials/straws of the samples as the recovery may not be very good especially in postchemotherapy and or radiotherapy patients. ICSI is the recommended treatment for such patients. Majority of the cancer patients do have sufficient time when they visit us. We should start the procedure on the day of referral itself if possible. Minimum six straws should be frozen from 2 to 3 samples collected 24 h apart. The patient and the parents should be counseled about the modality of treatment at latter date. The chances of offering ICSI should be disused at length. Semen banking is commonly carried out in patients with testicular cancers, Hodgkin's disease and leukemia [46].

Conclusions

Sperm banking has become a widely accepted compendium of infertility treatment and in recent times mooted as having new role as fertility insurance. Once looked upon with improbability, this practice has established time and time again to be a successful technique of keeping the anticipation of a family alive for countless families. The motives for storage are as diverse as humans themselves. So far, no limit has been established for how long human semen can be frozen when maintained and stored in appropriate liquid nitrogen storage. Scientific literature shows conclusively that sperm motility, viability, and morphology are not affected by proper long-term cryopreservation. Cryo thaw semen pregnancies have been reported after 2–3 decades of semen banking [47, 48]. Appropriate screening should be carried out before semen banking available guidelines. Currently, acceptable guidelines include those by the British Andrology Society [49] and the Practice Committee of the American Society for Reproductive Medicine [50].

See Appendix D

References

1. Parkes AS. Preservation of human spermatozoa at low temperatures. Br Med J. 1945;2:212–3.
2. Sherman JK, Buge RG. Observations on preservation of human spermatozoa at low temperatures. Proc Soc Exp Biol Med. 1953;82:686–8.
3. Hammerstedt RH, Graham JK, Nolan JP. Cryopreservation of mammalian sperm; what we ask them to survive. J Androl. 1990;11:73–88.
4. Spallanzani L, Opuscoli di Fisca, Amimale, e Vegetgabile, Opuscolo II. Osservazioni, e Speriencze inotrno ai Vermicelli Spermatici dell'Uomo e degli Amimali. Modena; 1776.
5. Montegazza J. Fisologia sullo sperma umano. Rendic reale Instit Lomb. 1866;3:183–5.
6. Shettels LB. The respiration of human spermatozoa and their response to various gases and low temperatures. Am J Physiol. 1940;128:404–15.
7. Hoagland H, Pincus G. Revival of mammalisan sperm after immersion in liquid nitrogen. J Gen Physiol. 1942;25:337–44.
8. Polge C, Smith AU, Parkes AS. Revival of spermatozoa after vitrification and dehydration at low temperatures. Nature. 1949;164:666–70.
9. Sherman JK. Research on frozen human semen. Past, present, and future. Fertil Steril. 1964;15:485–99.
10. Meryman HT. Mechanics of freezing in living cell and tissues. Science. 1956;124:515–21.
11. Quinn PH. Principles of membrane stability and phase behaviour under extreme conditions. J Bioenerg Biomemb. 1989;21:3–19.
12. Carruthers A, Melchior DL. Role of bilayer lipids in governing membrane transport processes. In: Aloia RC, Curtin CC, Gordon LM, editors. Lipid domains and the relationship to membrane function. New York: Liss; 1988. p. 201–25.
13. Smith AU, Ploge C, Smiles J. Microscopic observation of living cell during freezing and thawing. J Roy Microsc Soc. 1951;71:186–95.
14. Drobnis EZ, Crowe LM, Berger T, Anchordoguy TJ, Overstreet JW, Crowe JH. Cold shock damage is due to lipid phase transitions in cell membranes: a demonstration using sperm as a model. J Exp Zool. 1993;265:432–7.
15. Holt WV, North RD. Effects of temperature and restoration of osmotic equilibrium during thawing on the induction of plasma membrane damage in cryopreserved ram spermatozoa. Biol Reprod. 1992;46:1086–94.
16. Farrant J. Mechanism of cell damage during freezing and thawing and its prevention. Nature. 1965;205: 1284–7.
17. Shapiro SS. Strategies to improve efficiency of therapeutic donor insemination. In: Diamond MP, DeCherney AH, editors. Infertility and reproductive medicine clinics in North America. Male infertility. Philadelphia: WB Saunders; 1992. p. 469–85.
18. Bordson BL, Ricci E, Dicky RP, Dunway H, Taylor SN, Curole DW. Comparison of fecundability with fresh and frozen semen in therapeutic donor insemination. Fertil Steril. 1986;46:466–9.
19. Barthelemy D, Royere D, Hammahah S, Lebos C, Tharanne MJ, Lansac J. Ultrastructural changes in membrane and acrosome of human sperm during cryopreservation. Arch Androl. 1990;25:29–40.
20. Lasso JL, Noiles EE, Alvaraz JG, Storey BT. Mechanism of superoxide dismutase loss from human sperm cell during cryopreservation. J Androl. 1994;15: 255–65.
21. Van den Berg L, Soliman FS. Composition and pH changes during freezing of solutions containing calcium and magnesium phosphate. Cryobiology. 1969; 6:10–4.
22. Mortimer D. Semen cryopreservation. In: Mortimer D, editor. Practical laboratory andrology. Oxford: Oxford University Press; 1994.
23. Hammit DG, Walker DL, Willamson RA. Concentration of glycerol required for optimal survival and in vitro fertilization capacity of frozen sperm is dependent on cryopreservation medium. Fertil Steril. 1988;49:680–7.
24. Zimmerman SJ, Maude MB, Modldawer M. Freezing and storage of human semen in 50 healthy medical students. A comparative study of glycerol and dimethylsulfoxide as a preservative. Fertil Steril. 1964; 15:505–10.
25. Mahadevan M, Trounson A. Effects of cryopreservation media and dilution methods on the preservation of human spermatozoa. Andrologia. 1983;15: 355–66.
26. Prins GS, Weidel L. A comparative study of buffer systems as cryoprotectants for human spermatozoa. Fertil Steril. 1986;46:147–9.
27. Ramirez JP, Carreras A, Mendoza C. Sperm plasma membrane integrity in fertile and infertile men. Andrologia. 1992;24:141–4.
28. Jeyendran RS, Van der Ven HH, Kennedy W, Perez-Pelaez M, Zanelveld LJD. Comparison of glycerol and a zwitterions buffer system as cryoprotective media for human spermatozoa: effect on motility, penetration of zona-free hamster oocytes, and acrosin/proacrosin. J Androl. 1984;5:7.
29. Richardson DW, Sadlier RM. The toxicity of various non-electrolytes to human spermatozoa and their protective effects during freezing. J Reprod Fertil. 1967;14:439–44.
30. Mazur P. Freezing of living cells: mechanisms and implications. Am J Physiol. 1974;247:C125–42.
31. Tyler JPP, Kime L, Cooke S, Driscoll GL. Temperature change in cryo-containers during short exposure

to ambient temperatures. Hum Reprod. 1996;11: 1510–2.
32. Graham EF, Crabo BG. Some methods of freezing and evaluating human spermatozoa. Proc Natl Acad Sci U S A. 1978;4:274–304.
33. Bunge RG, Keetel WC, Sherman JL. Fertilization capacity of frozen human spermatozoa. Nature. 1953;172:767–8.
34. Mclaughlin EA, Ford WCL, Hull MGR. The contribution of the toxicity of a glycerol-egg yolk-citrate cryopreservative to the decline in human sperm motility during cryopreservation. J Repord Fertil. 1992;95:749–54.
35. Human Fertilization and Embryology Act. Schedule 3 para 2(2)(b). London: HMSO; 1990.
36. Oates RD, Lobel SM, Harris D, et al. Efficacy of intracytoplasmic sperm injection using intentionally cryopreserved epididymal sperm. Hum Reprod. 1996;11:133–8.
37. Elnaser TA, Rashwan H. Testicular sperm extraction and cryopreservation in patients with non-obstructive azoospermia prior to ovarian stimulation for ICSI. Middle East Fertil Soc J. 2004;9(2):128–35.
38. Oates RD, Mulhall J, Burges C, Cunniingham D, Carson R. Fertilization and pregnancy using intentionally cryopreserved testicular tissue as the sperm source for intracytoplasmic sperm injection in 10 men with non-obstructive azoospermia. Hum Reprod. 1997;12:734–9.
39. Gil-Salom M, Romero J, Minguez Y, et al. Pregnancies after intracytoplasmic sperm injection with cryopreserved testicular tissue. Hum Reprod. 1996;11: 1309–13.
40. Cohen J, Garrisi GJ, Congedo-Ferrara TA, Kieck KA, Schimmel TW, Scott RT. Cryopreservation of single human spermatozoa. Hum Reprod. 1997;12: 994–1001.
41. Isachenko V, Isachenko E, Katkov II, Montag M, Dessole S, Nawroth F, et al. Cryoprotectant-free cryopreservation of human spermatozoa by vitrification and freezing in vapor: effect on motility, DNA integrity, and fertilization ability. Biol Reprod. 2004;71(4):1167–73.
42. Isachenko E, Isachenko V, Katkov II, et al. DNA integrity and motility of human spermatozoa after standard slow freezing versus cryoprotectant-free vitrification. Hum Reprod. 2004;9(4):932–9.
43. Tomlinson MJ, Pacey AA. Practical aspects of sperm banking for cancer patients. Hum Fertil. 2003;6(3): 100–5.
44. Wallace WH, Anderson RA, Irvine DS. Fertility preservation for young patients with cancer: who is at risk and what can be offered? Lancet Oncol. 2005;6(4): 209–18.
45. Saito K, Suzuki K, Iwasaki A, Yumura Y, Kubota Y. Sperm cryopreservation before cancer chemotherapy helps in the emotional battle against cancer. Cancer. 2005;104(3):521–4.
46. Schmidt KL, Carlsen E, Andersen AN. Fertility treatment in male cancer survivors. Int J Androl. 2007; 30(4):413–9.
47. Horne G, Atkinson AD, Pease EH, et al. Live birth with sperm cryopreserved for 21 years prior to cancer treatment: case report. Hum Reprod. 2004;19(6): 1448–9.
48. Feldschuh J, Brassel J, Durso N, Levine A. Textbook of assisted reproductive technologies. Successful sperm storage for 28 years. Fertil Steril. 2005;84(4):1017.
49. British Andrology Society. British Andrology Society guidelines for the screening of semen donors for donor insemination. Hum Reprod. 1999;14(7):1823–6.
50. Guidelines for Gamete and Embryo Donation. The American society for reproductive medicine. Fertil Steril. 1998;70(4 Suppl 3):1S–3.

23 Preservation of Sperm Isolates or Testicular Biopsy Samples

Bhushan K. Gangrade

Abstract

Retrieval and cryopreservation of testicular sperm has become a cornerstone of infertility treatment in azoospermic men. In obstructive azoospermia, sperm may be retrieved from testis in almost 100% patients and can be frozen in multiple aliquots to avoid repeated surgeries. In approximately 50% of patients with nonobstructive azoospermia, enough sperm may be retrieved for IVF–intracytoplasmic sperm injection (ICSI). Any excess sperm remaining after ICSI can be frozen for future IVF attempts. Frozen testicular sperm are as effective in achieving successful pregnancies as freshly isolated testicular or ejaculated sperm. Treatment with motility-enhancing agents, such as pentoxifylline, induces motility in frozen-thawed testicular sperm and facilitates selection of viable sperm for ICSI. Special techniques, such as cryopreservation of single sperm within evacuated hamster zona or on a cryoloop, allow storage of extremely poor sperm samples. In USA, at this time, there is no legislative directive with respect to the posthumous procurement of sperm.

Keywords

Testicular biopsy samples • Sperm isolates • Cryopreservation of sperm • Infertility in men • Azoospermic men

The pioneering work of Steptoe and Edwards [1] in human reproduction leading to the birth of Louise Brown not only made history, but was also instrumental in developing a new field of reproductive technology for the treatment of infertility. While in vitro fertilization by coincubation of sperm and oocytes together works well when the sperm parameters (e.g., count, motility, morphology, etc.) are relatively within normal range, fertilization of oocytes is often compromised if the male partner has suboptimal semen characteristics. Microinjection-assisted fertilization techniques, such as zona pellucida drilling, subzona insemination, and direct sperm injection, have been historically employed to overcome the male factor infertility with varying success. The development of the technique of intracytoplasmic sperm

B.K. Gangrade (✉)
IVF Laboratory, Center for Reproductive Medicine, 3435 Pinehurst Avenue, Orlando, FL 32804, USA
e-mail: bkgangrade@hotmail.com

E. Seli and A. Agarwal (eds.), *Fertility Preservation: Emerging Technologies and Clinical Applications*, DOI 10.1007/978-1-4419-1783-6_23, © Springer Science+Business Media, LLC 2012

injection (ICSI) by Palermo et al. [2] has revolutionized the treatment of male factor infertility. Theoretically, this technique (ICSI) requires one viable sperm per oocyte. Originally, ICSI was employed to treat male factor infertility using ejaculated sperm; however, soon thereafter, successful pregnancies were reported following the injection of freshly retrieved testicular [3, 4] and epididymal [5] sperm and to treat infertility secondary to nonobstructive azoospermia (NOA) using testicular sperm [6]. Attempts at the retrieval and cryopreservation of testicular sperm in men with azoospermia have become integral practice in most if not all assisted reproductive technology programs.

Azoospermia

The complete absence of sperm in the ejaculate is referred to as azoospermia. This condition is observed in approximately 1% of the general population and approximately 10% of infertile men [7, 8]. The initial observation of the absence of sperm in the liquefied ejaculate is confirmed by centrifugation (300×*g* for 15 min) of the semen followed by a complete systematic microscopic search of the resuspended pellet. Failure to observe any sperm in the ejaculate on two occasions following at least 2–3 days of abstinence is suggestive of the diagnosis of azoospermia. Absolute absence of sperm in the ejaculate may be caused by either an obstruction in the reproductive tract (obstructive azoospermia) or by actual failure of sperm production in the seminiferous tubules (NOA). Occasionally, in some NOA patients, small isolated portions of seminiferous tubules may show some spermatogenesis, but the number of sperm released is so few that no sperm are found in the ejaculate. Such patients are, therefore, essentially given a diagnosis of NOA.

Obstructive Azoospermia

Obstructive azoospermia may be congenital (congenital bilateral absence of the vas deferens or CBAVD) or acquired (vasectomy, infection in the reproductive tract, or surgery-related). Men with obstructive azoospermia usually have normal spermatogenesis and sperm may be directly retrieved surgically from the testis or epididymis.

Nonobstructive Azoospermia

Some of the common causes of the NOA are Klinefelter's syndrome, testicular failure secondary to radiation or chemotherapy, primary germ cell failure, Sertoli cell-only syndrome, or hypophyseal (hypogonadotropic) hypogonadism. Other causative factors may include testicular torsion, mumps orchitis, and cryptorchidism. Some men with NOA exhibit focal spermatogenesis in isolated seminiferous tubules. When less than six mature spermatids are observed in the cross section by histological analysis, usually sperm are absent in the ejaculate [9], but a thorough examination of testicular biopsy tissue may yield few sperm in isolated tubules that may be enough to fertilize occytes by ICSI and achieve pregnancy. A variation of the technique of open testicular biopsy (*Te*sticular *s*perm *e*xtraction, TESE), sometimes referred to as micro-TESE, involves bisection of the testis and thorough examination of the seminiferous tubules by the urologist under optical magnification for the identification and excision of portions of seminiferous tubules with active spermatogenesis. Successful sperm retrieval following micro-TESE in almost 50% of men diagnosed with NOA has been reported [10, 11]. A comparison of standard TESE with micro-TESE shows that the latter surgical procedure is more effective in successful sperm retrieval in NOA patients [12]. A positive correlation between testis volume and the probability of retrieving sperm has been reported [13]. In men with small or atrophied testis (volume <5 mL), micro-TESE offers a better chance of retrieving sperm as compared with TESE. Furthermore, FSH levels in men with NOA negatively correlate with the probability of finding spermatozoa in testis; however, this correlation has recently been refuted by some researchers [13].

In NOA, in order to find sufficient sperm for ICSI, it may be necessary to remove relatively larger amount of testicular tissue. Sometimes, multiple biopsies at different loci are needed which in turn may negatively affect testicular androgen production. The decrease in androgen production following removal of testicular tissue may be more pronounced in men with relatively small or atrophied testis. Even though micro-TESE is a more invasive surgical procedure, the amount of tissue excised in micro-TESE is significantly less as only a few seminiferous tubules that show active spermatogenesis are removed.

In men with Klinefelter's syndrome, sperm can be retrieved in 21–69% of patients [13, 14]. Since these men have a high incidence of testicular atrophy, micro-TESE might be a better option for sperm retrieval. In order to avoid repeat testicular biopsy surgeries for future IVF–ICSI attempts, sperm may be frozen in multiple aliquots as described in a later section.

The role of Y chromosome in spermatogenesis is well-recognized. Microdeletions in the proximal regions of the long arm (AZFa and/or AZFb) of the Y chromosome have been shown to cause partial or complete failure of spermatogenesis. The extent of severity in terms of spermatogenesis depends on the size of the microdeletion in AZFa or AZFb region. In contrast, microdeletions in the distal AZFc region may cause mild to severe oligospermia and in some cases even azoospermia [15]. Approximately 10–15% of the patients of NOA may have microdeletions in one or more regions of the Y chromosome [16]. These couples if undergoing IVF–ICSI with testicular sperm should be advised extensively regarding the transmission of defective Y chromosome to the male progeny.

In men with NOA, even though it is possible to find sperm in isolated seminiferous tubules following testicular biopsy, the yield (of sperm) is significantly less than that usually retrieved in normospermic patients with obstruction. Testicular sperm from NOA patients can be frozen and used for ICSI at a later time. Several techniques for sperm retrieval from azoospermic men have been described. In men with obstructive azoospermia, sperm can be retrieved in almost all cases by testicular fine needle aspiration (TEFNA), *t*esticular *s*perm *a*spiration (TESA), *pe*rcutaneous epididymal *s*perm *a*spiration (PESA), or microepididymal sperm aspiration (MESA). In some cases, open conventional testicular biopsy or TESE may be preferred especially if the intent is to cryopreserve multiple aliquots of sperm for future use or other retrieval techniques have been unsuccessful in sperm retrieval.

Reasons for the Cryopreservation of Testicular Sperm

Cryopreservation of surgically retrieved testicular sperm from male partner prior to initiating the treatment cycle of the female almost ensures the availability of the male gamete on the day of the egg retrieval. This reason for cryopreservation may not be of as much significance in men with obstructive azoospermia since such patients usually have normal spermatogenesis. On the other hand, in patients with NOA, advance cryopreservation of sperm offers a psychological relief from the anxiety and uncertainty related to sperm retrieval on the same day when the female partner is undergoing oocyte retrieval. In addition, testicular sperm cryopreservation avoids the logistical difficulties of coordinating the surgeries for both the male and the female on the same day [17]. Administration of anesthesia to both the male and the female partner on the same day for gamete retrieval makes it necessary on patient's part to ask for outside help from friend or relative to accompany and transport them from the clinic/hospital after the surgery. This puts extra burden on the couple and also makes it difficult to keep the privacy, if they so desire, regarding their infertility treatment. Repeated testicular aspirations or biopsies have been shown to adversely affect testicular function [18], and cryopreservation of multiple aliquots of sperm avoids the need for repeated surgeries if the couple desires additional IVF/ICSI attempts.

Men who have had vasectomy, but decide to regain their fertility, often opt for vas reversal as the first line of treatment. Even though the vas patency rate for an obstructive interval of 10–15 years is quite high (approximately 74%), the reconstruction surgery is unsuccessful in approximately one quarter of the patients [19]. Between 50 and 80% of the patients who have undergone vasectomy reversal exhibit normal semen parameters (sperm concentration, morphology, motility, and progression) after the reconstruction surgery, but fail to achieve pregnancy due to the development of antisperm antibodies [20, 21]. Since the spermatozoa come into contact with the antibodies in the epididymis or in the ejaculatory duct, sperm retrieval from the site of origin, i.e., testis, avoids contact of spermatozoa with necrotic milieu. Cryopreservation of testicular/epididymal sperm in men at the time of vas reversal surgery avoids the need for a repeat surgical intervention if either vas reconstruction is unsuccessful or if the couple fails to achieve a successful pregnancy despite normal semen parameters. Use of cryopreserved testicular sperm has been documented in the case of necrospermia (a condition in which ejaculate contains only dead sperm) secondary to antisperm antibodies [22]. Exploratory or diagnostic testicular biopsies in men seeking infertility treatment if scheduled as "possible testicular sperm freeze" may in some cases avoid the need for another testicular biopsy [23]. Cryopreservation of testicular sperm has also been successfully used in men with spinal cord injury [24].

Cryopreservation of testicular sperm to preserve fertility in prepubertal boys undergoing gonadotoxic therapy for cancer has also been suggested [25, 26]. Even if sperm are not found in the ejaculate, there may be focal spermatogenesis in isolated tubules and enough testicular sperm may be retrieved and cryopreserved in prepubertal (10–13 years old) boys.

Retrieval of Sperm from Testicular Tissue

Testicular parenchymal biopsies are taken from a nonvascular region in or near equatorial plane. Approximately 5×5 mm biopsy samples from obstructive azoospermic men provide enough sperm that can be frozen in 3–4 (or more) aliquots, thus allowing several IVF–ICSI attempts. A small (1–2 mm) piece of seminiferous tubule in a drop of culture medium (HEPES-HTF) is transferred to a glass slide for a squash preparation and observed under the microscope and 200–400× magnification. It is usually not feasible to perform a quantitative sperm analysis in terms of the concentration, motility, vitality, and morphology in the testicular biopsy samples because of the extremely low sperm concentration. A subjective assessment of the number of spermatozoa, however, should be made to determine the number of aliquots that may be frozen. The testicular parenchyma is transferred to a petri dish and minced using a pair of fine-tipped sterile scissors or alternatively teased with a pair of 27-gauze needles attached to tuberculin syringes and squeezed to release the contents of the seminiferous tubules. A comparison of four different mechanical methods (shredding between two glass slides, mincing, vortexing, and crushing using electrical Potter) showed identical yields of sperm [27]. Since human sperm can survive at room temperature for prolonged duration [28], heated temperature-controlled (37°C) surface is not needed. Some investigators, however, prefer to maintain the temperature of contents at 37°C [29]. If the testicular cell suspension exhibits excessive red blood cells, short (5 min) exposure to erythrocyte lysis buffer (155 mM NH_4Cl, 10 mM $KHCO_3$, 2 mM EDTA; pH 7.2) effectively removes blood cells [27] without causing any detrimental effect to sperm [27, 30].

Isolation and Cryopreservation of Testicular Sperm

Thawing of Testicular Sperm

Frozen testicular sperm is thawed on the day of the oocyte retrieval approximately 3 h before performing ICSI. It is prudent to ascertain that mature (Metaphase II) oocytes are available for ICSI prior to thawing sperm. A vial of frozen sperm is removed from the liquid nitrogen and thawed at room temperature for 10 min. The contents of the vial are diluted by slowly adding equal volume of HSA (10 mg/mL)-supplemented

HEPES-HTF medium. The suspension is transferred to a 15-mL conical bottom tube and centrifuged at 300×*g* for 10 min. The supernatant is discarded and the crude sperm pellet is washed once by resuspending in one ml HEPES-HTF and centrifuged again as above. The final sperm pellet is resuspended in 0.1 mL medium and stored at room temperature until ICSI.

Selection of Viable Sperm in Testicular Biopsy Samples

With the introduction of ICSI, now it is possible to achieve high fertilization rate with sperm irrespective of the source. Testicular sperm are as effective as ejaculated sperm provided that viable sperm are selected for injection [31]. Testicular sperm usually are immature and immotile and achieve motility during the passage through the epididymis. The main characteristic that easily differentiates viable sperm from nonviable is the motion or slight tail movement. There is a further reduction in sperm viability after freeze and thaw. Since the testicular sperm are largely immotile, it is important to identify the viable sperm for ICSI. Several methods have been proposed to select viable sperm from fresh or frozen testicular specimen. Prolonged in vitro culture of fresh [32–35] and frozen [36] testicular sperm for 48–72 h significantly improves the proportion of motile sperm in the sample. The in vitro culture appears to mimic the maturation of spermatozoa and acquisition of motility in the epididymis. In fact, fresh testicular sperm may survive for up to 2 weeks at 37°C, whereas at slightly lower temperature (32°C) that is in line with scrotal temperature, their survival may prolong even longer. In case of obstructive azoospermia, in vitro culture of testicular sperm for 72 h followed by cryopreservation offers a high rate of recovery of motile spermatozoa at thaw [36].

Identification of the physiological integrity of the plasma membrane has been used to differentiate between viable and nonviable spermatozoa for ICSI. Live sperm when exposed to a hypo-osmotic solution (75 mmol/L fructose and 25 mmol/L sodium citrate dehydrate) exhibit swelling/curling of the tail, whereas dead sperm remain unaffected [37]. Successful pregnancies were achieved by incorporating hypo-osmotic swelling (HOS) test as the method to differentiate and select viable sperm for ICSI. A comparison of three different hypo-osmotic solutions (original Jeyendran solution containing fructose and sodium citrate as above, a mixture of 50% culture medium and 50% MilliQ water, and MilliQ water) showed that 50% culture medium with 50% MilliQ water offers better sperm quality than original HOS solution or pure MilliQ water [38]. Use of hypo-osmotic solution consisting of culture medium diluted (1:1) with MilliQ water is a simple and effective method for the selection of viable testicular and ejaculated spermatozoa [39].

Pentoxifylline, an inhibitor of phosphodiesterase activity, enhances sperm motility by increasing the accumulation of intracellular cAMP. In vitro exposure to pentoxifylline also causes an increase in acrosome reaction [40], sperm penetration in zona-free hamster oocytes [41], and preserves functional membrane integrity of human spermatozoa [42]. Pentoxifylline has been shown to enhance motility in normozoospermic as well as asthenozoospermic samples [43]. In vitro exposure of sperm to pentoxifylline prior to intrauterine insemination significantly increases pregnancy rate as compared to standard sperm preparation [44, 45]. The inability of pentoxifylline to improve in vitro fertilization in patients with history of failed or reduced fertilization [46, 47] is of limited significance in the era of ICSI. There is no doubt that pentoxifylline treatment significantly improves sperm motility in vitro and can be used to select viable but otherwise nonmotile testicular sperm. The average time to search for viable (motile) sperm following pentoxifylline exposure in fresh and frozen azoospermic TESE samples is significantly less than in its absence [48].

For induction of motility in testicular (fresh or frozen) sperm, a 30 μL flattened drop of pentoxifylline (3.6 mM) is made in the ICSI dish. In addition, the ICSI dish also contains drops of HEPES-HTF for holding oocytes and a drop of polyvinylpyrrolidone (PVP; 10%) for immobilizing the sperm. For ICSI, 5 μL fresh or thawed testicular sperm sample is added to the pentoxifylline drop. Motility is induced in testicular sperm in approximately 10 min. The motile hyperactivated sperm swim to the periphery of the pentoxifylline drop and are picked up using the ICSI pipette and transferred to the PVP drop for washing and immobilization. ICSI is then performed according to the standard protocol. The direct addition of sperm to the pentoxifylline drop is a simple and quick procedure to select viable but otherwise nonmotile sperm for ICSI.

Cryopreservation of Testicular Sperm Inside Evacuated Zona Pellucida

Cryopreservation of testicular sperm as described earlier works quite well (such as in case of obstructive azoospermia) when several hundred to a few thousand sperm are available for freezing. There is always a decrease in the proportion of viable spermatozoa after thawing the frozen testicular preparation. This is usually not of much concern in frozen-thawed obstructive azoospermic TESE samples since sufficient numbers of viable spermatozoa are present for ICSI. However, in patients with NOA, finding sperm in post-thaw TESE sample at the time of ICSI could be challenging and frustrating. In our experience, in frozen-thawed TESE samples from NOA patients, it may take up to 30 min or more to find a single twitching sperm suitable for ICSI. The prolonged exposure of oocytes to external environment while searching for a viable spermatozoon compromises the embryo quality and may adversely affect the outcome of the procedure.

In 1997, Cohen et al. [49] reported a unique method of cryopreserving a single sperm or few spermatozoa within an evacuated zona from hamster or mouse egg. This method can be used to freeze sperm in extreme cases when less than a few hundred sperm are available. This approach requires extensive preparation, expertise, and time and has been sparingly utilized in clinical setting. The empty zona envelop serves as a small container for spermatozoa that is easy to locate and handle. Locating the sperm inside the zona is easy and relatively less time consuming. It almost guarantees the availability of sperm at the time of ICSI. A brief description of the method of sperm cryopreservation inside evacuated zona is presented below.

If TESE reveals the presence of sperm in the sample, a 50-mm dish (Falcon 35-1006) is used to isolate and retrieve spermatozoa from the crude TESE preparation. Incubation of TESE preparation at room temperature for several (up to 24) hours may induce in vitro maturation and motility in viable sperm. A large, surface flat drop (called sedimentation layer) of HEPES-HTF with HSA (10 mg/mL) is made in the upper center portion of the dish. Approximately 5 μL of testicular sample preparation is added to the center of the sedimentation layer carefully without disturbing the contents. The cellular debris falls to the bottom in the sedimentation layer while rare sperm slowly swim to outer peripheral edges.

Frozen hamster oocytes purchased from a commercial vendor (Charles River Inc., Wilmington, MA, USA) are emptied of the ooplasm as follows. Cryostraws containing eggs are thawed as per supplier's instructions. Morphologically healthy oocytes are washed in the culture medium and transferred to 5 μL droplet of HEPES-HTF with HSA (10 mg/mL) in a micromanipulation dish (Falcon, 35-1006). The drop is covered with mineral oil to prevent evaporation. The oocyte is held against the holding pipette and penetrated with an ICSI-like micropipette (15 μm). The ooplasm is aspirated by moving the micropipette around the egg and applying suction until the zona is empty. Once the zona envelop has been cleared of ooplasm, spermatozoa in sedimentation drop are aspirated using an ICSI pipette and transferred to a 5 μL droplet of PVP (10%). Depending on the total number of sperm available for freezing, one to five sperm are loaded into the ICSI pipette and injected in the empty zona. The spermatozoa are not immobilized at this time. The zona containing sperm is then loaded into a cryostraw (one zona per straw) and frozen according to laboratory's standard sperm cryopreservation procedure.

In order to recover the sperm on the day of ICSI, straws are thawed by immersing in a water bath at 30°C. The contents of the straw are expelled into a drop in a culture dish. The zona is retrieved, washed in HTF-HEPES medium, and transferred to an ICSI dish. The ICSI dish contains a PVP drop surrounded by several drops of HTF-HEPES. The zona envelop is held onto a holding pipette and an ICSI micropipette preloaded with 10% PVP solution is inserted into the zona. It is preferable to enter the zona through the original hole that was made to deposit the sperm inside. The sperm are aspirated and transferred to the PVP drop, immobilized, and used for ICSI. The HTF-HEPES drops surrounding the PVP drop are used to contain cumulus-free oocytes.

Successful pregnancies and live birth were reported using post-thawed spermatozoa frozen inside hamster zonae [50]. Modifications of the technique of sperm cryopreservation in evacuated zona have been reported and claimed to be superior to conventional freezing methods [51, 52]; however, the feasibility of its widespread use and acceptance in clinical context remain to be seen.

Use of cryoloop (Hampton Research, Laguana, CA) to cryopreserve single (or few) sperm has also been attempted [53], but has not gained wide acceptance for clinical use.

Cryopreservation of Intact Seminiferous Tubules and Testicular Tissue

Cryopreservation of excess testicular sperm after ICSI has been a common practice in IVF laboratories. Some health care providers and/or patients may prefer to cryopreserve testicular sperm before the retrieval. Cryopreservation of intact testicular tissue on the other hand has achieved little attention and acceptance. There are a few reports describing the feasibility of testicular tissue cryopreservation in obstructive and NOA. One of the earlier attempts to cryopreserve testicular tissue involved freezing of isolated seminiferous tubules from men with obstructive and NOA [54]. The freezing protocol for tubule cryopreservation is presented below.

Intact tubules obtained by PESA are frozen in sperm-freezing medium containing 15% glycerol. Briefly, pieces of seminiferous tubules are washed in culture medium and transferred to cryovials in 0.5 mL of HTF-HEPES. Equal volume (0.5 mL) of sperm-freezing medium is added slowly to the vial and allowed to equilibrate for 30 min at room temperature. Cryovials are placed in a −20°C standard freezer (30 min), exposed to the liquid nitrogen vapor (30 min) in a wire basket, and then plunged in liquid nitrogen for storage.

The seminiferous tubules are thawed at room temperature for 15 min and then washed in HTF-HEPES medium. Spermatozoa are squeezed out of the tubules by 27-gauze tuberculin needle, washed by centrifugation, and used for ICSI.

Cryopreservation of testicular tissue from men followed by successful extraction of spermatozoa [55] is also reported. This method of freezing tissue is similar to that used for cryopreservation of seminiferous tubules described above. The biopsy tissue is incubated for 30 min in sperm-freeze medium and then slow cooled for 10 min at 4°C. The cryovials are then vapor cooled by hanging in liquid nitrogen vapor for 30 min and plunged in liquid nitrogen.

The thawing procedure involves warming the tissue in a water bath (37°C). The tissue pieces are washed with warm (37°C) HEPES-HTF and minced in a petri dish using 27-gauze syringe needles or a pair of sharp scissors. The testicular tissue suspension is then processed as described earlier for isolating sperm and performing ICSI.

The reason for the lack of interest on the part of reproductive laboratory personnel in freezing the testicular tissue appears to be the extra time and effort needed for preparation of sperm on the day of ICSI. In comparison, thawing of previously frozen testicular sperm is a relatively simple and less time-consuming procedure. Moreover, there is no additional advantage of freezing the testicular tissue when compared with freezing the sperm in azoospermic men.

Successful freezing and thawing of testicular tissue in prepubertal boys, who lack sperm in the ejaculate but require chemo- or radiation therapy for cancer treatment, would open a major avenue in fertility preservation [56]. It has been proposed [57] that cryopreservation of testicular cell suspension may be a better option than tissue cryopreservation in young boys who have not yet achieved spermatogenesis. It is, however, important to note that the basic architecture of testicular tissue is important in the development of spermatozoa from spermatogonial cells since physical contact with other cells (e.g., Sertoli cells) is required for spermatogenesis [58]. Keeping this in view, attempts have been made to freeze testicular tissue from young prepubertal males undergoing gonadotoxic therapy. Keros et al. [56] reported successful cryopreservation of prepubertal testicular tissue in boys undergoing cancer therapy. Small (2–5 mm^3) pieces of testicular tissue were frozen using 5% dimethyl sulfoxide (DMSO) and a slow freezing protocol. The tissue was cooled at a rate of −1°C/min from room temperature to 0°C and held for 5 min followed by cooling at −0.5°C/min until temperature dropped to −8°C. Manual seeding was performed at this time and cryovials were held at this temperature for 10 min. The cooling ramp then continued at the rate of −5°C/min until −40°C, held for 10 min, and the freezing continued to −70°C at −7°C/min. Thereafter, tissue vials were plunged into liquid nitrogen.

Thawing of testicular tissue was carried out by immersing the vials in a 37°C water bath until the ice melted. Tissue pieces were washed in culture medium and analyzed for structural changes by light and electron microscopy and also by immunohistochemistry. A comparison of fresh and frozen prepubertal testicular sample tissue revealed that frozen-thawed tissue maintained normal structure. Frozen tissue also exhibited good survival of spermatogonia in seminiferous tubules. The cell-to-cell contact and attachment of spermatogonia to the basal membrane was comparable to that noted in the fresh biopsy tissue. In this study, no attempt was made to analyze the endocrine status and responsiveness (i.e., androgen production by Leydig cells in response to gonadotropins) of the frozen-thawed sample. Further studies are required to explore and develop techniques and assess the feasibility of the whole tissue (testicular) freezing for preservation of fertility.

Posthumous Reproduction

Posthumous reproduction refers to the birth of a child after either one or both the genetic parents have died. Needless to say that posthumous conception and reproduction by IVF is a topic of ethics and law and is beyond the scope of this chapter and the expertise of the author. A brief discussion on this topic, however, deserves a place since now it is technically possible to cryopreserve both the sperm and eggs and to achieve conception by using the cryostored gametes of the dead person. Cryopreservation of oocytes for autologous use is still considered experimental and is rarely practiced. In contrast, sperm cryopreservation has been in practice for decades. Intentional cryopreservation of sperm by consenting male is a standard practice, where the intent of the male is clear, i.e., to achieve pregnancy with the spouse or female partner. Nowadays, some reproductive clinics require the male partner to record his wish regarding the fate of his sperm in the event of his death. In contrast, the practice of harvesting sperm from a deceased man (posthumous procurement of sperm) [59], usually at the request of his female partner, is quite controversial. It is obvious that in almost all such instances there is no directive from the deceased regarding the use of his sperm upon his death. In USA, the existing law does not provide clear guidance to whether the sperm may be retrieved form a dead man. The Ethics committee of the American Society for Reproductive Medicine issued a report [60] in September 2004 recognizing that "posthumous reproduction will be employed in instances when a couple faced with the imminent death of a partner or in anticipation of radiation or chemotherapy for cancer will ask to have gametes obtained and stored. Should death occur, posthumous reproduction using the stored gametes may be requested by the surviving partner." The Ethics committee further stated that "it is the responsibility of the health care provider (Reproductive Endocrinologist in this instance) to ascertain that appropriate informed consents are obtained and to ensure adequate screening and counseling of all concerned parties."

There are significant differences in beliefs and laws on posthumous reproduction depending on the country and religion. In Israel, women can have the sperm harvested and cryopreserved from their dead husbands even if the husband had not given any prior consent. Only in case the husband has clearly made his wish that he is not willing his sperm to be used for insemination after his death, the sperm cannot be retrieved postmortem on wife's request. The same guidelines apply to life partners (even if not married), but bar parents and any other relative to request sperm preservation from the dead man. In Japan, posthumous reproduction has been practiced several times in the absence of any legislative regulation. A survey of over 3,000 Japanese people revealed that there is strong support from the public in favor of posthumous conception [61]. In Belgium, posthumous reproduction is permitted. In several eastern European countries like Cyprus, the Czech Republic, Latvia, Lithuania, Malta, Poland, and Slovakia, there is no legislation concerning posthumous reproduction. Posthumous sperm retrieval is prohibited in Hungary, Slovenia, and Malaysia [62]. Australia (State of Victoria), Canada, Germany, and Sweden prohibit posthumous sperm procurement [63] while in France posthumous insemination is prohibited [64]. In terms of religion, posthumous procurement of sperm (or oocytes) is strictly prohibited in Islam [65].

Conclusions

Retrieval and cryopreservation of testicular sperm has become a cornerstone of infertility treatment in azoospermic men. In obstructive azoospermia, sperm may be retrieved from testis in almost 100% patients and can be frozen in multiple aliquots to avoid repeated surgeries. In approximately 50% of patients with NOA, enough sperm may be retrieved for IVF–ICSI. Any excess sperm remaining after ICSI can be frozen for future IVF attempts. Frozen testicular sperm are as effective in achieving successful pregnancies as freshly isolated testicular or ejaculated sperm. Treatment with motility-enhancing agents, such as pentoxifylline, induces motility in frozen-thawed testicular sperm and facilitates selection of viable sperm for ICSI. Special techniques, such as cryopreservation of single sperm within evacuated hamster zona or on a cryoloop, allow storage of extremely poor sperm samples. In USA, at this time, there is no legislative directive with respect to the posthumous procurement of sperm.

See Appendix E

References

1. Steptoe PC, Edwards RG. Birth after the reimplantation of a human embryo. Lancet. 1978;2:366.
2. Palermo G, Joris H, Devroey P, et al. Pregnancies after intracytoplasmic sperm injection of single spermatozoon into an oocyte. Lancet. 1992;340:17–8.
3. Schoysman R, Vanderzwalmen P, Nijs M, et al. Pregnancy after fertilization with human testicular spermatozoa. Lancet. 1993;342:1237.
4. Craft I, Bennett V, Nicholson N. Fertilizing ability of testicular spermatozoa. Lancet. 1993;342:864.
5. Tournaye H, Devroey P, Liu J, et al. Microsurgical epididymal sperm aspiration and intracytoplasmic sperm injection: a new effective approach to infertility as a result of congenital absence of the vas deferens. Fertil Steril. 1994;61:1045–51.
6. Devroey P, Liu J, Nagy ZP, et al. Pregnancies after testicular sperm extraction and intracytoplasmic sperm injection in non-obstructive azoospermia. Hum Reprod. 1995;10:1457–60.
7. Willott GM. Frequency of azoospermia. Forensic Sci Int. 1982;20:9–10.
8. Jarow JP, Espeland MA, Lipschultz LI. Evaluation of the azoospermic patients. J Urol. 1989;142:62.
9. Silber SJ. Microsurgical TESE and distribution of spermatogenesis in non-obstructive azoospermia. Hum Reprod. 2000;15:2278–84.
10. Tallarini A, Borini A, Bonu MA, et al. Testicular fine needle aspiration in non-obstructive azoospermia. Fert Steril. 2002;78(Suppl):S209.
11. Ishikawa T, Nose R, Yamaguchi K, et al. Learning curves of microdissection of testicular sperm extraction for nonobstructive azoospermia. Fert Steril. 2010;94:1008–11.
12. Colpi GM, Colpi EM, Piediferro G. Microsurgical TESE versus conventional TESE for ICSI in non-obstructive azoospermia: a randomized controlled study. Reprod Biomed Online. 2009;18:315–9.
13. Turunc T, Gul U, Haydardedeoglu B, et al. Conventional testicular sperm extraction combined with the microdissection technique in nonobstructive azoospermic patients: a prospective comparative study. Fertil Steril. 2010;94(6):2157–60.
14. Schiff JD, Palermo GD, Veeck LL, et al. Success of testicular sperm extraction and intracytoplasmic sperm injection in men with Klinefelter's syndrome. J Clin Endocrinol Metab. 2005;90:6263–7.
15. O'Brien KL, Varghese AC, Agarwal A. The genetic cause of male factor infertility: a review. Fertil Steril. 2010;93:1–12.
16. Pryor JL, Kent-First M, Muallem A, et al. Microdeletions in the Y chromosome of infertile men. New Engl J Med. 1997;336:534–40.

17. Ben-Yosef D, Yogev L, Hauser R, Yavetz I, et al. Testicular sperm retrieval and cryopreservation prior to initiating ovarian stimulation as the first line approach in patients with non-obstructive azoospermia. Hum Reprod. 1999;14:1794–801.
18. Schlegel P. Physiological consequences of TESE. Hum Reprod. 1996;11:159.
19. Kolettis PN, Sabanegh ES, D'Amico AM, et al. Outcome for vasectomy reversal performed after obstructive intervals of at least 10 years. Urology. 2002;60:885–9.
20. Kay DJ, Clifton V, Taylor JS, et al. Anti-sperm antibodies and sperm profiles in re-anastomosed men. Reprod Fertil Dev. 1993;5:135–9.
21. Royle MG, Parslow JM, Kingscott MM, et al. Reversal vasectomy: the effects of sperm antibodies on subsequent fertility. Br J Urol. 1981;53:654–9.
22. Chavez-Badiola A, Drakeley AJ, Finney V, et al. Necrospermia, antisperm antibodies, and vasectomy. Fertil Steril. 2008;89:723.e5–7.
23. Verheyen G, Vernaeve V, Van Landuyt L, et al. Should diagnostic testicular sperm retrieval followed by cryopreservation for later ICSI be the procedure of choice for all patients with non-obstructive azoospermia. Hum Reprod. 2004;19:2822–30.
24. Kanto S, Uto H, Mayumi T, et al. Fresh testicular sperm retrieved from men with spinal cord injury retains equal fecundity to that from men with obstructive azoospermia via intracytoplasmic sperm injection. Fertil Steril. 2004;92:1333–6.
25. Hovatta O. Cryopreservation of testicular tissue in young cancer patients. Hum Reprod Update. 2001;7: 378–83.
26. Keros V, Hultenby K, Borhstrom B, et al. Method of cryopreservation of testicular tissue with viable spermatogonia in pre-pubertal boys undergoing gonadotoxic canter treatment. Hum Reprod. 2007;22:1384–95.
27. Verheyen G, De Croo I, Tournaye H, et al. Comparison of four mechanical methods to retrieve spermatozoa from testicular tissue. Hum Reprod. 1995;10: 2956–9.
28. Petrella C, Hsieh J, Blake E, et al. Human sperm can survive at room temperature for weeks: Measure by motility and viability of sperm maintained under various conditions. Fertil Steril. 2003;80(Suppl):S210.
29. Habermann H, Seo R, Cislak J, et al. In vitro fertilization outcomes after intracytoplasmic sperm injection with fresh or frozen-thawed testicular spermatozoa. Fertil Steril. 2000;73:955–60.
30. Nagy ZP, Verheyen G, Tournaye H, et al. An improved treatment procedure for testicular biopsy specimens offers more efficient sperm recovery: case series. Fertil Steril. 1997;68:376–9.
31. Palermo GD, Schlegel PN, Hariprashad J, et al. Fertilization and pregnancy outcome with intracytoplasmic sperm injection for azoospermic men. Hum Reprod. 1999;14:741–8.
32. Craft I, Tsirigotis M, Zhu J. In vitro culture of testicular sperm. Lancet. 1995;346:1438.
33. Edirisinghe WR, Junk SM, Matson PL, et al. Case report: changes in motility pattern during in vitro culture of fresh and frozen thawed testicular and epididymal spermatozoa: implications for planning treatment by intracytoplasmic sperm injection. Hum Reprod. 1996;11:2474–6.
34. Liu J, Garcia E, Baramki A. The difference in outcome of in vitro culture of human testicular spermatozoa between obstructive and non-obstructive azoospermia. Hum Reprod. 1996;11:1587–8.
35. Zhu J, Tsirigotis M, Pelekanos M, et al. In vitro maturation of human testicular spermatozoa. Hum Reprod. 1996;11:231–2.
36. Emiliani S, Van den Bergh M, Vannin AS, et al. Increased sperm motility after in-vitro culture of testicular biopsies from obstructive azoospermic patients results in better post-thaw recovery rate. Hum Reprod. 2000;15:2371–4.
37. Jeyendran RS, Van der Ven HH, Perez-Pelaez M, Crabo BG, Zaneveld LJD, et al. Development of an assay to assess the functional integrity of the human sperm plasma membrane and its relationship to other semen characteristics. J Reprod Fertil. 1984;70: 219–28.
38. Verheyen G, Joris H, Crits K, et al. Comparison of different hypo-osmotic swelling solutions to select viable immotile spermatozoa for potential use in intracytoplasmic sperm injection. Hum Reprod. 1997;3:195–203.
39. Sallam HN, Farrag A, Agameya AF, et al. The use of a modified hypo-osmotic swelling test for the selection of viable ejaculated and testicular immotile spermatozoa in ICSI. Hum Reprod. 2001;16:272–6.
40. Cummins JM, Pember SM, Jequier AM, et al. A test of the human sperm acrosome reaction following ionophore challenge (ARIC)-Relationship to fertility and other semen parameters. J Androl. 1991;12:98–103.
41. Lambert HL, Steinleitner A, Eiserman J, et al. Enhanced gamete interaction in the sperm penetration assay after coincubation with pentoxifylline and human follicular fluid. Fertil Steril. 1992;58: 1205–8.
42. Stanic P, Sonicki Z, Suchanek E. Effect of pentoxifylline on motility and membrane integrity of cryopreserved human spermatozoa. Int J Androl. 2002;25: 186–90.
43. McKinney KA, Lewis SEM, Thompson W. Persistent effects of pentoxifylline on human sperm motility after drug removal in normozoospermic and asthenozoospermic individuals. Andrologia. 1994;26:235–40.
44. Negri P, Grechi E, Tomasi A, et al. Effectiveness of pentoxifylline in semen preparation for intrauterine insemination. Hum Reprod. 1996;6:1236–9.
45. Mehrannia T. The effect of pentoxifylline in semen preparation for intrauterine insemination. Pak J Med Sci. 2009;25:359–63.
46. Tournaye H, Janssens R, Camus M, et al. Pentoxifylline is not useful in enhancing sperm function in cases with previous in-vitro fertilization failure. Fertil Steril. 1993;59:210–5.

47. Tournaye H, Janssens R, Verheyen G, et al. An indiscriminate use of pentoxifylline does not improve in-vitro fertilization in poor fertilizers. Hum Reprod. 1994;9:1289–92.
48. Griveau JF, Lobel B, Laurent MC, et al. Interest of pentoxifylline in ICSI with frozen-thawed testicular spermatozoa from patients with non-obstructive azoospermia. Reprod Biomed Online. 2007;12: 14–8.
49. Cohen J, Garrisi GJ, Congedo-Ferrara TA, et al. Cryopreservation of single human spermatozoa. Hum Reprod. 1997;12:994–1001.
50. Walmsley R, Cohen J, Ferrara-Congedo T, et al. The first births and ongoing pregnancies associated with sperm cryopreservation within evacuated egg zonae. Hum Reprod. 1998;13:61–70.
51. Hsieh YY, Tsai HD, Chang CC, Lo HY. Cryopreservation of human spermatozoa within human or mouse empty zona pellucidae. Fertil Steril. 2000;73:694–8.
52. Ye Y, Xu C, Qian Y, et al. Evaluation of human sperm function after being cryopreserved within the zona pellucidae. Fertil Steril. 2009;92:1002–8.
53. Desai NN, Blackmon H, Goldfarb J. Single sperm cryopreservation on cryoloops: an alternative to hamster zona for freezing individual spermatozoa. Reprod Biomed Online. 2004;9:47–53.
54. Allan JA, Cotman AS. A new method for freezing testicular biopsy sperm: three pregnancies with sperm extracted from cryopreserved sections of seminiferous tubules. Fertil Steril. 1997;68:741–4.
55. Scholtes MCW, van Hoogstraten DG, Schmoutziguer A, et al. Extraction of testicular sperm from previously cryopreserved tissue in couples with or without transport of oocytes and testicular tissue. Fertil Steril. 1999;72:785–91.
56. Keros V, Hultenby K, Borgstrom B, et al. Methods of cryopreservation of testicular tissue with viable spermatogonia in pre-pubertal boys undergoing gonadotoxic cancer treatment. Hum Reprod. 2007;22: 1384–95.
57. Brook PF, Radford JA, Shalet SM, et al. Isolation of germ cells from human testicular tissue for low temperature storage and autotransplantation. Fertil Steril. 2001;75:269–74.
58. Ehmcke J, Hubner K, Scholer HR, et al. Spermatogonia: origin, physiology and prospects for conservation and manipulation of the male germ line. Reprod Fertil Dev. 2006;18:7–12.
59. Batzer FR, Hurwitz JM, Caplan A. Postmortem parenthood and the need for a protocol with posthumous sperm procurement. Fertil Steril. 2003;79: 1263–9.
60. The Ethics committee of the American Society for Reproductive Medicine. Posthumous reproduction. Fertil Steril. 2004;82(Suppl):S260–2.
61. Ueda N, Kushi N, Nakatsuka M, et al. Study of views on posthumous reproduction, focusing on its relation with views on family and religion in modern Japan. Acta Med Okayama. 2008;62:285–96.
62. Dostal J, Utrata R, Loyka S, et al. Post-mortem sperm retrieval in new European union countries: case report. Hum Reprod. 2005;20:2359–61.
63. Bahadur G. Death and conception. Hum Reprod. 2002;17:2769–75.
64. Lansac J. French law concerning medically-assisted reproduction. Hum Reprod. 1996;11:1843–7.
65. Samani RO, Ashrafi M, Alizadeh L, et al. Posthumous assisted reproduction from Islamic perspective. Int J Fertil Steril. 2008;2:96–100.

Germ Cell Transplantation and Neospermatogenesis

24

Queenie V. Neri, Zev Rosenwaks,
and Gianpiero D. Palermo

Abstract

Germ cells isolated from testicular tissue or derived from embryonic stem cells represent a possible treatment option for children who have survived cancer and for azoospermic men. Self-renewal of spermatogonial stem cells (SSCs) is the foundation for sustaining spermatogenesis throughout life. In recent years, spermatogonia transplantation in mammals has been successfully carried out and, more recently, the appearance of germ cell lines derived from embryonic stem cells has rendered the treatment or prevention of azoospermia conceivable. In our laboratory, we have identified and labeled human spermatogenic cells at various steps of differentiation and carried out their xenogeneic transplantation into recipient gonads. In addition, by appropriately modifying culture conditions, we have been able to propagate mouse SSCs in vitro.

Keywords

Fertility preservation • Transplantation • Spermatogonia stem cell • Neogametogenesis • In vitro spermatogenesis • Male infertility • Meiosis • Embryonic stem cells • Cryopreservation • Azoospermia • Cancer • Antineoplastic treatment

Among the general population, the ability of man to procreate appears to have decreased progressively during the past half century [1]. An estimated 15% of couples in their reproductive age are afflicted by a range of causes pertaining to their ability to procreate, and about half of all infertility cases are directly attributed to the male partner. Approximately 6% of males between the ages of 15 and 44 are deemed infertile or have their fecundity

G.D. Palermo (✉)
The Ronald O. Perelman and Claudia Cohen Center for Reproductive Medicine, Weill Cornell Medical College, 1305 York Avenue, 6th Floor, New York, NY 10021, USA

E. Seli and A. Agarwal (eds.), *Fertility Preservation: Emerging Technologies and Clinical Applications*,
DOI 10.1007/978-1-4419-1783-6_24,

severely compromised [2]. Evaluation of endocrine data implies that the main causes of male infertility are hormonal disturbances and aberrations in the production of semen [1]. Absence of spermatozoa in the ejaculate is termed azoospermia. About 50% of all azoospermic cases are due to hypo-spermatogenesis while complete germ cell aplasia accounts for about 25%, the remainder are due to spermatogenic arrest [3].

Intracytoplasmic sperm injection (ICSI), focused at selecting a single viable spermatozoon to inseminate a mature oocyte, is now considered the best procedure to generate consistent fertilization and satisfactory pregnancy outcome [4, 5]. The ability of virtually any spermatozoon, regardless of its morphology or functional characteristics to induce oocyte activation and consequent pronuclear development has meant that even some azoospermic patients can be treated successfully with this procedure. Yet, in spite of these accomplishments in treating many forms of male factor infertility, there are still limitations since the fully developed spermatozoon remains the *conditio sine qua non* for treatment of male infertility, leaving those men afflicted by Sertoli-cell-only (SCO) syndrome or spermatogenic arrest as currently untreatable.

Recent research has aimed at the in vitro proliferation and maturation of germ cells, at least in animals [6]. However, spermatogonial stem cells (SSCs) are the only cell type in the male gonad that retain the ability to self-renew and under certain stimuli to be driven into meiotic and post-meiotic differentiation [7]. Thus, SSCs are unique in that respect – they represent the progenitor cells leading to spermatozoa production, while at the same time displaying the ability to revert to a more totipotent state similar to the embryonic stem cell in order to reclaim their precursor state [8]. For this reason, mastering ability to culture and propagate SSCs in vitro may pave the way to an array of treatments aimed at alleviating several disorders characterizing male infertility.

Antineoplastic Treatment

Adults with Cancer

In 2009, approximately 62,000 individuals in the US between the ages of 20–39 years are expected to be diagnosed with cancer [9]. Many of them will require chemotherapy and/or radiotherapy regimens, antineoplastic, and cytotoxic agents that being highly gonadotoxic compromise the ability to procreate. Gonadal failure is one of the major consequences of cancer therapy, remarkably so after ionizing radiation. Many surveys on cancer survivors have found that they often have an increased risk of emotional distress related to infertility as a result of their treatment [10].

Depending on the underlying disease, the age of the oncological patient, the type of therapeutic agent used to treat the cancer, the cumulative radiation doses used or the duration of the chemotherapy, between 10 and 100% of surviving cancer patients will show impaired semen parameters following treatment [11]. It is estimated that between 15 and 30% of these cured cancer patients become permanently sterile [12]. There is the concern that many patients undergoing potentially sterilizing antineoplastic cancer treatments have not been appropriately counseled about their future fertility. Even though assisted reproductive technologies (ART) can successfully restore the reproductive capacity of many of these cancer patients, oncologists and their practitioners are still not fully aware of its limitations [13–15]. In fact, this lack of awareness in the medical community seems to be the main culprit for causing untimely cryostorage of gametes in these patients.

In the ICSI era, any semen sample, even those comprising only a few motile spermatozoa, may be considered for sperm cryopreservation, and this should be performed prior to any cytotoxic treatment. Since most patients will develop azoospermia 2 or 3 months following chemotherapy [12], cryobanking semen during the first month should still be advocated to overcome this major collateral effect [16]. However, patients should be informed that chemotherapy may induce genetic and epigenetic abnormalities in both the short- [17] and long-term [18, 19].

Azoospermia can sometimes be encountered in patients during the diagnosis of testicular cancer – this as a direct result of factors related to the malignancy on spermatogenesis [20]. At orchiectomy, these patients may be offered sperm harvesting by vasal, epididymal aspiration [21], or by testicular sperm extraction (TESE) [12] in order to bank their gametes prior to chemotherapy.

At our Center, the efficacy of standard IVF and ICSI has been evaluated in 118 couples who cryobanked semen before cancer treatment and then underwent 169 ART cycles [22]. These included men with cancer of the testis, prostate, brain, lung, pancreas, and bladder, as well as leukemia, lymphomas, sarcomas, and multiple myelomas. The mean age of the females at the time of their ART procedure was 34.8 ± 4 years and their husbands were 38.5 ± 8 years. The post-thaw semen analysis had an average concentration of 40.9×10^6/mL and motility of 14.2%. The overall clinical pregnancy rate was 56.8% (96/169) with 11 (5.6%) undergoing spontaneous abortions while the remainder went to term delivery (85/169). Patients with prostate cancer displayed the worst semen parameters prior to banking and the lowest clinical pregnancy rates. In a subanalysis that compared the clinical pregnancy according to the insemination method, it appears that ICSI proved to be more efficacious than standard in vitro insemination ($P < 0.001$). Overall, ARTs particularly with ICSI is recommended for couples whose male partners are diagnosed with cancer and cryobanked their specimen before, after, and even during antineoplastic treatment.

Prepubertal Age Cancer

About one in every 600 children will develop cancer before the age of 15. However, as a result of better treatment options, today the cancer death rate in infants and children has decreased more than that for any other age group, with up to 75% of all cases now being curable. At present, one in 1,000 adults in the second or third decade of life is a childhood cancer survivor [23], and it has been estimated that this proportion will rise to one in 250 [24].

With childhood cancer being treated more successfully, the focus of treatment has shifted toward the quality of life after treatment, and one aspect commonly overlooked is the prevention of sterility. In adults and adolescents, semen banking or cryopreservation of testicular tissue before radiation/chemotherapy are valuable preventive measures that circumvent sterility after treatment. However, no such prevention is possible prior to puberty since no active spermatogenesis has occurred, and so an average of 30% of children cured of cancer remain sterile in the long term [12].

One possible remedy consists of SSC transplantation, the procedure in which either testis germ cells or stem cell colonies are transferred from a fertile donor into the seminiferous tubules of an infertile recipient. This is a technique currently being perfected in mice before attempts to apply it to man.

The first experimental attempt at autologous spermatogonial transplantation was conducted by Brinster and Avarbock in 1994 when mouse testicular cells obtained after enzymatic digestion were directly injected into seminiferous tubules of a recipient mouse previously made sterile with injections of busulfan (a strong alkylating agent) [25]. A month after post-germ cell transplantation, nascent germ cell colonies were identified as chains of cells on the tubular basement membrane. These spermatogonia stem cells (SSCs) implanted in the tubules then differentiated into later spermatogenic stages. In a further experiment, higher colonization rates of germ cells were reported by enhancing the transplanted donor cell population [26].

Following these demonstrations, xenogeneic germ cell transplantation was conducted with testicular cells from fertile transgenic rats injected into the seminiferous tubules of immunosuppressed mice [27]. In this case, mouse sertoli cells clearly were able to serve as supporting somatic cells in rat germ cell recolonization and

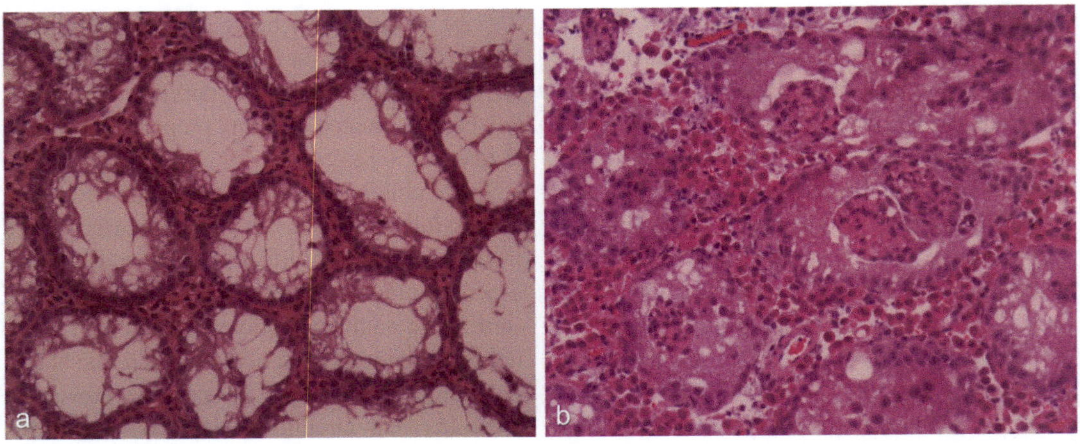

Fig. 24.1 Seminiferous tubules of SCID mouse (**a**) 4 weeks after busulfan treatment and (**b**) 1 week after injection of human spermatogenetic cells

eventual differentiation. Other species used for donor cell injection into mouse seminiferous tubules that achieved successful colonization and differentiation were hamster [28], rabbits, and dogs [29].

The first xenogeneic germ cell transplantation in which human testicular cells were injected into mouse seminiferous tubules was performed by our group in 2000 [30] that resulted in a lack of recolonization of SSCs in the recipient tubules of SCID or W/W^v mice after 150 days (Fig. 24.1). We have postulated then that lack of success may have been due to inter-species variability, to non-compatible cell adhesion molecules and/or to immunological rejection. Nagano et al. transplanted baboon germ cells into mice and found that germ cells were able to repopulate and survive for 6 months [31].

Investigators have reported the survival of at least some undifferentiated spermatogonia during distant xeno-germ cell transplantations [30–32]. In Reis et al., however, we did not find evidence of donor tissue survival following a human-to-immunodeficient mouse testicular tissue transplantation [30]. In this study, the antibody stain for proacrosin was used to detect successful tubular implantation. Proacrosin is a marker of differentiated human spermatogonia (primary spermatocytes and spermatids) and would not have detected transplanted cells that had survived or propagated without further differentiation. On the other hand, Sofikitis et al. reported successful spermatogenesis from transplantation of human tissue into rat and mouse recipients [33]. In 2002, Nagano and colleagues reported on the use of anti-baboon testes antibody to identify the survival of human spermatogonia in mouse recipients for up to 6 months post-transplantation [32]. The same group then used human SSCs but found that these cells had not differentiated beyond spermatogonia 1 month after transplantation to the mouse testes. These interspecies transplantation studies seem to suggest, at present, that the vast phylogenetic distance between species (baboon to mouse and human to mouse) will likely translate into host testicular environment and donor spermatogonia incompatibilities that prohibit completion of spermatogenesis.

An alternative to germ cell transplantation is tissue grafting. Successful tissue xenografting of cryopreserved testicular specimen has been reported between various species [34–36] and non-human primates [37]. Testicular tissue grafting entails placing a piece of the tissue, approximately 1 mm^3 size, subcutaneously under the dorsal skin or the testicular bursa. However, the differentiation capacity of frozen spermatogonia showed xenografts of testicular tissues from immature rhesus monkeys as having a blockage in the spermatocyte stage [38]. Similar results were observed in frozen-thawed immature human testicular tissue grafted on the peritoneal bursa of mice testicles [39] where numerous pre-meiotic spermatocytes were observed; however, spermatid-like cells did not express meiotic or post-meiotic markers.

Severely Compromised Spermatogenesis

Spermatogeneic Arrest

Spermatogenic arrest in men occurs most frequently at the spermatogonium stage or during arrested meiosis, the latter being most common at the primary spermatocyte stage [40, 41]. Meiotic arrest, at the primary spermatocyte, is the most common in men with non-obstructive azoospermia [40].

Successful in vitro differentiation of spermatogenic cells into spermatids would be an extremely attractive option to treat some forms of male infertility caused by spermatogenic arrest [42, 43]. Various approaches to the challenge of in vitro spermatogenesis have been attempted, these mainly focusing on testicular cell cultures that are directed at achieving male germ cell differentiation. Among them, tissue culture, organ culture, and co-culture systems have been investigated extensively to assess which system can obtain meiotic or post-meiotic differentiation of these male germ cells [42–48]. Despite the report of a normal child being born after fertilization with germ cells from a man with maturation arrest at the primary spermatocyte stage [42], an effective procedure for the completion of spermatogenesis in vitro still remains elusive.

Treatment: In Vitro Spermatogenesis

In the early embryo, primordial germ cells (PGCs), the progenitors of gametes migrate from their yolk sac origin to nest within the somatic component of the gonad. They are one of the first discernable cell types in the mouse embryo. While their progress is essentially amebic in nature, several growth factors have been shown to influence the appearance, migration, survival, and proliferation. However, the signaling through which these cells localize in the genital ridges has not been identified.

Germ cells appear in mammals at the onset of gastrulation, soon after implantation of the embryo. In the mouse, PGCs are first distinguishable at the base of the allantois on embryonic day (E) 7.25 [49]. Lineage studies of epiblast cells show that mouse PGCs are specified by inductive interactions at the onset of gastrulation [50, 51]. Moreover, primary cultures of epiblast fragments from embryos at E5.5–E6.0 generate migrating PGCs when they are cocultured with extraembryonic ectoderm [52], and whole epiblasts from E6.0 embryos cultured on feeder cells expressing both bone morphogenic protein 4 (BMP4) and BMP8b will give rise to PGCs [53, 54]. These results reveal that bone morphogenic proteins (BMPs) derived from the extraembryonic ectoderm play crucial roles in PGC appearance in the proximal epiblast. PGCs will then migrate in the developing embryo toward the gonadal ridge, proliferate, and differentiate into gonocytes between E8.5–12.5 (Fig. 24.2).

To study the morphology, appearance, and migratory behavior of PGCs, in preliminary experiments we assessed embryo development as early as 6.5 dpc (Fig. 24.3a, b) with the embryo being 120×250 μm in size with an implantation site of 2×3 mm. After immunohistochemistry on 4–7 sections of 5 μm, a total of 14.6% (12/82) of the cells per embryo section expressed *fragilis* – a marker for premigratory PGCs. In a 12.5 dpc embryo (Fig. 24.4a) with a distinguishable gonadal ridge measuring 25.0×87.5 μm, sections of this region evidenced 25.9% (45/174) VASA positive (Fig. 24.4b) post-migratory PGCs.

Gonocytes/SSCs self-renew and produce daughter cells that commit to differentiate into spermatozoa throughout adult life. SSCs can be identified unequivocally by a functional assay using a transplantation technique in which donor testis cells are injected into the seminiferous tubules of infertile recipient males [26, 55]. Under these conditions, only SSCs are able to generate foci of germ cells undergoing spermatogenesis [7].

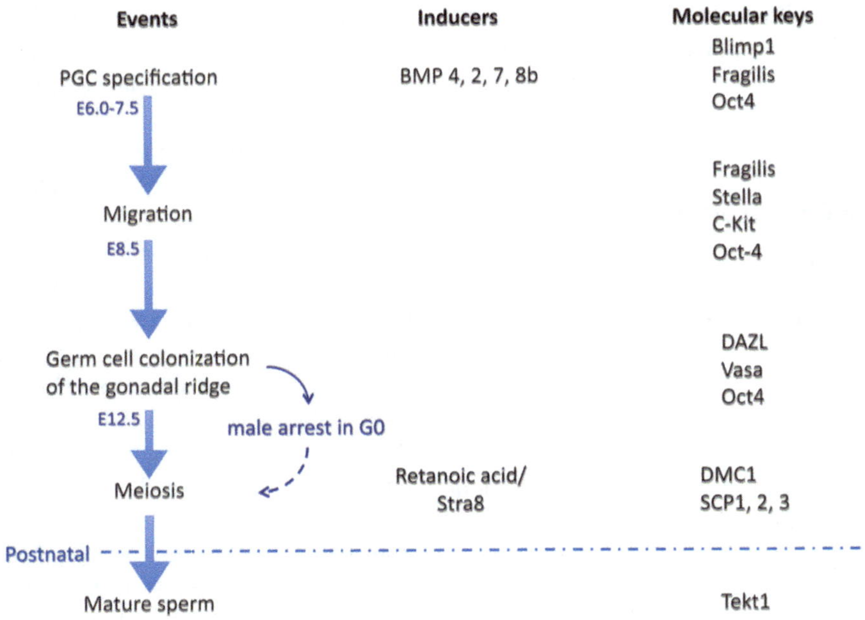

Fig. 24.2 The germline development: events and molecular keys. During in vivo development, BMP family stimulates a subset of epiplast cells that begin to express Fragilis and Blimp1 and become PGCs that migrate toward the gonadal ridge

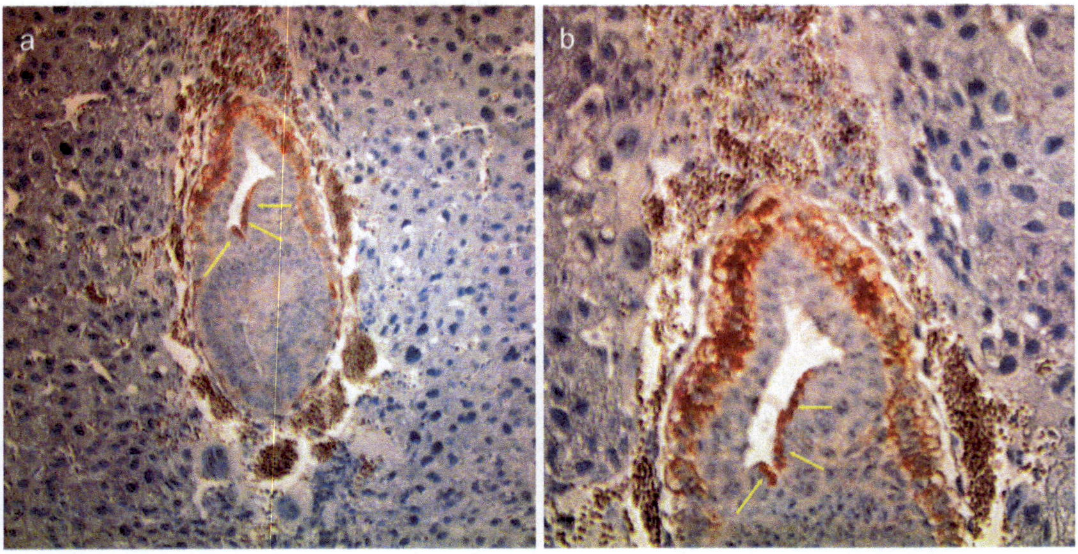

Fig. 24.3 Perfused embryo at 6.5 dpc. Primordial germ cells (PGCs) are labeled with VASA (*yellow arrows*) at (**a**) 100× and (**b**) 200×

One approach to the problem is cultivating SSCs under conditions that allow self-renewal and possibly induce differentiation. For this purpose, it is essential to establish a culture system with defined, experimentally modifiable characteristics. Glial cell line-derived neurotrophic factor (GDNF) is responsible for stimulating SSC renewal in rodents [56]. In addition, several BMP

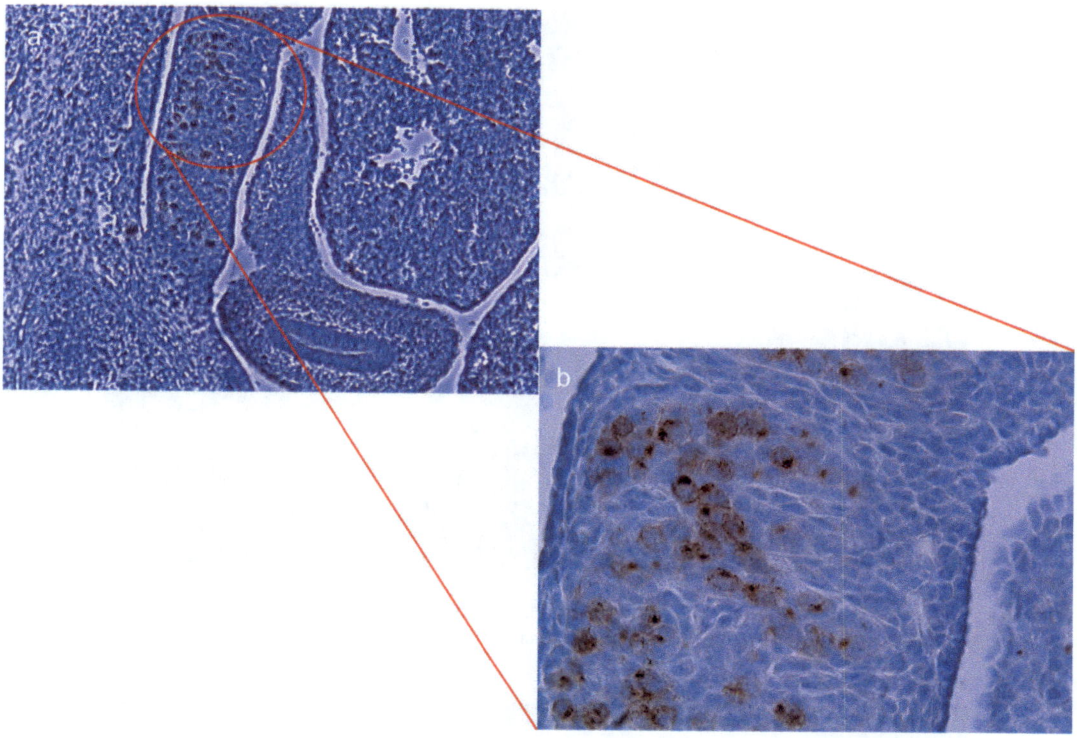

Fig. 24.4 Perfused embryo at 12.5 dpc. PGCs are labeled with VASA at (**a**) 100× and (**b**) 400×

genes are expressed in the testis, in particular, BMP7 and BMP8 a/b transcripts [57, 58], suggesting that not only Sertoli cells, but also germ cells secrete molecules that can mediate paracrine interactions in the testis. In vitro co-culture experiments have shown that both BMP4 and BMP8B proteins function synergistically to induce PGC specification from the epiblast [54]. BMP2, another member of the BMP family, acts in cooperation with BMP4 as a third signal in PGC formation [53].

Induction of meiosis by retinoic acid (RA) released from the mesonephros helps determine the timing of meiosis according to the gender. In males, induction is delayed until the post-natal period and RA has been postulated as being secreted by Sertoli cells [59]. RA plays a role in the regulation of meiosis and is required to be present to have the *Stra8* (stimulated by RA 8) gene to become expressed. The Stra8 encodes a cytoplasmic protein and is expressed specific to the developing male gonad during mouse embryogenesis and in the adult is restricted to the premeiotic germ cells.

Another important key to the full establishment of SSC in vitro is the incorporation of an appropriate feeder layer [60]. In the 1960s, Steinberger and associates found in numerous experimental trials that culturing germ cells that were mechanically or enzymatically dissociated from the testis had their survival directly dependent upon sertoli cell co-culture [61–63]. Recently, differentiation of rat germ cells into spermatids from preleptotene spermatocytes has been documented using Sertoli cells as a feeder layer [64].

As another variant, we attempted to isolate and select SSCs by utilizing magnetic activated cell sorting (Fig. 24.5). In order to obtain the maximal yield of SSCs, testes were retrieved from both neonate and adult B_6D_2-F_1 mice. The testicular cell suspension were layered onto a 30% density gradient and centrifuged for 8 min. CD90 (Thy1.2) antibody was elected as a marker to identify SSC fractions. The excess antibody

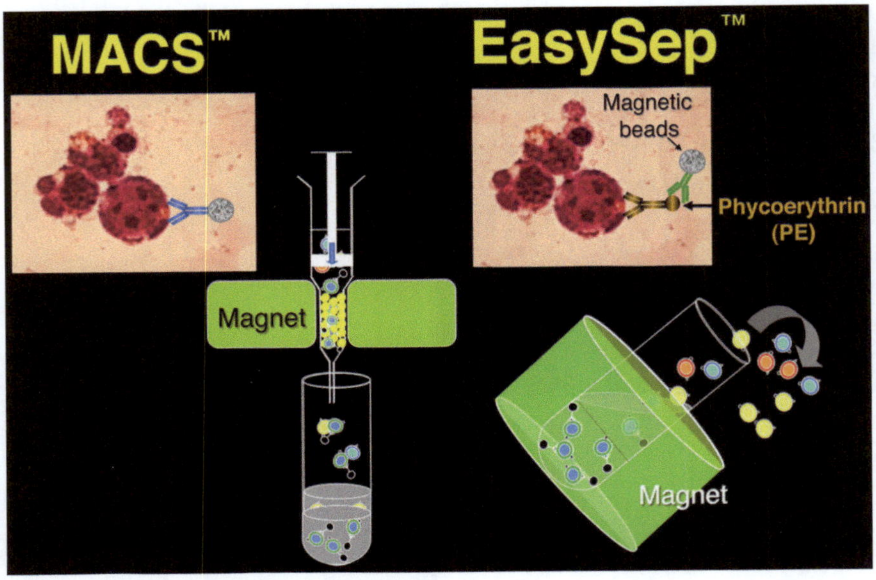

Fig. 24.5 Magnetic cell sorting by MACS™ or EasySep™. Cell suspensions were first processed by density gradient centrifugation and then sorted by incubation with monoclonal Thy-1 (CD90) magnetic-bead conjugated antibody

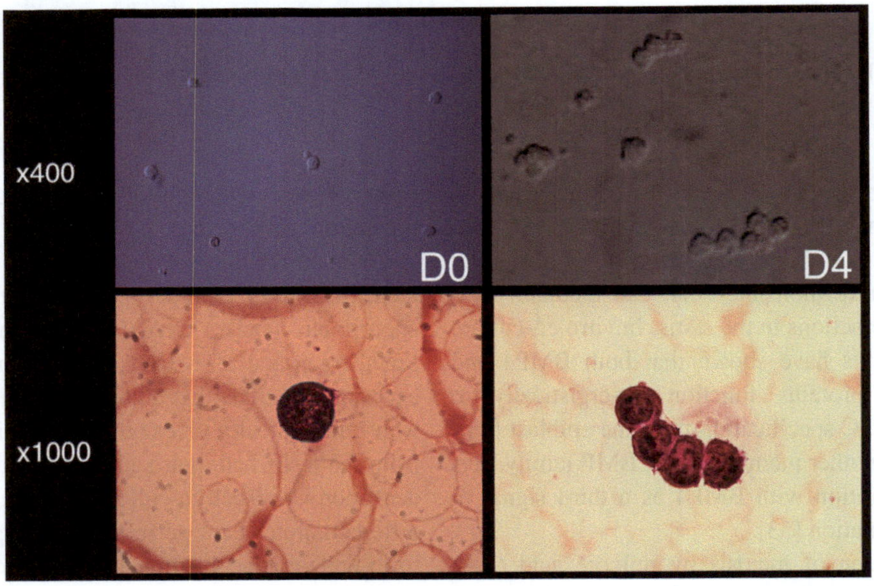

Fig. 24.6 After germ cells enrichment, cells were seeded (D0 of culture) while others were stained for morphological analysis. At D4 of culture, some germ cell proliferation was observed

was removed and the resulting cell suspension was loaded onto a separation column within a magnetic field. Fourteen testes retrieved from seven-6 dpc mice had an average weight of 5.0 mg±0.001. Sorting of six dissected testes (795,000 testicular cells) by magnetic separation yielded ~43,500 CD90 positive cells with a >80% enrichment rate ($P<0.0001$). After seeding the germ cells on a feeder layer (Fig. 24.6), one putative SSC colony, confirmed by alkaline phosphatase (AP) activity/VASA positivity, proliferated up to day 9 of culture (Fig. 24.7).

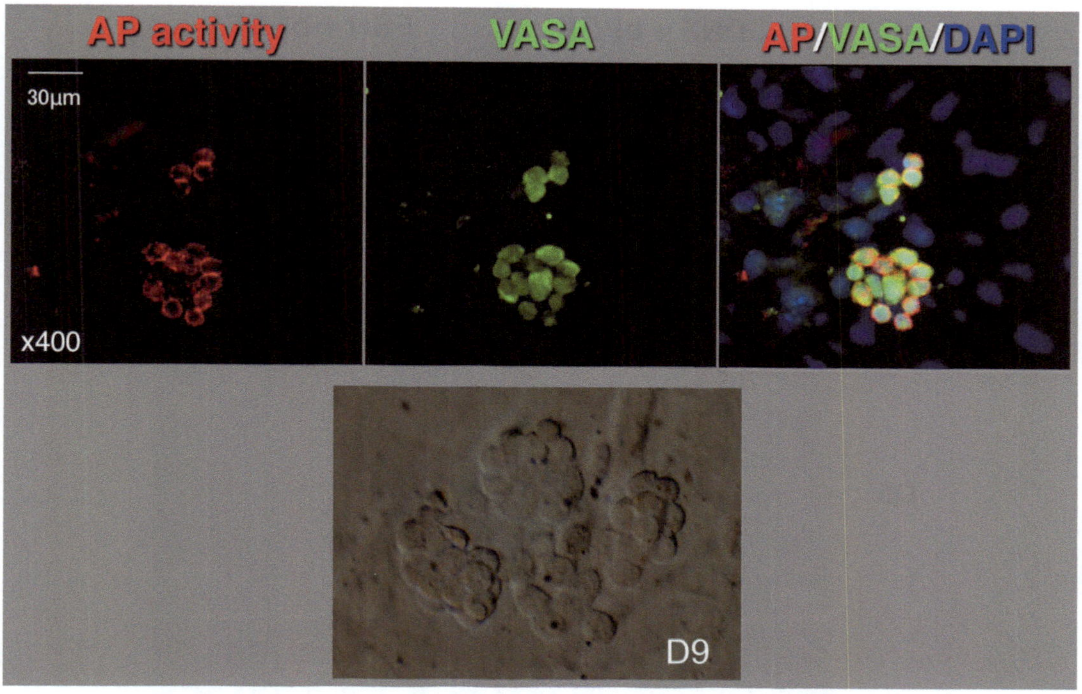

Fig. 24.7 Double expression of pluripotent markers alkaline phosphatase activity and VASA were assessed on plating D4 of SSC colonies. Nuclei were counterstained with DAPI. Putative colonies proliferated up to day 9 of culture

The remaining eight testes provided about 1.1×10^6 cells with ~3% germ cells retrieved after enzymatic digestion. The unselected suspensions of testicular cells consisting of a mixture of SSCs, sertoli cells, and interstitial cells were plated under three culture conditions: (1) GDNF+bFGF or (2) FSH+bFGF or (3) GDNF+bFGF+LIF. The following day, epithelial cells had become attached to the dish to form a confluent monolayer with scattered adherent germ cells (Fig. 24.8). On D3, cell clumps of ~25 μm diameter were uniformly distributed in the dish. Over the subsequent days, there was a spontaneous and progressive migration of the somatic cells around each germ cell, creating dispersed cell clumps. In every media combination, the size of the cell aggregates gradually increased to reach a diameter of ~100 μm by D10 (Fig. 24.9). In GDNF+bFGF and GDNF+bFGF+LIF, the aggregates remained as tightly packed entities, whereas the clumps in FSH+bFGF had started to spread. When LIF was removed, cell aggregates behaved similarly to those in FSH+bFGF. The putative germ cells in these tridimensional seminiferous tubule-like structures were then monitored for up to 75 days. Such cellular aggregates had a steady linear growth up to D56 and then plateaued thereafter (Fig. 24.10). Aggregates were fixed at different times to assess presence of germ cells. VASA expression and AP activity were present mainly on the outside periphery of the aggregates and on some individual cells within them, indicating that the germ cells were maintained for an extended time. The GDNF+bFGF supplemented with LIF appeared to provide the best support for the germ cell population. Interestingly, after 2 months in culture, VASA staining revealed only a few PGCs within the aggregates, while the majority coating them, presumably SSCs, adopted the form of an inverted seminiferous tubule (as described by Schlatt, ASRM Course 21, 2007). Very few cells positive for FE-J1 (~1%; post-meiotic stage) were observed [65].

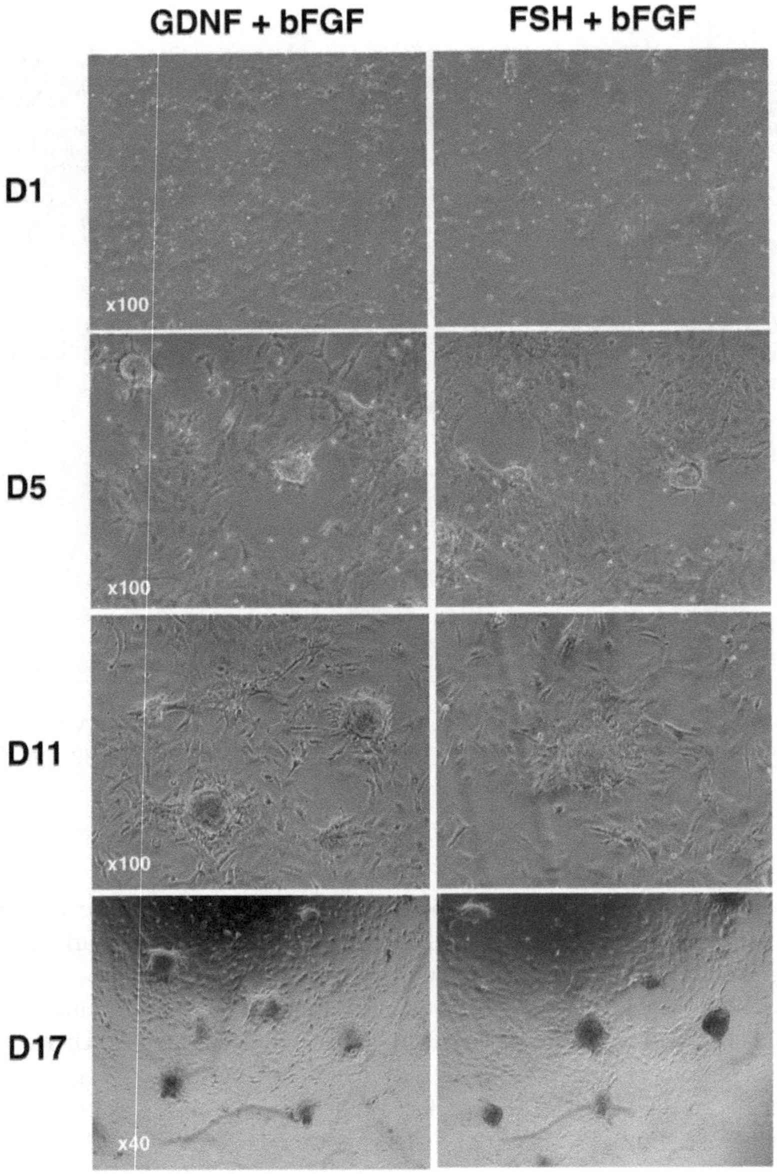

Fig. 24.8 Neonatal testicular cells started to form cell clumps on D1 of culture in a serum-free medium supplemented with growth factors. Cells began to adhere to the dish and form scattered clusters (D5). These cell aggregates progressively grew over time as seen on frames D11 and D17

The possibility to propagate and immortalize SSCs in vitro would allow the creation of colonies for a specific individual suffering from germ cell maturation arrest, and more importantly, patients suffering from childhood cancer. These cells can be maintained, cryopreserved, and made available for transplantation to the testes of individuals ready to start a family.

Germ Cell Aplasia

SCO syndrome, also called germ cell aplasia, describes a condition of the testes in which germ cells are absent and only Sertoli cells populate the seminiferous tubules. SCO syndrome presents during the evaluation of azoospermia in couples having difficulty in initiating a pregnancy.

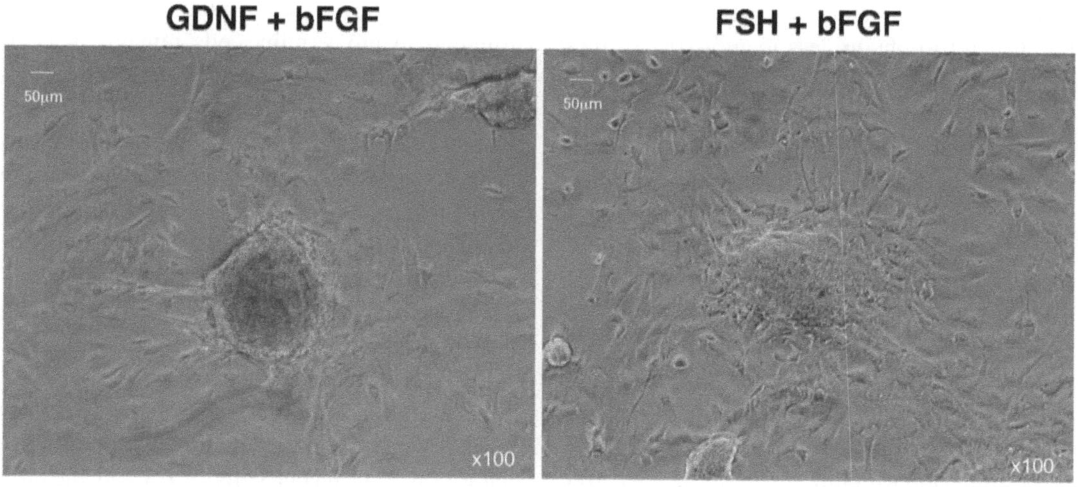

Fig. 24.9 Extended culture of unselected testicular cells in serum-free medium supplemented different growth factors. On D17, in GDNF+bFGF were tightly packed while FSH+bFGF began to spread over the dish

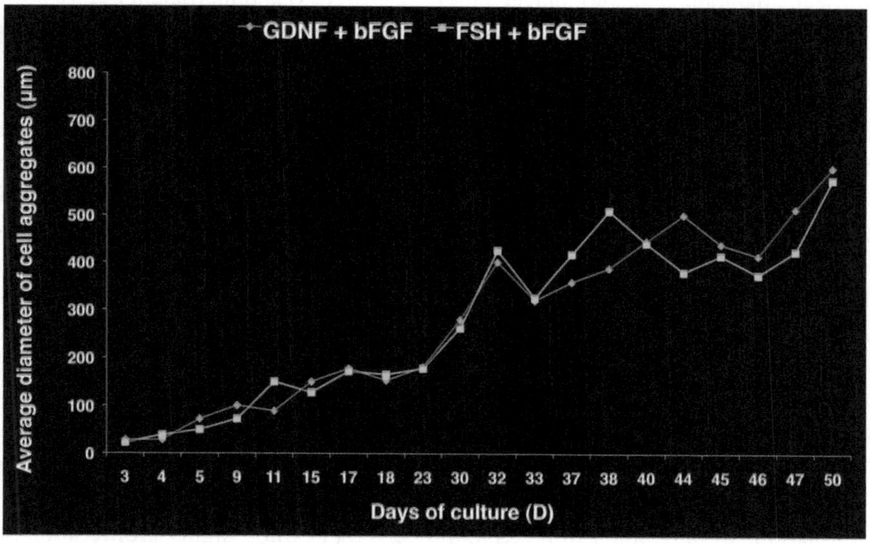

Fig. 24.10 Size of the cell aggregates progressively increased with time

These men typically present with infertility as the only symptom, and most present between age 20 and 40 years, and are found to have no sperm in the ejaculate. The male urologic physical examinations are often unremarkable, and the diagnosis is made from a testicular biopsy. While investigation of the basis for the SCO syndrome is ongoing, the etiology and mechanism of this process are currently unknown. Some examples of diseases and disorders characterized by SCO and Leydig cell hyperplasia are Klinefelter's Syndrome, mumps orchitis, deletion of the Y chromosome at the AZFa or AZFb regions, and men/children that had anti-neoplastic therapy with radiation or chemotherapy.

The prevalence of the SCO syndrome in the overall population is extremely low. Of the 10% of US couples affected by infertility, approximately 30% have a pure male factor as the underlying cause, and another 20% have a combined male and female factor. Although precise figures are difficult to obtain, less than 5–10% of these infertile men have SCO syndrome. SCO syndrome has no known racial predilection; however, SCO is more common in white men. There are no known effective treatment(s) at the present time.

Future Options

During embryonic stem cell (ESC) differentiation through embryoid body (EB) development in vitro, tissue-specific genes and proteins are differentially expressed in patterns that in some way mimic mouse embryogenesis in vivo [66]. ESCs have shown the ability to recapitulate features of embryonic development by spontaneously forming somatic lineages in culture. While ESCs have been maintained in monolayer cultures, sporadic oocyte-like structures have been identified that presumably develop into a structure resembling a blastocyst [67]. Follicle-like structures were also obtained by feeding EBs with conditioned medium isolated from cultures of mouse neonatal testicular tissue [68]. These structures were not, however, capable of developing further.

Growth factors such as leukemia inhibiting factor (LIF) and the BMP derivatives [69] are considered essential to grant their survival to get germ cell proliferation and self-renewal, while other factors such as RA would help them to enter meiosis [70–73]. By changing the concentration of BMPs (BMP4, 7, and 8b; 1 or 10 μM RA) and using EB formation, we attempted to recreate a seminiferous tubule-like environment [74]. We observed that 10 μm RA had the highest proportion of germ cell appearance on day 3 and was able to maintain up to day 20. In a Brazilian study, using commercially made neurobasal medium supplemented with B27 were presumably able to obtain sperm in 10 days and oocytes in 25 days of culture in the same EB [75].

The inherent characteristics of ESCs – their capacity to differentiate into cells of all three germ layers and grow indefinitely [76–78], presented an enormous potential for the field of regenerative medicine. A reserve of patient-specific pluripotent cell cultures, with the potential to differentiate into germline cells, remains an idealized goal in the minds of ART researchers and clinicians. Though the possibilities for ESC-based therapy could be enmeshed in ethical controversy, alternative treatment options have shown promise in recent years. Specifically, overexpression of the transcription factors Oct4, Klf4, Sox2, and c-Myc can induce somatic cells to exhibit molecular and functional similarities to ESCs, aptly designated as induced pluripotent stem cells (iPSCs) [79]. Initial efforts promoted the introduction of candidate genes into mouse embryonic fibroblasts (MEFs) via retroviral transduction, with primary culture on STO feeder cells in ES cell medium [79]. However, further efforts have revealed a need for alternative sources, as the reprogramming of fibroblasts proves to be a relatively slow (2–3 weeks) and inefficient (<1%) process [80].

Derivation of iPSCs from human peripheral blood cells, specifically T lymphocytes and myeloid cells, is one such alternative showing recent promise. Utilizing a polycistronic vector encoding the four standard transcription factors, followed by transfer onto MEF feeder layers and ensuing culture incubation, blood cells can give rise to expandable and stable iPSC colonies [81]. These colonies stain for pluripotency markers such as Oct4, Nanog, and Tra1-81, and are capable of in vitro differentiation into cells of the three germ layers [81]. Exploring alternative candidate sources such as blood for pluripotent induction provides much of the contemporary focus in regenerative medicine [81–83]. Isolating cells that exhibit higher proliferation rates, better long-term growth potentials and convenient accessibility remains an important step to making iPSC technology more broadly applicable.

Conclusions

Aimed At: Azoospermic Men

ICSI is the most effective treatment currently employed for severe male factor infertility but this requires the presence of fully developed spermatozoa. However, for men suffering from complete spermatogenic arrest treatment is still elusive. In mammals, millions of spermatozoa are produced daily, ultimately from spermatogonia stem cells. These precursors of spermatogonia depend not only on specific growth factors for survival and proliferation, but also on the close relationship with the enveloping presumptive sertoli cells which seems to prevent apoptosis of these germ cells. The ability to propagate and immortalize SSCs in vitro would allow the creation of colonies, which on intra-testicular transplantation might be able to repopulate the germinal epithelium of azoospermic men with SCO syndrome.

Men with germ cell aplasia as well as men and pre-pubescent children surviving cancer who had undergone chemotherapy treatment require the transplantation of germ cells to re-initiate the spermatogenic process. Since chemotherapy involves the use of extremely cytotoxic chemicals, it can cause tremendous harm to germ cells, often rendering the patient sterile.

In this regard, significant strides have been accomplished in experimental animals through spermatogonia transplantation that offers an opportunity for future clinical application in the human. A handful of researchers have been able to recolonize the basal membrane of seminiferous tubules by injecting SSCs, therefore re-establishing spermatogenesis in autologous transfers. The so derived viable spermatozoa were capable of fertilizing mature oocytes resulting in the production of progeny [84]. As experiments involving human subjects are limited by ethical constraints, xenogeneic transplantation using experimental models has helped provide an adequate environment to study the different stages of germ cell differentiation. Attempts have been made to transplant human germ cell suspensions into the testes of immunodeficient mice, but unfortunately without success in the case of fresh cell suspensions [31] and with limited success in the case of cryopreserved cells [33]. Xenografting of human testicular tissue pieces has also met with limited success [39, 85, 86]. Both xeno-transplantation and -grafting are blocked at pre-meiotic stages and have no differentiation capacity to the later spermiogeneic stages.

What Can Be Done in the Future

At the present time, cryopreservation methods are limited to utilizing standard freezing methods either with DMSO or glycerol [87–89]. Another method that has been attempted to assess if spermatogonial cells have better survival rates is vitrification [90]. However, the main limitation of cryopreservation is the possible damage to the spermatogonia thereby limiting its survival as well as its ability to self renew. In this instance, the addition of growth factors may help germ cell survival, proliferation, and eventual differentiation to mature gametes.

Where Future Research Is Heading

With the availability of embryonic stem cells the possibility to recreate a male germ cell niche, through the tridimensional model represented by the embryoid bodies, appears to be concrete. While these studies indicate that alternative sources of gametes are not merely the result of science fiction but a reality, further advances on the epigenesis of the neo-gametes are welcome.

Moreover, the possibility to propagate a male genome would provide an alternative means through which to consistently obtain conceptuses. We attempted to replicate a haploid male genome in the mouse [75, 91] and generate conceptuses capable of undergoing full-term development, so as to enhance the reproductive performance of a single spermatozoon.

When all the techniques are refined, scientists and physicians can develop a possible treatment that will restore fecundity in a sterile man and reproductive medicine will forever be changed. Thus, SSCs and their subsequent culturing in vitro is the future of reproductive medicine, finally giving hope to infertile couples who wish for nothing more than starting a family of their own.

Acknowledgments We are grateful to the clinicians, scientists, embryologists, and nursing staff of The Ronald O. Perelman and Claudia Cohen Center for Reproductive Medicine. We are thankful to Dr. J. Michael Bedford for his critical review, Dr. Noriko Tanaka for establishing the magnetic cell sorting, and for Justin Kocent for literature search on iPS. Queenie V. Neri was funded by the grant ULI RR024996 of the Clinical and Translational Science Center at Weill Cornell Medical College.

References

1. Stefankiewicz J, Kurzawa R, Drozdzik M. Environmental factors disturbing fertility of men. Ginekol Pol. 2006;77:163–9.
2. Wright VC, Chang J, Jeng G, Macaluso M. Assisted reproductive technology surveillance – United States, 2003. MMWR Surveill Summ. 2006;55:1–22.
3. Chan PT, Schlegel PN. Nonobstructive azoospermia. Curr Opin Urol. 2000;10:617–24.
4. Palermo G, Joris H, Devroey P, Van Steirteghem AC. Pregnancies after intracytoplasmic injection of single spermatozoon into an oocyte. Lancet. 1992;340: 17–8.
5. Palermo GD, Schlegel PN, Colombero LT, Zaninovic N, Moy F, Rosenwaks Z. Aggressive sperm immobilization prior to intracytoplasmic sperm injection with immature spermatozoa improves fertilization and pregnancy rates. Hum Reprod. 1996;11:1023–9.
6. Tanaka H, Baba T. Gene expression in spermiogenesis. Cell Mol Life Sci. 2005;62:344–54.
7. Brinster RL. Germline stem cell transplantation and transgenesis. Science. 2002;296:2174–6.
8. Dym M, He Z, Jiang J, Pant D, Kokkinaki M. Spermatogonial stem cells: unlimited potential. Reprod Fertil Dev. 2009;21:15–21.
9. Bleyer A, Barr R. Cancer in young adults 20 to 39 years of age: overview. Semin Oncol. 2009;36: 194–206.
10. Schover LR, Rybicki LA, Martin BA, Bringelsen KA. Having children after cancer. A pilot survey of survivors' attitudes and experiences. Cancer. 1999;86: 697–709.
11. Tournaye H, Goossens E, Verheyen G, Frederickx V, De Block G, Devroey P, et al. Preserving the reproductive potential of men and boys with cancer: current concepts and future prospects. Hum Reprod Update. 2004;10:525–32.
12. Schrader M, Heicappell R, Muller M, Straub B, Miller K. Impact of chemotherapy on male fertility. Onkologie. 2001;24:326–30.
13. Zapzalka DM, Redmon JB, Pryor JL. A survey of oncologists regarding sperm cryopreservation and assisted reproductive techniques for male cancer patients. Cancer. 1999;86:1812–7.
14. Allen C, Keane D, Harrison RF. A survey of Irish consultants regarding awareness of sperm freezing and assisted reproduction. Ir Med J. 2003;96:23–5.
15. Schrader M, Muller M, Straub B, Miller K. Testicular sperm extraction in azoospermic patients with gonadal germ cell tumors prior to chemotherapy – a new therapy option. Asian J Androl. 2002;4:9–15.
16. Carson SA, Gentry WL, Smith AL, Buster JE. Feasibility of semen collection and cryopreservation during chemotherapy. Hum Reprod. 1991;6:992–4.
17. Robbins WA, Meistrich ML, Moore D, Hagemeister FB, Weier HU, Cassel MJ, et al. Chemotherapy induces transient sex chromosomal and autosomal aneuploidy in human sperm. Nat Genet. 1997;16: 74–8.
18. Brandriff BF, Meistrich ML, Gordon LA, Carrano AV, Liang JC. Chromosomal damage in sperm of patients surviving Hodgkin's disease following MOPP (nitrogen mustard, vincristine, procarbazine, and prednisone) therapy with and without radiotherapy. Hum Genet. 1994;93:295–9.
19. Foresta C, Bettella A, Marin P, Galeazzi C, Merico M, Scandellari C. Analysis of sperm aneuploidy in infertile subjects after chemotherapy treatment. Ann Ital Med Int. 2000;15:189–94.
20. Berthelsen JG, Skakkebaek NE. Gonadal function in men with testis cancer. Fertil Steril. 1983;39:68–75.
21. Baniel J, Sella A. Sperm extraction at orchiectomy for testis cancer. Fertil Steril. 2001;75:260–2.
22. Hourvitz A, Goldschlag DE, Davis OK, Gosden LV, Palermo GD, Rosenwaks Z. Intracytoplasmic sperm injection (ICSI) using cryopreserved sperm from men with malignant neoplasm yields high pregnancy rates. Fertil Steril. 2008;90:557–63.
23. Hawkins MM, Stevens MC. The long-term survivors. Br Med Bull. 1996;52:898–923.
24. Bleyer WA. The impact of childhood cancer on the United States and the world. CA Cancer J Clin. 1990; 40:355–67.
25. Brinster RL, Avarbock MR. Germline transmission of donor haplotype following spermatogonial transplantation. Proc Natl Acad Sci U S A. 1994;91:11303–7.
26. Brinster RL, Nagano M. Spermatogonial stem cell transplantation, cryopreservation and culture. Semin Cell Dev Biol. 1998;9:401–9.
27. Clouthier DE, Avarbock MR, Maika SD, Hammer RE, Brinster RL. Rat spermatogenesis in mouse testis. Nature. 1996;381:418–21.
28. Ogawa T, Dobrinski I, Avarbock MR, Brinster RL. Xenogeneic spermatogenesis following transplantation of hamster germ cells to mouse testes. Biol Reprod. 1999;60:515–21.
29. Dobrinski I, Avarbock MR, Brinster RL. Transplantation of germ cells from rabbits and dogs into mouse testes. Biol Reprod. 1999;61:1331–9.
30. Reis MM, Tsai MC, Schlegel PN, Feliciano M, Raffaelli R, Rosenwaks Z, et al. Xenogeneic transplantation of human spermatogonia. Zygote. 2000;8: 97–105.
31. Nagano M, Brinster CJ, Orwig KE, Ryu BY, Avarbock MR, Brinster RL. Transgenic mice produced by retroviral transduction of male germ-line stem cells. Proc Natl Acad Sci U S A. 2001;98:13090–5.
32. Nagano M, Watson DJ, Ryu BY, Wolfe JH, Brinster RL. Lentiviral vector transduction of male germ line stem cells in mice. FEBS Lett. 2002;524:111–5.
33. Sofikitis N, Ono K, Yamamoto Y, Papadopoulos H, Miyagawa I. Influence of the male reproductive tract on the reproductive potential of round spermatids abnormally released from the seminiferous epithelium. Hum Reprod. 1999;14:1998–2006.

34. Honaramooz A, Snedaker A, Boiani M, Scholer H, Dobrinski I, Schlatt S. Sperm from neonatal mammalian testes grafted in mice. Nature. 2002;418:778–81.
35. Schlatt S, Kim SS, Gosden R. Spermatogenesis and steroidogenesis in mouse, hamster and monkey testicular tissue after cryopreservation and heterotopic grafting to castrated hosts. Reproduction. 2002;124: 339–46.
36. Shinohara T, Inoue K, Ogonuki N, Kanatsu-Shinohara M, Miki H, Nakata K, et al. Birth of offspring following transplantation of cryopreserved immature testicular pieces and in-vitro microinsemination. Hum Reprod. 2002;17:3039–45.
37. Orwig KE, Schlatt S. Cryopreservation and transplantation of spermatogonia and testicular tissue for preservation of male fertility. J Natl Cancer Inst Monogr. 2005;34:51–6.
38. Jahnukainen K, Ehmcke J, Hergenrother SD, Schlatt S. Effect of cold storage and cryopreservation of immature non-human primate testicular tissue on spermatogonial stem cell potential in xenografts. Hum Reprod. 2007;22:1060–7.
39. Wyns C, Van Langendonckt A, Wese FX, Donnez J, Curaba M. Long-term spermatogonial survival in cryopreserved and xenografted immature human testicular tissue. Hum Reprod. 2008;23:2402–14.
40. Martin-du Pan RC, Campana A. Physiopathology of spermatogenic arrest. Fertil Steril. 1993;60:937–46.
41. Sasagawa I, Yazawa H, Suzuki Y, Tateno T, Ichiyanagi O, Kobayashi T, et al. Reevaluation of testicular biopsies of males with nonobstructive azoospermia in assisted reproductive technology. Arch Androl. 2001;46:79–83.
42. Tesarik J, Bahceci M, Ozcan C, Greco E, Mendoza C. Restoration of fertility by in-vitro spermatogenesis. Lancet. 1999;353:555–6.
43. Staub C. A century of research on mammalian male germ cell meiotic differentiation in vitro. J Androl. 2001;22:911–26.
44. Hue D, Staub C, Perrard-Sapori MH, Weiss M, Nicolle JC, Vigier M, et al. Meiotic differentiation of germinal cells in three-week cultures of whole cell population from rat seminiferous tubules. Biol Reprod. 1998;59:379–87.
45. Cremades N, Bernabeu R, Barros A, Sousa M. In-vitro maturation of round spermatids using co-culture on Vero cells. Hum Reprod. 1999;14:1287–93.
46. Sousa M, Cremades N, Alves C, Silva J, Barros A. Developmental potential of human spermatogenic cells co-cultured with Sertoli cells. Hum Reprod. 2002;17:161–72.
47. Tesarik J, Nagy P, Abdelmassih R, Greco E, Mendoza C. Pharmacological concentrations of follicle-stimulating hormone and testosterone improve the efficacy of in vitro germ cell differentiation in men with maturation arrest. Fertil Steril. 2002;77:245–51.
48. Tanaka A, Nagayoshi M, Awata S, Mawatari Y, Tanaka I, Kusunoki H. Completion of meiosis in human primary spermatocytes through in vitro coculture with Vero cells. Fertil Steril. 2003;79 Suppl 1:795–801.
49. Ginsburg M, Snow MH, McLaren A. Primordial germ cells in the mouse embryo during gastrulation. Development. 1990;110:521–8.
50. Lawson KA, Hage WJ. Clonal analysis of the origin of primordial germ cells in the mouse. Ciba Found Symp. 1994;182:68–84. discussion 84–91.
51. Saitou M, Barton SC, Surani MA. A molecular programme for the specification of germ cell fate in mice. Nature. 2002;418:293–300.
52. Yoshimizu T, Obinata M, Matsui Y. Stage-specific tissue and cell interactions play key roles in mouse germ cell specification. Development. 2001;128:481–90.
53. Ying Y, Zhao GQ. Cooperation of endoderm-derived BMP2 and extraembryonic ectoderm-derived BMP4 in primordial germ cell generation in the mouse. Dev Biol. 2001;232:484–92.
54. Ying Y, Qi X, Zhao GQ. Induction of primordial germ cells from murine epiblasts by synergistic action of BMP4 and BMP8B signaling pathways. Proc Natl Acad Sci U S A. 2001;98:7858–62.
55. Brinster RL, Zimmermann JW. Spermatogenesis following male germ-cell transplantation. Proc Natl Acad Sci U S A. 1994;91:11298–302.
56. Schmidt JA, Avarbock MR, Tobias JW, Brinster RL. Identification of glial cell line-derived neurotrophic factor-regulated genes important for spermatogonial stem cell self-renewal in the rat. Biol Reprod. 2009;81:56–66.
57. Zhao GQ, Deng K, Labosky PA, Liaw L, Hogan BL. The gene encoding bone morphogenetic protein 8B is required for the initiation and maintenance of spermatogenesis in the mouse. Genes Dev. 1996;10:1657–69.
58. Zhao GQ, Chen YX, Liu XM, Xu Z, Qi X. Mutation in Bmp7 exacerbates the phenotype of Bmp8a mutants in spermatogenesis and epididymis. Dev Biol. 2001;240:212–22.
59. Bowles J, Knight D, Smith C, Wilhelm D, Richman J, Mamiya S, et al. Retinoid signaling determines germ cell fate in mice. Science. 2006;312:596–600.
60. Kanatsu-Shinohara M, Ogonuki N, Inoue K, Miki H, Ogura A, Toyokuni S, et al. Long-term proliferation in culture and germline transmission of mouse male germline stem cells. Biol Reprod. 2003;69:612–6.
61. Steinberger A, Steinberger E, Perloff WH. Mammalian testes in organ culture. Exp Cell Res. 1964;36:19–27.
62. Steinberger E, Steinberger A, Perloff WH. Studies on growth in organ culture of testicular tissue from rats of various ages. Anat Rec. 1964;148:581–9.
63. Rassoulzadegan M, Paquis-Flucklinger V, Bertino B, Sage J, Jasin M, Miyagawa K, et al. Transmeiotic differentiation of male germ cells in culture. Cell. 1993; 75:997–1006.
64. Ogawa T, Kita K, Kubota Y. Proliferation of spermatogonial stem cells and spermatogenesis in vitro. Reprod Med Biol. 2006;5:169–74.
65. Neri QV, Tanaka N, Takeuchi T, Toschi M, Rosenwaks Z, Palermo GD. Propagation and maturation of male gonocytes in vitro. Fertil Steril. 2006;86:s14.
66. Rohwedel J, Guan K, Wobus AM. Induction of cellular differentiation by retinoic acid in vitro. Cells Tissues Organs. 1999;165:190–202.

67. Hubner K, Fuhrmann G, Christenson LK, Kehler J, Reinbold R, De La Fuente R, et al. Derivation of oocytes from mouse embryonic stem cells. Science. 2003;300:1251–6.
68. Lacham-Kaplan O, Chy H, Trounson A. Testicular cell conditioned medium supports differentiation of embryonic stem cells into ovarian structures containing oocytes. Stem Cells. 2006;24:266–73.
69. Clark AT, Bodnar MS, Fox M, Rodriquez RT, Abeyta MJ, Firpo MT, et al. Spontaneous differentiation of germ cells from human embryonic stem cells in vitro. Hum Mol Genet. 2004;13:727–39.
70. Geijsen N, Horoschak M, Kim K, Gribnau J, Eggan K, Daley GQ. Derivation of embryonic germ cells and male gametes from embryonic stem cells. Nature. 2004;427:148–54.
71. Guan K, Nayernia K, Maier LS, Wagner S, Dressel R, Lee JH, et al. Pluripotency of spermatogonial stem cells from adult mouse testis. Nature. 2006;440: 1199–203.
72. Nayernia K, Nolte J, Michelmann HW, Lee JH, Rathsack K, Drusenheimer N, et al. In vitro-differentiated embryonic stem cells give rise to male gametes that can generate offspring mice. Dev Cell. 2006;11: 125–32.
73. Izadyar F, Pau F, Marh J, Slepko N, Wang T, Gonzalez R, et al. Generation of multipotent cell lines from a distinct population of male germ line stem cells. Reproduction. 2008;135:771–84.
74. Palermo GD, Neri QV, Takeuchi T, Rosenwaks Z. ICSI: where we have been and where we are going. Semin Reprod Med. 2009;27:191–201.
75. Kerkis A, Fonseca SA, Serafim RC, Lavagnolli TM, Abdelmassih S, Abdelmassih R, et al. In vitro differentiation of male mouse embryonic stem cells into both presumptive sperm cells and oocytes. Cloning Stem Cells. 2007;9:535–48.
76. Evans MJ, Kaufman MH. Establishment in culture of pluripotential cells from mouse embryos. Nature. 1981;292:154–6.
77. Martin GR. Isolation of a pluripotent cell line from early mouse embryos cultured in medium conditioned by teratocarcinoma stem cells. Proc Natl Acad Sci U S A. 1981;78:7634–8.
78. Thomson JA, Itskovitz-Eldor J, Shapiro SS, Waknitz MA, Swiergiel JJ, Marshall VS, et al. Embryonic stem cell lines derived from human blastocysts. Science. 1998;282:1145–7.
79. Takahashi K, Yamanaka S. Induction of pluripotent stem cells from mouse embryonic and adult fibroblast cultures by defined factors. Cell. 2006;126: 663–76.
80. Hochedlinger K, Plath K. Epigenetic reprogramming and induced pluripotency. Development. 2009;136: 509–23.
81. Staerk J, Dawlaty MM, Gao Q, Maetzel D, Hanna J, Sommer CA, et al. Reprogramming of human peripheral blood cells to induced pluripotent stem cells. Cell Stem Cell. 2010;7:20–4.
82. Loh YH, Hartung O, Li H, Guo C, Sahalie JM, Manos PD, et al. Reprogramming of T cells from human peripheral blood. Cell Stem Cell. 2010;7:15–9.
83. Seki T, Yuasa S, Oda M, Egashira T, Yae K, Kusumoto D, et al. Generation of induced pluripotent stem cells from human terminally differentiated circulating T cells. Cell Stem Cell. 2010;7:11–4.
84. Kubota H, Avarbock MR, Schmidt JA, Brinster RL. Spermatogonial stem cells derived from infertile Wv/Wv mice self-renew in vitro and generate progeny following transplantation. Biol Reprod. 2009;81: 293–301.
85. Geens M, De Block G, Goossens E, Frederickx V, Van Steirteghem A, Tournaye H. Spermatogonial survival after grafting human testicular tissue to immunodeficient mice. Hum Reprod. 2006;21:390–6.
86. Schlatt S, Honaramooz A, Ehmcke J, Goebell PJ, Rubben H, Dhir R, et al. Limited survival of adult human testicular tissue as ectopic xenograft. Hum Reprod. 2006;21:384–9.
87. Kvist K, Thorup J, Byskov AG, Hoyer PE, Mollgard K, Yding Andersen C. Cryopreservation of intact testicular tissue from boys with cryptorchidism. Hum Reprod. 2006;21:484–91.
88. Keros V, Hultenby K, Borgstrom B, Fridstrom M, Jahnukainen K, Hovatta O. Methods of cryopreservation of testicular tissue with viable spermatogonia in pre-pubertal boys undergoing gonadotoxic cancer treatment. Hum Reprod. 2007;22:1384–95.
89. Wyns C, Curaba M, Martinez-Madrid B, Van Langendonckt A, Francois-Xavier W, Donnez J. Spermatogonial survival after cryopreservation and short-term orthotopic immature human cryptorchid testicular tissue grafting to immunodeficient mice. Hum Reprod. 2007;22:1603–11.
90. Curaba M, Verleysen M, Amorim CA, Dolmans MM, Van Langendonckt A, Hovatta O, et al. Cryopreservation of prepubertal mouse testicular tissue by vitrification. Fertil Steril. 2011;95(4):1229–34.
91. Takeuchi T, Neri QV, Palermo GD. Male gamete empowerment. Ann N Y Acad Sci. 2008;1127:64–6.

Testicular Tissue Transplantation for Fertility Preservation

25

Jose R. Rodriguez-Sosa, Stefan Schlatt, and Ina Dobrinski

Abstract

Spermatogenesis is a complex process of cell proliferation and differentiation, sustained by a pool of stem cells in the testis. While our knowledge of spermatogenesis and the biology of spermatogonial stem cells is constantly increasing, it is still far from being complete, especially in the primate testis. Ectopic xenografting of testis tissue from larger animals into immunodeficient mice results in donor spermatogenesis, with production of fertilization-competent sperm. Since its first report in 2002, testis tissue xenografting has been evaluated in many mammalian species, including primates. This technique represents a powerful tool to study testis development and spermatogenesis of diverse species in a mouse host. The current chapter focuses on transplantation of testis tissue as a potential alternative for fertility preservation in prepubertal boys that are subjected to potentially sterilizing cytotoxic cancer treatment. The chapter provides a review of recent developments in testis tissue xenografting in primates, including humans. Second, we focus on the cryopreservation of testis tissue as an essential prerequisite to make testis xenografting a feasible and practical approach under clinical settings. Finally, we provide a detailed description of the methodology involved in cryopreservation and xenotransplantation of testis tissue.

Keywords

Fertility preservation in men • Testicular tissue transplantation • Xenografting in primates • Cryopreservation of testis tissue • Male fertility

I. Dobrinski (✉)
Department of Comparative Biology and Experimental Medicine, Faculty of Veterinary Medicine, University of Calgary, 3300 Hospital Drive, Calgary, AB, Canada T2N 4N1
e-mail: idobrins@ucalgary.ca

E. Seli and A. Agarwal (eds.), *Fertility Preservation: Emerging Technologies and Clinical Applications*,
DOI 10.1007/978-1-4419-1783-6_25,

Male fertility is based on the efficient production of sperm during the adult life of a male. Sperm are generated through a complex process of cell proliferation and differentiation known as spermatogenesis. In this process, spermatogonia undergo successive divisions to give rise to spermatocytes, in which genetic material is recombined and segregated through meiosis. Resulting haploid spermatids initiate a series of morphological changes to finally transform into sperm structurally equipped for the efficient delivery of the genetic material at fertilization [1]. Spermatogenesis occurs as a continuous process, assuring virtually unlimited sperm production during adult life. This process is sustained by spermatogonial stem cells (SSCs), which not only give rise to differentiating spermatogonia during the proliferative phase of spermatogenesis, but also self-renew to maintain the reserve stem cell pool [2]. Whether SSCs undergo asymmetric division resulting in stem cells and differentiated daughter cells is still a subject of investigation [3]. Recent evidence also points to the possibility that the decision to remain a stem cell or to enter the differentiation pathway can occur through fragmentation of clones of spermatogonia that will enter different pathways. This phenomenon has been observed in rodent and primate species [4–6].

Germ cell differentiation occurs in close interaction with testicular somatic cells, particularly Sertoli cells in the seminiferous epithelium of the testis. These SSCs reside in a particular microenvironment, the stem cell niche. In this niche, Sertoli cells, and presumably also cells from the interstitial compartment, modulate SSCs renewal and differentiation. The existence of a testicular stem cell niche is well documented but its cellular and molecular components are poorly understood and might vary between species [7–9]. However, transplantation of germ cells from a large variety of species into the mouse testis revealed that stem cell recognition and niche colonization are highly conserved among different species [10–13]. The plasticity of the niche to receive and host other germ line cells has been demonstrated; in addition to SSCs, primordial germ cells or teratocarcinoma cells have the potential to enter the niche and generate spermatogenesis [14, 15].

The SSC niche is established during the remodeling of the seminiferous epithelium that occurs during the postnatal phase of testis development. After an initial period of Sertoli cell proliferation, the seminiferous cords transform into tubules, and germ cells actively divide and differentiate to give rise to the first wave of spermatogenesis [16]. This first spermatogenic wave sets the framework for the future sperm production in adult life [2].

In primate adult testes, three types of spermatogonia are distinguished by morphological criteria: A_{dark}, A_{pale} and B spermatogonia [17–19]. The number of subsequent divisions of B-spermatogonia differ between primate species. For example, one division is described in men and four divisions (B1–B4) in macaques. Various models for spermatogonial kinetics have been described and are currently under debate signifying that the exact details of spermatogonial turnover in the primate testis are still largely unresolved [20, 21]. However, it is generally agreed that A_{dark} spermatogonia are mitotically quiescent and act as reserve stem cells since they become proliferatively active during pubertal expansion [22] and following depletion of spermatogonia due to irradiation or toxic exposure [23, 24]. On the other hand, A_{pale} spermatogonia proliferate regularly and are considered self-renewing progenitors [4, 5].

Rapidly dividing spermatogonia are highly sensitive to irradiation and toxins in adult and immature monkeys [24, 25]. Low doses of cytotoxic drugs or irradiation deplete the differentiating spermatogonia while less sensitive SSCs as well as spermatocytes and spermatids survive. Recovery of spermatogenesis occurs from the remaining SSCs and relies on the type, dose and fractionation of cytotoxic drugs and irradiation [23]. During recovery, testicular histology reveals an all-or-nothing pattern with areas of full spermatogenesis and areas with a Sertoli-cell-only pattern. This histological pattern during spermatogenic recovery indicates a critical role of

SSCs for re-initiation of spermatogenesis and shows that with doses used in these primate studies the somatic environment is not heavily affected by chemotherapy or radiation exposure [26].

Cytotoxic drugs and irradiation are commonly used as therapeutic alternatives in oncological patients, and due to their toxicity for germ cells, development of alternatives that avoid testicular damage is highly desirable. Several strategies for protection of spermatogonial cells in these patients have been developed for application in a clinical setting (for review see Jahnukainen et al. [27]; Schlatt et al. [26]). While cryopreservation of sperm offers a standardized and routine option for fertility preservation in adults and pubertal patients, prepubertal children cannot donate spermatozoa for cryostorage. Moreover, cryopreserved sperm represent a finite source of gametes. Several alternatives have been discussed based on the high regenerative potential of the seminiferous epithelium that is supported by SSCs. Although the somatic environment can also be affected by irradiation and toxic exposures, this appears to be more resistant than germ cells [28]. Therefore, autologous transplantation of SSCs could provide an option for fertility preservation for oncological patients, regardless of their age [27].

Another alternative is xenografting of testicular tissue into immunodeficient mice. Prior to puberty, when the testis consists of seminiferous cords and the only germ cells present are spermatogonia, the developing testis appears to be more tolerant to transient ischemia making it possible to manipulate cells and tissue fragments while maintaining their full developmental potential. Since the first report in 2002, it has been demonstrated that testis tissue from a wide variety of larger species can initiate and complete development after transplantation into immunodeficient adult-castrated mice, and that the resulting sperm are capable of fertilizing and triggering normal embryo development [29–32]. This approach has been explored in primates, including humans, as a potential alternative for fertility preservation in prepubertal boys undergoing cytotoxic treatment during oncological processes [33–36]. It is the aim of this chapter to initially summarize the advances achieved in cryopreservation and ectopic xenografting of testis tissue as an alternative for fertility preservation, with emphasis in primates. Second, this chapter focuses on the methodology involved in both cryopreservation and xenotransplantation of testis tissue xenografting.

Testis Tissue Xenografting

Because of the architecture of its vascular and duct systems and the complexity of the seminiferous epithelium, the testis does not appear to be a suitable tissue for grafting. However, transplantation of testicular tissue has been performed for two centuries, and has provided important insights into testicular function. Autologous and homologous transplantation of testicular tissue has been reviewed by Gosden and Aubard [37, 38]. The generation of immunodeficient lines of mice allowed xenotransplantation of testicular and other tissues from large animals into rodent host [39]. Xenotransplantation of testicular tissue into immunodeficient mice (human fetal testis into the abdominal wall of adult nude mice) was first performed in 1974, resulting in initial development of the donor tissue [40]. Subsequently, Hochereau-de-Reviers and Perreau [41] transplanted ovine fetal testis into the scrotum of intact nude mice and reported differentiation of gonocytes into spermatogonia and primary spermatocytes. Complete cross species spermatogenesis was first reported in 2002 [29]. In that report, pieces of testis tissue from newborn pigs and goats were able to survive and displayed complete development with production of sperm.

These studies made clear that the immature testis was capable of initiating development if exposed to adult levels of gonadotropins, as lack of testis development during the prepubertal phase is largely due to low gonadotropin support. This immediate response of the testis tissue to initiate development results in an apparent shortening of the time required until first appearance of testicular sperm in immature testicular xenografts from species such as pigs and monkeys [29, 33].

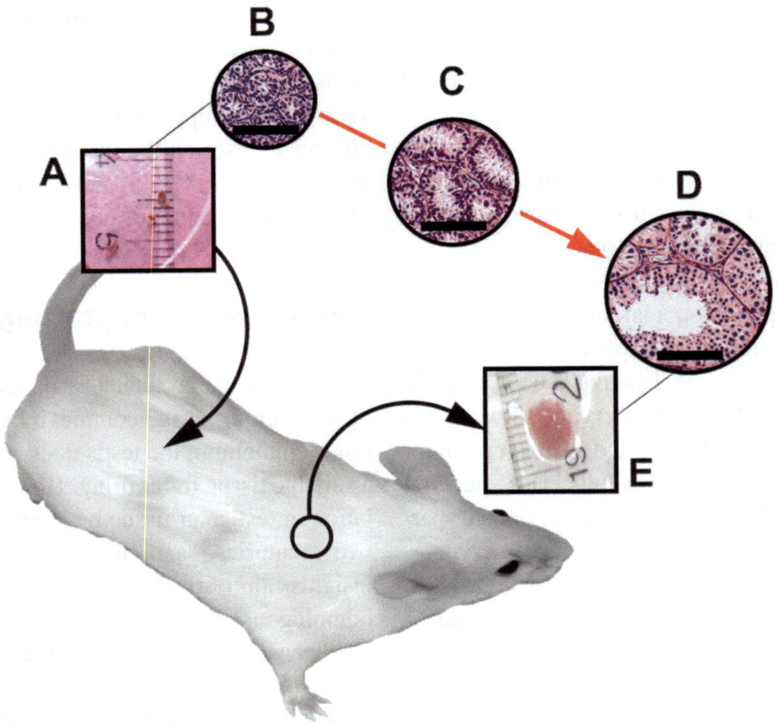

Fig. 25.1 Ectopic xenografting of immature testicular tissue from large animals into immunodeficient mice. Testis fragments (~1 mm^3) from immature donors are transplanted under the back skin of immunodeficient mice (**a**), in this case a SCID mouse. Once transplanted, the testis fragments are able to survive and respond to mouse gonadotropins. The tissue develops from its immature stage containing typical seminiferous cords and gonocytes (**b**, **c**) to a mature-like stage, including the formation of fertilization competent sperm (**d**). Then, testis xenografts are collected for analysis or harvesting sperm (**e**). Bars = 50 μm

In other species such as cats and cattle, this shortening is not evident mainly due to a developmental arrest of the tissue as grafts undergo meiosis [42–44].

Once testis tissue has been transplanted in host mice, they rely on the re-establishment of the blood flow to survive and function [45]. The re-established blood supply provides murine LH and FSH to the xenografts. Grafts respond to this stimulation by recovering their endocrine function and establishing a feedback loop restoring the hypothalamic–pituitary–gonadal axis. Graft Leydig cells secrete testosterone in response to host LH, which along with recipient FSH induce the entire development of graft seminiferous cords similar to pubertal development (Fig. 25.1) [29, 31]. Graft testosterone supports spermatogenesis and maintains the integrity of the recipient reproductive tract; size and histological appearance of seminal vesicles, an androgen-dependent secondary sex gland in recipient mice, are indistinguishable from those of noncastrated mice [29, 31, 46].

Testis tissue xenografting has been shown to result in spermatogenesis in all domestic species evaluated, although at variable efficiency (reviewed by Rodriguez-Sosa and Dobrinski [47]). From the studies performed so far, one important factor affecting the survival and function of testis xenografts is the age or developmental stage of the donor. It is clear that use of tissue from prepubertal males leads to better survival and spermatogenesis outcome than that of pubertal or adult donors, and the capability of survival and development decreases markedly as the donor tissue has undergone meiosis [31, 44, 48–50].

Cryopreservation and Xenografting of Testicular Tissue for Fertility Preservation in Humans

An important finding that originated from the testis xenografting studies carried out in domestic species was the fertilizing capacity of the resulting sperm. This capacity was initially explored using intracytoplasmic sperm injection (ICSI). Injection of sperm from goat and pig xenografts into mouse oocytes resulted in initial stages of embryo development [29]. Similarly, when the procedure was performed with fresh or cryopreserved sperm from neonatal mouse testis xenografts, normal embryos were obtained, which generated normal offspring after being transferred into pseudo-pregnant females [32]. The potential to generate fertile sperm from prepubertal testes created interest to explore primate testis xenografting as a clinically relevant strategy for fertility preservation in boys. To explore this option, sperm obtained from testis xenografts of rhesus macaques were injected into rhesus oocytes, leading to the generation of morphologically normal embryos at the morulae and blastocyst stage [33]. Clinical use of cryopreservation and xenografting of testis tissue for fertility preservation in boys undergoing cancer treatment has been postulated. In this approach (Fig. 25.2), it is envisioned that prior to treatment, biopsies from

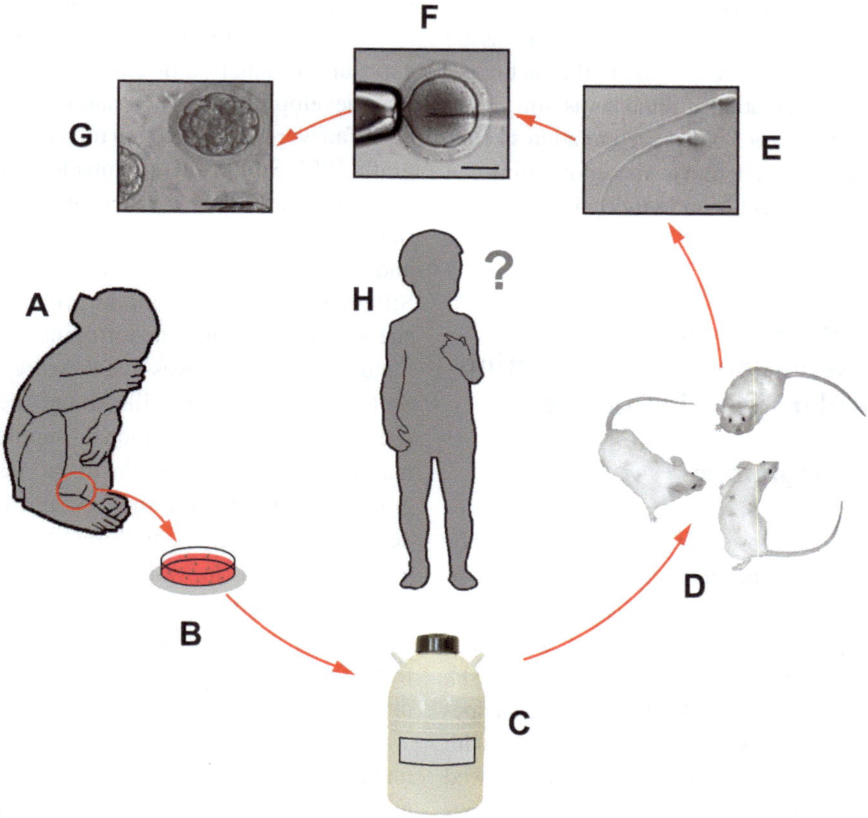

Fig. 25.2 Potential applications of testicular tissue xenografting for fertility preservation. Under this approach, testicular biopsies from immature males (**a**) are obtained and processed to produce fragments of suitable size for transplantation (**b**). The pieces are frozen and stored in liquid nitrogen (**c**). The fragments are thawed at a later point and ectopically grafted into immunodeficient mice (**d**), where they are able to mature. Once tissue has matured, grafts are recovered for sperm harvesting (**e**), which is subsequently used for microinjection of oocytes (**f**) to generate embryos (**g**) ready to be transferred. In the case illustrated here, a rhesus monkey serves as a donor since it has been a suitable model to start establishing these techniques in humans (**h**). However, feasible and safe application of testis tissue xenografting for human fertility preservation has not yet been achieved. Bar = 10 μm in (**e**), and 50 μm in (**f, g**)

the patient testes could be obtained to generate suitable fragments for xenografting that can be frozen and stored. Then, when the patient has reached maturity, has recovered from the original disease but has been rendered infertile by cytotoxic treatment, the tissue could be thawed and subsequently grafted into immunodeficient mice. Once the testis tissue has fully developed, grafts can be collected to harvest sperm to be used in ICSI to produce embryos for transfer. One advantage of this approach over autologous SSCs transplantation is the possibility to avoid the risk of reintroducing cancer cells into the patient testes that may be present in the donor cell suspension at the moment of collection [27, 51]. However, the possibility of host pathogens having an effect on the developing testis tissue and its resulting sperm must be considered in the case of testis tissue xenografting [27]. Therefore, use of this technique for fertility preservation in humans is still hypothetical and rigorous studies need to address ethical and safety issues before testis xenografting can be considered a feasible and safe option for fertility preservation in boys.

Recent Developments in Cryopreservation and Xenografting of Testicular Tissue in Primates

Ectopic Xenografting

One of the first reports of testis tissue xenografting from a large species explored the developmental potential of marmoset testis. Testis xenografts of immature marmosets showed a blockade in early germ cell differentiation [31]. Initially, it was not clear whether this was a species-specific event or an inherent problem with primate testis tissue. Subsequent studies showed that testis tissue from infant rhesus monkeys showed full developmental potential when transplanted to immunodeficient mice [33], and that gonadotropin supplementation promoted its maturation [52]. It is now known that the inability of marmoset testis tissue to develop in mice is due to an insensitivity to mouse LH caused by a deletion in exon 10 of the luteinizing hormone-receptor gene [53] and a hypersensitivity to the increased temperature at the subcutaneous site [54, 55]. Xenografting of human testis tissue has also been explored. The studies performed so far with human tissue show that the immature testis is more suitable for transplantation and differentiates once it is transplanted into host mice. However, so far complete spermatogenesis using immature human testis as xenografts has not been accomplished. Table 25.1 summarizes the strategies and outcome of studies which were performed using testicular xenografting from primate tissue donors.

Besides providing a potential alternative for fertility preservation, xenografting of testis tissue is an elegant experimental strategy to evaluate the effects of hormonal manipulation or exposure to radiation or gonadotoxins on testicular development. The gonadotoxic effects of busulfan were reproduced in monkey testis xenografts [25]. Similarly, the gonadotoxic effect of irradiation was shown when testis tissue from rhesus monkeys was exposed to irradiation doses and subsequently transplanted in mice [56]. Stimulation of grafted infant primate testis tissue with exogenous gonadotropins supported its maturation, and this was due to a direct effect on Sertoli cells [52]. This illustrated the potential use of exogenous hormonal supplement for manipulation of development of testis xenografts. As experimental strategy, an advantage of xenografting is the small number of donors needed to perform valid comparative studies. Instead of exposing groups of monkeys to different treatments, groups of mice carrying xenografts from few juvenile monkeys can be exposed to different regimens.

Although in the current chapter we have mainly focused on xenotransplantation of testis tissue, it is worth mentioning that autologous transplantation of primate testis tissue has also been performed. Autologous transplantation of testis tissue was performed in newborn marmosets to explore whether the developmental blockade in the mouse host was only due to the hormonal milieu provided by the rodent host.

Table 25.1 Outcome of testicular xenografting using primate testis tissue

Species	Donor age and developmental status	Cryopreservation	Recipient treatment	Collection time	Main outcome	References
Rhesus monkey	3 and 6 months, seminiferous cords	No	(A) PMSG, 10 IU twice a week (B) hCG, PMSG, 10 IU twice a week (C) None	7 months	Differentiation of cords into tubules relied on hormone treatment, and no difference was noticed between PMSG and hCG treatments. At collection, treated xenografts of 6-month-old donors showed mature sperm, whereas those of 3-month-old donors contained spermatocytes as the most advanced type of germ cells	[52]
Human	10–11 years, seminiferous cords/ tubules containing spermatogonia	No	None	4 and 9 months	No difference was observed between xenografts of both collection points. No active spermatogenesis was obtained, and spermatogonia were the most advanced germ cell type	[62]
Human	7, 12 and 14 years, seminiferous cords/tubules with spermatogonia (7, 12 and 14 years) and some tubules with focal spermatogenesis (14 years)[a]	Yes	None	6 months	Spermatocytes and spermatids were detected as the most advanced germ cells. No sperm with normal morphology were observed, but xenografts from the 14-year-old donor contained some sperm-like cells	[36]
Rhesus monkey	3, 6, 11 and 12 years, full spermatogenesis in 67–87% of tubules	No	None	<3, 3–6, and >6 months	Mostly atrophic tissue in xenografts from 11- and 12-year-old donors, very few tubules with spermatocytes in xenografts from the 6-year-old donor, and spermatocytes and spermatids in grafts of the 3-year-old donor at <3, 3–6 months	[50]
Rhesus monkey	16 and 19 months, seminiferous cords with spermatogonia	No	None, but testicular fragments were exposed to irradiation (0, 0.5, 1, and 4 Gy)	4 months	Type B spermatogonia were the most advanced germ cells present in all xenografts. However, spermatogonial numbers were reduced after exposure of donor tissue fragments to 0.5 Gy of irradiation. Spermatogonia were almost depleted in xenografts exposed to 1 and 4 Gy	[56]
Human	2–12 years, cryptorchid, seminiferous cords with spermatogonia[a]	Yes	None	21 days	Spermatogonia were the most advanced type of germ cells. In comparison to donor tissue the total number of spermatogonia was decreased in xenografts. However, number of proliferating spermatogonia was higher in xenografts than in donor tissue	[35]
Rhesus monkey	18 and 21 months, seminiferous cords with spermatogonia	Yes	None	3 and 5 months	Type B spermatogonia and spermatocytes were the most advanced type of germ cells at 3 and 5 months, respectively	[34]

(continued)

Table 25.1 (continued)

Species	Donor age and developmental status	Cryopreservation	Recipient treatment	Collection time	Main outcome	References
Human	Fetal, 20 and 26 weeks of gestation, seminiferous cords	No	None	116 and 135 days	Differentiation of cords into tubules was evident; increase in tubule diameter, initial stages lumen formation, and spermatogonia located on the basal lamina were observed	[63]
Rhesus monkey	16 and 19 months, seminiferous cords with spermatogonia	No	(A) None (B) Busulfan (38 mg/kg) in saline and DMSO (1:1) at 7 months (C) Saline and DMSO (1:1) at 7 months	7 months (A), and 8 months (B and C)	At 7 months (Group A), type B spermatogonia and spermatocytes were the most advanced germ cells. At 8 months, in vehicle-treated xenografts (Group B) round spermatids were detected, whereas in Busulfan-treated grafts no spermatogenesis was observed, and a decrease in the number of tubules with germ cells was evident	[25]
Human	Adult, complete spermatogenesis	No	None	30–195 days	Most xenografts (~60%) showed sclerosis. However, in ~22% of them spermatogonia were still observed, with increasing sclerosis in time	[48]
Human	Adult, intact and disturbed spermatogenesis	No	None	2–19 weeks	Xenografts from donors with obstructive azoospermia, hypospermatogenesis, and Sertoli cells only (SCO) showed complete atrophy and hyalinization, with no germ cells. Those from donors with testicular cancer showed SCO tubules and some with type A spermatogonia. Xenografts from transsexual donors (post estradiol treatment) contained some A- and few B-spermatogonia	[64]
Rhesus monkey	13 months, seminiferous cords	No	None	2–7 months	Increase in cord diameter and spermatogonia located on the basal lamina were evident at 2 months. Transformation of cords into tubules, with presence of spermatocytes and round spermatids were observed at 4 months. Complete spermatogenesis was observed at 7 months	[33]
Marmoset	Newborn, and 1, 3, and 7 months, seminiferous cords	No	None, but tissue was co-engrafted with hamster tissue	3 months	Spermatogonia were the most advanced type of germ cells in marmoset xenografts, and co-engraftment with hamster tissue did not induce any further differentiation of marmoset germ cells	[53]
Marmoset	Newborn, seminiferous cords	No	Gonadotropins (Pergonal, three injections of 1.5 IU/week) from day 100 to 135	100 and 135 days	Spermatogonia and spermatocytes were the most advanced types of germ cells at 100 and 135 days, respectively	[31]

[a]Xenografting was performed into the scrotum of castrated host mice

Partial differentiation of the transplanted tissue was obtained, with spermatocytes as the most advanced type of germ cell. This was attributed to the hyperthermia of the ectopic site [54]. Recently, testis tissue from immature and adult marmosets was transplanted in autologous recipients. Unlike adult tissue, immature tissue showed good survival and recovery, and similar to the previous study, ectopic placement of grafts resulted in partial differentiation, with spermatocytes as the most advanced germ cells. However, when the tissue was transplanted into the scrotum, mature spermatids were obtained [55]. Finally, partial differentiation of immature testis tissue after autologous transplantation and gonadotropin supplementation has also been reported in rhesus monkeys [57].

Cryopreservation

Long-term preservation of testicular tissue by cryostorage is an essential prerequisite for testis tissue xenografting to be widely applicable to clinical settings. The original studies performed in pigs and goats showed that freezing and thawing of testis tissue prior to transplantation did not affect its capability to undergo full development in the host mice [29]. Moreover, in vitro microinsemination using sperm harvested from frozen and thawed rabbit testis tissue orthotopically transplanted in noncastrated mice resulted in embryos and apparently normal offspring [31]. Since then, cryopreservation of testicular tissue has been evaluated in several species with varying degrees of germ cell differentiation after thawing and grafting (Table 25.2). In humans and other primates, the most promising results have been obtained with DMSO as the cryoprotectant component of the freezing medium. The structural attributes of human testicular tissue were maintained by freezing with a programmable freezer using 0.7 M DMSO. However, xenografting of the frozen tissue was not attempted [58, 59]. In the monkey study, 1.4 M DMSO resulted in better spermatogenesis outcome in comparison to 0.7 M DMSO [34]. More recently, immature testicular tissue frozen with 0.7 M DMSO and subsequently thawed and transplanted into the scrotum of castrated nude mice was able to differentiate. However, most of the xenografts arrested at the meiotic stage and except for few morphologically abnormal sperm in the samples of one donor, no normal sperm was obtained [36]. Importantly, studies performed in rhesus monkeys and pigs demonstrated that similarity of frozen-thawed and fresh tissue does not necessarily translate to equivalent development after grafting into mouse hosts [34, 60].

A complementary alternative to freezing testicular tissue is the possibility of freezing sperm that are harvested from testis xenografts [61]. Snap-frozen sperm harvested from mouse allografts led to normal embryos and progeny when microinjected into mouse oocytes [32]. Similarly, microinjection with snap-frozen sperm from pig testis xenografts resulted in generation of morphologically normal embryos, albeit of a lower efficiency in comparison to testicular, epididymal, or ejaculated sperm [30]. Freezing sperm recovered from pig testis xenografts resulted in viable sperm after thawing, but the number of viable sperm was reduced in comparison to frozen-thawed testicular sperm [60]. Differences in the characteristics or fertilizing ability of the sperm obtained from frozen-thawed testis xenografts has been attributed, at least in part, to the senescense of the resulting sperm. Since testis xenografts lack the excurrent ducts of in situ testes, resulting sperm accumulate in the lumen of the seminiferous tubules. Since so far there is no way to distinguish recently-formed sperm from those that have been accumulated, old sperm could be employed in ICSI leading to decreased fertilization outcome [30, 60]. This emphasizes the need for developing efficient methods not only for testis tissue and sperm cryopreservation, but also to determine optimal times for xenograft recovery and sperm harvesting. See Appendix F and [61] for a detailed protocol for xenografting.

Table 25.2 Testicular cryopreservation in non-rodent species

Species	Freezing method	Xenografting	Outcome	References
Human	0.7 M DMSO HBSS, supplemented with sucrose and human serum albumin, automated freezing	Yes[a]	Testis tissue showed minor degenerative signs. Incomplete spermatogenesis, with round spermatids as most advanced type of germ cells, with presence of few abnormal sperm-like cells in the xenografts of one donor	[36]
Human	(A) HBSS plus 5% DMSO and 5% human serum albumin. Automated freezing at cooling rate 1°C/ min to 0°C, 0.5°C/min to −40°C, 7°C to −70°C with final plunging into liquid nitrogen (B) Same medium as in (A), Automated freezing at cooling rate 1°C/ min to −8°C, 10°C/min to −80°C with final plunging into liquid nitrogen	No	After thawing, no structural differences were noticed between frozen and fresh tissue. Best method: B	[59]
Human	(A) Leibovitz L-15 plus 1.5 M ethylene glycol, 0.1 M sucrose and 0.1% human serum albumin, automated freezing (B) Same as in A, except that Leibovitz medium was replaced by PBS	No	After thawing, no structural and functional (in vitro culture) differences were noticed between frozen and fresh tissue. No difference was observed between the freezing methods used	[65]
Human	(A) Egg yolk-based medium containing 12% glycerol, supplemented with sucrose and patient serum, automated freezing (B) 1.5 M 1,2-propanediol HBSS supplemented with sucrose and patient serum, automated freezing (C) 0.7 M DMSO HBSS, supplemented with sucrose and patient serum, automated freezing	No	After thawing, no structural differences were noticed between frozen and fresh tissue. Best method: C	[58]
Rhesus monkey	(A) 1.4 M ethylene glycol Leibovitz L-15 plus 10% FCS, automated freezing (B) 1.4 M DMSO Leibovitz L-15 plus 10% FCS, automated freezing (C) Leibovitz L-15 plus 10% FCS, automated freezing	Yes	Initiation of spermatogenesis, with spermatocytes as the most advanced type of germ cells. Best method: B	[34]
Pig	(A) DMSO, DMEM, and FCS at 1:3:1 ratio, conventional slow-freezing in alcohol bath (B) Leibovitz L-15 plus 2% FCS and 10% DMSO, automated freezing (C) Vitrification with ethylene glycol containing 0.9% NaCl and 0.5 M raffinose	Yes	At collection spermatogenesis was delayed in comparison to fresh tissue xenografts, with spermatocytes as the most advanced type of germ cells. Best method: A	[60]
Pig	DMSO, DMEM, and FCS at 1:3:1 ratio, and conventional slow-freezing in alcohol bath	Yes	Complete spermatogenesis	[29]
Goat	DMSO, DMEM, and FCS at 1:3:1 ratio, and conventional slow-freezing	Yes	Complete spermatogenesis	[29]
Rabbit	DMSO, DMEM, and FCS, automated freezing	Yes[b]	Complete spermatogenesis. Normal progeny was obtained by ICSI with resulting sperm	[66]

[a]Xenografting was performed into the scrotum of castrated host mice

[b]Recipient mice were not castrated and donor testis tissue was transplanted into the testis

Conclusions

Based on results obtained from studies in domestic species and primates, testicular tissue xenografting represents a promising alternative for fertility preservation in prepubertal boys. Cryostorage of the patient testis tissue is essential for testis tissue xenografting to become a practical alternative under clinical settings. Important advances have been made towards that goal. However, additional studies need to be performed to address ethical and safety issues before testis tissue xenografting can be used in humans. So far, complete spermatogenesis has not been obtained from cryopreserved human testis tissue transplanted into immunodeficient mice. A key area to address in the future is the development of freezing methods that preserve not only the structure and developmental potential of prepubertal human tissue, but also the fertilization capability of the resulting sperm.

See Appendix F

References

1. Russell LD, Ettlin RA, SinhaHikim AP, et al. Mammalian spermatogenesis. In: Russell LD, Ettlin RA, SinhaHikim AP, Clegg ED, editors. Histological and histopathological evaluation of the testis. St. Louis: Cache River Press; 1990. p. 1–38.
2. de Rooij DG, Russell LD. All you wanted to know about spermatogonia but were afraid to ask. J Androl. 2000;21:776–98.
3. Luo J, Megee S, Dobrinski I. Asymmetric distribution of UCH-L1 in spermatogonia is associated with maintenance and differentiation of spermatogonial stem cells. J Cell Physiol. 2009;220:460–8.
4. Ehmcke J, Luetjens CM, Schlatt S. Clonal organization of proliferating spermatogonial stem cells in adult males of two species of non-human primates, *Macaca mulatta* and *Callithrix jacchus*. Biol Reprod. 2005;72: 293–300.
5. Ehmcke J, Simorangkir DR, Schlatt S. Identification of the starting point for spermatogenesis and characterization of the testicular stem cell in adult male rhesus monkeys. Hum Reprod. 2005;20:1185–93.
6. Nakagawa T, Sharma M, Nabeshima Y, et al. Functional hierarchy and reversibility within the murine spermatogenic stem cell compartment. Science. 2010;328:62–7.
7. Orwig KE, Ryu BY, Master SR, et al. Genes involved in post-transcriptional regulation are overrepresented in stem/progenitor spermatogonia of cryptorchid mouse testes. Stem Cells. 2008;26:927–38.
8. Kostereva N, Hofmann MC. Regulation of the spermatogonial stem cell niche. Reprod Domest Anim. 2009;43 Suppl 2:386–92.
9. Oatley JM, Oatley MJ, Avarbock MR, et al. Colony stimulating factor 1 is an extrinsic stimulator of mouse spermatogonial stem cell self-renewal. Development. 2009;136:1191–9.
10. Dobrinski I, Avarbock MR, Brinster RL. Transplantation of germ cells from rabbit and dogs into mouse testes. Biol Reprod. 1999;61:1331–9.
11. Dobrinski I, Avarbock MR, Brinster RL. Germ cell transplantation from large domestic animals into mouse testes. Mol Reprod Dev. 2000;57:270–9.
12. Nagano N, McCarrey JR, Brinster RL. Primate spermatogonial stem cells colonize mouse testis. Biol Reprod. 2001;64:1409–16.
13. Nagano N, Patrizio P, Brinster RL. Long-term survival of human spermatogonial stem cells in mouse testes. Fertil Steril. 2002;78:1225–33.
14. Nayernia K, Li M, Jaroszynski L, Khusainov R, et al. Stem cell based therapeutical approach of male infertility by teratocarcinoma derived germ cells. Hum Mol Genet. 2004;13:1451–60.
15. Chuma S, Kanatsu-Shinohara M, Inoue K, et al. Spermatogenesis from epiblast and primordial germ cells following transplantation into postnatal mouse testis. Development. 2005;132:117–22.
16. Orth JM. Cell biology of testicular development in the fetus and neonate. In: Desjardins C, Ewing LL, editors. Cell and molecular biology of the testis. New York: Oxford University Press; 1993. p. 3–42.
17. Clermont Y, Leblond CP. Differentiation and renewal of spermatogonia in the monkey, *Macaca rhesus*. Am J Anat. 1959;104:237–73.
18. Clermont Y. Spermatogenesis in man. A study of the spermatogonial population. Fertil Steril. 1966;17: 705–21.
19. Clermont Y. Two classes of spermatogonial stem cells in the monkey (*Cercopithecus aethiops*). Am J Anat. 1969;126:57–71.
20. Ehmcke J, Schlatt S. A revised model for spermatogonial expansion in man: lessons from non-human primates. Reproduction. 2006;132:673–80.
21. Amann RP. The cycle of the seminiferous epithelium in humans: a need to revisit? J Androl. 2008;29: 469–87.
22. Simorangkir DR, Marshall GR, Ehmcke J, et al. Prepubertal expansion of dark and pale type A spermatogonia in the rhesus monkey (*Macaca mulatta*) results from proliferation during infantile and juvenile development in a relatively gonadotropin independent manner. Biol Reprod. 2005;73:1109–15.
23. van Alphen MMA, van den Kant HJG, de Rooij DG. Repopulation of the seminiferous epithelium of the rhesus monkey after X-irradiation. Radiat Res. 1988; 113:487–500.

24. van Alphen MMA, van den Kant HJG, Davids JAG, et al. Dose-response studies on the spermatogonial stem cells of the Rhesus monkey (*Macaca mulatta*) after X-irradiation. Radiat Res. 1989;199:443–51.
25. Jahnukainen K, Ehmcke J, Schlatt S. Testicular xenografts: a novel approach to study cytotoxic damage in juvenile primate testis. Cancer Res. 2006;66: 3813–8.
26. Schlatt S, Ehmcke J, Jahnukainen K. Testicular stem cells for fertility preservations: preclinical studies on male germ cell transplantation and testicular grafting. Pediatr Blood Cancer. 2009;53:274–80.
27. Jahnukainen K, Ehmcke J, Soder O, et al. Clinical potential and putative risks of fertility preservation in children utilizing gonadal tissue or germ line stem cells. Pediatr Res. 2006;59:40R–7.
28. Zhang Z, Shao S, Meistrich ML. Irradiated mouse testes efficiently support spermatogenesis derived from donor germ cells of mice and rats. J Androl. 2006;27:365–75.
29. Honaramooz A, Snedaker A, Bioani M, et al. Sperm from neonatal testes grafted in mice. Nature. 2002;418:778–81.
30. Honaramooz A, Cui X, Kim N, et al. Porcine embryos produced after intracytoplasmic sperm injection using xenogeneic pig sperm from neonatal testis tissue grafted in mice. Reprod Fertil Dev. 2008;20:802–7.
31. Schlatt S, Kim SS, Gosden R. Spermatogenesis and steroidogenesis in mouse, hamster and monkey testicular tissue after cryopreservation and heterotopic grafting to castrated host. Reproduction. 2002; 124:339–46.
32. Schlatt S, Honaramooz A, Bioani M, et al. Progeny from sperm obtained after ectopic grafting of neonatal mouse testes. Biol Reprod. 2003;68:2331–5.
33. Honaramooz A, Li M, Penedo CT, et al. Accelerated maturation of primate testis by xenografting into mice. Biol Reprod. 2004;70:1500–3.
34. Jahnukainen K, Ehmcke J, Hergenrother SD, et al. Effect of cold storage and cryopreservation of immature non-human primate testicular tissue on spermatogonial stem cell potential in xenografts. Hum Reprod. 2007;22:1060–7.
35. Wyns C, Curaba M, Martinez-Madrid B, et al. Spermatogonial survival after cryopreservation and short-term orthotopic immature human cryptorchid testicular tissue grafting to immunodeficient mice. Hum Reprod. 2007;22:1603–11.
36. Wyns C, Van Langendonckt A, Wese FX, et al. Long-term spermatogonial survival in cryopreserved and xenografted immature human testicular tissue. Hum Reprod. 2008;23:2402–14.
37. Gosden RG, Aubard Y. Why transplant gonadal tissue? In: Gosden RG, Aubard Y, editors. Transplantation of ovarian and testicular tissues. Austin: Landes Company and Chapman & Hall; 1996. p. 1–15.
38. Gosden RG, Aubard Y. Transplantation of testicular tissue. In: Gosden RG, Aubard Y, editors. Transplantation of ovarian and testicular tissues. Austin: Landes Company and Chapman & Hall; 1996. p. 89–97.
39. Paris MCJ, Snow M, Cox S, et al. Xenotransplantation: a tool for reproductive biology and animal conservation? Theriogenology. 2004;61:277–91.
40. Skakkebaek NE, Jensen G, Povlsen CO, et al. Heterotransplantation of human foetal testicular and ovarian tissue to the mouse mutant Nude. Acta Obstet Gynecol Scand. 1974;53:73–5.
41. Hochereau-de-Reviers MT, Perreau C. Induced differentiation of ovine foetal gonocytes after grafting in the scrotum of nude mice. Reprod Nutr Dev. 1997;37:469–76.
42. Rathi R, Honaramooz A, Zeng W, et al. Germ cell fate and seminiferous tubule development in bovine testis xenografts. Reproduction. 2005;130:923–9.
43. Oatley JM, Reeves JJ, McLean DJ. Establishment of spermatogenesis in neonatal bovine testicular tissue following ectopic xenografting varies with donor age. Biol Reprod. 2005;72:358–64.
44. Kim Y, Selvaraj V, Pukazhenthi B, et al. Effect of donor age on success of spermatogenesis in feline testis xenografts. Reprod Fertil Dev. 2007;19:869–76.
45. Schlatt S, Westernströer B, Gassei K, et al. Donor-host involvement in immature rat testis xenografting into nude mouse hosts. Biol Reprod. 2010;82:888–95.
46. Rodriguez-Sosa JR, Foster RA, Hahnel A. Development of strips of ovine testes after xenografting under the skin of mice and co-transplantation of exogenous spermatogonia with grafts. Reproduction. 2010;139:227–35.
47. Rodriguez-Sosa JR, Dobrinski I. Recent developments in testis tissue xenografting. Reproduction. 2009;138:187–94.
48. Geens M, De Block G, Goossens E, et al. Spermatogonial survival after grafting human testicular tissue to immunodeficient mice. Hum Reprod. 2006;21:390–6.
49. Rathi R, Honaramooz A, Zeng W, et al. Germ cell development in equine testis tissue xenografted into mice. Reproduction. 2006;131:1091–8.
50. Arregui L, Rathi R, Zeng W, et al. Xenografting of adult mammalian testis tissue. Anim Reprod Sci. 2008;106:65–76.
51. Aslam I, Fishel S, Moore H, et al. Fertility preservation of boys undergoing anti-cancer therapy: a review of the existing situation and prospects for the future. Hum Reprod. 2000;15:2154–9.
52. Rathi R, Zeng W, Megee S, et al. Maturation of testicular tissue from infant monkeys after xenografting into mice. Endocrinology. 2008;149:5288–96.
53. Wistuba J, Mundry M, Luetjens CM, et al. Cografting of hamster (*Phodopus sungorus*) and marmoset (*Callithrix jacchus*) testicular tissues into nude mice does not overcome blockade of early spermatogenic differentiation in primate grafts. Biol Reprod. 2004;7:2087–91.
54. Wistuba J, Luetjens CM, Wesselmann R. Meiosis in autologous ectopic transplants of immature testicular

tissue grafted to *Callithrix jacchus*. Biol Reprod. 2006;74:706–13.
55. Luetjens CM, Stukenborg JB, Nieschlag E, et al. Complete spermatogenesis in orthotopic but not in ectopic transplants of autologously grafted marmoset testicular tissue. Endocrinology. 2008;149:1736–47.
56. Jahnukainen K, Ehmcke J, Nurmio M. Irradiation causes acute and long-term spermatogonial depletion in cultured and xenotransplanted testicular tissue from juvenile nonhuman primates. Endocrinology. 2007;148:5541–8.
57. Orwig KE, Schlatt S. Cryopreservation and transplantation of spermatogonia and testis tissue for preservation of male fertility. J Natl Cancer Inst Monogr. 2005;34:56.
58. Keros V, Rosenlund B, Hultenby K, et al. Optimizing cryopreservation of human testicular tissue: comparison of protocols with glycerol, propanediol and dimethylsulphoxide as cryoprotectants. Hum Reprod. 2005;20:1676–87.
59. Keros V, Hultenby K, Borgström B, et al. Methods of cryopreservation of testicular tissue with viable spermatogonia in pre-pubertal boys undergoing gonadotoxic cancer treatment. Hum Reprod. 2007;22:1384–895.
60. Zeng W, Snedaker AK, Megee S, et al. Preservation and transplantation of porcine testis tissue. Reprod Fertil Dev. 2009;21:489–97.
61. Rathi R, Dobrinski I. Ectopic grafting of mammalian testis tissue into mouse hosts. In: Hou SX, Singh SR, editors. Methods in molecular biology. Germ line stem cells. Totowa: Humana Press; 2008. p. 139–48.
62. Goossens E, Geens M, De Block G, et al. Spermatogonial survival in long-term human prepubertal xenografts. Fertil Steril. 2008;90:2019–22.
63. Yu J, Cai ZM, Wan HJ, et al. Development of neonatal mouse and fetal human testicular tissue as ectopic grafts in immunodeficient mice. Asian J Androl. 2006;8:393–403.
64. Schlatt S, Honaramooz A, Ehmcke J, et al. Limited survival of adult human testicular tissue as ectopic xenograft. Hum Reprod. 2006;21:384–9.
65. Kvist K, Thorup J, Byskov AG, et al. Cryopreservation of intact testicular tissue from boys with cryptorchidism. Hum Reprod. 2006;21:484–91.
66. Shinohara T, Inoue K, Ogonuki N, et al. Birth of offspring following transplantation of cryopreserved immature testicular pieces and in-vitro microinsemination. Hum Reprod. 2002;17:3039–45.

Stem Cells and Fertility Preservation in Males

26

Marcia Riboldi, Ana Isabel Marqués Marí, and Carlos Simón

Abstract

Recent advances in stem cell biology related to fertility preservation have contributed to improve our knowledge about the etiologies of male infertility and its potential reversion.

Several factors related with low fertility in males have been described, such as the effect of different drugs, several types of cancers, aging, hormonal treatments, and the increase in sexually transmitted diseases. Given these increasing hazards, different alternatives for the preservation of male fertility are currently suggested, for instance, cryopreservation of testicular tissue for patients with spermatogenesis activity. However, the number of prepubertal patients increases yearly due to more cancer cases and associated anomalies. Chemotherapy and radiotherapy treatments block or destroy spermatogenesis. Thus, the search for suitable sources of germ stem cells, unique thanks to their self-renewal and differentiation characteristics, is increasingly necessary. Recent studies have demonstrated the presence of stem cells in adult testes and have described the methodology to isolate and culture these germ stem cells. Here we describe the state of the art in the field and the work underway in male fertility preservation relating to stem cells.

Keywords

Men • Fertility • Infertility • Spermatogenesis • Germ cells • Stem cells

Currently, around 70 million couples suffer because of infertility worldwide [1, 2]. Around 15% of couples are not able to conceive within 1 year and treatments for infertility are necessary because of different causes [3], where the male factor is responsible for 50% of infertile couples. In general, infertility disorders affect a large percentage of men, roughly 30–40% [4].

Approximately 30–40% of men are survivors of malignant diseases such as cancer, and the consequences of this are subfertility or infertility as secondary effects. One in 650 children develops

C. Simón (✉)
Stem Cell Bank, Prince Felipe Research Centre, Autopista del Saler 16, Camino de las Moreras, Valencia 46012, Spain

E. Seli and A. Agarwal (eds.), *Fertility Preservation: Emerging Technologies and Clinical Applications*,
DOI 10.1007/978-1-4419-1783-6_26, © Springer Science+Business Media, LLC 2012

cancer by the age of 15, and 50–60% will be cured. Today, 1 in 1,000 young adults is a childhood cancer survivor [5]. Testicular germ cell cancer (TGCC) and lymphoma are the most frequent cancer types in young patients. In TGCC, for example, the cure rate is around 95% [6]. However, infertility is a huge burden for these cancer survivors.

The leading cause of male infertility is azoospermia which can be attributed to a variable number of causes. Male genital tract obstructions might result from infections, previous inguinal, and scrotal surgery (vasectomy), secondary genital duct obstruction due to inflammation, failed vasectomy reversal, congenital bilateral absence of the *vas deferens* [7], among others. The prevalence of azoospermia is approximately 1% among all men [8], and ranges between 10 and 15% among infertile men [9]. Whereas spermatozoa are abundant in the epididymis in male patients suffering tract obstructions, allowing a high rate of retrieval success, only a few *foci* with spermatogenesis are found in testes in secretory azoospermia [10–13]. However, the worst prognosis for infertility is for those in which no germ cells (GCs) are present [14, 15].

Stem cells (SCs) are a specific population of undifferentiated cells that are capable of self-renewal and differentiation into all the cell types of the organism. These cells were isolated for the first time in 1998 [16]. Thanks to their unique characteristics, they have opened a new horizon and have raised the possibility of their future use in regenerative medicine. Embryonic stem cells (ESC) are obtained from preimplantation stages of the embryo (blastocyst, morula, blastomere) [16–18] and have pluripotent characteristics since they are capable of differentiating into all cell types, including GCs. Adult SCs have been described in a wide range of tissues, including brain [19], blood [20, 21], bone marrow [22], fat [23] or skin [24, 25], among others, and also in testes [26]. These cells have pluripotent potential and can differentiate into various types of cells depending on their origin [27].

Male GCs develop into spermatogonial stem cells (SSCs) which are fundamental for the spermatogenesis process and have the ability to self-renew and to generate differentiated GCs [28, 29].

Male Germline

GCs are responsible for the transmission of individual genetic information to the next generation, and it is through this process that they assure the continuity of species [30]. GCs are derived from a population of primordial germ cells (PGC) [31, 32] that arise from the proximal epiblast [31] to then migrate into primitive gonads and proliferate by increasing in number (Fig. 26.1).

In men, once PGCs arrive at the genital ridge, they are enclosed by somatic SCs and become gonocytes [33]. Cells proliferate for a few days and then arrest in the G0/G1 phase during fetal development. After birth, PGCs migrate to the basement membrane of the seminiferous tubules and become SSCs. SSCs divide asymmetrically to give rise to one stem cell, and to one spermatogonia that initiates its differentiation into spermatozoa [34, 35] (Fig. 26.2).

Maintenance of the self-renewal and differentiation of the germ SCs in the testis depends on the niche that is located at the basal membrane of the seminiferous tubules [36], where only 1% of cells are SSCs [37]. The SSC niche is maintained

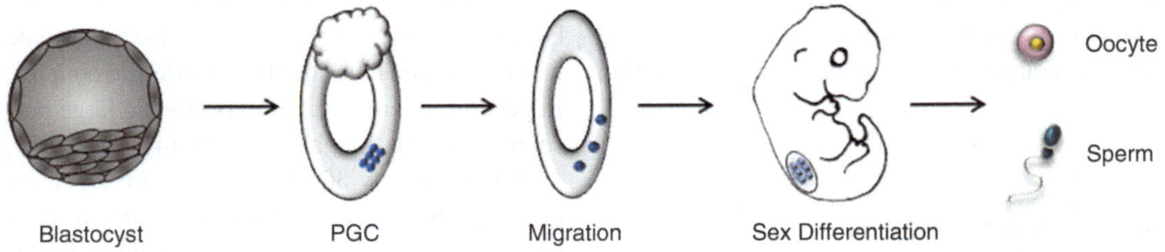

Fig. 26.1 PGC population and sex differentiation

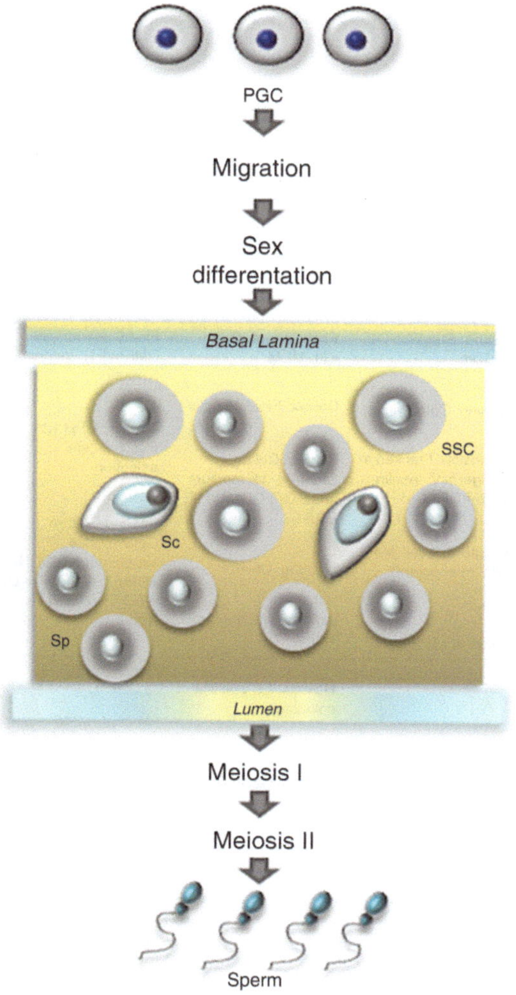

Fig. 26.2 Spermatogenesis. *SSC* spermatogonial stem cell; *Sc* Sertoli cell; *Sp* spermatogonial cell

due to the many factors produced and secreted by SCs. Of these, one of the most important is the glial cell line-derived neurotrophic factor (GDNF) as it is essential for the self-renewal and survival of SSCs [38, 39].

The presence in testes of a stem cell population (SSCs) responsible for continuous sperm production was demonstrated in 1994 by Brinster and Zimmermann in mice [30]. In 2008, Conrad and collaborators reported the isolation of SSCs from adult testicular tissue after generating human adult germline stem cells (haGSCs). These cell lines showed similar cellular and molecular characteristics to ESC and GCs. These cells are pluripotent and capable of giving rise to many different cell types in vitro and of forming teratomas when injected into immunosuppressed mice [40]. Very recently, however, these findings have been questioned by Scholer's group which demonstrated that haGSCs and fibroblasts have a similar gene-expression profile. Therefore, the pluripotency of the human SSCs population is currently under debate [41].

Methods of Isolation and Culture of Germ Cells

The first requisite to establish the culture of SSCs is to start with pure cell populations, while the second is to procure optimal conditions to support the survival and expansion of cells. Various methods and protocols have been described for the isolation, culture, and maintenance of the SSCs in vitro (Fig. 26.3).

Digestion of Testicular Tissue

Initially, testicular tissue is mechanically disrupted into small pieces, and then enzymatic techniques are used for the dispersion of seminiferous tubules to obtain a single cell suspension. Collagenase was the first enzyme utilized for the dispersion of seminiferous tubules, together with DNaseI, hyaluronidase, and trypsin. Various combinations of enzymes and concentrations have been described for different groups.

Conrad et al. used 0.5 g/mL collagenase type VI and 0.25 g/mL dispase II for 30 min at 37°C [40]. Dym's group used 1 mg/mL collagenase IV and 2 μg/μL DNase I [42]. Reijo-Pera reported a first incubation of 30 min at 37°C in 10 mg/mL collagenase [43].

Other types of enzymes that used to obtain individual cells were 4 mg/mL collagenase IV, 2.5 mg/mL hyaluronidase, 2 mg/mL trypsin, and 1 μg/μL DNase I for Dym's group, and 1 mg/mL collagenase IV, 1.5 mg/mL hyaluronidase, and 1 mg/mL trypsin for Reijo-Pera [40, 42, 44]. After digestion, the supernatant is filtered through 40, 50, and 70 μm filters [40, 45], and the cells obtained are cultured.

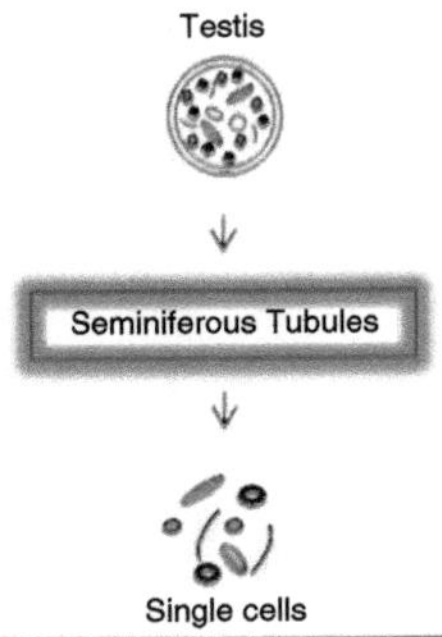

First Culture: Knouckout medium add GDNF; 96 hours, 32.5ºC, 5% CO_2 Maker: MACS CD49f positive cells selection Collagen matrix: negative cells selection Laminin matrix: positive cells selection Second Culture: Basic medium add LIF in 0.1% Gelatin Results: Pluripotent colonies *Conrad et al., 2008*	Marker: MACS Gfr α-1 positive cells selection Culture: MEM-α medium in gelatin coated dishes Forming: Colonies Expansion: Manual Passaging in MEF with hESC conditions (b-FGF) Results: hMGSCS (human multipotent germline stem cells) *Kossack et al., 2009*	Culture: Gelatin coated plate with ESC medium add b-FGF and TGF-β1 Forming: Small ES cell-like colonies Expansion: Mechanical in gelatin plate culture Results: Pluripotent ES-cell like cells *Golestaneh et al., 2009*	Culture: StemPro-34® medium Forming: Colonies Expansion: Piacked in Matrigel with Insulin and b-FGF Results: ES-cell like cells (multipotent) *Mizrak et al., 2009*	First Culture: O.N. with MEM Collect: Floating cells Second Culture: StemPro-34® medium and EGF; GDNF; LIF; EGF Forming: Cluster Expansion: Laminin matrix Results: ES-cell like colonies *Sadri-Ardekani et al., 2009*	First Culture: DMEM/F12 medium; 3 hours Marker: GPR 125 Positive cells selection Second Culture: Gelatin plate supplemented GDNF +Gfrα-1 + b-FGF + LIF + EGF + TGF-β + Nodal Results: SSC *He et al., 2010*

Fig. 26.3 Protocols for the isolation, culture and maintenance of the SSCs in vitro

Isolation of SSCs

Fluorescence-activated cell sorting (FACS) or magnetic-activated cell sorting (MACS) methods are coupled with a diverse number of markers to identify SSCs (Table 26.1). Extracellular matrices have been used for the cell purification process. Candidate cells are incubated for 4 h at 32.5°C onto a collagen matrix (5 mg cm^2). Then the supernatant is aspirated and incubated onto a laminin matrix (4.4 mg cm^2) for 45 min at 32.5°C [40]. This methodology was developed to specifically select the cell population attached to the laminin matrix [40, 46]. These cells showed the typical morphology of SSC, and loosely adherent clumps or small ES cell-like colonies were formed in the primary culture [42, 46].

Culture of SSCs

Three components are considered essential for the maintenance of SSCs in culture: feeders, medium, and addition of growth factors.

In order to maintain cells in an undifferentiated state, mouse embryonic fibroblasts (MEF) or mouse testicular stromal cells (MTS) are used as feeder cells [43, 46].

The most commonly used culture media are the embryonic stem cell (ESC) medium or the StemPro® medium [44]. A serum-free condition has been described as being important to provide a good environment for the initial culture in vitro [47, 48], but the addition of fetal bovine serum (FBS) to the medium is used for long-term cultures.

Table 26.1 Markers used to identify SSCs

Surface markers	Species	References
α6-integrin (CD49f)	Mouse Human Primate	[55] [40] [71]
β1-integrin	Mouse	[55]
Thy-1 (CD90)	Mouse Rat Primate Human	[47] [57] [56] [40]
CD9	Mouse Rat	[56]
Ep-CAM	Mouse Rat	[57]
E-cadherin	Mouse	[58]
Gfrα1	Mouse Human	[59] [60] [43]
CD24	Mouse	[72]
GPR125	Mouse Human	[61, 62] [44]

Table 26.2 Different exogenous agents added to the culture medium to induce cellular processes

Agent	In vivo	In vitro	References
GDNF	Proliferation Undifferentiated Maintenance Regulation Self-renewal	Survival Proliferation Self-renewal Expansion Proliferation	[48, 57, 63–67]
Bfgf		Survival Proliferation Expansion	[47]
EGF	Proliferation	Survival Proliferation Still unclear	[48, 67]
IGF-I		Increase in number	[48]
LIF	Maturation Self-renewal	Survive Proliferation Still unclear	[67–69]
GFRα-1		Expansion Proliferation	[47]
SCF	Proliferation Transition types spermatogonia		[67, 70]
FGF2	Regulation Behavior	Self-renewal Differentiation	[67]

Different exogenous agents are added to the culture medium to provide the culture with the appropriate signals in order to induce several cellular processes (Table 26.2).

Expansion of the SSCs Population In Vitro

The most critical period for the enrichment of SSCs is the first 2 weeks of culture. During this time, the formation of cell clumps loosely adhered to the plate is a good indication of GC morphology.

The first expansion should be mechanical or by gently pipetting up and down for few times to obtain a single cell suspension. The expansion ratio is then determined on the basis of the size and number of clumps. For subsequent expansions, cells are trypsinized and seeded into inactivated MEF or MTS.

The expansion rate for the second and third passages is about 1:1.5 or 1:2 (usually after 7 days). From three passages onward, the expansion rate ranges from 1:2 to 1:4 [49].

Transplantation of SSCs into In Vivo Systems

This step is essential for their functional characterization, but the stem cell potential depends on the recipient tissue. It is necessary to relocate these cells in the appropriate niche to maintain self-renewal and differentiation abilities [50].

Although different techniques exist, the basic procedure is the microinjection of the cell suspension into the seminiferous tubules of a receptive testis after GC depletion by busulfan treatment or chemotherapy. Under these conditions, SSCs are capable of colonizing the host testis by re-establishing the spermatogenesis process [39]. The efficacy of transplantation depends mainly on the number of SSCs injected into the testis and the quality of the recipient niche [51–53].

In the mouse model, most of the proliferating donor-derived cells are located on the basement membrane during the first post-transplant month. From the second month onward, GCs move from the basement membrane to the lumen and they differentiate into spermatozoa [54].

Therapy for Fertility Preservation

Currently, there are different established methods for fertility preservation in postpubertal patients:

- Gonadal prevention: using shields during radiotherapy treatments with cancer patients to avoid directly striking the reproductive organs.
- Sperm banking: freezing healthy sperm before cancer treatment to use it in the future to conceive.
- Testis cancer surgery: nowadays, a usual procedure in surgical oncology is to remove only the cancerous nodule instead of the whole testis in selected patients with testis cancer.
- Testis tissue preservation: freezing testicular tissue from adult men is a valuable method to potentially preserve fertility since sperm may be found in the testis tissue despite the lack of sperm in the ejaculate.

In prepubertal boys, some of the above-mentioned strategies are unworkable because spermatozoa are lacking. In the human testis, spermatogenesis starts after puberty, so retrieval of spermatozoa and preservation of gonadal tissue before puberty are not feasible options to preserve fertility. However, there is a potential fertility preservation method for these patients:

- SSC banking: ejaculated sperm is not present in prepubertal boys with cancer. Nonetheless, it may be possible to freeze the SSC collected from the testis by means of a testicular biopsy before they receive cancer treatment, thus enabling fertility to be restored during their adult life after autologous transplantation:

Young cancer patients who require chemotherapy and radiotherapy treatments present a blockade of the spermatogenesis process as a secondary effect.

The most helpful clinical strategy for retrieval of SSCs or GCs in such patients is to submit them to a testicular biopsy before starting the cancer treatment. This procedure consists in minor surgery to collect a piece of testicular tissue containing GCs and SSCs. These cells can be grown in vitro, and are then expanded and frozen in order to recover these patients' fertility in the future. Post-treatment would be the ideal time to transplant cells, as malignancy is absent, in order to provide them an appropriate niche for differentiation.

Conclusions

Isolation, maintenance and differentiation of SSCs form the next step in male fertility to restore spermatogenesis and to use SSCs in assisted reproduction. Since a small amount of tissue was collected by means of biopsy, the propagation of the SSCs in vitro to repopulate the testis becomes very important. The methodology developed and research conducted in human SSCs and GCs is very recent, and further research is needed.

References

1. Fathalla MF. Reproductive health: a global overview. Early Hum Dev. 1992;29:35–42.
2. Boivin J, Bunting L, Collins JA, Nygren KG. International estimates of infertility prevalence and treatment-seeking: potential need and demand for infertility medical care. Hum Reprod. 2007; 22:1506–12.
3. World Health Organization. WHO manual for the standardised investigation and diagnosis of the infertile couple. Cambridge: Cambridge University Press; 2000.
4. Thonneau P, Marchand S, Tallec A, Ferial ML, Ducot B, Lansac J, et al. Incidence and main causes of infertility in a resident population (1,850,000) of three French regions (1988–1989). Hum Reprod. 1991; 6:811–6.
5. Bahadur G, Ralph D. Gonadal tissue cryopreservation in boys with pediatric cancers. Hum Reprod. 1999;14:11–7.
6. Daugaard G, Hansen HH, Rùrth M. Treatment of malignant germ cell tumors. Ann Oncol. 1990;1: 195–202.
7. Pasqualotto FF, Rossi LM, Guilherme P, Ortiz V, Iaconelli Jr A, Borges Jr E. Etiology-specific outcomes of intracytoplasmic sperm injection in azoospermic patients. Fertil Steril. 2005;83(3):606–11.
8. Stephen EH, Chandra A. Declining estimates of infertility in the United States: 1982–2002. Fertil Steril. 2006;86:516–23.
9. Jarow JP, Espeland MA, Lipshultz LI. Evaluation of the azoospermic patient. J Urol. 1989;142:62–5.

10. Silber SJ, Nagy Z, Devroey P, Tournaye H, Van Seirteghem AC. Distribution of spermatogenesis in the testicles of azoospermic men: the presence or absence of spermatids in the testes of men with germinal failure. Hum Reprod. 1997;12:2422–8.
11. Silber SJ, Rodrigues-Rigau LJ. Quantitative analysis of testicle biopsy: determination of partial obstruction and prediction of sperm count after surgery for obstruction. Fertil Steril. 1981;36:480–5.
12. Turek PJ, Cha I, Ljung B-M. Systematic fine-needle aspiration of the testis: correlation to biopsy and results of organ "mapping" for mature sperm in azoospermic men. Urology. 1997;49:743–8.
13. Gorgy A, Podsiadly BT, Bates S, Craft IL. Testicular sperm aspiration (TESA): the appropriate technique. Hum Reprod. 1998;13:1111–3.
14. Choi J, Koh E, Suzuki H, Maeda Y, Yoshida A, Namiki M. Alu sequence variants of the BPY2 gene in proven fertile and infertile men with Sertoli cell-only phenotype. Int J Urol. 2007;14(5):431–5.
15. Mancini M, Carmignani L, Gazzano G, Sagone P, Gadda F, Bosari S, et al. High prevalence of testicular cancer in azoospermic men without spermatogenesis. Hum Reprod. 2007;22(4):1042–6.
16. Thomson JA, Itskovitz-Eldor J, Shapiro SS, Waknitz MA, Swiergiel JJ, Marshall VS, et al. Embryonic stem cell lines derived from human blastocysts. Science. 1998;282:1145–7.
17. Ilic D, Giritharan G, Zdravkovic T, Caceres E, Genbacev O, Fisher SJ, et al. Derivation of human embryonic stem cell lines from biopsied blastomeres on human feeders with minimal exposure to xenomaterials. Stem Cells Dev. 2009;18(9):1343–50.
18. Aguilar-Gallardo C, Poo M, Gomez E, Galan A, Sanchez E, Marques-Mari A, et al. Derivation, characterization, differentiation, and registration of seven human embryonic stem cell lines (VAL-3, -4, -5, -6M, -7, -8, and -9) on human feeder. In Vitro Cell Dev Biol Anim. 2010;46(3–4):317–26.
19. Katz J, Keenan B, Snyder EY. Culture and manipulation of neural stem cells. Adv Exp Med Biol. 2010;671:13–22.
20. Goodman JW, Hodgson GS. Evidence for stem cells in the peripheral blood of mice. Blood. 1962;19:702–14.
21. Barnes DW, Loutit JF. Haemopoietic stem cells in the peripheral blood. Lancet. 1967;2(7526):1138–41.
22. Pittenger MF, Mackay AM, Beck SC, Jaiswal RK, Douglas R, Mosca JD, et al. Multilineage potential of adult human mesenchymal stem cells. Science. 1999;284(5411):143–7.
23. Zuk PA. The adipose-derived stem cell: looking back and looking ahead. Mol Biol Cell. 2010;21(11): 1783–7.
24. Toma JG, Akhavan M, Fernandes KJ, Barnabé-Heider F, Sadikot A, Kaplan DR, et al. Isolation of multipotent adult stem cells from the dermis of mammalian skin. Nat Cell Biol. 2001;3(9):778–84.
25. Alonso L, Fuchs E. Stem cells of the skin epithelium. Proc Natl Acad Sci USA. 2003;100 Suppl 1: 11830–5.
26. Guan K, Nayernia K, Maier LS, Wagner S, Dressel R, Lee JH, et al. Pluripotency of spermatogonial stem cells from adult mouse testis. Nature. 2006;440 (7088): 1199–203.
27. Fortier LA. Stem cells: classifications, controversies, and clinical applications. Vet Surg. 2005;34(5):415–23.
28. Olive V, Cuzin F. The spermatogonial stem cells: from basic knowledge to transgenic technology. Int J Biochem Cell Biol. 2005;37:246–50.
29. Nayernia K, Li M, Engel W. Spermatogonial stem cells. In: Schatten H, editor. Germ cell protocols: methods in molecular biology, vol. 253. Totowa: Humana Press; 2003. p. 105–20.
30. Brinster RL, Zimmermann JW. Spermatogenesis following male germ-cell transplantation. Proc Natl Acad Sci USA. 1994;9:11298–302.
31. Lawson KA, Hage WJ. Clonal analysis of the origin of primordial germ cells in the mouse. Ciba Found Symp. 1994;182:68–91.
32. McLaren A. Germ and somatic cell lineages in the developing gonad. Mol Cell Endocrinol. 2000;163: 3–9.
33. Donovan PJ, Stott D, Cairns LA, et al. Migratory and postmigratory mouse primordial germ cells behave differently in culture. Cell. 1986;44:831–8.
34. de Rooij DG, Grootegoed JA. Spermatogonial stem cells. Curr Opin Cell Biol. 1998;10:694–701.
35. Nayernia K, Li M, Engel W. Spermatogial stem cells. Methods Mol Biol. 2004;253:105–20.
36. Russell LD, Ettlin RA, Hikim APS, Clegg ED. Histological and histopathological evaluation of the testis. Clearwater, FL: Cache River Press; 1990.
37. Tegelenbosch RA, de Rooij DG. A quantitative study of spermatogonial multiplication and stem cell renewal in the C3H/101 F1 hybrid mouse. Mutat Res. 1993;290:193–200.
38. Hofmann MC. Gdnf signaling pathways within the mammalian spermatogonial stem cell niche. Mol Cell Endocrinol. 2008;288:95–103.
39. Oatley JM, Brinster RL. Regulation of spermatogonial stem cell self-renewal in mammals. Annu Rev Cell Dev Biol. 2008;24:263–86.
40. Conrad S, Renninger M, Hennenlotter J, Wiesner T, Just L, Bonin M, et al. Generation of pluripotent stem cells from adult human testis. Nature. 2008;456(7220): 344–9.
41. Ko K, Araúzo-Bravo MJ, Tapia N, Kim J, Lin Q, Bernemann C, et al. Human adult germline stem cells in question. Nature. 2010;465(7301):E1. discussion E3.
42. Golestaneh N, Kokkinaki M, Pant D, Jiang J, DeStefano D, Fernandez-Bueno C, et al. Pluripotent stem cells derived from adult human testes. Stem Cells Dev. 2009;18(8):1115–26.
43. Kossack N, Meneses J, Shefi S, Nguyen HN, Chavez S, Nicholas C, et al. Isolation and characterization of pluripotent human spermatogonial stem cell-derived cells. Stem Cells. 2009;27(1):138–49.
44. He Z, Kokkinaki M, Jiang J, Dobrinski I, Dym M. Isolation, characterization, and culture of human spermatogonia. Biol Reprod. 2010;82(2):363–72.

45. Mizrak SC, Chikhovskaya JV, Sadri-Ardekani H, van Daalen S, Korver CM, Hovingh SE, et al. Embryonic stem cell-like cells derived from adult human testis. Hum Reprod. 2010;25(1):158–67.
46. Guan K, Wolf F, Becker A, Engel W, Nayernia K, Hasenfuss G. Isolation and cultivation of stem cells from adult mouse testes. Nat Protoc. 2009;4(2): 143–54.
47. Kubota H, Avarbock MR, Brinster RL. Culture conditions and single growth factors affect fate determination of mouse spermatogonial stem cells. Biol Reprod. 2004;71:722–31.
48. Kubota H, Avarbock MR, Brinster RL. Growth factors essential for self-renewal and expansion of mouse spermatogonial stem cells. Proc Natl Acad Sci USA. 2004;101:16489–94.
49. Oatley JM, Brinster RL. Spermatogonial stem cells. Meth Enzymol. 2006;419:259–82.
50. Wyns C, Curaba M, Vanabelle B, Van Langendonckt A, Donnez J. Options for fertility preservation in prepubertal boys. Hum Reprod Update. 2010;16(3): 312–28.
51. Dobrinski I, Ogawa T, Avarbock MR, Brinster RL. Computer assisted image analysis to assess colonization of recipient seminiferous tubules by spermatogonial stem cells from transgenic donor mice. Mol Reprod Dev. 1999;53:142–8.
52. Ogawa T, Dobrinski I, Avarbock MR, Brinster RL. Transplantation of male germ line stem cells restores fertility in infertile mice. Nat Med. 2000;6:29–34.
53. Ohta H, Yomogida K, Dohmae K, Nishimune Y. Regulation of proliferation and differentiation in spermatogonial stem cells: the role of c-kit and its ligand SCF. Development. 2000;127:2125–31.
54. Nagano M, Avarbock MR, Brinster RL. Pattern and kinetics of mouse donor spermatogonial stem cell colonization in recipient testes. Biol Reprod. 1999;60(6):1429–36.
55. Shinohara T, Avarbock MR, Brinster RL. Beta1- and alpha6-integrin are surface markers on mouse spermatogonial stem cells. Proc Natl Acad Sci USA. 1999;96(10):5504–9.
56. Kanatsu-Shinohara M, Toyokuni S, Shinohara T. CD9 is a surface marker on mouse and rat male germline stem cells. Biol Reprod. 2004;70(1):70–5.
57. Ryu BY, Orwig KE, Kubota H, Avarbock MR, Brinster RL. Phenotypic and functional characteristics of spermatogonial stem cells in rats. Dev Biol. 2004;274(1): 158–70.
58. Tolkunova EN, Malashicheva AB, Chikhirzhina EV, Kostyleva EI, Zeng W, Luo J, et al. E cadherin as a novel surface marker of spermatogonial stem cells. Cell Tissue Biol. 2009;3(2):103–9.
59. Buageaw A, Sukhwani M, Ben-Yehudah A, Ehmcke J, Rawe VY, Pholpramool C, et al. GDNF family receptor alpha1 phenotype of spermatogonial stem cells in immature mouse testes. Biol Reprod. 2005;73(5):1011–6.
60. Ebata KT, Zhang X, Nagano MC. Expression patterns of cell-surface molecules on male germ line stem cells during postnatal mouse development. Mol Reprod Dev. 2005;72(2):171–81.
61. Seandel M, Falciatori I, Shmelkov SV, Kim J, James D, Rafii S. Niche players: spermatogonial progenitors marked by GPR125. Cell Cycle. 2008;7(2):135–40.
62. Seandel M, James D, Shmelkov SV, Falciatori I, Kim J, Chavala S, et al. Generation of functional multipotent adult stem cells from GPR125+ germline progenitors. Nature. 2007;449(7160):346–50.
63. Tadokoro Y, Yomogida K, Ohta H, et al. Homeostatic regulation of germinal stem cell proliferation by the GDNF/FSH pathway. Mech Dev. 2002;113:29–39.
64. Hamra FK, Chapman KM, Nguyen DM, Williams-Stephens AA, Hammer RE, Garbers DL. Self renewal, expansion, and transfection of rat spermatogonial stem cells in culture. Proc Natl Acad Sci USA. 2005;102(48):17430–5.
65. Brinster RL. Germline stem cell transplantation and transgenesis. Science. 2002;296(5576):2174–6. Review.
66. Hofmann MC, Braydich-Stolle L, Dym M. Isolation of male germ-line stem cells; influence of GDNF. Dev Biol. 2005;279:114–24.
67. de Rooij DG. The spermatogonial stem cell niche. Microsc Res Tech. 2009;72:580–5.
68. Kanatsu-Shinohara M, Miki H, Inoue K, Ogonuki N, Toyokuni S, Ogura A, et al. Long-term culture of mouse male germline stem cells under serum-or feeder-free conditions. Biol Reprod. 2005;72(4): 985–91.
69. Kanatsu-Shinohara M, Inoue K, Ogonuki O, Miki H, Yoshida S, Toyokuni S, et al. Leukemia inhibitory factor enhances formation of germ cell colonies in neonatal mouse testis culture. Biol Reprod. 2007;76:55–62.
70. Pellegrini M, Grimaldi P, Rossi P, et al. Developmental expression of BMP4/ALK3/SMAD5 signaling pathway in the mouse testis: a potential role of BMP4 in spermatogonia differentiation. J Cell Sci. 2003;116: 3363–72.
71. Maki CB, Pacchiarotti J, Ramos T, Pascual M, Pham J, Kinjo J, et al. Phenotypic and molecular characterization of spermatogonial stem cells in adult primate testes. Hum Reprod. 2009;24(6):1480–91.
72. Kubota H, Avarbock MR, Brinster RL. Spermatogonial stem cells share some, but not all, phenotypic and functional characteristics with other stem cells. Proc Natl Acad Sci USA. 2003;100(11):6487–92.

Approach to Fertility Preservation in Adult and Pre-pubertal Males

27

Fnu Deepinder and Ashok Agarwal

Abstract

Recent advances in medicine have led to ever increasing number of pre-pubertal and adult males surviving cancer treatment. This has increased the need to improve the existing technology for cryopreservation of gametes and search for new fertility preservation options. As of today, only sperm cryopreservation is considered accepted standard clinical practices. Fertility preservation options in pre-pubertal males are still experimental. Cryopreservation of testicular tissue and spermatogonial stem cell transplantation should only be offered within IRB-approved clinical protocol after thorough counseling of patients and their family members as there are still many unresolved issues related to these technologies.

Keywords

Fertility preservation in boys and men • Cancer treatment and fertility preservation in boys and men • Infertility in men • Ethical issues in prepubertal male fertility preservation • Sperm cryopreservation

In the past two decades, major strides have been made in the curability of cancers leading to striking improvements in the chances of long-term survival, particularly in young men including children and adolescents. The use of chemotherapeutic and radio-therapeutic interventions has led to 70% survival rates in children with malignancies [1]. However, one of the major complications of these advanced treatment modalities is sterility and loss of gonadal function [2]. The effects might be transient or permanent depending upon the individual variability in the sensitivity to reproductive damage [3]. The severity of damage is dependent on the type of chemotherapy or radiotherapy, the treatment protocol, and the age of patients [4].

Although, future fertility of young males is very low on the relative quality of life parameters list at the time of anti-cancer treatment, infertility becomes an important issue following cure from cancer. According to a recent survey, 51% of men with cancer wanted children in future, including 77% of men who were childless when their cancer was first

A. Agarwal (✉)
Director, Center for Reproductive Medicine, Cleveland Clinic, Cleveland, OH, USA

Department of Obstetrics-Gynecology and Women's Health Institute, Cleveland Clinic, Cleveland, OH, USA
e-mail: agarwaa@ccf.org

E. Seli and A. Agarwal (eds.), *Fertility Preservation: Emerging Technologies and Clinical Applications*,
DOI 10.1007/978-1-4419-1783-6_27,

diagnosed [5]. Since it is difficult to predict who will survive or become infertile after anti-cancer treatment, fertility conservation is an important issue for those young cancer patients who have not yet started or completed their family size.

Although, fertility preservation in adult men by sperm cryopreservation is already established, options in pre-pubertal males are still experimental. There has been a tremendous progress in the development of strategies for germ cell banking. This chapter discusses the approach for fertility preservation in pre-pubertal and adult males including a number of technical and ethical concerns requiring careful evaluation.

Ethical and Legal Prospects

Informed Consent/Assent

The question of preserving fertility beyond a cancer patient's current treatment raises the need for an informed consent. This may be complicated because of the paucity of meaningful options and very little time available to most of the patients for taking any decision. Informed consent process for minor needs the involvement of patient's parents or legal guardians. Assent (permission less than full consent) is required in case of minors who are able to understand the issue, such as post-pubertal boys and girls, together with the parental consent. However, for children too young to give an assent, parents may consent to experimental procedures only if the expected benefits are sufficient to justify the risks involved. In a recent study done in Netherlands, asking opinion of 318 parents of boys surviving cancer, sperm collection was approved by 70%; whereas spermatogonial stem cell collection by biopsy and hemicastration got only 61 and 33% approval, respectively [6]. The principles of beneficence and non-maleficence are considered paramount in such cases and oftentimes hospital ethics committee are asked to review parental decisions. Informed consent requires that the patient is given information that a reasonable person would want to know, and in enough detail that a reasonable person would be able to understand the procedure. Failure to offer all the existing methods of fertility preservation and accurately explaining the associated risks with such procedures may give rise to medical malpractice claims against health care providers. Hence it is essential to keep written documentation of such a process [7]. During assent, age appropriate information about sexual reproduction should be reviewed with patients as per their level of understanding, preferably by a psychologist or psychiatrist with expertise in children. In children enrolled in experimental fertility preservation protocols, the consent process should be performed in two stages. The decision to harvest immature germ cells made at the time of cancer treatment would rely on the parents or guardians. However, the decision of how to use the gametes could be made at a future time by the patient when he attains adulthood [8].

Risks Involved

A common cause of concern for both fertility specialists and cancer survivors seeking fertility preservation is whether their offspring are at higher risks for physical defects and cancer because of the effects of their disease, anti-cancer therapy, and cryopreservation techniques. Children born with disabilities may allege medical negligence in connection with their parent's fertility preservation during cancer treatment that preserved their life [9]. Men should be advised of a possible, not yet quantifiable, higher risk of genetic damage in sperm stored after diagnosis of cancer or initiation of cancer therapy. In non-cancer populations, there is no evidence of an increased risk of adverse outcomes if cryopreserved rather than fresh sperm are used for assisted reproduction. Pre-pubertal children may inadvertently get castrated because of loss of gonadal tissue during the collection of gametes in addition to the gonadotoxic chemotherapy and radiation. Hence children and their parents must be made aware of the risk of premature gonadal failure and delayed sexual development before proceeding with the fertility preservation options. Some experts have also questioned if it is ethical to enable cancer patients to reproduce as they face a greatly lowered life span, thus leaving a minor child bereft of one patient [10]. Furthermore, providers who store human genetic material for future use may face liability for damages in the event of loss or destruction of the cryopreserved tissue [9].

Religious Beliefs

Religious prohibitions against collection of genetic tissue for future use by natural or artificial methods may present another obstacle for fertility preservation. Beliefs and opinions vary across different religions and even among members of the same religion. In case of children, parental views should be respected, although a child's religious views may be different or may change over the time [11].

Financial Costs

Patients also need to be made aware of the financial costs involved as the insurance companies do not always cover the costs of cryopreservation.

Legal Parenthood

Challenges may arise with respect to legal parentage of the children resulting from cryopreserved tissues. The need to determine legal paternity arises in context of inheritance and federal benefits. Most courts recognize that if a man dies before placement of his gametes; he is not a parent of the child resulting from posthumous assisted reproduction, unless he consented otherwise. Hence it is important to include in the informed consent the patient's intended legal relationship to any resulting child.

Future Disposition in Case of Death

Discussing fertility preservation with young males and their families takes patience and sensitivity. In addition to addressing patient's own future use of his gamete, it also involves the consent for future disposition of that tissue or its use by a specific partner or parent in case of patient's death. Because of the important ethical and emotional issues raised, it is advisable to have a bioethicist available to talk to patients along with their families, and formulate institutional policies to provide limits and guidance [12]. In the absence of prior written directives, it is recommended that ownership of the tissue should not be transferred to the patient's relatives in case the patient dies prior to use of the stored germ cell material. However, many women have successfully gained access to their deceased husbands' sperms convincing the courts that the storage of gametes by adults itself implied a decision to procreate. Furthermore, complicated issues may also arise in situations such as when the patient dies without banking the semen and the surviving partner or parent requests the health care provider for posthumous extraction of sperm. As compared to adults, a child's rights get violated if his gametes are used posthumously as he had a limited understanding of the process at the time of collection of his gametes. The two-step consent process ensures that the stored genetic material cannot be used for reproductive purposes until the child becomes an adult and is able to give full consent [9].

Semen Collection

Although post-pubertal males are ordinarily capable of ejaculation and provide sperm for storage, teenagers may be embarrassed to discuss the option of masturbation in front of their parents. A mental health expert, oncology nurse, or a social worker can minimize the embarrassment by discussing it outside the presence of their parents. With parental permission, having some non-violent erotic magazines or videos in the collection room may be helpful. For boys who cannot ejaculate, invasive procedures can be done with their assent and parental consent [12]. If a young teen objects to any of the above procedures, they should not be done, despite parental wishes [13].

Established Options for Fertility Preservation

Modification of Treatment Regime

The first line of fertility protection is to reduce the exposure of gonads to cytotoxic chemotherapy and radiotherapy. Wherever possible, radiation exposure to the gonads should be limited by using radiation shields. Gonadal damage can also be reduced by reduction in the dose and frequency of cytotoxic chemotherapy, or by substitution with less gonadotoxic agents.

In a recent study of 355 adult patients treated for Hodgkin's disease, impairment of spermatogenesis was seen in 8% of patients receiving non-alkylating chemotherapy in contrast to 60% of patients who received alkylating agents. Moreover, recovery of spermatogenesis occurred in 82% of patients treated with non-alkylating chemotherapy as compared to just 30% in the alkylating group [14]. Another trial in Hodgkin's disease patients compared gonadal toxicity of combination chemotherapy with alkylating agents containing MOPP regime (mechlormethine, procarbazine, vincristine, and prednisone) vs. non-alkylating chemotherapy ABVD (doxorubicin, bleomycin, vinblastine, and dacarbazine). In this trial, 86% of the patients treated with MOPP demonstrated persistent azoospermia as compared to ABVD regime, in which all patients showed recovery of spermatogenesis [15]. This substantiates that cancer treatment can potentially be modified in some circumstances to reduce gonadal toxicity.

Sperm Banking

Traditionally, sperm banking by cryopreservation of at least three semen samples with an abstinence period of at least 48 h in between the samples has been recommended for adult males desiring to preserve their fertility [16]. However, many young cancer patients, especially those with testicular carcinoma or Hodgkin's disease already have decreased semen quality at the time of diagnosis and start of anti-cancer therapy [17]. Freezing and thawing semen further reduces sperm's count, motility, and viability [18]. Additional samples and longer abstinence periods may be used to achieve higher total sperm counts. Moreover, recent advances in assisted reproductive technology especially the advent of intracytoplasmic sperm injection (ICSI) have made it possible for a man to become a father even if only a few spermatozoa remain alive after cryopreservation [19, 20].

Semen banking should ideally be done before the start of cancer treatment. Theoretically semen collection and storage is feasible after the initiation of chemotherapy and radiation therapy, at least until azoospermia ensues. However, it is advisable to wait for 12–18 months because of the time taken for the recovery of spermatogenesis and significant increase in the frequency of sperm aneuploidy persisting for 18 months or more after initiation of anti-cancer treatment [21–23].

The semen collection process is achieved by masturbation. The patient should be provided a sterile specimen collection cup and ample time and privacy to produce sample. It is important to avoid lubricants such as petroleum jelly and saliva as these substances may be spermatotoxic. If no ejaculate is expelled on climax, then a post-ejaculate urinalysis should be done to assess for retrograde ejaculation. If retrograde ejaculation is observed, alpha agonist medications may be administered to convert retrograde to antegrade ejaculation. If this is not successful, then alkalization of the urine and subsequent collection and processing of the post-ejaculate urine sample may facilitate isolation of viable sperm. If the patient is unable to reach climax, care should be taken to ensure that he has had ample privacy and time [24].

Although post-pubertal males are generally able to ejaculate, some young cancer patients may not be able to produce a sample by masturbation. A strong vibrator or a rectal electric probe can be used to stimulate ejaculation in these boys; however it should be used under anesthesia to avoid pain [25]. Advanced methods for sperm retrieval include microsurgical epididymal sperm aspiration (MESA) for patients with un-reconstructable obstructive azoospermia [26] and testicular sperm extraction (TESE) for patients suffering from non-obstructive azoospermia [27–29]. Although TESE in combination with ICSI has shown some recent promise in patients with post-chemotherapy azoospermia, long-term potential for genetic risks among children born with this technique are still unknown (Fig. 27.1).

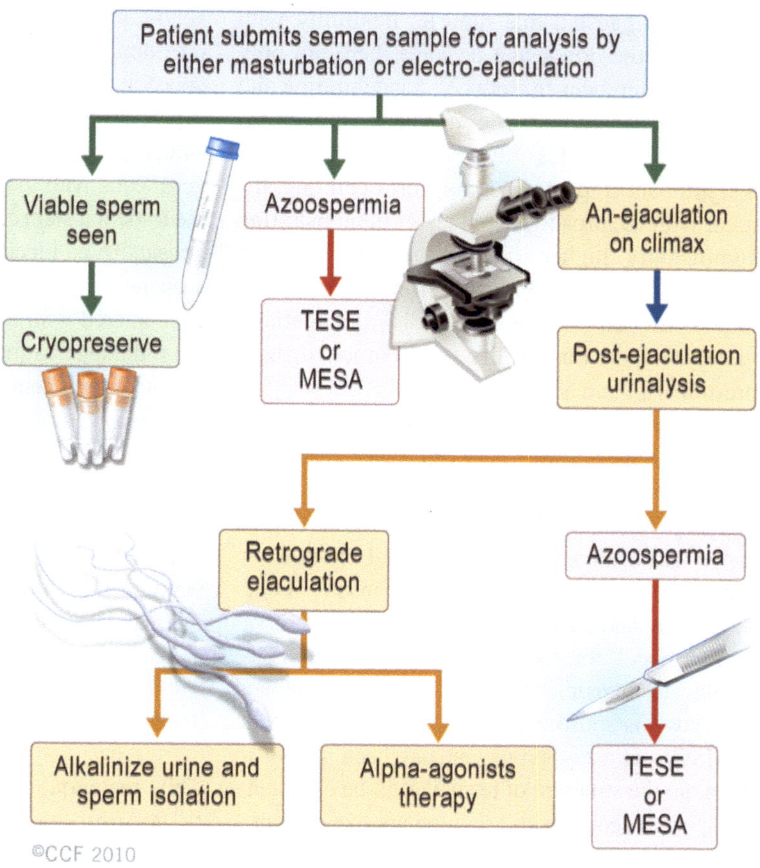

Fig. 27.1 Approach for sperm banking in post-pubertal and adult males

Experimental Options for Fertility Preservation

Absence of spermatozoa and spermatids in testis prevents pre-pubertal boys to benefit from the technique of sperm freezing. Spermarche, the production of sperm, occurs at approximately 13–14 years of age. Although few spermatids might be present in some seminiferous tubules in the late pre-pubertal period; such spermatocytes never seem to give rise to spermatozoa [30]. Furthermore, cryopreserved semen samples are a finite source and do not offer the patients a chance to achieve natural fertility. Hence, with the recent advances in assisted reproductive techniques, mature and immature sperm extraction and maturation [31, 32], focus has been shifted to the possibility of testicular tissue and germ cell preservation in young cancer patients.

Spermatogonial Stem Cell Transplantation

The testicular tissue contains testicular precursors called spermatogonial stem cells. These cells are the genuine totipotent population of cells in the adult body and they undergo self-renewal throughout the life. These are also more resistant than other testicular cells to variety of toxic insults [33]. The spermatogonial stem cells are present in pre-pubertal testicular tissue, and can be isolated and successfully cryopreserved with almost 70% cells surviving freezing and thawing as demonstrated in animal experiments [34]. These stem cells can be re-transplanted autologously into the testis, where they recolonize the seminferous tubules, generating complete spermatogenesis and mature germ cells thus restoring natural fertility (spermatogonial stem cell autotransplantation) [35–38]. Animal studies done so far have suggested that there is flexibility with respect to the donor age because there are only modest differences in stem cell concentration and colony expansion among newborn, preadolescent, and adult testes. However, recipient age has a significant impact on donor germ cell engraftment with the preadolescent testes demonstrating significantly higher spermatogenesis as compared to adults [38, 39]. Therefore, spermatogonial stem cell transplantation is a potentially viable method for fertility preservation in both pre-pubertal and adult males.

Although, successful establishment of fertility has been achieved with stem cell transplantation in mice, rats, and goats, the technique is still experimental in humans [35–38]. One of the crucial steps for re-fertilization is the safe retrieval of sufficient testicular tissue before the cytotoxic insults of chemotherapy and radiotherapy. A safe cryopreservation protocol for the spermatogenic stem cells, similar to freezing of human spermatozoa and embryos is also needed. Dimethylsulfoxide (DMSO), as a cryoprotectant has been shown to maintain the structure of testicular tissue especially the spermatogonia better than others like propanediol and glycerol. In a study involving 16 infertile men, after freezing with 5% DMSO, 70±6% of seminiferous tubules were found to be good as compared to 37±3% with propanediol [40]. However, DMSO itself is a potential carcinogen which restricts its application in clinical practice [41, 42]. Although DMSO is usually used in concentrations of 10–20% [32, 43, 44], Keros et al. have shown promising results with 5% DMSO in human testicular tissue cryopreservation [40, 45]. Another potential concern is the contamination of spermatogonia with tumor cells leading to retransmission of cancer back to the recipient. Systemic malignancies can metastasize to the testis, and transplantation of testicular cells can re-expose the recipients to the same problem. Presence of malignant cells in the gonadal tissue can be screened by various cytogenetic and molecular markers. However, if there is clinical evidence of testicular tissue involvement, or when the potential for occult metastasis is high, auto-transplantation of germ cells is not recommended. Transplantation of human spermatogonial stem cells into animals may be performed instead (spermatogonial stem cell xeno-transplantation) [46, 47]. However, animal infectious agents such as retroviruses may be introduced in human germ line when these cells are used to procure conception [48]. Furthermore, it is impossible to generate haploid male gametes from diploid germ cells with the existing in vitro approaches as the testicular stem cells expansion and meiotic entry appears to be blocked in cultures of testicular cell suspension [49]. Some of the other challenges encountered with this technology include the ischemic damage to the transplanted testicular tissue, in vitro enrichment of stem cell spermatogonia, and non-invasive transfer of germ cell suspensions into the rete testis. Hence, in spite of latest developments in the field suggesting bright prospects for fertility preservation in young cancer patients who have not achieved puberty, techniques still need to be developed for isolation, storage, and re-infusion of spermatogonial stem cells in humans.

Testicular Tissue Cryopreservation

This technique involves removal of testicular tissue from the patient before cytotoxic therapy and transplantation of testicular tissue pieces to an ectopic site such as under the skin of patients after successful completion of cancer therapy (ectopic auto-grafting of testicular tissue) or into animals (ectopic xeno-grafting of testicular tissue). The grafted testicular tissue revascularizes in the ectopic site producing complete spermatogenesis, first demonstrated by Brinster and Zimmermann in mice [50, 51]. Sperm retrieval from grafted tissue can be used to generate healthy offspring with assisted fertilization techniques. Grafting has been successfully demonstrated in hamsters, pigs, goats, calves, rabbits, and monkeys [52–54]. In a recently published on-going trial, Children's hospital of Philadelphia has been successful in enrolling 16 pre-pubertal boys aged 3 months to 14 years for testicular cryopreservation as a potential future use in an experimental protocol. In this study, 76% (16/21) of the parents of pre-pubertal boys with cancer consented for testicular biopsy. Of the 16 patients enrolled, 14 underwent the procedure without any negative intra- or post-operative sequelae [55]. A Belgian study has earlier demonstrated survival and proliferative capacity of spermatogonia and sertoli cells in cryopreserved immature human cryptorchid tissue transplanted into mice scrotum through orthotopic xenografting model [56]. Lately, Sato et al. from Japan demonstrated successful spermatogenesis in the xenograft of human infant testicular tissue grafted into nude mice until pachytene spermatocyte stage [57]. In addition to spermatogenic, the steroidogenic function of the testicular tissue has also been shown to restore with testicular grafting [53]. However, grafting of adult testicular tissue has shown only limited success in contrast to immature tissue. Furthermore, autografting of cryopreserved testicular tissue share similar concerns of cancer recurrence as spermatogonial stem cell auto-transplantation. Likewise, as previously discussed, xenografting of human gonadal tissue poses significant risks of introducing various animal infectious agents in human germ line.

Hormonal Gonado-Protection

This strategy of fertility preservation in cancer patients involves manipulation of the hypothalamic pituitary gonadal axis (HPG axis) to render the testes less susceptible to cytotoxic therapy. An initial hypothesis was that suppression of the HPG axis prior to cytotoxic therapy using agents such as sex steroids, gonadotropin releasing hormone (GnRH) agonists or antagonists would protect the gonad from cytotoxic damage by preventing the germ cells from actively proliferating thus inducing resting or a less sensitive state. Lately, it has been shown that GnRH agonists release the germ cells from the block on differentiation that would otherwise occur [58]. Furthermore, it has also been proposed that GnRH antagonists reduce the high levels of intra-testicular testosterone caused by radiation therapy, leading to a reduction in interstitial volume and testicular edema and thus allowing early resumption of spermatogenesis [59].

In rats, spermatogenesis and fertility has shown to be restored following treatment with radiation or chemotherapy by suppressing testosterone with GnRH agonists or antagonists either before or after the cytotoxic insult [60, 61]. However, hormonal therapy has not been successful in preserving fertility or speed recovery of spermatogenesis in men and other primates [62–64]. The only successful human study evaluated testosterone alone in men without cancer treated with cyclophosphamide for nephrotic disorders [65]. Thus despite the success in restoring fertility by hormone treatment of rats rendered azoospermic by chemotherapy and radiation, the application of this procedure to humans is uncertain [66].

Selection of Appropriate Technique for Adult and Pre-pubertal Males

Studies have revealed that most young male patients with cancer do not seek fertility preservation [67, 68]. Reasons for this apparent underutilization are either lack of timely information and referral by physicians, or psychological, logistic and financial constraints of the patients [5, 12]. Even when sperm is banked, less than 30% of men return to use their stored specimens [69, 70]. Preserving the fertility of young men requires coordinated efforts and attention to a host of ethical and legal issues by oncologists and fertility specialists. Physicians have a responsibility to inform patients that infertility is a potential risk of their cancer treatment. Basic questions such as whether fertility preservation options decrease their chance of successful cancer treatment, increase the risk of maternal or perinatal complications, or compromise the health of offspring should be answered. Informed consent should be obtained from the patients regarding the type of fertility preservation treatment and legal status of any resulting child. The more specifically and comprehensively a written directive is executed by the patient during his lifetime, the less likely it will be that disputes will arise regarding future use of the stored genetic material.

Literature suggests that sperm banking is the only effective method of fertility preservation in men. Cryopreservation of sperm before initiation of cytotoxic therapy is the best method of preserving fertility in post-pubertal males. Testicular or spermatogonial stem cell cryopreservation and auto-transplantation or xenografting are still investigational. However, these are the only options available for pre-pubertal boys. Gonadal protection through hormone manipulation is ineffective [49, 71, 72]. Table 27.1 summarizes available

Table 27.1 Summary for fertility preservation options in pre-pubertal and adult males

	Sperm cryopreservation	Spermatogonial stem cell autotransplantation	Spermatogonial stem cell xenotransplantation	Testicular tissue cryopreservation and ectopic autografting	Testicular tissue cryopreservation and ectopic xenografting
Eligible candidates	Adult male	Pre-pubertal male	Pre-pubertal male	Pre-pubertal male	Pre-pubertal male
Method of pregnancy	Assisted reproduction	Natural intercourse or assisted reproduction	Assisted reproduction	Assisted reproduction	Assisted reproduction
Critical comments	Unfeasible in pre-pubertal males, finite source, cannot achieve natural fertility	Invasive procedure, risk of retransmission of cancer	Invasive procedure, risk of transfer of animal infections into human germ line	Invasive procedure, risk of retransmission of cancer	Invasive procedure, risk of transfer of animal infections into human germ line
Status of procedure	Established	Investigational	Investigational	Investigational	Investigational

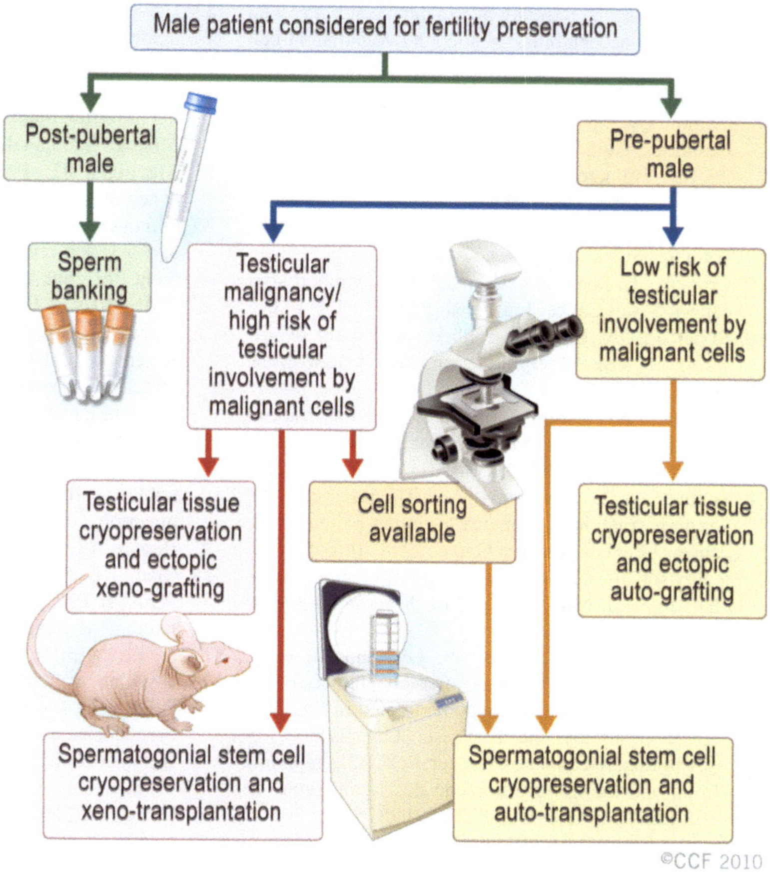

Fig. 27.2 Selection of appropriate established or experimental technique for fertility preservation in adult and pre-pubertal males

fertility preservation options for pre- and postpubertal males. Figure 27.2 provides a flowchart for selection of appropriate fertility preservation technique among the available established and experimental options.

Conclusions

Recent advances in medicine have led to ever increasing number of pre-pubertal and adult males surviving cancer treatment. This has increased the need to improve the existing technology for cryopreservation of gametes and search for new fertility preservation options. As of today, only sperm cryopreservation is considered accepted standard clinical practices. Fertility preservation options in pre-pubertal males are still experimental. Cryopreservation of testicular tissue and spermatogonial stem cell transplantation should only be offered within IRB-approved clinical protocol after thorough counseling of patients and their family members as there are still many unresolved issues related to these technologies.

References

1. Robison LL. Methodologic issues in the study of second malignant neoplasms and pregnancy outcomes. Med Pediatr Oncol Suppl. 1996;1:41–4.
2. Mackie EJ, Radford M, Shalet SM. Gonadal function following chemotherapy for childhood Hodgkin's disease. Med Pediatr Oncol. 1996;27(2):74–8.
3. Blumenfeld Z, Haim N. Prevention of gonadal damage during cytotoxic therapy. Ann Med. 1997;29(3): 199–206.
4. Howell S, Shalet S. Gonadal damage from chemotherapy and radiotherapy. Endocrinol Metab Clin North Am. 1998;27(4):927–43.
5. Schover LR, Brey K, Lichtin A, Lipshultz LI, Jeha S. Knowledge and experience regarding cancer, infertility, and sperm banking in younger male survivors. J Clin Oncol. 2002;20(7):1880–9.

6. van den Berg H, Repping S, van der Veen F. Parental desire and acceptability of spermatogonial stem cell cryopreservation in boys with cancer. Hum Reprod. 2007;22(2):594–7.
7. Committee on Drugs, American Academy of Pediatrics. Guidelines for the ethical conduct of studies to evaluate drugs in pediatric populations. Pediatrics. 1995;95(2):286–94.
8. Bahadur G. Ethics of testicular stem cell medicine. Hum Reprod. 2004;19(12):2702–10.
9. Crockin SL. Legal issues related to parenthood after cancer. J Natl Cancer Inst Monogr. 2005;34:111–3.
10. Robertson JA. Procreative liberty and harm to offspring in assisted reproduction. Am J Law Med. 2004;30(1):7–40.
11. Stegmann BJ. Unique ethical and legal implications of fertility preservation research in the pediatric population. Fertil Steril. 2010;93(4):1037–9.
12. Schover LR, Agarwal A, Thomas Jr AJ. Cryopreservation of gametes in young patients with cancer. J Pediatr Hematol Oncol. 1998;20(5):426–8.
13. Robertson JA. Cancer and fertility: ethical and legal challenges. J Natl Cancer Inst Monogr. 2005;34: 104–6.
14. van der Kaaij MA, Heutte N, Le Stang N, et al. Gonadal function in males after chemotherapy for early-stage Hodgkin's lymphoma treated in four subsequent trials by the European Organisation for Research and Treatment of Cancer: EORTC Lymphoma Group and the Groupe d'Etude des Lymphomes de l'Adulte. J Clin Oncol. 2007;25(19):2825–32.
15. Viviani S, Santoro A, Ragni G, Bonfante V, Bestetti O, Bonadonna G. Gonadal toxicity after combination chemotherapy for Hodgkin's disease. Comparative results of MOPP vs ABVD. Eur J Cancer Clin Oncol. 1985;21(5):601–5.
16. Meseguer M, Molina N, Garcia-Velasco JA, Remohi J, Pellicer A, Garrido N. Sperm cryopreservation in oncological patients: a 14-year follow-up study. Fertil Steril. 2006;85(3):640–5.
17. Agarwal A, Shekarriz M, Sidhu RK, Thomas Jr AJ. Value of clinical diagnosis in predicting the quality of cryopreserved sperm from cancer patients. J Urol. 1996;155(3):934–8.
18. Gandini L, Lombardo F, Lenzi A, Spano M, Dondero F. Cryopreservation and sperm DNA integrity. Cell Tissue Bank. 2006;7(2):91–8.
19. Hallak J, Hendin BN, Thomas Jr AJ, Agarwal A. Investigation of fertilizing capacity of cryopreserved spermatozoa from patients with cancer. J Urol. 1998;159(4):1217–20.
20. Kuczynski W, Dhont M, Grygoruk C, Grochowski D, Wolczynski S, Szamatowicz M. The outcome of intracytoplasmic injection of fresh and cryopreserved ejaculated spermatozoa – a prospective randomized study. Hum Reprod. 2001;16(10):2109–13.
21. De Mas P, Daudin M, Vincent MC, et al. Increased aneuploidy in spermatozoa from testicular tumour patients after chemotherapy with cisplatin, etoposide and bleomycin. Hum Reprod. 2001;16(6):1204–8.
22. Howell SJ, Shalet SM. Spermatogenesis after cancer treatment: damage and recovery. J Natl Cancer Inst Monogr. 2005;34:12–7.
23. Shin D, Lo KC, Lipshultz LI. Treatment options for the infertile male with cancer. J Natl Cancer Inst Monogr. 2005;34:48–50.
24. Brannigan RE. Fertility preservation in adult male cancer patients. Cancer Treat Res. 2007;138:28–49.
25. Ohl DA, Wolf LJ, Menge AC, et al. Electroejaculation and assisted reproductive technologies in the treatment of anejaculatory infertility. Fertil Steril. 2001;76(6):1249–55.
26. Janzen N, Goldstein M, Schlegel PN, Palermo GD, Rosenwaks Z, Hariprashad J. Use of electively cryopreserved microsurgically aspirated epididymal sperm with IVF and intracytoplasmic sperm injection for obstructive azoospermia. Fertil Steril. 2000;74(4):696–701.
27. Devroey P, Liu J, Nagy Z, et al. Pregnancies after testicular sperm extraction and intracytoplasmic sperm injection in non-obstructive azoospermia. Hum Reprod. 1995;10(6):1457–60.
28. Chan PT, Palermo GD, Veeck LL, Rosenwaks Z, Schlegel PN. Testicular sperm extraction combined with intracytoplasmic sperm injection in the treatment of men with persistent azoospermia postchemotherapy. Cancer. 2001;92(6):1632–7.
29. Palermo G, Joris H, Devroey P, Van Steirteghem AC. Pregnancies after intracytoplasmic injection of single spermatozoon into an oocyte. Lancet. 1992;340(8810): 17–8.
30. Paniagua R, Nistal M. Morphological and histometric study of human spermatogonia from birth to the onset of puberty. J Anat. 1984;139(Pt 3):535–52.
31. Fishel S, Green S, Bishop M, et al. Pregnancy after intracytoplasmic injection of spermatid. Lancet. 1995;345(8965):1641–2.
32. Hovatta O, Foudila T, Siegberg R, Johansson K, von Smitten K, Reima I. Pregnancy resulting from intracytoplasmic injection of spermatozoa from a frozen-thawed testicular biopsy specimen. Hum Reprod. 1996;11(11):2472–3.
33. Russell LD, Ettlin RA, Hakim AP. Mammalian spermatogenesis. Clearwater, FL: Cache River Press; 1999.
34. Izadyar F, Matthijs-Rijsenbilt JJ, den Ouden K, Creemers LB, Woelders H, de Rooij DG. Development of a cryopreservation protocol for type A spermatogonia. J Androl. 2002;23(4):537–45.
35. Brinster RL, Avarbock MR. Germline transmission of donor haplotype following spermatogonial transplantation. Proc Natl Acad Sci USA. 1994;91(24):11303–7.
36. Ogawa T, Dobrinski I, Avarbock MR, Brinster RL. Transplantation of male germ line stem cells restores fertility in infertile mice. Nat Med. 2000;6(1):29–34.
37. Honaramooz A, Behboodi E, Megee SO, et al. Fertility and germline transmission of donor haplotype following germ cell transplantation in immunocompetent goats. Biol Reprod. 2003;69(4):1260–4.
38. Ryu BY, Orwig KE, Avarbock MR, Brinster RL. Stem cell and niche development in the postnatal rat testis. Dev Biol. 2003;263(2):253–63.

39. Shinohara T, Orwig KE, Avarbock MR, Brinster RL. Remodeling of the postnatal mouse testis is accompanied by dramatic changes in stem cell number and niche accessibility. Proc Natl Acad Sci USA. 2001;98(11):6186–91.
40. Keros V, Rosenlund B, Hultenby K, Aghajanova L, Levkov L, Hovatta O. Optimizing cryopreservation of human testicular tissue: comparison of protocols with glycerol, propanediol and dimethylsulphoxide as cryoprotectants. Hum Reprod. 2005;20(6):1676–87.
41. Avarbock MR, Brinster CJ, Brinster RL. Reconstitution of spermatogenesis from frozen spermatogonial stem cells. Nat Med. 1996;2(6):693–6.
42. Aslam I, Fishel S, Moore H, Dowell K, Thornton S. Fertility preservation of boys undergoing anti-cancer therapy: a review of the existing situation and prospects for the future. Hum Reprod. 2000;15(10):2154–9.
43. Shinohara T, Inoue K, Ogonuki N, et al. Birth of offspring following transplantation of cryopreserved immature testicular pieces and in-vitro microinsemination. Hum Reprod. 2002;17(12):3039–45.
44. Jezek D, Schulze W, Kalanj-Bognar S, Vukelic Z, Milavec-Puretic V, Krhen I. Effects of various cryopreservation media and freezing-thawing on the morphology of rat testicular biopsies. Andrologia. 2001;33(6):368–78.
45. Keros V, Hultenby K, Borgstrom B, Fridstrom M, Jahnukainen K, Hovatta O. Methods of cryopreservation of testicular tissue with viable spermatogonia in pre-pubertal boys undergoing gonadotoxic cancer treatment. Hum Reprod. 2007;22(5):1384–95.
46. Nagano M, Patrizio P, Brinster RL. Long-term survival of human spermatogonial stem cells in mouse testes. Fertil Steril. 2002;78(6):1225–33.
47. Sofikitis N. Transplantation of human spermatogonia into the seminiferous tubules (STs) of animal testicles results in the completion of the human meiosis and the generation of human motile spermatozoa. Fertil Steril. 1999;72 suppl 1:S83–4.
48. Patience C, Takeuchi Y, Weiss RA. Infection of human cells by an endogenous retrovirus of pigs. Nat Med. 1997;3(3):282–6.
49. Orwig KE, Schlatt S. Cryopreservation and transplantation of spermatogonia and testicular tissue for preservation of male fertility. J Natl Cancer Inst Monogr. 2005;34:51–6.
50. Brinster RL, Zimmermann JW. Spermatogenesis following male germ-cell transplantation. Proc Natl Acad Sci USA. 1994;91(24):11298–302.
51. Ogawa T. Spermatogonial transplantation technique in spermatogenesis research. Int J Androl. 2000;23 Suppl 2:57–9.
52. Honaramooz A, Snedaker A, Boiani M, Scholer H, Dobrinski I, Schlatt S. Sperm from neonatal mammalian testes grafted in mice. Nature. 2002;418(6899): 778–81.
53. Schlatt S, Kim SS, Gosden R. Spermatogenesis and steroidogenesis in mouse, hamster and monkey testicular tissue after cryopreservation and heterotopic grafting to castrated hosts. Reproduction. 2002;124(3):339–46.
54. Oatley JM, de Avila DM, Reeves JJ, McLean DJ. Spermatogenesis and germ cell transgene expression in xenografted bovine testicular tissue. Biol Reprod. 2004;71(2):494–501.
55. Ginsberg JP, Carlson CA, Lin K, et al. An experimental protocol for fertility preservation in prepubertal boys recently diagnosed with cancer: a report of acceptability and safety. Hum Reprod. 2010;25(1): 37–41.
56. Wyns C, Curaba M, Martinez-Madrid B, Van Langendonckt A, Francois-Xavier W, Donnez J. Spermatogonial survival after cryopreservation and short-term orthotopic immature human cryptorchid testicular tissue grafting to immunodeficient mice. Hum Reprod. 2007;22(6):1603–11.
57. Sato Y, Nozawa S, Yoshiike M, Arai M, Sasaki C, Iwamoto T. Xenografting of testicular tissue from an infant human donor results in accelerated testicular maturation. Hum Reprod. 2010;25(5):1113–22.
58. Meistrich ML, Wilson G, Kangasniemi M, Huhtaniemi I. Mechanism of protection of rat spermatogenesis by hormonal pretreatment: stimulation of spermatogonial differentiation after irradiation. J Androl. 2000;21(3):464–9.
59. Porter KL, Shetty G, Meistrich ML. Testicular edema is associated with spermatogonial arrest in irradiated rats. Endocrinology. 2006;147(3):1297–305.
60. Meistrich ML, Kangasniemi M. Hormone treatment after irradiation stimulates recovery of rat spermatogenesis from surviving spermatogonia. J Androl. 1997;18(1):80–7.
61. Meistrich ML, Wilson G, Huhtaniemi I. Hormonal treatment after cytotoxic therapy stimulates recovery of spermatogenesis. Cancer Res. 1999;59(15):3557–60.
62. Johnson DH, Linde R, Hainsworth JD, et al. Effect of a luteinizing hormone releasing hormone agonist given during combination chemotherapy on posttherapy fertility in male patients with lymphoma: preliminary observations. Blood. 1985;65(4):832–6.
63. Waxman JH, Ahmed R, Smith D, et al. Failure to preserve fertility in patients with Hodgkin's disease. Cancer Chemother Pharmacol. 1987;19(2):159–62.
64. Thomson AB, Anderson RA, Irvine DS, Kelnar CJ, Sharpe RM, Wallace WH. Investigation of suppression of the hypothalamic-pituitary-gonadal axis to restore spermatogenesis in azoospermic men treated for childhood cancer. Hum Reprod. 2002;17(7):1715–23.
65. Masala A, Faedda R, Alagna S, et al. Use of testosterone to prevent cyclophosphamide-induced azoospermia. Ann Intern Med. 1997;126(4):292–5.
66. Shetty G, Meistrich ML. Hormonal approaches to preservation and restoration of male fertility after cancer treatment. J Natl Cancer Inst Monogr. 2005;34:36–9.
67. Wallace WH, Anderson RA, Irvine DS. Fertility preservation for young patients with cancer: who is at risk and what can be offered? Lancet Oncol. 2005;6(4):209–18.
68. Chung K, Irani J, Knee G. Sperm cryopreservation for male patients with cancer: an epidemiological analysis at the University of Pennsylvania. Eur J Obstet Gynecol Reprod Biol. 2004;113 Suppl 1:S7–11.

69. Audrins P, Holden CA, McLachlan RI, Kovacs GT. Semen storage for special purposes at Monash IVF from 1977 to 1997. Fertil Steril. 1999;72(1):179–81.
70. Blackhall FH, Atkinson AD, Maaya MB, et al. Semen cryopreservation, utilisation and reproductive outcome in men treated for Hodgkin's disease. Br J Cancer. 2002;87(4):381–4.
71. Lee SJ, Schover LR, Partridge AH, et al. American Society of Clinical Oncology recommendations on fertility preservation in cancer patients. J Clin Oncol. 2006;24(18):2917–31.
72. Jeruss JS, Woodruff TK. Preservation of fertility in patients with cancer. N Engl J Med. 2009;360(9):902–11.

Part IV

Protocols

Appendix A: Embryo Cryopreservation and Alternative Strategies for Controlled Ovarian Hyperstimulation

Bulent Urman, Ozgur Oktem, and Basak Balaban

Cryopreservation of Embryos with Slow Freezing Method

I. Equipment

Programmable freezing machine for slow freezing (Planer Series III, England).
LN2 storage tank with stainless steel canes for straws.
Sterile serological pipettes and culture dishes, NUNC ART products, Denmark.
Cryo Bio System (Paris, France) high security embryo freezing straws.
Colored cryomarkers.
Cryogloves.
Slow freezing media for cleavage stage embryos.

B. Urman (✉)
Department of Obstetrics and Gynecology, American Hospital, Guzelbahce Sok No 20 Nisantasi, 34365, Istanbul, Turkey
e-mail: burman@superonline.com

II. Reagents

Slow freezing medium (Freeze Kit-1 Ref: 10066 Vitrolife, Sweden Ready to use kit).

III. Specimen

Human embryos.

IV. Procedure

The duration of the slow freezing procedure including cryoprotectant equilibration is approximately 2 h. Freeze kit-I (Vitrolife, Sweden) is used for slow freezing, whereas Thaw Kit-I (Vitrolife, Sweden) for thawing procedure.

The slow freezing program used by Planer Series III (England) is as follows:
Starting temperature: +18.0 to +25°C
Step 1: –2.0°C/min to –7.0°C
Step 2: HOLD AT –7.0°C for 10 min, SEED after 2 min

E. Seli and A. Agarwal (eds.), *Fertility Preservation: Emerging Technologies and Clinical Applications*, DOI 10.1007/978-1-4419-1783-6, © Springer Science+Business Media, LLC 2012

Step 3: −0.3°C/min to −30.0°C

Step 4: −30.0°C to below −80.0°C (at least 10°C/min)

The straws are removed and plunged into LN2, and store submerged in LN2 (not in the vapor phase).

A. Essentials

Only embryos of high quality are cryopreserved since the thaw survival rate is related to initial quality of the embryo.

Patient's identity is checked and all paperwork is prepared before starting the procedure.

B. Hands on

1. Label dishes for freezing solutions, pipette appropriate volumes of Cryo-PBS, FS1, and FS2 into respective dishes (Cryo-PBS, FS1, FS2 are freezing solutions included in Freeze Kit-1). Equilibrate to ambient temperature.
2. Observe and record detailed morphology. Rinse embryos for freezing in Cryo-PBS.
3. Gently place embryos into FS1 for 10 min (never more than 20 min). The cells of the embryo will shrink and then re-equilibrate during this step.
4. Move embryos across to FS2 and then load into straws (High Security Straws, Cryo Bio System Paris, France) by attaching the straw to a 1-mL syringe (syringes are connected to the straws by using 1 cm of silastic tubing). Remove the straws and seal so that liquid nitrogen will not leak inside the straw.
5. Place into freezing chamber at ambient temperature and commence freezing program.
6. Manually seed the straws at −7°C with liquid nitrogen (LN2) cooled forceps close to the cotton plug. Do not seed the straws close to the embryos. Do not drop straw or shake it. If there are air bubbles in the straw it may reduce cell survival. Continue the freezing program.
7. The freezing takes approximately 2 h. Attention must be paid to the handling of the straws at low temperatures as they may break very quicly. Plunge into liquid nitrogen and arrange for storage at −196°C.

Thawing Procedure for Embryos Cryopreserved by Slow Freezing Method

I. Equipment

37°C water bath for warming.

Sterile serological pipettes and culture dishes, NUNC Art products, Denmark.

Colored cryomarkers.

Cryogloves.

Sterile stainless steel scissor.

Thawing media for cleavage stage embryos.

II. Reagents

A. Thawing medium (Thaw Kit-1 Ref: 10067 Vitrolife, Sweden Ready to use kit).

III. Specimen

Human embryos

IV. Procedure

A. Essentials

Straws are thawed one at a time.

All steps are performed at ambient temperature.

Patient's identity is checked carefully and all paperwork is prepared accordingly.

B. Hands on

1. Identify patient and location of straw. Prepare all paperwork. Keep the straw under LN2 in small Dewar vessel until actual thawing. Label dishes for thaw solutions and pipette appropriate amounts of TS1, TS2, TS3 and Cryo-PBS (TS1, TS2, TS3 and Cryo-PBS are included in Thaw-Kit I).
2. Remove straw and air thaw for 30 s. During this time, handle the straws carefully and examine it for air bubbles, cracks in the seal, and any leakage of LN2.

3. Place in a +30°C water bath for 30 s. Remove and carefully wipe. Cut the plug end of the straw with sterile scissors, and attach in to a 1-mL syringe. Cut the other end carefully, do not shake straw or make air bubbles.
4. Gently expel the embryos into TS1. Observe the embryos coming out of the straw and if you do not see them all, quickly refill the straw and gently flush. Occasionally the embryos will stick to the sides of the straw.
5. Incubate the embryos for 5 min in TS1.
6. Gently move the embryos to TS2 for 5 min (approximately). Remember that these embryos are osmotically STRESSED and need to be handled VERY CAREFULLY.
7. Move the embryos to TS3 for 5–10 min. The embryos will be stressed and the memranes will be very fragile.
8. Place embryos in Cryo-PBS at ambient temperature for 6 min, followed by 4 min at 37°C on heated stage. Do not place in the CO_2 incubator, as the buffer capacity of Cryo-PBS is insufficient for the 6%CO_2 atmosphere.
9. Embryos may be placed into equilibrated supplemented G-1/G-2 PLUS (or G-2/G-2 PLUS (G1/G2 PLUS culture media, Vitrolife Sweden) if the embryos were cryopreserved on day 3). Assess and record survival and morphology compared to prefreeze. Embryos may be transferred immediately or left for further culture. Pronuclear oocytes should be cultured overnight in G-1/G-1 PLUS and only transferred if normal cleavage has occured.
10. The thaw-cycle is very important to the success of embryo cryopreservation. Ensure that the embryo transfer is performed on the appropriate day. Embryos may be replaced in a natural or artificial cycle.

Cryopreservation of Embryos with Vitrification Method

I. Equipment

LN2 dewar
Sterile serological Pipettes and Dishes, NUNC ART products, Denmark
Sterile cryoloops (Vitrolife, Sweden)
Colored cryomarkers
Magnetic Wand to manuplate the cryoloop (Vitrolife, Sweden)
Stainless steel canes for cryovials
Cryogloves
Vitrification media

II. Reagents

A. Vitrification media (RapidVit Cleave Ref: 10117 Vitrolife, Sweden Ready to use)

III. Specimen

A. Human embryos

IV. Procedure

A. Place 0.5–1 mL of each vitrification solution into separate wells of a labeled 4-well plate.
B. Allow the solutions to warm to 37°C.
C. Label the cryoloops and canes to allow subsequent identification.
D. Embryos are then moved from their culture medium to Vitri 1™ Cleave solution using a pipette.
E. Move 1–3 embryos into the Vitri 2™ Cleave solution. Start the Timer. The embryos are incubated in the Vitri 2™ Cleave for a total time of 2 min.
F. Once 1:30 s have elapsed, pipette 20 μL of the Vitri 3™ Cleave solution onto a culture dish. NOTE: This restricts the movement of the embryos and allows them to be quickly mixed with the Vitri 3™ Cleave solution and then gathered.
G. Attach the cryoloop to the magnetic wand and then dip the nylon loop into the Vitri 3™ Cleave solution. The magnetic wand can then be left flat on the microscope stage and subsequently rotated until the cryoloop is in the correct plane for loading.
H. Once 1:50 s have elapsed, begin to collect the embryos from the Vitri 2™ Cleave using the pipette.

I. The embryos should leave the Vitri 2™ Cleave solution and enter the 20 μL droplet of the Vitri 3™ Cleave solution as the 2 min expires.
J. Use the tip of the pipette to mix the solution once the embryos have been released.
K. The total time embryos are exposed to the Vitri 3™ Cleave solution before being plunged into liquid nitrogen is 25–30 s. Therefore, once the embryos have been exposed to the Vitri 3™ Cleave solution for 20 s, collect the embryos and pipette them onto the film of solution held across the cryoloop.
L. Using the magnetic wand, screw the cryoloop into the cryovial that is attached to the cryocane submerged in a Dewar of liquid nitrogen.

Warming of Embryos Cryopreserved by Vitrification Method

I. Equipment

LN2 dewar
Sterile serological pipettes and dishes, NUNC ART products, Denmark.
Colored cryomarkers.
Magnetic Wand to manipulate the cryoloop (Vitrolife, Sweden).
Cryogloves.
Warming media.

II. Reagents

Warming media (RapidWarm Cleave Ref: 10118 Vitrolife, Sweden Ready to use).

III. Specimen

Human embryos.

IV. Procedure

A. Place 0.5–1 mL of each warming solution into separate wells of a labeled 4-well plate.
B. Allow the solutions to warm to 37°C.
C. Remove the cryocanes from the long-term liquid nitrogen storage vessel. Place the cryocane into the Dewar-containing liquid nitrogen.
D. Use the magnetic wand to unscrew the lid from the cryovial. Rapidly, move the cryoloop into Warm 1™ Cleave solution.
E. By touching the cryoloop to the surface of the solution, the embryos will be removed from the cryoloop. Allow the embryos to sink and collect within 10–30 s.
F. Move the embryos into Warm 2™ Cleave solution. The embryos will remain in this second solution for 1 min.
G. Once the 1 min has elapsed move the embryos into Warm 3™ Cleave solution. The embryos will remain in this third solution for 2 min.
H. Once the 2 min have elapsed, move the embryos into Warm 4™ Cleave solution. The embryos will remain in this third solution for 5 min.
I. Move the embryos into appropriate culture media if in vitro development is required.

Appendix B: Oocyte Cryopreservation

Andrea Borini and Veronica Bianchi

Oocyte Cryopreservation: Slow Cooling

I. Equipment

Sterile straws (Paillettes Crystal 133 mm; Cryo Bio System, France)
Sterile serological pipettes (Falcon 1, 2, 10 mL)
Adapted syringes
Micropipette tips (170 μm) to transfer the oocytes
Impulse heat sealer
Stopwatch or timer
Automated Kryo 10 series biological vertical freezer (Planer Kryo GB)
Liquid nitrogen reservoir (dewar or equivalent)
Liquid nitrogen
LN2 dewar with ten canisters for long-term storage
Cryo tape
Multicolor goblets
Stainless steel canes
Cryogloves
Protective goggles
Latex gloves
37°C incubator
Freezing media (PROH, sucrose, PBS, protein supplement, or albumin)
Culture media
Sterile Nunclon surface 4 well plate (Nunc)

II. Reagents

Freezing

Equilibration solution 1.5 M PROH in PBS
Loading solution 1.5 M PROH plus 0.2 M sucrose in PBS
Albumin or synthetic protein supplement from Irvine is added in all the solution (20%)
The solutions are filtered and used for a maximum of 3 days

Thawing

1.0 M PROH plus 0.3 M sucrose in PBS
0.5 M PROH plus 0.3 M sucrose in PBS
0.3 M sucrose in PBS

A. Borini (✉)
Tecnobios Procreazione Centre for Reproductive Health, Via Dante 15, 40125 Bologna, Italy
e-mail: borini@tecnobiosprocreazione.it

E. Seli and A. Agarwal (eds.), *Fertility Preservation: Emerging Technologies and Clinical Applications*, DOI 10.1007/978-1-4419-1783-6, © Springer Science+Business Media, LLC 2012

Albumin or synthetic protein supplement from Irvine is added in all the solution (20%).

All the reagents are homemade filtered and used for a maximum of 3 days. They are stored in the refrigerator at 4°C and have to be warmed to room temperature prior to use.

III. Specimen

Oocytes are retrieved transvaginally under ultrasound guidance 36 h after HCG subministration. Specimen are collected into a clean sterile petri dish with 30 μL drops of fertilization media (SAGE) and labeled with the patient's name.

Oocytes are decumulated enzymatically and mechanically around 2 h after the retrieval and prior to freezing. Only metaphase II eggs are frozen.

The denuded oocytes are transferred into a 20 μL drop of cleavage medium (SAGE) at 37°C and 5% CO_2.

IV. Procedure

Freezing

A. In a 4 well Nunc labeled with patient's name, 0.5 mL of the equilibrium solution is placed in wells #1 and 3 while 0.5 mL of the loading solution is placed in wells #2 and 4. Each row corresponds to one straw. The oocytes are transferred from the cleavage medium to the equilibration solution for 10 min and then to the loading solution for 5 min.

B. The samples are then loaded and placed in the biological freezer. The starting temperature is around 20°C; then, it is slowly cooled from 20 to −8°C at a rate of 2°C/min. Manual seeding of oocytes within straws is performed at near −7°C and this temperature is maintained for 10 min in order to allow uniform ice propagation. Temperature is then decreased to −30°C at a rate of 0.3°C/min and then (relatively rapidly) brought to −150°C at a rate of 50°C/min. Straws are then directly plunged into liquid nitrogen at −196°C and stored.

C. Record the location on the Laboratory Worksheet.

Thawing

A. Label a 4 well Nunc with patients' name.

B. Warm the solutions to room temperature prior to start.

C. The thawing solutions are dispensed with decreasing PrOH concentration (1.0, 0.5 M, and sucrose alone) in a 4 well Nunc; the last well just contains PBS and albumin.

D. Thawing consists of rapid rewarming (30 s air and 40 s in a 30°C water bath), the subsequent stepwise dilution of the cryoprotectants (5 min in the 1.0 M PROH solution; 5 min in 0.5 M PROH, 10 min in sucrose solution, and 10 min in PBS) and finally, the return of oocytes to cleavage culture media at 37°C prior to insemination.

E. ICSI is performed in all the frozen/thawed eggs as the elective choice for the insemination maximum within 2 h after thawing.

Oocyte Cryopreservation: Vitrification

I. Equipment

Irvine scientific sterile CryoTip (Cat # 40709)
Irvine scientific connector (Cat # 40736)
Sterile petri dishes (50×9 mm, Falcon 351006 or equivalent)
Cryotubes (4.5 mL) or goblets and cryocanes
Modified HTF – HEPES (Cat# 90126) culture medium supplemented with DSS or equivalent
Disposable gloves
Hamilton GASTIGHT® Syringe
Micropipette tips (170 μm) to transfer the oocytes
Tweezers
Impulse heat sealer
Stopwatch or timer
Liquid nitrogen reservoir (dewar or equivalent)
Liquid nitrogen
N2 dewar with ten canisters for long-term storage
Gilson pipette with sterile tips (100 μl)
Cryo tape and straws for patient's identification
Multicolor goblets

Stainless steel canes
Cryogloves
Protective goggles
Latex gloves
37°C incubator
Freezing media (Irvine catalog No. 90133) equilibration solution (ES) and vitrification solution (VS)
Culture media

II. Reagents

Freezing

A. *Equilibration solution* (*ES*): It is a HEPES buffered solution of Medium-199 containing gentamicin sulfate, 7.5% (v/v) of each DMSO and ethylene glycole, and 20% (v/v) dextran serum supplement.
B. *Vitrification solution VS*: It is a HEPES buffered solution of Medium-199 containing gentamicin sulfate, 15% (v/v) of each DMSO and ethylene glycole, 20% (v/v) dextran serum supplement, and 0.5 M sucrose.

Thawing

Thawing solution (*TS*): It is a HEPES buffered solution of Medium-199 containing gentamicin sulfate, 1.0 M sucrose, and 20% (v/v) dextran serum supplement.

Dilution solution (*DS*): It is a HEPES buffered solution of Medium-199 containing gentamicin sulfate, 0.5 M sucrose, and 20% (v/v) dextran serum supplement.

Washing solution (*WS*): It is a HEPES buffered solution of Medium-199 containing gentamicin sulfate and 20% (v/v) dextran serum supplement.

Bring all the solutions to room temperature (20–27°C) for at least 30 min prior to vitrification.

III. Specimen

Oocytes are retrieved transvaginally under ultrasound guidance 36 h after HCG subministration. Specimen are collected into a clean sterile petri dish with 30 μL drops of fertilization media (SAGE) and labeled with the patient's name.

Oocytes are decumulated enzymatically and mechanically around 2 h after the retrieval and prior to freezing. Only metaphase II eggs are frozen.

The denuded oocytes are transferred into a 20 μL drop of cleavage medium (SAGE) at 37°C and 5% CO_2.

IV. Procedure

Freezing

A. Place close to the microscope the liquid nitrogen reservoir full of LN, the heat sealer, and the Hamilton Syringe with the CryoTip attached.
B. Label each sterile petri dish (or lid) and CryoTip with patient information and sample #.
C. Dispense one drop (20 μL) of culture media (H) and two drops of ES solution (20 μL) in close proximity on the petri dish.
D. Dispense another 20 μL ES drop on the same petri dish.
E. Transfer the oocyte(s) with minimal volume from the culture media to the H drop for 1 min.
F. Merge the drop of H to ES1 with the tip of the transfer pipette and allow spontaneous mixing for 2 min.
G. Merge with the second ES2 drop and leave for 2 more minutes.
H. Transfer the oocyte(s) to the new ES3 drop and leave for 3 min.

   In the meantime, set up four drops of VS solution and, when the equilibration phase is over, transfer the oocyte(s) to the first VS drop for 5 s, then to the second VS drop for 5 s, and to the third VS drop for 10 s.
I. Transfer the oocyte(s) to the last VS drop and load them in the CryoTip according to the instruction; seal the fine end first and then the thick end of the CryoTip.
J. Plunge the covered CryoTip directly into liquid nitrogen and place it into a goblet with

patient name and then in the cryo bank for the long-term storage.

K. Record the location on the Laboratory Worksheet.

Thawing

A. Label with the patient name the lid of a petri dish.
B. Dispense a sequence of three microdrops (20 μL): one of TS and two of DS (DS1 and DS2).
C. Place the 37°C water bath close to the microscope together with sharp scissors, transfer flexi pipette, tips, and Hamilton syringe.
D. Remove the CryoTip from the liquid nitrogen and immerse it directly in the water bath for 3 s.
E. Cut both ends of the CryoTip, release the content, and merge the drop with the oocyte(s) to the first TS drop for 1 min.
F. Transfer to the second TS drop for 1 min and then to the first and second DS drops for 2 min each.
G. In the meantime, dispense three drops of washing solution and transfer the oocyte(s) in each drop for 2 min.
H. Finally, transfer the specimens to preequilibrated culture media.

Appendix C: Ovarian Tissue Cryopreservation

Mohamed A. Bedaiwy, Gihan M. Bareh, Katherine J. Rodewald, William W. Hurd, and Ashok Agarwal

Ovarian Cortical Strip Cryopreservation Procedure

I. Introduction

Success in cryopreservation of ovarian tissue in animal models has been applied to human female ovarian tissue cryopreservation. Investigation of this type of tissue freezing provides hope to young women facing a cancer diagnosis.

The preservation of female fertility has relatively few successful options for those diagnosed with cancer. Very little success in pregnancy outcomes has been experienced in oocyte cryopreservation. This is probably due to the hardening of the zone pellucida during the process of freezing. Cryopreservation of embryos has been very successful for full-term babies, but has its limitations. Usually, the patient has very little time to undergo an IVF cycle prior to chemotherapy. Some single women may reject the acceptance of donor sperm for fertilization prior to freezing.

Therefore, ovarian tissue cryopreservation may be an alternative to the more conventional cryopreservation of embryos. This type of procedure is considered experimental because little information is available on the success of the autotransplanted tissue producing follicles that are matured in vitro and fertilized to produce embryos that are carried to term. Obtaining ovarian tissue is invasive to the patient and, in reality, may not be an option for those patients with more advanced stages of cancer.

II. Principle

The ovarian tissue is retrieved, generally, by laparoscopic surgery but may be retrieved by other methods dependent on the medical condition of the patient.

M.A. Bedaiwy (✉)
Department of Obstetrics and Gynecology, University Hospitals Case Medical Center, Case Western Reserve University, 11100 Euclid Avenue, Cleveland, OH 44106, USA
e-mail: bedaiwymmm@yahoo.com

E. Seli and A. Agarwal (eds.), *Fertility Preservation: Emerging Technologies and Clinical Applications*, DOI 10.1007/978-1-4419-1783-6, © Springer Science+Business Media, LLC 2012

Thin sections of the ovary are cut in 1×1×3 cm by the surgeon and placed in cryoprotectant media poured over into a sterile specimen jar obtained in OR.

Each thin section of ovarian tissue is cryopreserved in a cryoprotectant using a controlled rate of freezing to a final temperature of −196°C.

III. Presurgery Patient Preparation

A. The Reproductive Tissue Bank has a 1–2-day notice that an ovarian tissue case may occur.
B. The lab is notified by Gynecology as to the patient name, CCF#, doctor, and day of surgery. The time of surgery must be scheduled as the *first case* because of the length of time needed for processing.
C. Create a chart for the patient and assign the next consecutive freeze number. The chart consists of the following:
   1. The Frozen Ovarian Tissue Agreement: Ask the patient to read through the whole agreement, including the fee schedule on the back page, and sign where the Patient Signature is indicated under the Director's signature. This should be done on duplicate copies of the Frozen Ovarian Tissue Agreement so that one can be kept in the client's file and one can be sent home with/to the client.
   2. The demographics of the ovarian bank questionnaire should be completed by the patient.
   3. During this initial visit, the patient is scheduled to meet with the Reproductive Tissue Bank Director to answer any questions or concerns.
   4. The final step for positive identification in the future is made by taking the patient's picture with the Polaroid camera. Once the picture is developed, the client's name, CCF#, and freeze number are written at the bottom. The picture is stored in the patient's cryobank folder.
   5. Label an index card for the ovarian tissue box with patient name, CCF#, freeze #, date, etc.
   6. The patient is now required to have blood drawn.
      a. Label a white requisition form for a venipuncture for this patient; use the memo account number as well as the patient name and CCF#. Three 7-mL red top tubes, as well as one 7-mL EDTA tube and one 7-mL plasma preparation tube (PPT), are to be drawn by PLC and returned to us by the patient in a biohazard bag sealed with tamper evident tape.
      b. When we receive the blood, we then login to the client depositor book: the patient name, CCF#, freeze #, bar code label, date blood sent, date received results, date of blood draw, tech initials.
      c. Place a barcode label with the test tube icon on the 7-mL EDTA tube and the PPT tube, as well as the two 7-mL red top tubes. Be absolutely sure that you are using the bar codes with the same accession numbers for the proper barcode label placement on the test tubes. If there is room on the tube, please initial and date the tubes.
      d. The tubes are placed into a biohazard bag with a copy of the Blood Bank requisition. This is hand carried to Blood Bank and given either to a specific person assigned to it. Give an extra set of barcode labels to blood bank with the Blood Requisition. Include one label with and without test tube icon and one with accession number and no barcode.
      e. The remaining 7-mL red top tube should be centrifuged and the serum removed. Aliquot the serum into 1–2 cryovials. The cryovials should be labeled with patient name, CCF#, freeze #, date, and serum, and then placed in a Puffer-Hubbard freezer. These client depositor serum banks can be used for retesting or for future new tests. These serum specimens are archived for 10 years after the date of distribution.

f. The blood tests performed by the Red Cross are: HIV 1/2, HBcAb, HTLV 1/2, HIV Ag, RPR, HbsAg, HCV, HCVNAT, WNV, and HIVNAT.
   i. If any of the test results are positive, the Director of the laboratory is contacted by the Blood Bank.
   ii. Positive test results are confirmed by retesting.
   iii. The Director contacts the patient's ordering physician, as well as the patient, either by phone or letter when any of the tests are positive.
   iv. Occasionally, the Blood Bank is able to retest positive specimens by forwarding them as STATS to Microbiology. Copy the original blood sheet and a note explaining which test needs to be retested. Put the accession number on the serum vial and call appropriate personnel to inquire if and when any STATS will be done so that our specimen will be ready at the same time. STATS are done once or twice a week.
   v. If any of the blood results are positive, refer to Frozen Ovarian Tissue Agreement, as to the proper procedure. The cryopreserved specimens remain in a quarantined area.
   vi. All blood reports on client depositors are reviewed and initialed by the Director.

IV. Equipment and Materials

- Sterile 5-mL serological pipettes
- Sterile 10-mL serological pipettes
- 5 mL cryovials
- LN_2 Dewars
- Cryocanes
- Cryosleeves
- Culture media flasks
- Cryomarkers
- Sterile forceps
- Planer temperature-controlled freezer
- Culture plates
- Nalgene sterile filters
- Ovarian dissection kit
- Crushed ice
- Sterile hood

V. Equipment Preparation

Ovarian Dissection Kit for OR – Sterile surgery tray should be prepared by person in charge of the particular OR suite prior to case containing the following:

1. Forceps
2. Pickup forceps
3. Scalpel
4. Hemostats
5. Crushed ice

VI. Reagents

A. Dimethyl sulfoxide (DMSO), 1.5 M (Sigma-Aldrich, St. Louis, MO) – Store at room temperature until expiration date.
B. Leibovitz L-15 Medium (Irvine Scientific, Santa Ana, CA, Cat #9082), sterile. Store at 2–4°C until expiration date.
C. 10% fetal calf serum (Irvine Scientific, Santa Ana, CA, Cat #6000). Aliquoted into 10-mL aliquots by sterile technique into sterile tubes. Store at −50°C until needed.
D. Sucrose (Sigma-Aldrich, St. Louis, MO, Cat #S-7903) – Store at room temperature until expiration date.

VII. Reagent Preparation: Use Sterile Technique

Cryopreservation media: 1.5 M DMSO, 10% fetal calf serum, 0.1 M sucrose in Leibovitz L-15 media, makes 250 mL.

A. One day prior to case, make two batches of 250 mL, one for OR and one for lab.
   1. Using the sterile hood, transfer 198.4 mL of Leibovitz L-15 media to a sterile flask.
   2. Add 8.56 g of sucrose. Allow sucrose to dissolve.

3. Label flask with contents and place in refrigerator (2–4°C) until next day.
4. Remove three tubes of aliquoted fetal calf serum from the freezer and place in refrigerator to thaw.

B. Day of case:
1. Remove media from refrigerator and place the flask on crushed ice.
2. Add 25 mL of well-mixed fetal calf serum to the flask and mix.
3. Slowly add 26.6 mL of DMSO to flask and mix.
4. Filter media through a Nalgene filter. Transfer media into another labeled sterile flask.
5. Place media in refrigerator (2°C) until the OR calls.
6. When OR calls, take one flask of 250 mL of cryoprotection media to OR and leave the other 250 mL flask in refrigerator. Label the flask with its contents. Place on crushed ice using the cryocontainer.
7. A medical technologist goes to OR to carry supplies and return with the tissue.
8. Upon return to the lab, start to precool the Planer Programmable freezer.

VIII. Procedure for Tissue Preparation: Use Sterile Technique

Note: The Planer Cryofreezer can only freeze 13 tissue pieces in a 3-h period.

A. In OR
1. The ovary is removed and cut into 1 × 1 × 3-mm strips/sections by the surgeon in OR.
2. The tissue strips are placed in a sterile specimen container containing the cold cryoprotectant.

B. In lab

Note: Use the sterile hood when processing.

1. The tissues are transported to the lab on ice, along with the OR Collection Form (Appendix A).
2. Transfer the tissue to a large sterile culture plate and cover with cryoprotectant.
3. The tissue should equilibrate at 2–4°C in the cryoprotectant for 30 min. The covered culture plate should be placed in the refrigerator (2–4°C) during this time.
4. Label one 5 mL cryovial for each tissue strip with patient name, CCF#, freeze #, date, and type of tissue, as follows, for example:
   Name
   CCF#
   V04-001A
   Date
   Ovarian Tx
5. Use a different colored cryomarker for each tissue strip:
   RED – for first tissue
   BLUE – for second tissue
   GREEN – for third tissue
   Black – for fourth tissue
   RED – for fifth tissue, etc.

   Note: Please have a second technologist review the labeling of each tube if available, initial, and date the worksheet.
6. Remove culture plate from refrigerator and place in the ice bath; use the small metal tray for the ice bath.
7. Place 3 mL of cold cryoprotectant into each labeled cryovial.
8. Transfer one ovarian stripper labeled Cryovial using sterile applicator sticks.
9. Add 1 mL of cold cryoprotectant to each tube, cap.
10. Transfer the cryovials carefully to the Planer Cryofreezer.

    Note: Wrap a piece of colored tape around each cryovial and the cane to prevent the vials from popping off the cane. Refer to the instrument manual for proper setup and loading.
11. Select the preloaded program #1 (OV).
12. Once the cryofreezer starts the cooling process, it alarms after 30 min or until the freezing chamber achieves the proper seeding temperature. Each cryovial must be "seeded."
13. Seeding
    a. Use proper safety equipment when working with liquid nitrogen.

b. Fill a small LN_2 Dewar with liquid nitrogen.
c. Place the tips of the long-length forceps into the LN_2 to freeze.
d. Open each ampoule chamber, one at a time, and lift the cane upward.
e. Place the frozen tips of the forceps straddling the cryovial near the top of the specimen for 5 s or more.
f. Reposition the cane in its holder, secure the tabs at the top of freezer, and move onto the next ampoule chamber.
g. Repeat steps c–f on the next cryovial. Continue these steps until all cryovials have been seeded.
h. Double check that all ampoule chambers are locked down before pressing the "Run" button to continue the freezing process.
i. Due to the size of the tissues, the cryofreezer takes some time to complete the freezing process. It may take 2–3 h depending on the size of the tissue and the number of vials.

14. Label a cryocane using a cryomarker with the freeze number and the last name of the patient. Up to two cryovials may be placed on one cryocane for the long-term storage.
15. At the end of the freezing program, the alarm buzzes and read "Program Finished Remove Samples – Run."
16. Carefully remove the samples and load onto the labeled cryocane. Place a cryosleeve over the cryocane.
17. Transfer the samples to the 35HVC#3 LN_2 tank. Record the location of the specimens on the worksheet.
18. Press "Run" on the cryofreezer so that the freezing chamber undergoes its warming cycle. This takes about 15 min.
19. The display reads "Do Not Switch Off." When the display changes to "Ready to Restart," press "Run" to return to the "Main Menu."
20. The unit may be powered off at this point.
21. Record information in the Ovarian Tissue logbook.
22. Charge the patient for Ovarian Tissue Cryo using MISYS Code "OVCRYO." Keep a copy in the patient chart.
23. Make a Xerox copy of the Planer Cryofreezer printout and put into the patient chart as well as the logbook.

IX. Calibration

The Planer Cryofreezer is on a yearly schedule for manufacturer's preventative maintenance and calibration.

X. Quality Control

There are no quality control standards available for this procedure.

Suggested Reading

1. Gosden RG, Baird DT, Wade JC, Webb R. Restoration of fertility to oophorectomized sheep by ovarian autografts stored at −196°C. Hum Reprod. 1994;9:597–603.
2. Oktay K, Newton H, Aubard Y, Salha O, Gosden RG. Cryopreservation of immature human oocytes and ovarian tissue: an emerging technology. Fertil Steril. 1998;69:1–7.
3. Oktay K. Ovarian tissue cryopreservation and transplantation. In: Dunitz M, editor. Textbook of assisted reproductive techniques, laboratory and clinical perspectives, Chapter 23. London: Martin Dunitz; 2001.

Appendix D: Sperm Preparation and Freezing for Banking

Pankaj Talwar

Semen Freezing Protocol

Here, we describe the *liquid nitrogen vapor technique* for cryofreezing the spermatozoa. The technique is simple, easy to learn, and does not require expensive equipment.

I. Sample Collection

The semen collection is achieved via masturbation in a healthy man.

A. The patient should be counseled and provided with a sterile specimen collection jar and ample time and privacy to do so.
B. Avoidance of lubricants (soap, jelly, and saliva) is important, as these are spermatotoxic.
C. If no ejaculate is achieved by the patient, then a postejaculate urinalysis should be examined to review for retrograde ejaculation.
D. If retrograde ejaculation is observed, alpha agonists may be administered in an effort to convert retrograde to antegrade ejaculation. Alkalinization of the urine may be carried out and urine post ejaculation may be used to harvest the sperms.
E. If the patient is unable to reach climax or have errection, audio–video aids may be used.
F. If this difficulty persists, then consideration should be given to vibratory stimulation, electroejaculation.
G. Surgical testicular sperm extraction techniques may have a potential role in such patients.

II. Liquid Nitrogen Vapor Cooling

Cryovials

The sample is collected in the designated sample collection room of the center to avoid contamination and temperature-related changes in the semen sample.

P. Talwar (✉)
ART Centre, Army Hospital Research and Referral, Dhaula Kuan, New Delhi 110010, India
e-mail: pankaj_1310@yahoo.co.in

E. Seli and A. Agarwal (eds.), *Fertility Preservation: Emerging Technologies and Clinical Applications*,
DOI 10.1007/978-1-4419-1783-6, © Springer Science+Business Media, LLC 2012

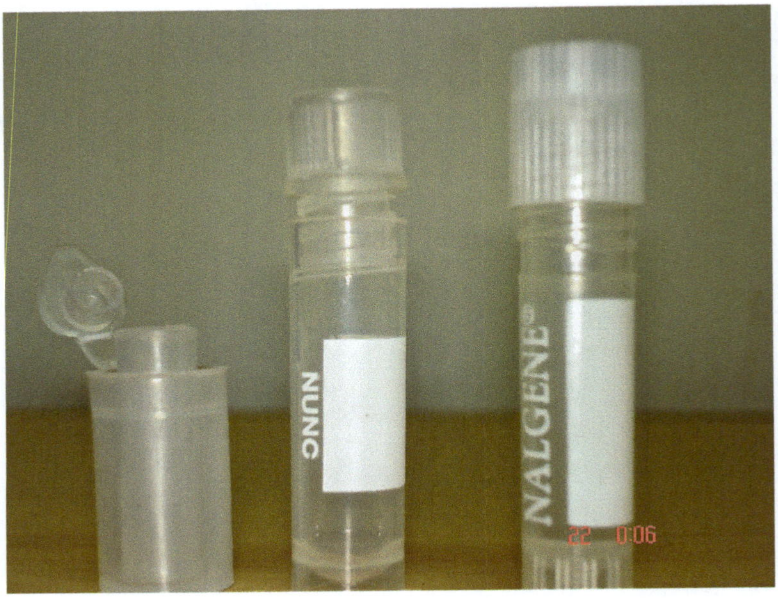

Fig. 31.1 Commercially available vials for semen packaging. Cryotubes may have internal or external threads. Few of these may have an external reservoir for liquid nitrogen for better temperature maintenance

A. Confirm the particulars of the patient and indication of the cryofreezing on receiving the sample from the patient. Keep the semen sample at room temp for 30 min for liquefaction. Label the cryovials and make the entries in the logbook.
B. After liquefaction, measure the total volume of the ejaculate and carry out semen analysis as per WHO guidelines. If the sample is satisfactory, then it can be frozen raw. Concentrating the sperm prior to freezing can enhance post-thaw recovery of semen samples with low sperm counts. Samples in cases of surgical sperm retrieval techniques or those with poor counts are prepared by double-density gradient or other suitable methods and frozen. Sperm pellet may also be frozen if the swim up does not have adequate number of sperms or sperms have poor morphological scoring.
C. Take out 2–3 mL of semen freezing media from the bottle at 4–8°C under laminar flow hood and keep in the incubator at room temperature.
D. Ensure that both semen sample and sperm freezing medium (SFM) are at room temperature. Dilute the semen 1:1 (v/v) with the SFM. The medium should be added dropwise to the semen and the solution carefully mixed after each additional drop of SFM. This is done as glycerol is toxic for the sperms. The procedure should be carried out under laminar flow hood. The mixture is left at room temperature for a period of 10–15 min to equilibrate.
E. Load the diluted semen into straws or cryovials and seal according to the manufacturer's recommendations (Figs. 31.1 and 31.2). It is very important to leave some air space in the lower part of the straw for sealing as well as to allow expansion of the solution during freezing.
F. Suspend the straws horizontally for 30 min, just above the surface of the liquid nitrogen. Cryovials should be attached to a cane and then suspended above the surface of the liquid nitrogen for the same period of time.
G. Finally, transfer the straws or cryovials into liquid nitrogen and store at −196°C (Fig. 31.3).

For Thawing the Sample

A. Remove straws or cryovials from liquid nitrogen and wash the vial under running water for few minutes. Now place them at room temperature till the time the sweating gets over and the sample has liquefied completely.

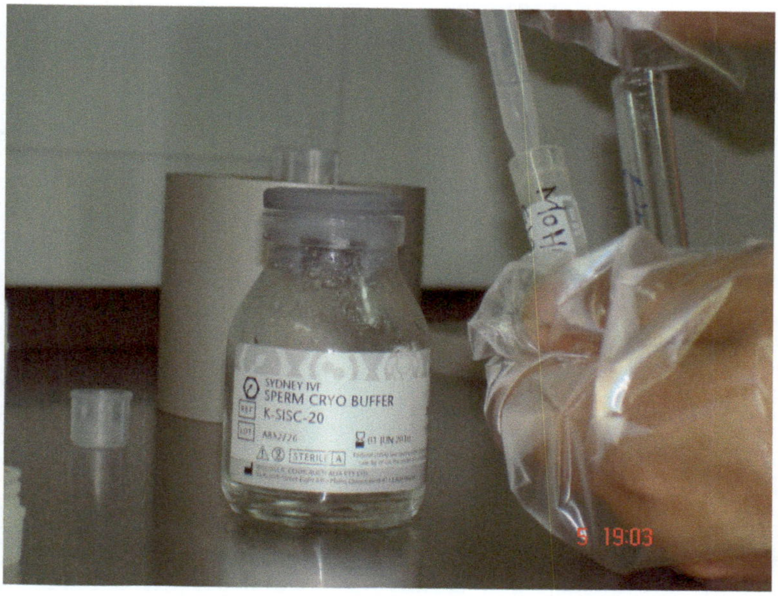

Fig. 31.2 Glycerolated semen being loaded in the cryovials

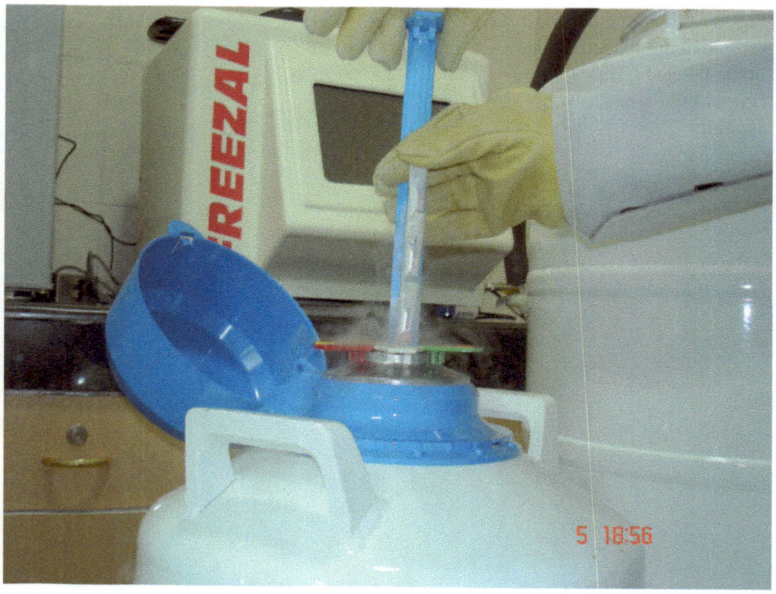

Fig. 31.3 Semen containing vials being loaded on the aluminum canes before being dipped in LN_2 in the cryocontainers

B. Wipe the vials totally dry and open the cap of the cryovials according to the manufacturer's instructions and remove the thawed semen.
C. Dilute the semen with sperm preparation medium (1:1) to reduce the toxic effect of glycerol.
D. Quickly evaluate the survival of the sperm. If necessary, thaw additional cryovials/straws.
E. Place the specimen in incubator for 15 min.
F. Immediately prepare sample by the density gradient method, swim up, or single wash technique with sperm preparation medium.
G. A final concentration of minimum 10×10^6/mL is recommended.

Cryostraws

A. Washed and prepared semen sample is filled in the prelabeled 0.5-mL, clear, flexible,

Fig. 31.4 Daisy goblets, visotubes, and straws for semen packaging. Color coding helps in easy identification of the sample

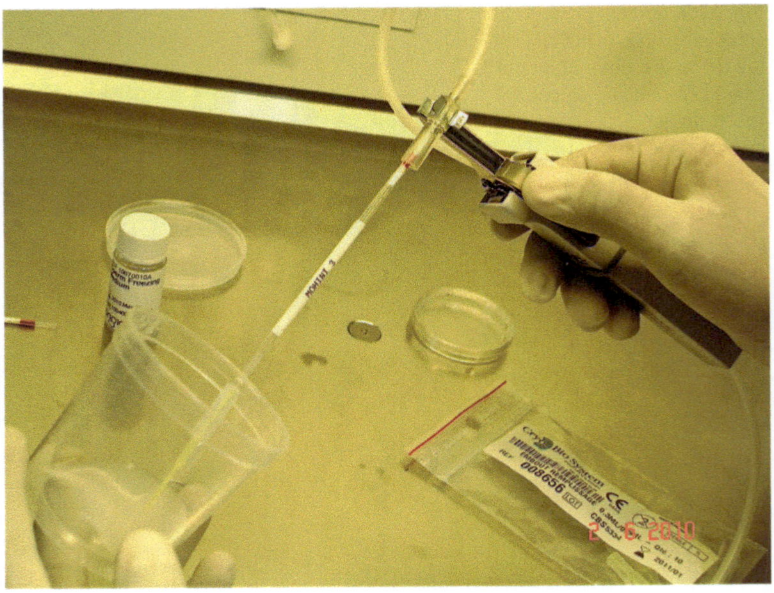

Fig. 31.5 Raw semen sample being loaded in the labeled CBS cryostraws using a manual aspirator

ionomeric resin straws (CBS) using aspirating device (Figs. 31.4 and 31.5).

B. Straws are sealed using SYMS sealer (Fig. 31.6).
C. Sealed straws are suspended 4–5 cm above liquid nitrogen level in a Styrofoam box horizontally (Fig. 31.7).
D. The straw should not be touching each other and they should be minimum 0.2–0.3 cm apart for proper circulation of the liquid nitrogen vapor around the straws.
E. Straws are cooled for 10–20 min. By this time, the contents of the straws have frozen.
F. Now plunge them into liquid nitrogen and transferred in the predecided goblets in LN_2 container and taken on inventory (Fig. 31.8).

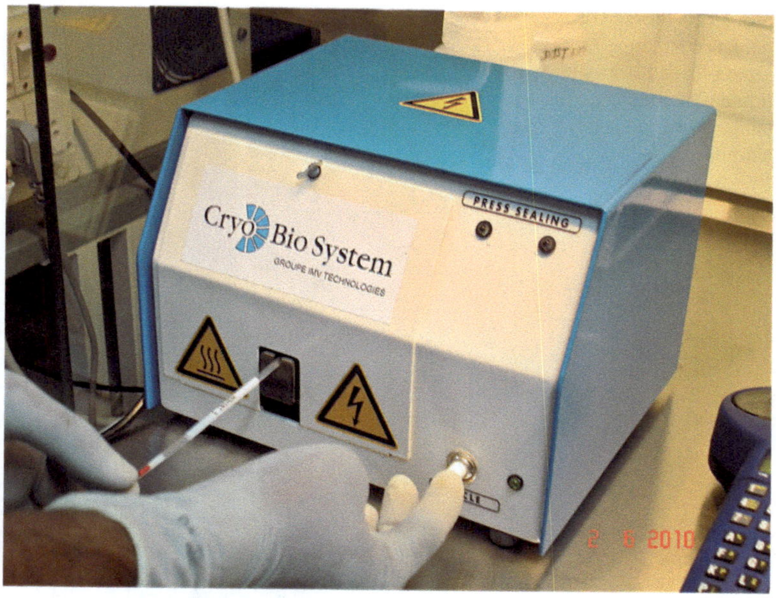

Fig. 31.6 SYMS CBS: Straw sealer being used to seal the labeled and loaded straws

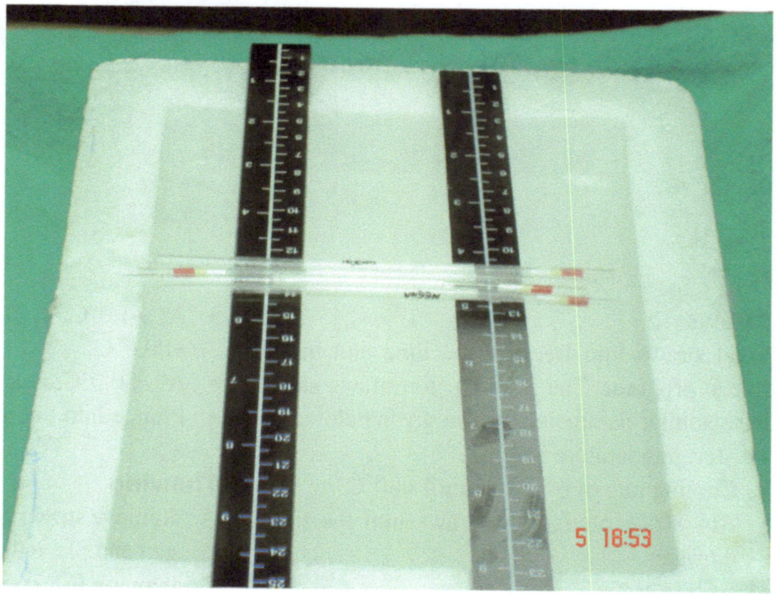

Fig. 31.7 Vapor-phase cooling of the semen-loaded straws being done using a Styrofoam box containing liquid nitrogen

III. Technique Using Controlled Rate Freezing Method

Computer-controlled freezers. (a) Liquid nitrogen vapor-filled chambers (e.g., Kryo-10, Planer, Sunbury, UK and Nicool Models, Air Liquide, Bussy-Saint-Georges, France); (b) cooled metal blocks – CryoLogic, Australia – are widely used cryoplanes.

A. Ensure that both the sample and sperm cryopreservation buffer are at room temperature.
B. Mix two volumes of sperm cryopreservation buffer to 1 volume of sample.
C. Leave mixture for 10 min at room temperature.
D. Label straws with relevant information.
E. Load the sample into a freezing straw or cryovial and seal according to manufacturer's instructions.

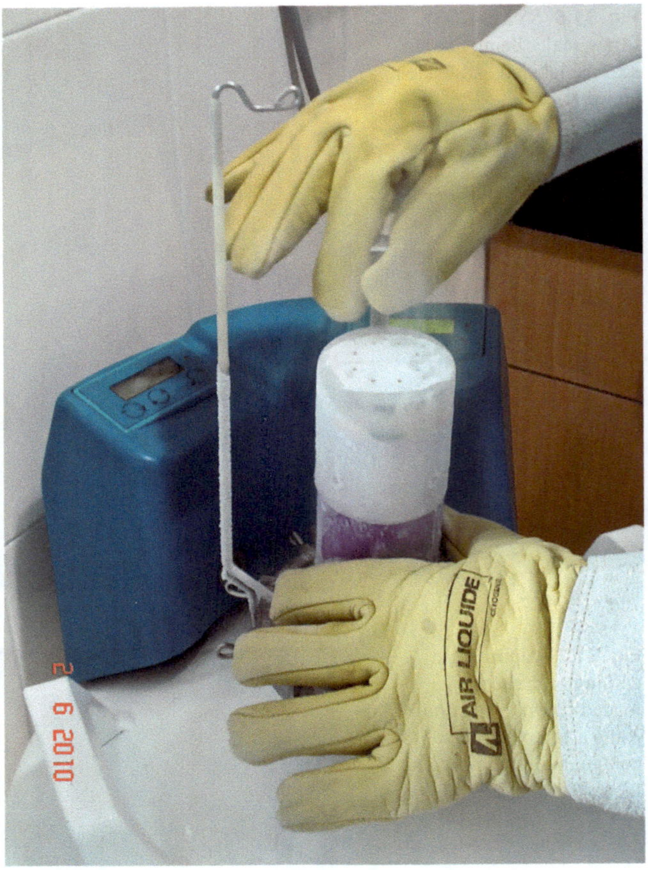

Fig. 31.8 Storage of goblets containing semen-loaded straws in liquid nitrogen tank with temperature and level alarms

Straws

Load straws into freezing machine and initiate freeze program. The program for straws should have similar parameters to that given below:

- Start temperature is 20°C.
- Cooling rate of 6°C/min until −80°C.
- At −80°C, plunge the straws into the liquid nitrogen.

Cryovials

Load cryovials into freezing machine and initiate freeze program. The freeze program for cryovials should have similar parameters to those given below:

- Start temperature is 20°C.
- Cooling rate of −0.5°C/min to +5.0°C.
- At +5.0°C, cool at a rate of −1°C/min to +4.0°C.
- At +4.0°C, cool at a rate of −2°C/min to +3.0°C.
- At +3.0°C, cool at a rate of −4°C/min to +2.0°C.
- At +2.0°C, cool at a rate of −8°C/min to +1.0°C.
- At +1.0°C, cool at a rate of −10°C/min to −80.0°C.
- At −80.0°C, hold for 10 min.
- Plunge into liquid nitrogen.

Thawing

A. Remove straws or cryovials from liquid nitrogen and place them at room temperature until thawing is complete.

B. Open the straws or cryotubes according to the manufacturer's instructions and remove the thawed semen.

C. Dilute the semen with HEPES buffer (1:1) to reduce the toxic effect of glycerol.

D. Quickly evaluate the survival of the sperm. If necessary, thaw additional straws for preparation.

E. Immediately prepare the thawed sample by the density gradient/single wash and swim up method.

Appendix E: Preservation of Sperm Isolates and Testicular Biopsy Samples for Banking

Bhushan K. Gangrade

Protocol for Isolation and Cryopreservation of Testicular Sperm

I. Materials and Equipment

Sterile culture dishes
Polystyrene conical tubes
Sterile glass pipettes
Syringes (50 mL)
Syringe (3 mL) with 21 gauze needle
Syringe filter (Nalgene, 0.22 μm)
Microscope glass slides
Cover glass
Gloves
Protective goggles
Sterile pair of scissors
Sterile pair of forceps
Centrifuge
Microscope
Refrigerator
Weighing chemical balance
Nunc Cryo tube vials
Aluminum canes
Plastic cryosleeves
Liquid nitrogen storage dewar
Liquid nitrogen

II. Reagents

Sperm Washing Medium (modified HTF with human serum albumin, 5 mg/mL; SAGE IVF Inc., Trumbull, CT, USA).

Sperm Freezing Medium (TEST yolk buffer with gentamycin sulfate; Irvine Scientific, Santa Ana, CA, USA).

Tissue Culture Grade Water (SAGE IVF Inc., Trumbull, CT, USA).

Erythrocyte (RBC) LYSIS BUFFEr (155 mM ammonium chloride, 10 mM potassium bicarbonate, 2 mM EDTA; pH 7.2).

Mineral oil.

B.K. Gangrade (✉)
IVF Laboratory, Center for Reproductive Medicine, 3435 Pinehurst Avenue, Orlando, FL 32804, USA
e-mail: bkgangrade@hotmail.com

E. Seli and A. Agarwal (eds.), *Fertility Preservation: Emerging Technologies and Clinical Applications*, DOI 10.1007/978-1-4419-1783-6, © Springer Science+Business Media, LLC 2012

III. Preparation of Erythrocyte (RBC) Lysis Buffer

Weigh the following chemicals separately.

NH_4Cl (Sigma Cat #A 0171)	0.829 g
$KHCO_3$ (Sigma Cat #P 9144)	0.100 g
EDTA (Sigma Cat #ED2SS)	0.074 g

Dissolve in 100 mL of tissue culture grade water and adjust the pH to 7.2. Filter sterilize using Nalgene syringe filters (0.22 μm) attached to 50-mL syringe. The RBC lysis buffer may be stored in refrigerator (4°C) for 1 month.

IV. Specimen

Note: Patient identification and labeling of the specimen container should be performed as per standard laboratory protocol. Sterile techniques and universal precautions should be exercised while processing the tissue sample.

The testicular biopsy sample is surgically retrieved by the surgeon or the urologist. The tissue is collected in a sterile petri dish in 2.0 mL of sperm washing medium. If the surgery is performed at a distant facility and the tissue needs to be transported to the Andrology Laboratory for processing, the testicular tissue may be transferred to a sterile container with tight lid. At all times, the tissue should be completely submerged in the sperm washing medium. Care should be taken to avoid exposure to extreme temperature during transport. The testicular tissue can be kept at room temperature for short (1–2 h) duration without any detrimental effect on sperm parameters.

V. Procedure

A. Wash the testicular tissue with 1–2 mL of sperm washing medium (hereafter referred as "culture medium") to remove red blood cells.
B. Transfer the tissue to a petri dish and add a few drops of culture medium to keep the tissue moist. Using a pair of sterile scissor, mince the tissue.
C. Add 1.0 mL of culture medium to the finely minced sample. Aspirate the suspension and gently pass it through a 21 gauze needle attached to a 3.0-mL syringe. Repeat this process 2–3 times. The passage through the needle breaks down the seminiferous tubules in small pieces and also dislodges the sperm from the lumen.
D. Place 5–10 μL of suspension in petri dish and overlay with mineral oil. Observe and record the number of spermatozoa per high power field (200×).
E. Transfer the suspension to a conical tube and let it stand at room temperature for 5 min. This allows the pieces of seminiferous tubules and large tissue clumps to settle down.
F. Transfer the supernatant to another conical tube and centrifuge (300 × *g*) for 10 min.
G. Discard the supernatant.
H. If the pellet exhibits the presence of red blood cells (as it invariably does), add 2.0 mL of RBC lysis buffer to the tube and resuspend the pellet.
I. Centrifuge at 400 × *g* for 5 min.
J. Discard the supernatant and wash the pellet with culture medium (1.0 mL).
K. Resuspend the pellet in 0.2-mL culture medium.
L. Add equal volume (0.2 mL) of TEST-Yolk buffer (freezing medium) slowly to the sperm suspension.
M. Transfer the above mixture (sperm suspension and freezing medium) to a prelabeled Nunc cryovial.
N. Place the cryovial in a refrigerator (4°C) for 45 min.
O. Expose the cryovial to the liquid nitrogen vapor phase (20–30 cm above the liquid nitrogen surface) for 45 min. This can be achieved by placing the vial in a wire mesh container hanging in the neck of a Dewar tank.
P. Place the cryovial on the aluminum cane and immerse in liquid nitrogen Dewar for storage.
Q. Record the location in the Dewar inventory.

Appendix F: Testis Tissue Xenografting

Jose R. Rodriguez-Sosa, Stefan Schlatt, and Ina Dobrinski

Testis tissue xenografting involves a series of procedures from donor tissue preparation to graft recovery for analysis and/or sperm harvesting. In the current section, we describe these protocols in detail (modified from Dobrinski and Rathi [1]).

Materials

Preparation of Donor Tissue

1. Phosphate-buffered saline (PBS).
2. Dulbecco's modified Eagle's medium (DMEM) (or other balanced cell culture medium).
3. Plastic tissue culture dishes (60×15 and 100×20 mm).
4. Dissection instruments: Iris forceps (~10 cm, 0.8-mm tips), small forceps (~11 cm, 0.1×0.6-mm tips), curved iris forceps (~10 cm, 0.8-mm tips, curved), dissecting scissors (~10 cm, straight), disposable scalpel (#10).

Cyropreservation of Donor Tissue

1. Freezing medium: Any suitable cyoprotectant medium and method can be used. Here, for its practical use and the results obtained previously, we describe the use of DMSO-based medium and a conventional slow freezing method. Reagents: Heat-inactivated fetal calf serum (FCS), DMSO, and DMEM or other cell culture medium.
2. 2-mL cryovials.
3. Nalgene freezing container designed to provide a controlled cooling rate of 1°C/min when placed into a −70°C freezer.

I. Dobrinski (✉)
Department of Comparative Biology and Experimental Medicine, Faculty of Veterinary Medicine, University of Calgary, 3300 Hospital Drive, Calgary, AB, Canada T2N 4N1
e-mail: idobrins@ucalgary.ca

E. Seli and A. Agarwal (eds.), *Fertility Preservation: Emerging Technologies and Clinical Applications*, DOI 10.1007/978-1-4419-1783-6, © Springer Science+Business Media, LLC 2012

4. 15-mL centrifuge tubes.
5. Plastic tissue culture dishes (60×15 and 100×20 mm).

Recipient Preparation and Ectopic Xenografting of Donor Tissue

1. Recipient mice: Immunodeficient (e.g., NCR nu/nu, SCID, or RAG) male mice, 6–8-weeks old. Female mice can also be used, but castration is easier in males.
2. Anesthesia reagents: Any suitable anesthetic can be used in accordance with animal care and use guidelines.
3. Tuberculin syringes (1 mL).
4. Injection needles (26½ gauge).
5. 70% ethanol.
6. Betadine solution.
7. Instruments suitable for mouse surgery.
8. Suture material with needle (6-0 Silk braid w/needle).
9. Wound clips (7.5 mm, e.g., Michel® clips).
10. Clip applying-removing forceps (12.7 cm, e.g., Michel®).
11. Heating pad.

Collection of Testis Xenografts for Analysis and Sperm Harvesting

1. Bouin's solution (or any other fixative, depending on the analysis method).
2. Sample vials.
3. Plastic tissue culture dishes (60×15 and 100×20 mm).
4. A pair of small forceps (~11 cm, 0.1×0.6-mm tips).
5. One pair of dissecting scissors (~10 cm, straight).
6. Culture medium or PBS.
7. Disposable scalpel (#10).
8. 40-μm cell strainer.
9. 15-mL centrifuge tubes.
10. 70% ethanol.

Methods

Collection of Donor Tissue

1. Obtain testis tissue by castration or biopsy from a donor male.
2. Place testis in PBS or biopsies into culture medium, maintaining sterile conditions.
3. Keep the collected tissue on ice and transport to the laboratory.

Preparation of Donor Tissue

Prepare testis tissue in the tissue culture hood.

A. Wash each testis in ice-cold PBS containing antibiotics two to three times before transferring into a culture dish with PBS. In the case of biopsies, wash testis fragments two to three times with ice-cold culture medium containing antibiotics by spinning them down at 150 g for 2 min and resuspending in fresh ice-cold culture medium.
B. For intact testes, remove tunica vaginalis by making an incision along surface and extrude the testis. Proceed, then, to remove from testis all annex structures (spermatic cord, epididymis, connective tissue). Wash testes once in cold PBS and transfer them into a culture dish with PBS.
   1. Carefully remove the tunica albuginea of the testis by using a scalpel blade and a pair of scissors. If the testis is very small, the tunica can be removed by squeezing the testicular tissue out of the tunica through a small incision made on one end while holding the tunica with a pair of small forceps on the other end.
   2. Depending on the size of the testis, either the whole testis tissue can be cut into small pieces of around 1–2 mm^3 in size using curved forceps and a scalpel blade or large pieces of testis tissue can first be removed from the testis and then cut into smaller pieces. All this should be done in ice-cold

culture medium and under sterile conditions in a small culture dish (60 × 15 mm).

3. Transfer the prepared tissue fragments to ice-cold culture medium in small culture dishes on ice until grafting.

C. For testis biopsies, these should be cut into 1–2 mm3 pieces and transferred to ice-cold culture medium in small culture dishes on ice until grafting.

Cryopreservation of Donor Tissue

Freezing

1. Prepare freezing medium by mixing FCS, DMEM, and DMSO at a ratio of 1:3:1.
2. Transfer approximately ten pieces of testis tissue into a 2-mL cryovial and add 0.5 mL of freezing medium at room temperature.
3. Place the vials into the Nalgene freezing container set at room temperature and introduce the container to a −70°C freezer. Tissue fragments are left overnight into the container and freezer before being transferred into liquid nitrogen.

Thawing

1. Remove cryovials from liquid nitrogen and hold them at room temperature for 1 min to evaporate any remaining liquid nitrogen.
2. Introduce cryovials in a 25°C water bath for 1 min.
3. Add ~1.5 mL into each vial, transfer content into a sterile 15-mL centrifuge tube, and 10 mL of fresh medium.
4. Centrifuge at 300 g for 2 min, discard supernatant, and resuspend tissue fragments in 10 mL of fresh medium.
5. Repeat wash to completely eliminate cryoprotectant.
6. Resuspend fragments in ~2 mL of fresh medium and transfer content to a tissue culture dish containing enough culture medium to maintain the tissue fragments in suspension. Maintain dish with fragments on ice until transplantation.

Recipient Preparation and Ectopic Xenografting of Donor Tissue

Anesthesia of Recipient Mouse

1. Weigh the mouse and anesthetize. Monitor anesthetic depth by the absence of voluntary and reflex movement.
2. Position the mouse in dorsal recumbency for castration (described below) on warm pad and keep mouse warm during the entire procedure.

Castration of Recipient Mouse

1. Prepare sterile surgical field by clipping or plucking the hair (not necessary in nude mice), wiping with 70% ethanol and betadine solution.
2. Make a 0.5–1-cm ventral midline skin incision to expose the abdominal wall.
3. Lift abdominal wall by using a small forceps at the point of the white line to avoid accidentally injuring abdominal organs and proceed to make a ~0.5-cm incision at the midline of the abdominal wall to expose the peritoneal cavity.
4. Use one iris forceps to hold the abdominal wall, and use another pair of iris forceps to search for the fat pads attached to the epididymis and testis in the peritoneal cavity. Gently pull the fat pad out until the testis is exteriorized and the testicular artery and epididymis are clearly visible.
5. Detach the tail of the epididymis from the gubernaculum by blunt dissection (Fig. 33.1a).
6. Ligate the testicular artery and the vas deferens together with the blood vessel with silk, and section the ligated structures by cutting between the testis and the ligature (Fig. 33.1b).
7. Repeat the procedure for the second testis.
8. Suture the abdominal wall with one or two surgical stitches.
9. Close the skin incision with one or two Michel clips.

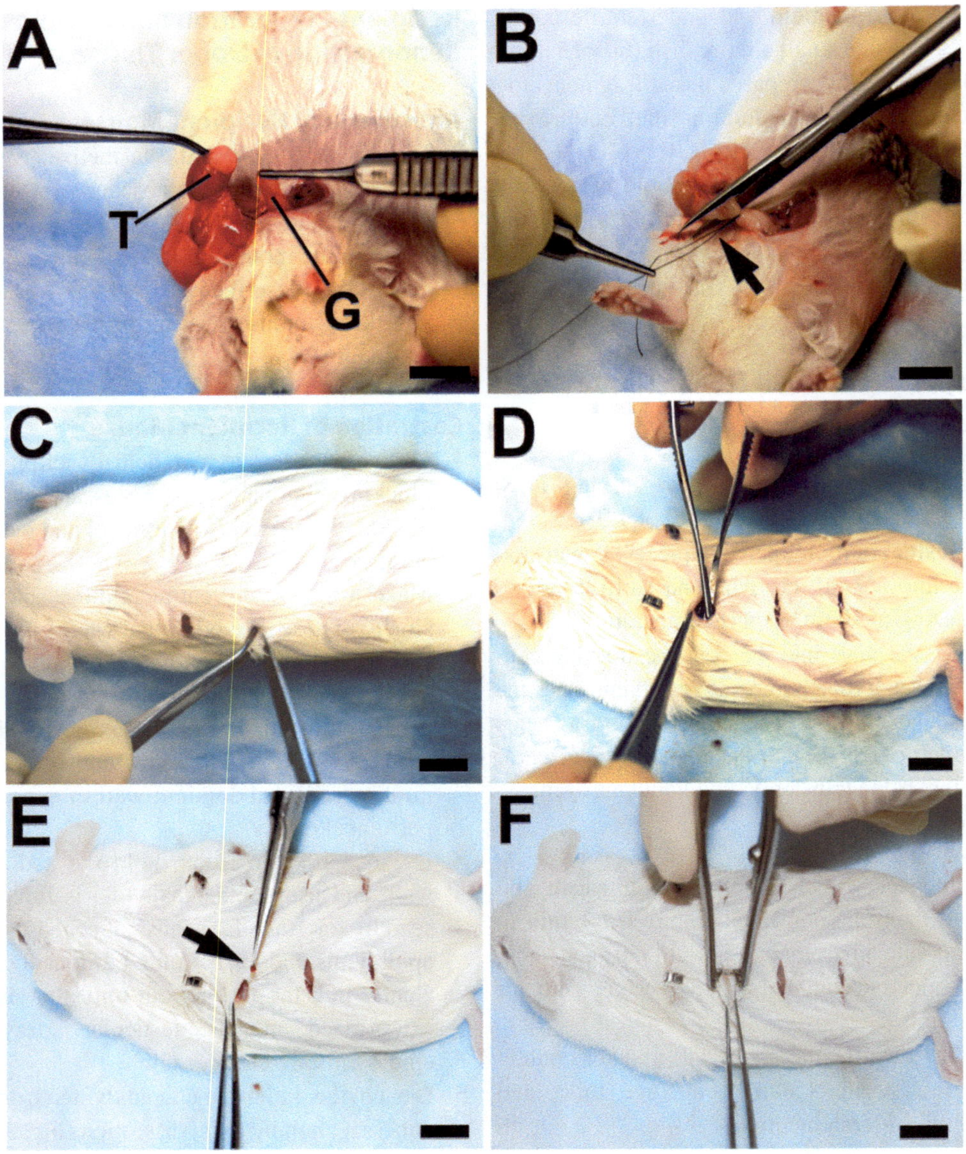

Fig. 33.1 Key methodological steps in recipient preparation and ectopic transplantation of testicular tissue. (**a**) Photograph illustrating the detachment of the testis (T) from the gubernaculum (G) prior to ligation of the testicular artery and annexed structures. (**b**) Once the testis has been detached from the gubernaculum, the testicular artery and vas deferens are ligated. In this photograph, the ligature has been already placed (*arrow*) and the sectioning of the ligated structures is about to be performed. (**c**) Once the mouse has been castrated and its back has been aseptically prepared, 0.5–1-cm incisions are made in the skin to introduce each testis fragment. (**d**) In order to place each testis fragment under the dorsal skin, the subcutaneous tissue is teased apart with small scissors to produce a small cavity. (**e**) The testis fragment is placed with fine forceps deep into the subcutaneous cavity by holding the border of the skin incision with small forceps to expose the cavity. (**f**) Once the testis fragment has been placed, the incision is closed with a Michel clip. Bars = 1 cm

Ectopic Xenografting

1. Position the mouse in ventral recumbency and prepare a sterile surgical field on its back as above.
2. Depending on how many grafts are to be inserted (generally 4–8/mouse), make ~0.5-cm-long skin incisions on each side of the midline of the back of the mouse (Fig. 33.1c).
3. Use forceps to hold a border of the skin incision, and make a subcutaneous cavity by teasing apart the connective tissue using scissors (Fig. 33.1d).
4. Using an iris forceps, place a piece of testis tissue deep into the subcutaneous cavity, holding the border of the skin incision with another iris forceps (Fig. 33.1e).
5. Close the skin incision with one Michel clip (Fig. 33.1f).

Postoperative Care

1. Keep the mouse on heating pad until it starts to recover from anesthesia.
2. Move the mouse to a cage with additional insulation and cover and monitor until mice are fully recovered.

Collection of Testis Xenografts for Analysis and Sperm Harvesting

Xenografts Recovery

1. Sacrifice the host mouse according to animal care and use guidelines.
2. Make a midline skin incision on the back skin running from the tail to the neck and open skin. This exposes the grafts which can be located either on the subcutaneous tissue or attached to the skin.
3. Carefully remove the grafts using a pair of forceps and a pair of scissors.
4. Record the number of grafts recovered and size and weight of individual grafts.
5. Retrieve the seminal vesicles from the abdomen of the mouse and record their weight as an indication of testosterone production by the grafted tissue.

For Histological Analysis

1. Suspend xenografts into sample vial containing Bouin's solution (or other fixative) in a volume ~10 timed that of the xenograft, and label vial appropriately.
2. Incubate overnight in the refrigerator followed by washing at least three times in 70% ethanol at intervals of 24 h preferably.
3. Proceed for processing and embedding in paraffin.

For Sperm Harvesting

1. Wash xenografts by spinning them down at 300 g for 1 min and resuspending them in culture medium containing antibiotics.
2. Cut grafts into small pieces and mince carefully with the forceps in a tissue culture dish containing 3–5 mL of culture medium.
3. Filter minced tissue through the 40-μm cell strainer.

Reference

1. Dobrinski I, Rathi R. Ectopic grafting of mammalian testis tissue into mouse hosts. In: Hou SX, Singh SR, editors. Methods in molecular biology germ line stem cells. Totowa: Humana Press; 2008. p. 139–48.

Index

D

MIX
Papier aus verantwortungsvollen Quellen
Paper from responsible sources
FSC® C105338

If you have any concerns about our products, you can contact us on
ProductSafety@springernature.com

In case Publisher is established outside the EU, the EU authorized representative is:
Springer Nature Customer Service Center GmbH
Europaplatz 3, 69115 Heidelberg, Germany

Printed by Libri Plureos GmbH
in Hamburg, Germany